Mathematical Methods in Physics, Engineering, and Chemistry

Mathematical Methods in Physics, Engineering, and Chemistry

BRETT BORDEN AND JAMES LUSCOMBE

Naval Postgraduate School
Monterey, CA, USA

WILEY

This edition first published 2020

Registered Office
John Wiley & Sons, Inc., 111 River Street, Hoboken, NJ 07030, USA

Editorial Office
111 River Street, Hoboken, NJ 07030, USA

For details of our global editorial offices, customer services, and more information about Wiley products visit us at www.wiley.com

Wiley also publishes its books in a variety of electronic formats and by print-on-demand. Some content that appears in standard print versions of this book may not be available in other formats.

Library of Congress Cataloging-in-Publication data applied for

ISBN: 9781119579656

Cover Design: Wiley
Cover Image: © FrankRamspott/Getty Images

Set in 11/13pt Computer Modern by SPi Global, Pondicherry, India

Printed in the United States of America

V10014409_100319

Contents

Preface *xi*

1 Vectors and linear operators **1**
1.1 The linearity of physical phenomena 1
1.2 Vector spaces 2
1.2.1 A word on notation 4
1.2.2 Linear independence, bases, and dimensionality 5
1.2.3 Subspaces 7
1.2.4 Isomorphism of N-dimensional spaces 8
1.2.5 Dual spaces 8
1.3 Inner products and orthogonality 10
1.3.1 Inner products 10
1.3.2 The Schwarz inequality 11
1.3.3 Vector norms 12
1.3.4 Orthonormal bases and the Gram–Schmidt process 12
1.3.5 Complete sets of orthonormal vectors 15
1.4 Operators and matrices 16
1.4.1 Linear operators 17
1.4.2 Representing operators with matrices 18
1.4.3 Matrix algebra 20
1.4.4 Rank and nullity 22
1.4.5 Bounded operators 23
1.4.6 Inverses 24
1.4.7 Change of basis and the similarity transformation 25
1.4.8 Adjoints and Hermitian operators 27
1.4.9 Determinants and the matrix inverse 29
1.4.10 Unitary operators 33
1.4.11 The trace of a matrix 35
1.5 Eigenvectors and their role in representing operators 36
1.5.1 Eigenvectors and eigenvalues 36
1.5.2 The eigenproblem for Hermitian and unitary operators 39
1.5.3 Diagonalizing matrices 40
1.6 Hilbert space: Infinite-dimensional vector space 43
Exercises 47

2 Sturm–Liouville theory **51**
2.1 Second-order differential equations 52
2.1.1 Uniqueness and linear independence 52
2.1.2 The adjoint operator 55
2.1.3 Self-adjoint operator 56

2.2 Sturm–Liouville systems 57
2.3 The Sturm–Liouville eigenproblem 60
2.4 The Dirac delta function 64
2.5 Completeness 66
2.6 Recap 68
Summary 68
Exercises 69

3 Partial differential equations 71
3.1 A survey of partial differential equations 71
3.1.1 The continuity equation 71
3.1.2 The diffusion equation 72
3.1.3 The free-particle Schrödinger equation 73
3.1.4 The heat equation 73
3.1.5 The inhomogeneous diffusion equation 74
3.1.6 Schrödinger equation for a particle in a potential field 74
3.1.7 The Poisson equation 74
3.1.8 The Laplace equation 75
3.1.9 The wave equation 75
3.1.10 Inhomogeneous wave equation 76
3.1.11 Summary of PDEs 76
3.2 Separation of variables and the Helmholtz equation 76
3.2.1 Rectangular coordinates 78
3.2.2 Cylindrical coordinates 80
3.2.3 Spherical coordinates 82
3.3 The paraxial approximation 83
3.4 The three types of linear PDEs 84
3.4.1 Hyperbolic PDEs 85
3.4.2 Parabolic PDEs 87
3.4.3 Elliptic PDEs 87
3.5 Outlook 88
Summary 88
Exercises 89

4 Fourier analysis 91
4.1 Fourier series 91
4.2 The exponential form of Fourier series 96
4.3 General intervals 98
4.4 Parseval's theorem 103
4.5 Back to the delta function 105
4.6 Fourier transform 107
4.7 Convolution integral 111
Summary 115
Exercises 116

5 Series solutions of ordinary differential equations **121**
5.1 The Frobenius method 122
5.1.1 Power series 122
5.1.2 Introductory example 123
5.1.3 Ordinary points 125
5.1.4 Regular singular points 130
5.2 Wronskian method for obtaining a second solution 137
5.3 Bessel and Neumann functions 137
5.4 Legendre polynomials 142
Summary 144
Exercises 145

6 Spherical harmonics **147**
6.1 Properties of the Legendre polynomials, $P_l(x)$ 148
6.1.1 Rodrigues formula 148
6.1.2 Orthogonality 150
6.1.3 Completeness 151
6.1.4 Generating function 152
6.1.5 Recursion relations 155
6.2 Associated Legendre functions, $P_l^m(x)$ 157
6.3 Spherical harmonic functions, $Y_l^m(\theta, \phi)$ 158
6.4 Addition theorem for $Y_l^m(\theta, \phi)$ 160
6.5 Laplace equation in spherical coordinates 166
Summary 167
Exercises 168

7 Bessel functions **173**
7.1 Small-argument and asymptotic forms 173
7.1.1 Limiting forms for small argument 173
7.1.2 Asymptotic forms for large argument 174
7.1.3 Hankel functions 174
7.2 Properties of the Bessel functions, $J_n(x)$ 175
7.2.1 Series associated with the generating function 175
7.2.2 Recursion relations 177
7.2.3 Integral representation 178
7.3 Orthogonality 180
7.4 Bessel series 182
7.5 The Fourier-Bessel transform 185
7.6 Spherical Bessel functions 186
7.6.1 Reduction to elementary functions 186
7.6.2 Small-argument forms 188
7.6.3 Asymptotic forms 188
7.6.4 Orthogonality and completeness 189
7.7 Expansion of plane waves in spherical harmonics 190
Summary 192
Exercises 192

8 Complex analysis **195**
8.1 Complex functions 195
8.2 Analytic functions: differentiable in a region 197
8.2.1 Continuity, differentiability, and analyticity 197
8.2.2 Cauchy–Riemann conditions 198
8.2.3 Analytic functions are functions only of $z = x + iy$ 201
8.2.4 Useful definitions 201
8.3 Contour integrals 202
8.4 Integrating analytic functions 206
8.5 Cauchy integral formulas 210
8.5.1 Derivatives of analytic functions 211
8.5.2 Consequences of the Cauchy formulas 212
8.6 Taylor and Laurent series 213
8.6.1 Taylor series 213
8.6.2 The zeros of analytic functions are isolated 215
8.6.3 Laurent series 215
8.7 Singularities and residues 217
8.7.1 Isolated singularities, residue theorem 217
8.7.2 Multivalued functions, branch points, and branch cuts 220
8.8 Definite integrals 221
8.8.1 Integrands containing $\cos\theta$ and $\sin\theta$ 222
8.8.2 Infinite integrals 223
8.8.3 Poles on the contour of integration 226
8.9 Meromorphic functions 228
8.10 Approximation of integrals 230
8.10.1 The method of steepest descent 233
8.10.2 The method of stationary phase 235
8.11 The analytic signal 236
8.11.1 The Hilbert transform 237
8.11.2 Paley–Wiener and Titchmarsh theorems 239
8.11.3 Is the analytic signal, analytic? 241
8.12 The Laplace transform 242
Summary 245
Exercises 245

9 Inhomogeneous differential equations **251**
9.1 The method of Green functions 251
9.1.1 Boundary conditions 252
9.1.2 Reciprocity relation: $G(x, x') = G(x', x)$ 253
9.1.3 Matching conditions 254
9.1.4 Direct construction of $G(x, x')$ 255
9.1.5 Eigenfunction expansions 257
9.2 Poisson equation 260
9.2.1 Boundary conditions and reciprocity relations 261
9.2.2 So, what's the Green function? 263

9.3 Helmholtz equation 266
9.3.1 Green function for two-dimensional problems 267
9.3.2 Free-space Green function for three dimensions 270
9.3.3 Expansion in spherical harmonics 270
9.4 Diffusion equation 272
9.4.1 Boundary conditions, causality, and reciprocity 272
9.4.2 Solution to the diffusion equation 274
9.4.3 Free-space Green function 275
9.5 Wave equation 279
9.6 The Kirchhoff integral theorem 283
Summary 284
Exercises 284

10 Integral equations 287
10.1 Introduction 287
10.1.1 Equivalence of integral and differential equations 287
10.1.2 Role of coordinate systems in capturing boundary data 288
10.2 Classification of integral equations 290
10.3 Neumann series 291
10.4 Integral transform methods 293
10.4.1 Difference kernels 293
10.4.2 Fourier kernels 294
10.5 Separable kernels 295
10.6 Self-adjoint kernels 297
10.7 Numerical approaches 302
10.7.1 Matrix form 302
10.7.2 Measurement space 303
10.7.3 The generalized inverse 306
Summary 314
Exercises 315

11 Tensor analysis 319
11.1 Once over lightly: A quick intro to tensors 319
11.2 Transformation properties 327
11.2.1 The two types of vector: Contravariant and covariant 327
11.2.2 Coordinate transformations 328
11.2.3 Contravariant vectors and tensors 332
11.2.4 Covariant vectors and tensors 336
11.2.5 Mixed tensors 339
11.2.6 Covariant equations 339
11.3 Contraction and the quotient theorem 340
11.4 The metric tensor 342
11.5 Raising and lowering indices 344
11.6 Geometric properties of covariant vectors 347
11.7 Relative tensors 350
11.8 Tensors as operators 353
11.9 Symmetric and antisymmetric tensors 356

11.10 The Levi-Civita tensor 357
11.11 Pseudotensors 360
11.12 Covariant differentiation of tensors 363
Summary 373
Exercises 374

A Vector calculus **377**
A.1 Scalar fields 377
A.1.1 The directional derivative 377
A.1.2 The gradient 378
A.2 Vector fields 379
A.2.1 Divergence 379
A.2.2 Curl 380
A.2.3 The Laplacian 380
A.2.4 Vector operator formulae 381
A.3 Integration 382
A.3.1 Line integrals 382
A.3.2 Surface integrals 383
A.4 Important integral theorems in vector calculus 384
A.4.1 Green's theorem in the plane 384
A.4.2 The divergence theorem 386
A.4.3 Stokes' theorem 386
A.4.4 Conservative fields 387
A.4.5 The Helmholtz theorem 389
A.5 Coordinate systems 390
A.5.1 Orthogonal curvilinear coordinates 390
A.5.2 Unit vectors 391
A.5.3 Differential displacement 392
A.5.4 Differential surface and volume elements 393
A.5.5 Transformation of vector components 393
A.5.6 Cylindrical coordinates 394

B Power series **401**

C The gamma function, $\Gamma(x)$ **403**
Recursion relation 403
Limit formula 404
Reflection formula 405
Digamma function 405

D Boundary conditions for Partial Differential Equations **409**
Summary 417

References *419*

Index *421*

Preface

Mathematics is deeply ingrained in physics, in how it's taught and how it's practiced. Courses in mathematical methods of physics are core components of physics curricula, at the advanced undergraduate and beginning graduate levels. In our experience, textbooks that seek to provide a comprehensive coverage of mathematical methods tend not to mesh well with the needs of today's students, who face curricula continually under pressure to squeeze more content into the allotted time. The amount of mathematics one *could* be called upon to know in a scientific career is daunting – hence, the temptation to try to cover it all in textbooks. We have developed, over years of teaching, a set of notes outlining the essentials of the subject, which has turned into this book.[1] Our goal has been to emphasize topics that the majority of students will require in the course of their studies. Not every student will go on to a career in theoretical physics, and thus not every student need be exposed to specialized topics at this point in their education.

Following is a sketch of the contents of this book.

- **Linear algebra**: What's more important in the education of scientists and engineers, calculus or linear algebra? We opt for the latter. Students are assumed to have had vector calculus (a standard component of Calculus III), a summary of which is provided in Appendix A. We start with vector spaces – a robust, unifying concept that pertains to much of the mathematics students will encounter in their studies. We introduce Dirac notation – never too early to see in a scientific education – but only when it makes sense. It's not a law of nature that we use Dirac notation, yet students must be well versed in it. An overarching theme of the book is *operators*. We develop linear transformations and matrices in Chapter 1, even though it's assumed that students already have a certain dexterity with matrices. A key topic is that of function spaces, which underlies much of what's done throughout the book in terms of representing functions with various expansions – Fourier series, for example. Chapter 1 is a review of a full course in linear algebra; students sufficiently prepared could begin with Chapter 2.
- **Partial differential equations**: A pervasive concept in physics is that of *fields*, the behavior of which in space and time is described by partial differential equations (PDEs). To study physics at the advanced undergraduate level, one must be proficient in solving PDEs and that provides another overarching theme: Boundary value

[1] A condensed and partial version of these notes was published as [1].

problems. Chapters 2–7 form the backbone of a fairly standard one-quarter, upper-division course in math methods. Chapters 8–11 cover separate topics that could form the basis of a follow-on course – or a graduate-level course appropriate for some instructional programs. Ambitious programs could cover the entire book in one semester.

Chapter 2 starts with second-order differential equations with variable coefficients. We find that an early introduction to Sturm–Liouville theory provides a unifying framework for discussing second-order differential equations. The key notion of the *completeness* of solutions to the Sturm–Liouville eigenvalue problem is of vital importance to solving boundary value problems. We restrict ourselves to homogeneous boundary conditions at this point (inhomogeneous problems are considered in Chapter 9). In Chapter 3 we review the three types of PDEs most commonly encountered – wave equation, diffusion equation, and Laplace equation. We introduce the method of separation of variables in the three main coordinate systems used in science and engineering (Cartesian, spherical, cylindrical) and how PDEs separate into systems of ordinary differential equations (ODEs). In Chapter 5, we develop the Frobenius method for solving ODEs with variable coefficients. Along the way, Fourier analysis is introduced in Chapter 4.

- **Special functions**: We cover the most commonly encountered special functions. The gamma function is treated in Appendix C, and the properties of Bessel functions, Legendre polynomials, and spherical harmonics are covered in Chapters 6 and 7. We omit special functions seen only in other courses, e.g. the Laguerre and Hermite polynomials.
- **Complex analysis**: It's never quite obvious where a chapter on complex analysis should go. We place the theory of analytic functions (Chapter 8) before the chapter on Green functions (Chapter 9). We cover the standard topics of contour integration and Cauchy's theorem. We develop the *nonstandard* topics of the approximation of integrals (steepest descent and stationary phase) and the analytic signal (Hilbert transform, the Paley–Weiner and Titchmarsh theorems). The latter is important in applications involving signal processing, which many students end up doing in their thesis work, and is natural to include in a discussion on analytic functions.
- **Green functions**: Inhomogeneous differential equations are naturally solved using the method of Green functions. We illustrate (in Chapter 9) the Green function method for the inhomogeneous Helmholtz, diffusion, and wave equations.
- **Integral equations**: Many key equations occur as integral equations – in scattering theory, for example, either quantum or electromagnetic. We cover integral equations in Chapter 10, a topic not always included in books at this level. Yet, it's important for students to understand that the separation-of-variables method (introduced

earlier in the book) is often not realistic – it relies on the boundaries of systems having rather simple shapes (spherical, cylindrical, etc.). Numerical solutions of PDEs are often based on integral-equation approaches.

- **Tensor analysis**: The book ends with an introduction to tensors, Chapter 11. Tensors are motivated as a mathematical tool for treating systems featuring anisotropy, and for their fundamental role in establishing covariant equations. We present tensors with sufficient depth so as to provide a foundation for their use in the special theory of relativity. The covariant derivative, developed at the end of the chapter, would be the starting point for more advanced applications.

Brett Borden
James Luscombe
Monterey, California

1

Vectors and linear operators

1.1 The linearity of physical phenomena

Much of the mathematics employed in science and engineering is predicated on the *linearity* of physical phenomena. As just one among many possible examples, linearity in quantum mechanics means that if two quantum wavefunctions, ψ_1 and ψ_2, separately satisfy the time-dependent Schrödinger equation, $H\psi = \mathrm{i}\hbar\partial\psi/\partial t$, then so does any *linear combination* $\psi \equiv \alpha\psi_1 + \beta\psi_2$, where α and β are constants. Underlying linearity is *superposition*: The net physical effect, or system response, due to two or more sources (stimuli) is the sum (superposition) of the responses that would have been obtained from each acting individually.

Example. The electrostatic potential φ at location $\boldsymbol{r}$, produced by a fixed continuous charge distribution described by charge density function $\rho(\boldsymbol{r}')$, is obtained as a superposition of the potentials produced by infinitesimal charges $\rho(\boldsymbol{r}')\,\mathrm{d}^3r'$,

$$\varphi(\boldsymbol{r}) = k\int \frac{\rho(\boldsymbol{r}')}{|\boldsymbol{r}-\boldsymbol{r}'|}\,\mathrm{d}^3r',$$

where k is a constant that depends on the system of units employed.

Not all physical phenomena are linear. Human senses, for example, are not responsive unless the strength of various stimuli exceeds a threshold value. Finding the self-consistent charge distribution in semiconductor devices is a nonlinear problem; in contrast to the *fixed* charge distribution in the aforementioned example, charges are able to move and respond to the potentials they generate. The general theory of relativity (the fundamental theory of gravitation) is nonlinear for the same reason – masses respond to the gravitational potential they generate. Nonlinear theories

Mathematical Methods in Physics, Engineering, and Chemistry, First Edition.
Brett Borden and James Luscombe.

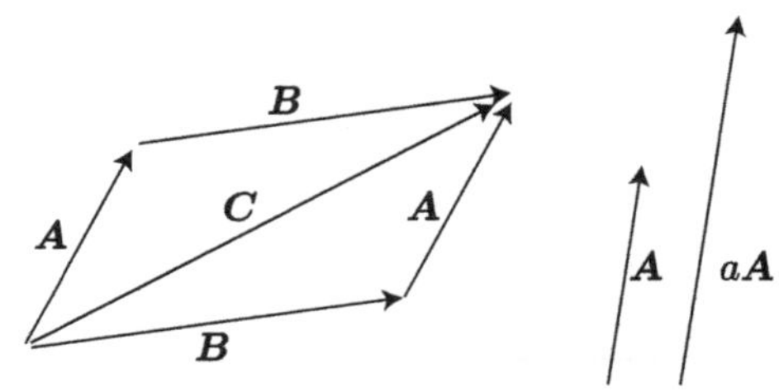

Figure 1.1 Vector addition and scalar multiplication.

are among the most difficult in physics. Let's learn to walk before we run. There are sufficiently many physical effects having linearity as a common feature that it's not surprising there is a common body of mathematical techniques to treat them – and so, the subject of this chapter. We note that nonlinear effects can be described in terms of linear theories when the strength of their sources is sufficiently small.[1]

1.2 Vector spaces

A basic idea in elementary mechanics is that of *vector*, quantities having two attributes: direction and magnitude. The position vector $\boldsymbol{r}$ (physical displacement relative to an origin) is the prototype vector. Anything called "vector" must share the properties of the prototype. From a mathematical perspective, vectors have two main properties: Vectors can be added to produce new vectors, and vectors can be multiplied by numbers to produce new vectors. These operations are depicted in Figure 1.1: the parallelogram rule of vector addition[2] $\boldsymbol{A}+\boldsymbol{B}=\boldsymbol{C}=\boldsymbol{B}+\boldsymbol{A}$; and scalar multiplication, $\boldsymbol{B}=a\boldsymbol{A}$, where a is a constant. Many physical quantities have a vector nature (which is why we study vectors): velocity $\boldsymbol{v}$, acceleration $\boldsymbol{a}$, electric field $\boldsymbol{E}$, etc. We'll refer to the vectors used in three-dimensional position space as *elementary vectors*.

In more advanced physics (such as quantum mechanics), we encounter quantities that *generalize* elementary vectors in two ways – as existing in spaces of dimension greater than 3, and/or for which their components

[1] One speaks of linearized hydrodynamics (the equations of fluid dynamics are nonlinear) or linearized gravity within the theory of general relativity.

[2] Physics is an experiment-based subject, and mathematics is a tool for modeling physical phenomena. When, through experimental work, it's found that certain concepts are no longer applicable for previously unobserved phenomena, new mathematical methods are developed that are suitable for the application at hand. The parallelogram rule reflects our experience with physical displacements $\boldsymbol{r}$. Velocities – time derivatives of $\boldsymbol{r}$ – obey the parallelogram rule in non relativistic physics (the body of physics applicable for speeds v much less than the speed of light, c). It's found in the theory of relativity, however, that non colinear velocities $\boldsymbol{v}_1$, $\boldsymbol{v}_2$ do *not* add like position vectors: $\boldsymbol{v}_1+\boldsymbol{v}_2\neq\boldsymbol{v}_2+\boldsymbol{v}_1$. Velocities, which involve a comparison of space and time, are found not to behave in ways that seem obvious in a limited realm of experience ($v\ll c$). If one generalizes the concepts of space and time to be aspects of a four-dimensional reality ("spacetime") then four-dimensional velocities (appropriately defined) have the properties of the four-dimensional spacetime position vector.

are not real valued (complex valued, for example). Such quantities may not have direction and magnitude in the traditional sense, yet they can be combined mathematically in the same way as elementary vectors.

A *vector space* $\mathbb{V}$ is a set of mathematical objects (called vectors) $\mathbb{V} = \{|\phi\rangle, |\chi\rangle, |\psi\rangle, \dots\}$ having the two main properties of vectors: The sum of vectors is a vector, $|\psi\rangle + |\phi\rangle = |\chi\rangle$, and each vector when multiplied by a *scalar* is a vector, $a|\phi\rangle = |\psi\rangle$. The scalar is chosen from another set, the *field*[3] $\mathbb{F}$. If a is a real (complex) number, then $\mathbb{V}$ is referred to as a real (complex) vector space.[4] Vector spaces, as collections of objects that combine linearly, provide a common mathematical "platform" to describe diverse physical phenomena.

Definition. *A* vector space $\mathbb{V}$ *over a field* $\mathbb{F}$ *is a set of elements* $\{|\psi\rangle, |\phi\rangle, |\chi\rangle, \dots\}$ *that satisfy the following requirements.*

(a) To every pair of vectors $|\phi\rangle$ and $|\psi\rangle$ in $\mathbb{V}$, there corresponds a vector $|\phi\rangle + |\psi\rangle \in \mathbb{V}$, called the sum of $|\phi\rangle$ and $|\psi\rangle$, such that
   1. addition is commutative, $|\psi\rangle + |\phi\rangle = |\phi\rangle + |\psi\rangle$,
   2. addition is associative, $|\psi\rangle + (|\phi\rangle + |\chi\rangle) = (|\psi\rangle + |\phi\rangle) + |\chi\rangle$,
   3. there exists a unique vector $|0\rangle \in \mathbb{V}$ (the additive identity or *null vector*) obeying $|\psi\rangle + |0\rangle = |\psi\rangle$ for every $|\psi\rangle \in \mathbb{V}$, and
   4. to every $|\psi\rangle \in \mathbb{V}$, there exists a unique vector $-|\psi\rangle \in \mathbb{V}$ (the *negative vector*) such that $|\psi\rangle + (-|\psi\rangle) = |0\rangle$.

(b) To every pair $a \in \mathbb{F}$ and $|\psi\rangle \in \mathbb{V}$, there corresponds a vector $a|\psi\rangle \in \mathbb{V}$, called the product of a and $|\psi\rangle$, such that
   1. scalar multiplication is associative, $a(b|\psi\rangle) = (ab)|\psi\rangle$ and
   2. there exists a multiplicative identity 1 such that $1|\psi\rangle = |\psi\rangle$ for every $|\psi\rangle \in \mathbb{V}$.

(c) For scalars $a, b \in \mathbb{F}$ and $|\psi\rangle \in \mathbb{V}$,
   1. scalar multiplication is distributive with respect to vector addition $a(|\psi\rangle + |\phi\rangle) = a|\psi\rangle + a|\phi\rangle$, and
   2. multiplication by vectors is distributive with respect to scalar addition $(a + b)|\psi\rangle = a|\psi\rangle + b|\psi\rangle$.

It may be shown as a result of these axioms [2, p. 32] that $0|\psi\rangle = |0\rangle$ holds for every $|\psi\rangle \in \mathbb{V}$, where 0 is the scalar zero.[5] The axioms in group

[3]The word *field* here has no relation to its usual use in physics. Is there a difference between scalars and numbers? Scalars are elements of $\mathbb{F}$, which can consist of different *types* of number (real or complex, for example). With the term *scalar*, we don't have to commit to any particular type of number. In physics, $\mathbb{F}$ is usually $\mathbb{R}$ or $\mathbb{C}$, the set of real numbers and the set of complex numbers.

[4]Quantum mechanics makes use of complex vector spaces, while the theory of relativity employs real vector spaces for the most part.

[5]Fields $\mathbb{F}$ are also defined through a set of axioms similar to those for vector spaces, including the existence of a unique scalar 0 such that $a + 0 = a$ for all $a \in \mathbb{F}$ and a nonzero scalar 1 such that $1a = a$ for all a [3, p. 1].

(A) prescribe how vectors add, those in (B) prescribe the product of scalars and vectors, and those in (C) prescribe the connection between the additive and multiplicative structures of a vector space.

Examples

The concept of vector space is quite robust. As the following examples show, diverse collections of mathematical objects meet the requirements of a vector space. To check that a given set is indeed a vector space, one should, to be meticulously correct, verify that each of the requirements are met. Often it's enough to check that the sum of two elements in the set remains in the set.

- The set of all elementary vectors $\boldsymbol{A}$, $\boldsymbol{B}$, etc., comprises a vector space under the parallelogram law of addition and multiplication by numbers.
- The set of all N-tuples of numbers (which could be complex) $|\psi\rangle \equiv (x_1, \ldots, x_N)$ is called N-dimensional Euclidean space, denoted $\mathbb{E}^N$. To show that N-tuples form a vector space, the operations of vector addition and scalar multiplication must be specified. These operations are defined *componentwise*, with $|\psi\rangle + |\phi\rangle \equiv (x_1 + y_1, \ldots, x_N + y_N)$ and $a|\psi\rangle \equiv (ax_1, \ldots, ax_N)$. For real-valued scalars, $\mathbb{E}^N$ is denoted $\mathbb{R}^N$; for complex-valued scalars, $\mathbb{E}^N$ is denoted $\mathbb{C}^N$.
- The set of all *infinite sequences* of numbers $\psi = (x_1, \ldots, x_k, \ldots)$ having the property that $\sum_{k=1}^{\infty} |x_k|^2$ is finite, with addition and scalar multiplication defined componentwise, is a *sequence space*, l^2. Convergent infinite series do not naturally come to mind as "vectors," but they satisfy the requirements of a vector space.[6]
- The set of all continuous functions of a real variable x, with addition and scalar multiplication defined pointwise, $(\psi + \phi)(x) = \psi(x) + \phi(x)$ and $(a\psi)(x) = a\psi(x)$, is called a *function space.*
- The set of all *square-integrable functions* $\psi(x)$ of a real variable x for which $\int_a^b |\psi(x)|^2 \, dx$ is finite (for specified limits of integration), with addition and scalar multiplication defined pointwise, is a vector space known as $L^2[a, b]$. (This is the "arena" of quantum mechanics.[7,8])
- The set of all polynomials $p_N(x) = a_0 + a_1 x + \cdots + a_N x^N$, in the variable x, of degree at most N, and with real coefficients, is called a *polynomial vector space.*

1.2.1 A word on notation

We've used the *abstract* symbols $|\phi\rangle$, $|\chi\rangle$, and $|\psi\rangle$ in defining the axioms of vector spaces. This notation – *Dirac notation* – is widely used in quantum

[6] Minkowski's inequality [4, p. 30], $(\sum_{k=1}^{n} |x_k + y_k|^2)^{1/2} \leq (\sum_{k=1}^{n} |x_k|^2)^{1/2} + (\sum_{k=1}^{n} |y_k|^2)^{1/2}$, guarantees that the sum of two elements of l^2 stays in l^2. Treating infinite series as vectors is further discussed in Section 1.6.

[7] The sum of two square-integrable functions is square integrable, see [5, p. 43]. See also Exercise 1.5b.

[8] A *closed interval*, denoted $[a, b]$, indicates the set of points $\{x\}$ such that $a \leq x \leq b$.

physics and quantum mechanics and is a prominent reason physical scientists study vector spaces. Dirac notation was invented by (you guessed it) P.A.M. Dirac, one of the founders of quantum mechanics.[9] In quantum mechanics, physical states of systems are represented as elements of vector spaces (often denoted $|\psi\rangle$) because it's been found through much experimentation that states of quantum systems combine like elements of vector spaces. As seen in the aforementioned examples, membership in vector spaces applies to a variety of mathematical objects, and hence the use of abstract notation is apt. Dirac could have used $\vec{\psi}$ to indicate vectors (elements of vector spaces) as applied to quantum states; arrows, however, imply a *spatial* nature that's not apropos of quantum states. Vectors symbolized as $|\phi\rangle$ are known as *kets*. Kets are another word for vectors (physics vocabulary lesson). The word *ket* is derived from "bracket"; Dirac also introduced another notation, $\langle\psi|$ (referred to as *bras*), that we'll discuss in Section 1.3. A benefit of Dirac notation is that kets indicate vectors without specifying a representation.[10] With all this said, Dirac notation is not universally used outside of quantum mechanics.[11] In many ways Dirac notation is clumsy, such as in its representation of the norm of a vector. Other notations for vectors, such as bold symbols $\boldsymbol{x}$ are also used. To work with vector-space concepts in solving problems, one must be able to handle a variety of notations for elements of vector spaces. In what follows, we'll interchangeably use different notations for vectors, such as $|\psi\rangle$, ψ, or $\boldsymbol{x}$.

1.2.2 Linear independence, bases, and dimensionality

The definition of vector space does not specify the *dimension* of the space. The key notion for that purpose is *linear independence.*

Definition. *A set of vectors* $\{\psi_1, \ldots, \psi_N\}$ *(elements of a vector space) is* linearly independent *if for scalars* $\{a_k\}$*, the* linear combination $\sum_{k=1}^{N} a_k \psi_k = 0$ *holds only for the trivial case* $a_1 = \cdots = a_N = 0$. *Otherwise the set is* linearly dependent.

Linear independence means that *every nontrivial linear combination of vectors is different from zero.* Thus, no member of a linearly independent set can be expressed as a linear combination of the other vectors in the set.

---

[9]It's not widely appreciated because today Dirac notation appears in ostensibly all books on quantum mechanics, that Dirac first proposed what's become known as Dirac notation in a research article [6].

[10]By *representation* is meant a vector expressed as a linear combination of basis vectors (Section 1.2.2). For example, the electric field vector $\boldsymbol{E}$ can be represented in either spherical or cylindrical coordinates. When we write a vector simply as $\boldsymbol{E}$ or $|\psi\rangle$, we're not committing ourselves to a particular coordinate system.

[11]It's not a law of nature that we must use Dirac notation, but you need to be versed in it.

Definition. *A vector space is N-dimensional if it contains N linearly independent vectors, but not $N+1$.*

The notation $\dim \mathbb{V}$ is used to indicate the dimension of $\mathbb{V}$. Vector spaces containing an infinite number of linearly independent vectors are said to be *infinite dimensional*; l^2 and L^2 are examples. We work for the most part in this chapter with finite-dimensional vector spaces; we touch on infinite-dimensional spaces in Section 1.6.

A set of vectors $\{\psi_1, \ldots, \psi_N\}$, each an element of the vector space $\mathbb{V}$, *spans the space* if any vector $\psi \in \mathbb{V}$ can be expressed as a linear combination, $\psi = \sum_{k=1}^{N} a_k \psi_k$.

Definition. *A set of vectors $\{\psi_1, \ldots, \psi_N\}$ is a* basis *if (i) it spans the space and (ii) is linearly independent.*

Thus, a vector space is N-dimensional if (and only if) it has a basis of N vectors.[12]

Definition. *The numbers $a_1, \ldots, a_N$ in the linear combination $\psi = \sum_{k=1}^{N} a_k \psi_k$ are the* components *(or* coordinates*) of ψ with respect to the basis.*

Vector components are *unique* with respect to a given basis. Suppose ψ has a representation in the basis $\psi_1, \ldots, \psi_N$, with $\psi = \sum_{k=1}^{N} a_k \psi_k$. Assume we can find another representation in the same basis, with $\psi = \sum_{k=1}^{N} b_k \psi_k$. Subtract the two formulas for ψ: $0 = \sum_{k=1}^{N} (a_k - b_k)\psi_k$. Because the basis set $\{\psi_k\}$ is linearly independent, it must be that $b_k = a_k$. If we change the basis, however (a generic task, frequently done), the vector components will change as a result; see Section 1.4.7.

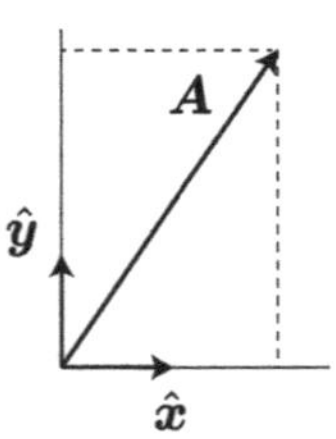

Figure 1.2 Vectors $\hat{\boldsymbol{x}}$ and $\hat{\boldsymbol{y}}$ span the space of vectors confined to a plane.

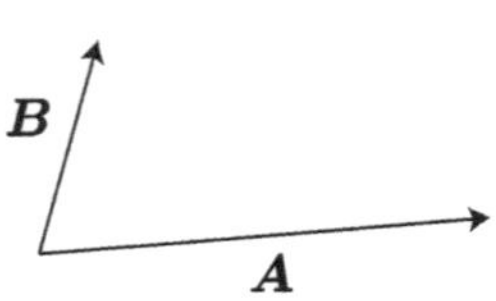

Figure 1.3 Vectors $\boldsymbol{A}$ and $\boldsymbol{B}$ are a basis for the space of vectors confined to a plane.

Examples

- Elementary vectors confined to a plane are elements of a two-dimensional vector space. We can use as a basis the unit vectors $\hat{\boldsymbol{x}}$ and $\hat{\boldsymbol{y}}$ shown in Figure 1.2 Any vector $\boldsymbol{A}$ in the plane can be constructed from a linear combination, $\boldsymbol{A} = a_x \hat{\boldsymbol{x}} + a_y \hat{\boldsymbol{y}}$; $\hat{\boldsymbol{x}}$ and $\hat{\boldsymbol{y}}$ span the space. They're also linearly independent: $\hat{\boldsymbol{y}}$ can't be expressed in terms of $\hat{\boldsymbol{x}}$. Basis vectors, however, don't have to be orthogonal (such as they are in Figure 1.2). *Any* two non colinear vectors $\boldsymbol{A}$ and $\boldsymbol{B}$, such as shown in Figure 1.3, can serve as a basis. No linear combination of $\boldsymbol{A}$ and $\boldsymbol{B}$ can sum to zero: $a\boldsymbol{A} + b\boldsymbol{B} \neq 0$. (Try it!) The only way we can force $a\boldsymbol{A} + b\boldsymbol{B} = 0$ with $a \neq 0$ and $b \neq 0$ is if $\boldsymbol{B} = -(a/b)\boldsymbol{A}$, i.e. if $\boldsymbol{B}$ is not independent of $\boldsymbol{A}$. We can easily add *three* vectors confined

[12]We omit a formal proof of this statement, which in other books might be given as a theorem.

to a plane to produce zero, $\boldsymbol{A}+\boldsymbol{B}+\boldsymbol{C}=0 \Longrightarrow \boldsymbol{C}=-(\boldsymbol{A}+\boldsymbol{B})$, in which case $\boldsymbol{C}$ is linearly dependent on $\boldsymbol{A}$ and $\boldsymbol{B}$. The maximum number of linearly independent vectors restricted to a plane is two, and hence, the set of all vectors confined to a plane is a two-dimensional vector space. Note that while there are an unlimited number of possible vectors confined to a plane, a plane is a two-dimensional space.

- For the space l^2 of infinite sequences $|\psi\rangle = (x_1, x_2, \dots)$ such that $\sum_{k=1}^{\infty} |x_k|^2$ is finite, a basis (called the *standard basis*) is $|\psi_1\rangle = (1, 0, \dots)$, $|\psi_2\rangle = (0, 1, 0, \dots)$, $|\psi_i\rangle = (0, \dots, 0, 1_i, 0, \dots)$ with 1 in the ith slot. Each $|\psi_i\rangle$ belongs to the space (the sum of the squares of its elements is finite). The vectors $\{|\psi_i\rangle\}$ span the space, $|\psi\rangle = \sum_{i=1}^{\infty} x_i |\psi_i\rangle$, and they are linearly independent. Thus, l^2 is infinite dimensional: It has an infinite number of basis vectors.
- For the space of Nth-degree polynomials in x, $p_N(x)$, a basis is $\phi_0(x) \equiv 1$, $\phi_1(x) \equiv x$, $\phi_2(x) \equiv x^2$, $\cdots$, $\phi_N(x) \equiv x^N$. The vectors $\{\phi_i(x)\}_{i=0}^{N}$ span the space, $p_N(x) = \sum_{k=0}^{N} a_k \phi_k(x)$, and they're linearly independent.[13] The space is therefore $(N+1)$-dimensional.
- The set of all functions $\psi(x)$ that satisfy the differential equation $\psi'' + \omega^2 \psi = 0$ is a two-dimensional function space: Any solution $\psi(x)$ can be expressed as a linear combination of basis functions (vectors) $\cos \omega x$ and $\sin \omega x$ so that $\psi(x) = a \cos \omega x + b \sin \omega x$.

1.2.3 Subspaces

Definition. *A subset* $\mathbb{V}'$ *of a vector space* $\mathbb{V}$ *is called a* subspace *of* $\mathbb{V}$ *if* $\mathbb{V}'$ *is a vector space under the same rules of addition and scalar multiplication introduced in* $\mathbb{V}$.[14]

Examples

- The zero vector $|0\rangle$ of $\mathbb{V}$ is a subspace of $\mathbb{V}$.
- The whole space $\mathbb{V}$ can be considered a subspace of $\mathbb{V}$ if a set can be considered a subset of itself. It's easier to allow the whole space to be a subspace (especially when we come to infinite-dimensional spaces).
- For $\mathbb{V}$ the three-dimensional position space, any plane in $\mathbb{V}$ passing through the origin is a subspace of $\mathbb{V}$.
- For the space of N-tuples, for any vector $|\psi\rangle = (\psi_1, \psi_2, \dots, \psi_N)$, all vectors obtained by setting $\psi_1 = 0$ form a subspace.

---

[13] Can one, for example, express x^2 as a linear combination, $x^2 = a + bx$, *for all* x? Such an equality might hold for specific values of x, but not for arbitrary x.

[14] Defining subspaces of infinite-dimensional spaces (see Section 1.6) involves an extra consideration, that a set of vectors be *closed*.

1.2.4 Isomorphism of N-dimensional spaces

Two vector spaces over the same field and having the same dimension are instances of the "same" vector space. Consider elementary vectors $\boldsymbol{A}$ in three-dimensional space. Once a basis $\hat{\boldsymbol{x}}$, $\hat{\boldsymbol{y}}$, and $\hat{\boldsymbol{z}}$ has been chosen, so that $\boldsymbol{A} = a_x\hat{\boldsymbol{x}} + a_y\hat{\boldsymbol{y}} + a_z\hat{\boldsymbol{z}}$, there is a *correspondence* between $\boldsymbol{A}$ and the three-tuple (a_x, a_y, a_z), and this correspondence *exists between every element of each set.* This "sameness" of vector spaces is called an *isomorphism.*

Definition. *Two vector spaces* $\mathbb{V}$ *and* $\mathbb{V}'$ *are* isomorphic *if it's possible to establish a one-to-one correspondence* $\psi \leftrightarrow \psi'$ *between* $\psi \in \mathbb{V}$ *and* $\psi' \in \mathbb{V}'$ *such that if* $\psi \leftrightarrow \psi'$ *and* $\phi \leftrightarrow \phi'$*,*

1. the vector which this correspondence associates with $\psi + \phi$ is $\psi' + \phi'$, and
2. the vector which this correspondence associates with $a\phi$ is $a\phi'$.

All vector spaces of dimension N are isomorphic. Consider that for isomorphic spaces $\mathbb{V}$ and $\mathbb{V}'$, if $\phi, \chi, \ldots$ are in $\mathbb{V}$ and $\phi', \chi', \ldots$ are their counterparts in $\mathbb{V}'$, the equation $a\phi + b\chi + \cdots = 0$ corresponds to $a\phi' + b\chi' + \cdots = 0$. Thus the counterparts in $\mathbb{V}'$ of linearly independent vectors in $\mathbb{V}$ are also linearly independent. The maximum number of linearly independent vectors in $\mathbb{V}'$ is therefore the same as that in $\mathbb{V}$, i.e. the dimensions of $\mathbb{V}$ and $\mathbb{V}'$ are the same.

1.2.5 Dual spaces

Definition. *Linear scalar-valued functions of vectors are called* linear functionals.[15]

Linear functionals F assign a scalar $F(\phi)$ to each vector ϕ such that for any vectors ϕ and ψ and scalars a and b, the property of linearity holds:

$$F(a\phi + b\psi) = aF(\phi) + bF(\psi).$$

Let $\boldsymbol{e}_1, \ldots, \boldsymbol{e}_N$ be a basis for $\mathbb{V}_N$. Any vector $\psi \in \mathbb{V}_N$ can be expressed as a linear combination, $\psi = \sum_{k=1}^{N} a_k \boldsymbol{e}_k$. The action of a linear functional is (by definition)

$$F(\psi) = F\left(\sum_k a_k \boldsymbol{e}_k\right) = \sum_{k=1}^{N} a_k F(\boldsymbol{e}_k) \equiv \sum_{k=1}^{N} a_k f_k, \tag{1.1}$$

[15] The action of F is indicated as a *mapping* $F : \mathbb{V} \to \mathbb{F}$, which informs us that F associates an element of $\mathbb{V}$, the *domain* of the mapping, with an element of $\mathbb{F}$, the *codomain* or the *target space* of the mapping.

where $f_k = F(\boldsymbol{e}_k)$. The value of F acting on $\psi \in \mathbb{V}_N$ is determined by the values of F acting on the basis vectors for $\mathbb{V}_N$. Because bases are not unique, one would seemingly have reason to worry that the value of $F(\psi)$ is basis dependent. Such is not the case, however, but to prove that here would be getting ahead of ourselves.[16]

As a consequence of linearity, consider that

$$F(|0\rangle) = F(0|0\rangle) = 0F(|0\rangle) = 0. \tag{1.2}$$

Thus, F maps the zero vector onto the scalar zero. This property rules out adding a constant to Eq. (1.1), as in $F(\psi) = \sum_{k=1}^{N} a_k f_k + \beta$; $F(|0\rangle) = 0$ implies $\beta = 0$.

Linear functionals form a vector space in their own right, called the *dual space*, often denoted[17] $\mathbb{V}_N^*$, which is closely connected with $\mathbb{V}_N$. The set of all linear functionals forms a vector space because they have the properties of a linear space:

(i) $(F_1 + F_2)(\phi) = F_1(\phi) + F_2(\phi)$,

(ii) $(aF)(\phi) = aF(\phi)$.

Definition. *For $\mathbb{V}_N$ an N-dimensional vector space, the* dual space $\mathbb{V}_N^*$ *is the vector space whose elements are linear functionals on $\mathbb{V}_N$. Addition and scalar multiplication in $\mathbb{V}_N^*$ follow from the rules of addition and scalar multiplication in $\mathbb{V}_N$.*

Relative to a given basis $\{\boldsymbol{e}_i\}_{i=1}^N$ in $\mathbb{V}_N$, every linear functional F is uniquely determined by the N-tuple $(f_1, \ldots, f_N)$, where $f_k = F(\boldsymbol{e}_k)$. This correspondence is preserved under vector addition and multiplication of vectors by scalars; it follows that $\mathbb{V}_N^*$ is isomorphic to $\mathbb{V}_N$. The dual space associated with $\mathbb{V}_N$ is N-dimensional.

---

[16]Under a change of basis (see Section 1.4.7), the quantities f_k transform in a way that "undoes" how the vector components a_k transform, with the result that $\sum_k a_k f_k$ is invariant, independent of basis – a *second* use of the word "scalar," one more familiar to physicists than scalars are elements of a field $\mathbb{F}$; scalars are technically zeroth-order tensors (Chapter 11). Physics is concerned with statements that are independent of the choice of basis, i.e. independent of coordinate system. Coordinates are a necessary evil in physics – we need them to make measurements and to do calculations, but they don't exist in nature; they're artifacts of our thinking. Some concepts, such as the dimension of a vector space, explicitly depend on a basis. Yet, vector spaces having the same dimension are all isomorphic. In the same way, the goal of physics is to formulate laws of nature that if true in one coordinate system, are true in all coordinate systems. To vanquish coordinates, transcend them!

[17]The asterisk in $\mathbb{V}_N^*$ has nothing to do with complex conjugation – it simply denotes the dual space. The dual space is not used very much in this chapter; it becomes important when we discuss tensors, Chapter 11.

1.3 Inner products and orthogonality

Thus, vectors – elements of vector spaces – generalize elementary vectors. Vectors in N-dimensional spaces generalize the notion of *direction*. Direction is a concept associated with our experience in three-dimensional space, and is specified by the components of a vector relative to a given set of basis vectors. We can't intuitively speak of direction in spaces of arbitrary dimension, but we can continue to speak of the components of vectors relative to bases in spaces of any dimension. What about the other half of the maxim that vectors have direction and magnitude? We've so far ignored the notion of the *length* of a vector. We address that point now.

1.3.1 Inner products

Can vectors be multiplied by vectors? Nothing in the definition of vector space tells us how to do that. One way is with the *inner product*,[18] a rule that associates a scalar to a pair of vectors. Vector spaces equipped with the additional property of an inner product are known as *inner product spaces*.[19]

Definition. *An* inner product *for vector space* $\mathbb{V}$ *over field* $\mathbb{F}$ *associates a scalar, denoted* (ϕ, ψ)*, with every pair of vectors* $\phi, \psi \in \mathbb{V}$ *(for* $a \in \mathbb{F}$*), such that it satisfies:*

(i) $(\psi, \phi) = (\phi, \psi)^*$,

(ii) $(\psi, a\phi) = a(\psi, \phi)$,

(iii) $(\psi, \phi + \chi) = (\psi, \phi) + (\psi, \chi)$,

(iv) $(\psi, \psi) \geq 0$ with equality holding for $\psi = 0$, (positive definite)

where $(*)$ denotes complex conjugation.[20,21] Requirements (i) and (ii) imply that $(a\phi, \psi) = a^*(\phi, \psi)$.

Definition. *The* norm *(or* length*) of a vector* ϕ*, denoted* $\|\phi\|$*, is the square root of the inner product of a vector with itself,* $\|\phi\| \equiv \sqrt{(\phi, \phi)}$.

---

[18] An *outer product* also exists as a way of multiplying vectors; see Section 1.4.8.

[19] That is, inner products are not intrinsic to vector spaces. Adding a rule for multiplying vectors in the form of an inner product is a new mathematical structure, an inner-product space.

[20] The rules for inner product as specified here are used in many areas of physics and especially in quantum mechanics. The inner product used in special relativity is defined differently; notably the requirement of positive definiteness is relaxed, apropos of the difference between Euclidean and non-Euclidean geometries.

[21] Note how mathematics works. In the definition of vector space (Section 1.2), nowhere is it stated *how* its elements are to be added and multiplied by numbers; each instance of a vector space comes with its own rules of combination. Similarly with the inner product – nowhere is it specified what the form of the function (ϕ, ψ) should be so that it satisfies the requirements.

Examples

- The *dot product* of elementary vector analysis satisfies the definition of inner product, $\boldsymbol{A}\cdot\boldsymbol{B} \equiv |A||B|\cos\theta$, where θ is the angle between the vectors when their "tails" are joined together (see Figure 1.4) and $|A| = \sqrt{\boldsymbol{A}\cdot\boldsymbol{A}}$ denotes the magnitude of $\boldsymbol{A}$.
- For $\mathbb{E}^N$, the space of N-tuples, let $\psi = (\xi_1, \xi_2, \ldots, \xi_N)$ and $\phi = (\eta_1, \eta_2, \ldots, \eta_N)$. The operation $(\psi, \phi) = \sum_{k=1}^{N} \xi_k^* \eta_k$ satisfies the requirements of an inner product.
- For the space l^2 of infinite sequences, if $|\psi\rangle = (x_1, x_2, \ldots, x_k, \ldots)$ and $|\phi\rangle = (y_1, y_2, \ldots, y_k, \ldots)$, then $(\psi, \phi) = \sum_{k=1}^{\infty} x_k^* y_k$ satisfies the requirements for an inner product. The condition for l^2, that $\sum_{k=1}^{\infty} |x_k|^2$ is finite, is the requirement that the norm $\|\psi\|$ is finite.
- For the space $L^2[a,b]$ of square-integrable functions $\phi(x)$ and $\psi(x)$, the following rule meets the requirements of an inner product:

$$\langle\phi|\psi\rangle \equiv \int_a^b \phi^*(x)\psi(x)\,\mathrm{d}x. \tag{1.3}$$

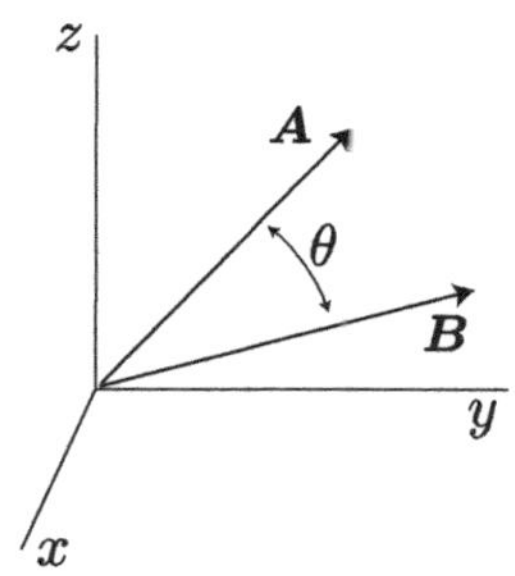

Figure 1.4 Dot product of elementary vectors, $\boldsymbol{A}\cdot\boldsymbol{B} = |A||B|\cos\theta$.

The condition for L^2, that $\int_a^b |\psi(x)|^2\,\mathrm{d}x$ is finite, is again the requirement that the norm $\|\psi\|$ is finite. That $\|\psi\|$ is finite (for all ψ) implies that the inner product of two vectors is finite.[22] Note that the complex conjugate in Eq. (1.3) ensures a positive result for $\langle\phi|\phi\rangle = \int_a^b |\phi(x)|^2\,\mathrm{d}x$.

Dirac notation $\langle\phi|\psi\rangle$ has been used in Eq. (1.3) to indicate the inner product, instead of (ϕ, ψ). The notation arises from the fact that a linear functional F_ψ can be defined for each[23] ψ by the requirement that $F_\psi(\phi) = (\psi, \phi)$ for every vector ϕ. Linear functionals $F_\psi(\phi) = (\psi, \phi)$ are written in Dirac notation simply as $\langle\psi|$ (termed *bras*), so that $(\psi, \phi) \equiv \langle\psi|\phi\rangle$; $\langle\psi|$ operates on $|\phi\rangle$ to produce (ψ, ϕ). For the inner product of L^2 (Eq. (1.3)), the functional $\langle\psi|$ is the operator $\langle\psi| \equiv \int \mathrm{d}x\, \psi^*(x)\, \cdot$, where $(\cdot)$ is a placeholder for the vector ϕ, so that $\langle\psi|\phi\rangle = \int \mathrm{d}x\, \psi^*(x)\phi(x)$.

1.3.2 The Schwarz inequality

From the dot product, the angle between elementary vectors can be inferred from the relation $\cos\theta = \boldsymbol{A}\cdot\boldsymbol{B}/(|\boldsymbol{A}||\boldsymbol{B}|)$. For θ to be calculated in this way, it must be the case that $-1 \le \boldsymbol{A}\cdot\boldsymbol{B}/(|\boldsymbol{A}||\boldsymbol{B}|) \le 1$ for all $\boldsymbol{A}$ and $\boldsymbol{B}$. This inequality finds its generalization for any inner product in the *Schwarz inequality*:

$$|(\phi, \psi)|^2 \le (\phi, \phi)(\psi, \psi), \tag{1.4}$$

where equality holds when ϕ and ψ are linearly dependent.

[22]For a proof, see [5, p. 42]; see also Exercise 1.5a.

[23]The inner product is a function involving pairs of vectors, whereas linear functionals are functions of a single vector. Hence the careful statement, for each vector ψ, a linear functional is defined by $F_\psi(\phi) = (\psi, \phi)$. The fact that there is a functional F_ψ for each vector ψ underscores the isomorphism between $\mathbb{V}_N$ and its dual space, $\mathbb{V}_N^*$.

To prove the Schwarz inequality, introduce $\psi + t\phi$, where t is any complex number. Because the inner product is positive definite, we must have $(\psi + t\phi, \psi + t\phi) \geq 0$, or

$$(\phi, \phi)|t|^2 + t(\psi, \phi) + t^*(\phi, \psi) + (\psi, \psi) \geq 0. \tag{1.5}$$

The inequality (1.5) is valid for all t, and thus it remains valid for any particular t. The Schwarz inequality follows under the substitution $t = -(\phi, \psi)/(\phi, \phi)$ in (1.5).[24]

1.3.3 Vector norms

The norm has the following properties, for any vectors ψ and ϕ, and for any scalar a,

(i) $\|\psi\| \geq 0$, with equality holding only for $\psi = |0\rangle$,

(ii) $\|a\psi\| = |a| \, \|\psi\|$, and

(iii) $\|\psi + \phi\| \leq \|\psi\| + \|\phi\|$. (triangle inequality)

Properties (i) and (ii) follow by definition. Property (iii) can be derived using the Schwarz inequality (see Exercise 1.7): If ψ, ϕ, and $\psi + \phi$ are envisioned as forming the sides of a triangle, the length of the side $\psi + \phi$ is less than the sum of the lengths of the other two sides (hence, the "triangle inequality").

1.3.4 Orthonormal bases and the Gram–Schmidt process

With vector spaces endowed with an inner product, we can define orthogonality.

Definition. *Vectors ϕ, ψ in an inner-product space are* orthogonal *if* $(\phi, \psi) = 0$.

Elementary vectors are orthogonal when the angle between them is 90°; when $\boldsymbol{A} \cdot \boldsymbol{B} = 0$. The definition of orthogonality generalizes $\boldsymbol{A} \cdot \boldsymbol{B} = 0$ to any inner product, $(\phi, \psi) = 0$.

Any set of linearly independent vectors that spans a space is a basis for that space. Nothing requires basis vectors to be orthogonal. Bases comprised of orthogonal vectors, however, have many convenient properties.[25] In this section, we show that an *orthonormal* basis (orthogonal vectors

---

[24] Note that the Schwarz inequality has been derived using only the defining properties of the inner product. We don't have to prove the Schwarz inequality every time we encounter a new inner product space.

[25] In Exercise 1.8 it's shown that a set of mutually orthogonal vectors is linearly independent.

having unit norm) can always be constructed from an arbitrary basis. No loss of generality is implied therefore by working with orthonormal bases.

Definition. *A set of vectors* $\{\hat{\mathbf{e}}_1, \hat{\mathbf{e}}_2, \ldots, \hat{\mathbf{e}}_N\}$ *is called an* orthonormal set *if*

$$\langle \hat{\mathbf{e}}_i | \hat{\mathbf{e}}_j \rangle = \delta_{ij} \equiv \begin{cases} 1 & \text{if } i = j \\ 0 & \text{if } i \neq j, \end{cases}$$

where δ_{ij} is the *Kronecker delta symbol.*

Example. The basis for $\mathbb{E}^N$, $\{\hat{\mathbf{e}}_i = (0, \ldots, 0, 1_i, 0, \ldots, 0)\}_{i=1}^N$ is an orthonormal set.

Let $\{\hat{\mathbf{e}}_1, \hat{\mathbf{e}}_2, \ldots, \hat{\mathbf{e}}_N\}$ be an orthonormal basis for $\mathbb{V}_N$. We can express any vector $\boldsymbol{a} \in \mathbb{V}_N$ as the linear combination $\boldsymbol{a} = a_1 \hat{\mathbf{e}}_1 + a_2 \hat{\mathbf{e}}_2 + \cdots + a_N \hat{\mathbf{e}}_N = \sum_{i=1}^N a_i \hat{\mathbf{e}}_i$. The components a_j are easily found when the basis is orthonormal because

$$\langle \hat{\mathbf{e}}_j | \boldsymbol{a} \rangle = \left\langle \hat{\mathbf{e}}_j \middle| \sum_{i=1}^N a_i \hat{\mathbf{e}}_i \right\rangle = \sum_{i=1}^N a_i \langle \hat{\mathbf{e}}_j | \hat{\mathbf{e}}_i \rangle = \sum_{i=1}^N a_i \delta_{ji} = a_j. \tag{1.6}$$

In a general basis, a given set of scalars $\{a_j\}$ determines a vector, in an orthonormal basis, the coefficients $\{a_j\}$ are determined as in Eq. (1.6), $a_j = \langle \hat{\mathbf{e}}_j | \boldsymbol{a} \rangle$. In this case, a_j is simply the *projection* of $|\boldsymbol{a}\rangle$ along the "direction" $|\hat{\mathbf{e}}_j\rangle$. From Eq. (1.6), any vector $|\psi\rangle$ can be expressed as a linear combination of orthonormal basis vectors in the form

$$|\psi\rangle = \sum_{i=1}^N \langle \hat{\mathbf{e}}_i | \psi \rangle | \hat{\mathbf{e}}_i \rangle. \tag{1.7}$$

Moreover, in terms of this orthonormal basis, with $\boldsymbol{a} = \sum_{i=1}^N a_i \hat{\mathbf{e}}_i$ and $\boldsymbol{b} = \sum_{j=1}^N b_j \hat{\mathbf{e}}_j$,

$$\begin{aligned} \langle \boldsymbol{a} | \boldsymbol{b} \rangle &= \left\langle \sum_{i=1}^N a_i \hat{\mathbf{e}}_i \middle| \sum_{j=1}^N b_j \hat{\mathbf{e}}_j \right\rangle = \sum_{i=1}^N \sum_{j=1}^N a_i^* b_j \langle \hat{\mathbf{e}}_i | \hat{\mathbf{e}}_j \rangle \\ &= \sum_{i=1}^N \sum_{j=1}^N a_i^* b_j \delta_{ij} = \sum_{i=1}^N a_i^* b_i. \end{aligned} \tag{1.8}$$

This result is sometimes used to define the inner product (but it requires that $\boldsymbol{a}$ and $\boldsymbol{b}$ have a known expansion in terms of an orthonormal basis). Equation (1.8) is the inner product for $\mathbb{E}^N$. Any N-dimensional space can therefore be identified with $\mathbb{E}^N$ through the use of an orthonormal basis.[26]

[26] In a sense, there is only one finite-dimensional vector space, $\mathbb{E}^N$ – all spaces of the same dimension are isomorphic, and arbitrary bases can be made orthonormal. One might

Gram–Schmidt process

Let $\{\boldsymbol{v}_1, \boldsymbol{v}_2, \dots, \boldsymbol{v}_N\}$ be a set of linearly independent vectors in $\mathbb{V}_N$. The $\boldsymbol{v}_i$ need not be orthogonal or normalized.[27] The basic idea behind the Gram–Schmidt process is to note that $\langle \boldsymbol{v}_i | \boldsymbol{a} \rangle$ is the projection of $\boldsymbol{a}$ in the "direction" of $\boldsymbol{v}_i$. The vector

$$\boldsymbol{u} \equiv \boldsymbol{a} - \frac{\langle \boldsymbol{v}_i | \boldsymbol{a} \rangle}{\langle \boldsymbol{v}_i | \boldsymbol{v}_i \rangle} \boldsymbol{v}_i$$

is orthogonal to $\boldsymbol{v}_i$ – a fact that can be readily verified by direct calculation:

$$\langle \boldsymbol{v}_i | \boldsymbol{u} \rangle = \left\langle \boldsymbol{v}_i \middle| \boldsymbol{a} - \frac{\langle \boldsymbol{v}_i | \boldsymbol{a} \rangle}{\langle \boldsymbol{v}_i | \boldsymbol{v}_i \rangle} \boldsymbol{v}_i \right\rangle = \langle \boldsymbol{v}_i | \boldsymbol{a} \rangle - \frac{\langle \boldsymbol{v}_i | \boldsymbol{a} \rangle}{\langle \boldsymbol{v}_i | \boldsymbol{v}_i \rangle} \langle \boldsymbol{v}_i | \boldsymbol{v}_i \rangle = \langle \boldsymbol{v}_i | \boldsymbol{a} \rangle - \langle \boldsymbol{v}_i | \boldsymbol{a} \rangle = 0.$$

We can construct an orthogonal basis from $\{\boldsymbol{v}_i\}$ as follows. Let $\boldsymbol{u}_1 = \boldsymbol{v}_1$. Now set

$$\boldsymbol{u}_2 = \boldsymbol{v}_2 - \frac{\langle \boldsymbol{u}_1 | \boldsymbol{v}_2 \rangle}{\langle \boldsymbol{u}_1 | \boldsymbol{u}_1 \rangle} \boldsymbol{u}_1$$

($\boldsymbol{u}_2$ is orthogonal to $\boldsymbol{u}_1$). Next, set

$$\boldsymbol{u}_3 = \boldsymbol{v}_3 - \frac{\langle \boldsymbol{u}_2 | \boldsymbol{v}_3 \rangle}{\langle \boldsymbol{u}_2 | \boldsymbol{u}_2 \rangle} \boldsymbol{u}_2 - \frac{\langle \boldsymbol{u}_1 | \boldsymbol{v}_3 \rangle}{\langle \boldsymbol{u}_1 | \boldsymbol{u}_1 \rangle} \boldsymbol{u}_1$$

($\boldsymbol{u}_3$ is orthogonal to $\boldsymbol{u}_2$ and $\boldsymbol{u}_1$). The process is to subtract from each vector $\boldsymbol{v}_k$ its projections in the direction of the previously orthogonalized vectors so that the set $\{\boldsymbol{u}_1, \boldsymbol{u}_2, \dots, \boldsymbol{u}_N\}$ is orthogonal. If we further divide each element of this set by its norm so that

$$\hat{\boldsymbol{e}}_i = \frac{\boldsymbol{u}_i}{\sqrt{\langle \boldsymbol{u}_i | \boldsymbol{u}_i \rangle}}$$

then the set $\{\hat{\boldsymbol{e}}_1, \hat{\boldsymbol{e}}_2, \dots, \hat{\boldsymbol{e}}_N\}$ is orthonormal. We've defined the Gram–Schmidt process using a finite-dimensional vector space, but the method can also be applied to separable Hilbert spaces (see Section 1.6) [5, p. 67].

Example. Consider vectors $\boldsymbol{v}_1 = 3\hat{\boldsymbol{x}} + \hat{\boldsymbol{y}}$ and $\boldsymbol{v}_2 = 2\hat{\boldsymbol{x}} + 2\hat{\boldsymbol{y}}$, which are non orthogonal ($\boldsymbol{v}_1 \cdot \boldsymbol{v}_2 = 8$). The Gram–Schmidt process can be used to

think, therefore, there is no point in studying any particular vector space – just study $\mathbb{E}^N$. There's a catch, however. The most important properties of vectors and vector spaces are those that are independent of coordinate system. If we studied only $\mathbb{E}^N$, we'd be tied to a particular basis. While all N-dimensional vector spaces are isomorphic, any particular application will involve a vector space best suited for that application.

[27] A *normalized* vector $\hat{\boldsymbol{X}}$, associated with vector $\boldsymbol{X}$, is obtained by dividing the vector by its magnitude, $\hat{\boldsymbol{X}} = \boldsymbol{X}/\|\boldsymbol{X}\|$. Any vector may be represented as $\boldsymbol{X} = \|\boldsymbol{X}\|\hat{\boldsymbol{X}}$

construct orthonormal vectors from $\boldsymbol{v}_1$ and $\boldsymbol{v}_2$: Set $\boldsymbol{u}_1 = \boldsymbol{v}_1 = 3\hat{\boldsymbol{x}} + \hat{\boldsymbol{y}}$. Now set

$$\boldsymbol{u}_2 = \boldsymbol{v}_2 - \frac{\boldsymbol{u}_1 \cdot \boldsymbol{v}_2}{\boldsymbol{u}_1 \cdot \boldsymbol{u}_1}\boldsymbol{u}_1 = \frac{1}{10}(-4\hat{\boldsymbol{x}} + 12\hat{\boldsymbol{y}}),$$

which is orthogonal to $\boldsymbol{u}_1$ (check it!). Normalize each vector to produce an orthonormal set:

$$\hat{\mathbf{e}}_1 = \frac{1}{\sqrt{10}}(3\hat{\boldsymbol{x}} + \hat{\boldsymbol{y}}) \qquad \hat{\mathbf{e}}_2 = \frac{1}{\sqrt{10}}(-\hat{\boldsymbol{x}} + 3\hat{\boldsymbol{y}}).$$

As can be verified, $\hat{\mathbf{e}}_1 \cdot \hat{\mathbf{e}}_2 = 0$, $\hat{\mathbf{e}}_1 \cdot \hat{\mathbf{e}}_1 = 1$, and $\hat{\mathbf{e}}_2 \cdot \hat{\mathbf{e}}_2 = 1$.

1.3.5 Complete sets of orthonormal vectors

For an inner-product space $\mathbb{V}_N$, let $\mathbb{X}_k$ denote a set of k mutually orthonormal vectors, $\mathbb{X}_k \equiv \{\hat{\mathbf{e}}_1, \ldots, \hat{\mathbf{e}}_k\}$, where $1 \le k \le N$. That is, $(\hat{\mathbf{e}}_i, \hat{\mathbf{e}}_j) = \delta_{ij}$ for $i, j = 1, \ldots, k$.

Definition. *An orthonormal set of vectors is called* complete *if it is not contained in a larger orthonormal set.*

Example. Finite-dimensional inner-product spaces automatically have complete orthonormal bases. A basis is the maximum number of linearly independent vectors that span the space; an orthonormal basis (achievable through the Gram–Schmidt process) is therefore a complete set because it's not a subset of a larger set of orthonormal vectors.

The notion of completeness – not particularly sophisticated for finite-dimensional spaces – is quite important for infinite-dimensional vector spaces (Section 1.6); yet it's easier to explain for a finite number of dimensions. The idea is that there's no mutually orthogonal vector "outside" a complete set, and hence, there is no vector in the space not expressible as a linear combination of the vectors in a complete set. Let $\mathbb{X}_k = \{\hat{\mathbf{e}}_1, \ldots, \hat{\mathbf{e}}_k\}$ be a set of orthonormal vectors in inner-product space $\mathbb{V}_N$, with $k < N$. Mutually orthogonal vectors are linearly independent (Exercise 1.8), and the vectors in $\mathbb{X}_k$ span a k-dimensional subspace of $\mathbb{V}_N$. For $\psi \in \mathbb{V}_N$, and for $\alpha_i \equiv (\hat{\mathbf{e}}_i, \psi)$, the nonzero vector $\psi' \equiv \psi - \sum_{i=1}^k \alpha_i \hat{\mathbf{e}}_i$ is orthogonal to each vector $\hat{\mathbf{e}}_j \in \mathbb{X}_k$:

$$(\hat{\mathbf{e}}_j, \psi') = (\hat{\mathbf{e}}_j, \psi) - \sum_{i=1}^{k} \alpha_i (\hat{\mathbf{e}}_j, \hat{\mathbf{e}}_i) = \alpha_j - \sum_{i=1}^{k} \alpha_i \delta_{ij} = \alpha_j - \alpha_j = 0. \quad (1.9)$$

Thus, there is "more" to $\psi \in \mathbb{V}_N$ than what can be expressed in the basis formed by the vectors in $\mathbb{X}_k$. If $k = N$, then $\psi' = 0$ (see Eq. (1.7)), and Eq. (1.9) is trivially satisfied. The norm of the part of the vector expressed in the basis vectors of $\mathbb{X}_k$ is always less than the norm of the vector $\psi \in \mathbb{V}_N$ (*Bessel's inequality*):

$$\sum_{i=1}^{k} |\alpha_i|^2 \leq \|\psi\|^2, \tag{1.10}$$

with equality holding for $k = N$.

There are several equivalent conditions for an orthonormal set $\mathbb{X}$ to be complete, which we state as a theorem without proof [3, p. 124].

Theorem 1.3.1. *If $\mathbb{X} = \{\hat{\mathbf{e}}_i\}$ is any finite orthonormal set in an inner product space $\mathbb{V}_N$, the following six conditions are equivalent:*

1. $\mathbb{X}$ is complete.
2. If $(\hat{\mathbf{e}}_i, \psi) = 0$ for $i = 1, \ldots, N$, then $\psi = 0$ (i.e. there is nothing "outside" the set).
3. The subspace spanned by $\mathbb{X}$ is the whole space $\mathbb{V}_N$.
4. If $\psi \in \mathbb{V}_N$, then $\psi = \sum_{i=1}^{N} (\hat{\mathbf{e}}_i, \psi)\hat{\mathbf{e}}_i$.
5. For $\phi, \psi \in \mathbb{V}_N$, then $(\phi, \psi) = \sum_{i=1}^{N} (\phi, \hat{\mathbf{e}}_i)(\hat{\mathbf{e}}_i, \psi)$ (Parseval's identity).
6. For $\psi \in \mathbb{V}_N$, $\|\psi\|^2 = \sum_{i=1}^{N} |(\hat{\mathbf{e}}_i, \psi)|^2$.

1.4 Operators and matrices

Now that we've defined vector spaces as collections of objects (vectors) that combine linearly (which represent physical quantities), and how to assign lengths to vectors (by inner products), we want to *do* something with vectors. Physics is not concerned merely with the existence of vector quantities, but with processes that drive transitions between states of systems that are described by vectors. In the theory of quantum mechanics, any possible state of a system is represented by a vector $|\psi\rangle$. What can be measured on a system in state $|\psi\rangle$ (its "observables") is represented in the theory as a *linear operator* (to be defined momentarily) – i.e. the act of measurement is an operation applied to a system; one that affects the system. In a sense, our interest in vector spaces derives from the fact that that's where linear operators are defined! Quantum mechanics isn't the only reason we should study linear operators (rotations, for example), but it's a big one.[28]

[28]To quote Richard Feynman: "... nature isn't classical, dammit, and if you want to make a simulation of nature, you'd better make it quantum mechanical ..." [7].

1.4.1 Linear operators

An operator $\mathcal{A}$ acts on elements of a vector space and transforms them into other vectors, not necessarily in the same space. That is, for $\boldsymbol{x} \in \mathbb{V}$, $\mathcal{A}$ associates $\boldsymbol{x}$ with another, unique vector $\boldsymbol{y} \in \mathbb{W}$, the action of which we can write symbolically as $\boldsymbol{y} = \mathcal{A}\boldsymbol{x}$ (i.e. the operator operates to the right). Thus, $\mathcal{A}$ defines a mapping between vector spaces, $\mathcal{A} : \mathbb{V} \to \mathbb{W}$, where $\mathbb{V}$ and $\mathbb{W}$ are defined over the same set of scalars. In many cases of interest, $\mathbb{W} = \mathbb{V}$, but there are exceptions. When $\mathbb{W} = \mathbb{F}$, $\mathcal{A}$ is a linear functional; the set of scalars is a one-dimensional vector space.

Definition. *An operator is* linear *if for all scalars α and β and all vectors $\boldsymbol{a}$ and $\boldsymbol{b}$ in $\mathbb{V}$*

$$\mathcal{A}(\alpha\boldsymbol{a} + \beta\boldsymbol{b}) = \alpha\mathcal{A}\boldsymbol{a} + \beta\mathcal{A}\boldsymbol{b}.$$

Two operators $\mathcal{A}$ and $\mathcal{B}$ are equal, $\mathcal{A} = \mathcal{B}$, if $\mathcal{A}\psi = \mathcal{B}\psi$ for all ψ. We're often interested in the compound effect of two or more operators applied successively to a vector. The product of two operators acting in succession is denoted $\mathcal{AB}|\psi\rangle \equiv \mathcal{A}(\mathcal{B}|\psi\rangle)$. The order of operators is important – see the following.

Examples

- **Identity operator**: The *identity operator* $\mathcal{I}$ leaves any vector unchanged: $\mathcal{I}|\psi\rangle \equiv |\psi\rangle$. Scalar multiplication, $b|\psi\rangle$, is therefore the effect of the *multiplicative operator* $b\,\mathcal{I}$. An explicit expression for $\mathcal{I}$ emerges when the expansion of a vector in an orthonormal basis, Eq. (1.7), is expressed in Dirac notation: $|\psi\rangle = \left(\sum_{i=1}^{N} |\hat{\boldsymbol{e}}_i\rangle\langle\hat{\boldsymbol{e}}_i|\right)|\psi\rangle$. The construct $\sum_{i=1}^{N} |\hat{\boldsymbol{e}}_i\rangle\langle\hat{\boldsymbol{e}}_i|$ is therefore the identity operator,

$$\sum_{i=1}^{N} |\hat{\boldsymbol{e}}_i\rangle\langle\hat{\boldsymbol{e}}_i| = \mathcal{I}. \tag{1.11}$$

  Equation (1.11) is referred to as the *completeness relation* (see Section 1.4). Parseval's identity can be written[29] using Eq. (1.11): $\langle\phi|\psi\rangle = \langle\phi|\mathcal{I}|\psi\rangle = \langle\phi|\left(\sum_i |\hat{\boldsymbol{e}}_i\rangle\langle\hat{\boldsymbol{e}}_i|\right)|\psi\rangle$.
- **Integral operators**: For a space of continuous functions defined on $[a, b]$,

$$\mathcal{A}f(t) \equiv \int_a^b A(t,x)f(x)\,\mathrm{d}x \tag{1.12}$$

  defines a continuous function of t, and hence, $\mathcal{A}$ is a linear operator on this space, an *integral operator*. (Note that, generally, a and b may depend on t.) The operator $\mathcal{A}$ defined by Eq. (1.12) has the properties of a linear operator:

[29] The identity operator in the form of Eq. (1.11) may be placed in any formula wherever it's to our advantage to do so; a trick referred to as "inserting a complete set of states."

i. $\mathcal{A}(f_1 + f_2) = \int_a^b A(t,x)(f_1(x) + f_2(x))\,\mathrm{d}x$
$= \int_a^b A(t,x)f_1(x)\,\mathrm{d}x + \int_a^b A(t,x)f_2(x)\,\mathrm{d}x = \mathcal{A}f_1 + \mathcal{A}f_2$;

ii. $\mathcal{A}(\lambda f) = \int_a^b A(t,x)\lambda f(x)\,\mathrm{d}x = \lambda \int_a^b A(t,x)f(x)\,\mathrm{d}x = \lambda \mathcal{A}f$.

We'll encounter integral operators in Chapter 4 (Fourier transform) and in Chapter 10 (integral equations).

- **Differential operators**: Linear combinations of the differentiation operator naturally comprise linear operators – *differential operators* – on the space of smooth functions[30] of the real variable x. For example, for continuous functions $\{p_i(x)\}$,

$$\mathcal{A}\psi(x) \equiv \left(p_0(x)\frac{\mathrm{d}^n}{\mathrm{d}x^n} + p_1(x)\frac{\mathrm{d}^{n-1}}{\mathrm{d}x^{n-1}} + \cdots + p_{n-1}(x)\frac{\mathrm{d}}{\mathrm{d}x} + p_n(x)\right)\psi(x) \tag{1.13}$$

is a linear differential operator for $n \geq 1$. One can form linear operators from partial derivatives as well, e.g. the diffusion equation (Chapter 3)

$$\left(\frac{\partial}{\partial t} - D\frac{\partial^2}{\partial x^2}\right)\psi(x,t) = 0$$

specifies a differential operator (D is a constant). A significant portion of this book is devoted to linear differential equations. In Chapter 2, we consider the important case of second-order[31] differential operators, $n = 2$ in Eq. (1.13).

Example. Consider the product of two linear operators, a multiplicative operator $\mathcal{T}$ and the derivative operator $\mathcal{D}$. For the space of smooth functions of the real variable t, let $\mathcal{D}f(t) = \mathrm{d}f/\,\mathrm{d}t$ and $\mathcal{T}f(t) = tf(t)$. Then, $\mathcal{D}\mathcal{T}f(t) = \mathrm{d}(tf(t))/\,\mathrm{d}t = f(t) + t\,\mathrm{d}f/\,\mathrm{d}t$, whereas $\mathcal{T}\mathcal{D}f(t) = t\,\mathrm{d}f/\,\mathrm{d}t$. Clearly $\mathcal{D}\mathcal{T} \neq \mathcal{T}\mathcal{D}$; the order of the operations matters.

1.4.2 Representing operators with matrices

Specifying an operator as a mapping $\mathcal{A} : \mathbb{V} \to \mathbb{W}$ is an *abstract* statement. To get specific about the action of an operator, we need to know how it acts on basis vectors. Let $\mathbb{V}_N$ have an orthonormal basis $\{\hat{\mathbf{e}}_i\}_{i=1}^N$, and let $\mathbb{W}$ have orthonormal basis $\{\hat{\mathbf{e}}'_k\}_{k=1}^M$, where $M = \dim \mathbb{W}$. Let $\mathcal{A}$ operate

[30] *Smooth functions* are infinitely differentiable – they possess derivatives of all possible orders.

[31] The *order* of a differential equation is the order of the highest order derivative present in the equation.

on basis vector $\hat{\mathbf{e}}_j$. The result $\mathcal{A}\hat{\mathbf{e}}_j$ is a vector in $\mathbb{W}$, and as such can be expressed as a linear combination of basis vectors for $\mathbb{W}$:

$$\mathcal{A}\hat{\mathbf{e}}_j = \sum_{i=1}^{M} A_{ij}\hat{\mathbf{e}}'_i, \qquad (j = 1, \ldots, N) \tag{1.14}$$

where the expansion coefficients A_{ij} are to be determined (note the order of the indices in Eq. (1.14)). Form the inner product between $\hat{\mathbf{e}}'_k$ (an arbitrary element of the basis set for $\mathbb{W}$) with $\mathcal{A}\hat{\mathbf{e}}_j$ as given by Eq. (1.14),[32]

$$\langle \hat{\mathbf{e}}'_k | \mathcal{A}\hat{\mathbf{e}}_j \rangle = \langle \hat{\mathbf{e}}'_k | \sum_{i=1}^{M} A_{ij}\hat{\mathbf{e}}'_i \rangle = \sum_{i=1}^{M} A_{ij} \langle \hat{\mathbf{e}}'_k | \hat{\mathbf{e}}'_i \rangle = \sum_{i=1}^{M} A_{ij}\delta_{ki} = A_{kj}. \tag{1.15}$$

Thus, A_{ij} is the ith component of the vector $\mathcal{A}\hat{\mathbf{e}}_j$,

$$A_{ij} = \langle \hat{\mathbf{e}}'_i | \mathcal{A}\hat{\mathbf{e}}_j \rangle. \qquad (i = 1, \ldots, M; j = 1, \ldots, N) \tag{1.16}$$

The set of $M \times N$ scalars $\{A_{ij}\}$ is the *representation* of the abstract operator $\mathcal{A}$ in the orthonormal basis sets $\{\hat{\mathbf{e}}_j\}$ and $\{\hat{\mathbf{e}}'_i\}$. The numbers A_{ij} are called the *matrix elements* of $\mathcal{A}$, where the matrix $\boldsymbol{A}$ associated with $\mathcal{A}$ is the rectangular array:

$$\boldsymbol{A} = \begin{pmatrix} A_{11} & A_{12} & \cdots & A_{1N} \\ A_{21} & A_{22} & \cdots & A_{2N} \\ \vdots & \vdots & \ddots & \vdots \\ A_{M1} & A_{M2} & \cdots & A_{MN} \end{pmatrix}. \tag{1.17}$$

This array has M rows and N columns and is an $M \times N$ matrix. *There is a one-to-one correspondence $\mathcal{A} \leftrightarrow \boldsymbol{A}$ between linear operators $\mathcal{A} : \mathbb{V}_N \to \mathbb{W}_M$ and the $M \times N$ matrices $\boldsymbol{A}$ that represent them in particular basis sets for $\mathbb{W}$ and $\mathbb{V}$.*

Example. For $\mathbb{V} = \mathbb{R}^3$, consider the operator $\mathcal{A}$ that projects every vector $\psi \in \mathbb{V}$ onto the xy plane. To find the matrix elements of $\mathcal{A}$, we must know it's action on the basis vectors. For $\hat{\mathbf{e}}_1, \hat{\mathbf{e}}_2, \hat{\mathbf{e}}_3$ an orthonormal basis for $\mathbb{V}$, let $\mathcal{A}\hat{\mathbf{e}}_1 = \hat{\mathbf{e}}_1$, $\mathcal{A}\hat{\mathbf{e}}_2 = \hat{\mathbf{e}}_2$, and $\mathcal{A}\hat{\mathbf{e}}_3 = 0$. Using Eq. (1.16), $\mathcal{A}$ has the matrix representation $\begin{pmatrix} 1 & 0 & 0 \\ 0 & 1 & 0 \\ 0 & 0 & 0 \end{pmatrix}$.

---

[32] Forming the inner product of $\hat{\mathbf{e}}'_k$ with $\mathcal{A}\hat{\mathbf{e}}_j$ in Eq. (1.15) is a process known as "taking matrix elements."

If we know the representations of $\mathcal{A}$, $\boldsymbol{x}$, and $\boldsymbol{y}$, the abstract statement $\boldsymbol{y} = \mathcal{A}\boldsymbol{x}$ can be written as a matrix equation. For $\boldsymbol{y} = \sum_{i=1}^{M} y_i \hat{\mathbf{e}}'_i$ and $\boldsymbol{x} = \sum_{j=1}^{N} x_j \hat{\mathbf{e}}_j$, $\boldsymbol{y} = \mathcal{A}\boldsymbol{x} \implies$

$$\sum_{i=1}^{M} y_i \hat{\mathbf{e}}'_i = \mathcal{A} \sum_{j=1}^{N} x_j \hat{\mathbf{e}}_j = \sum_{j=1}^{N} x_j \mathcal{A} \hat{\mathbf{e}}_j = \sum_{j=1}^{N} x_j \sum_{i=1}^{M} A_{ij} \hat{\mathbf{e}}'_i$$
$$= \sum_{i=1}^{M} \left(\sum_{j=1}^{N} x_j A_{ij} \right) \hat{\mathbf{e}}'_i, \tag{1.18}$$

where we've used Eq. (1.14). Identifying the coefficients of the basis vectors,[33]

$$y_i = \sum_{j=1}^{N} A_{ij} x_j. \qquad (i = 1, \ldots, M) \tag{1.19}$$

Equation (1.19) is sometimes used to define the action of a linear operator – but it must be understood that it's a basis-dependent formula: The components x_i, y_i, and A_{ij} will generally be different for a different basis set.

The only nontrivial exception is the matrix representation of the identity operator, denoted $\boldsymbol{I}$. For any orthonormal basis, its matrix representation has the same form: $I_{ij} = \langle \hat{\mathbf{e}}_i | \mathcal{I} \hat{\mathbf{e}}_j \rangle = \langle \hat{\mathbf{e}}_i | \hat{\mathbf{e}}_j \rangle = \delta_{ij}$. The matrix $\boldsymbol{I}$ is referred to as the *unit matrix*. The unit matrix has the unit scalar 1 as its *diagonal* elements, the diagonal running from the top left corner of the matrix to the bottom right corner. For $N = 3$,

$$\boldsymbol{I} = \begin{pmatrix} 1 & 0 & 0 \\ 0 & 1 & 0 \\ 0 & 0 & 1 \end{pmatrix}.$$

More generally, a matrix $\boldsymbol{D}$ is called *diagonal* if $D_{ij} = 0$ unless $i = j$. That is, only the terms on the diagonal are nonzero, $D_{ii} \neq 0$. The unit matrix is diagonal with $I_{ii} = 1$.

1.4.3 Matrix algebra

Matrices are one of the most important mathematical tools in the use of linear operators on finite-dimensional vector spaces. Matrices have their own algebra, with rules derived from the behavior of abstract operators. Matrices have the same properties of addition and scalar multiplication that linear operators have. The sum $\boldsymbol{A} + \boldsymbol{B}$ of two $M \times N$ matrices and the scalar product of a matrix $\boldsymbol{A}$ and a number λ obey the simple rules:

$$(\boldsymbol{A} + \boldsymbol{B})_{ij} = A_{ij} + B_{ij} \quad \text{and} \quad (\lambda \boldsymbol{A})_{ij} = \lambda A_{ij}.$$

[33]Passing from Eq. (1.18) to Eq. (1.19) relies on the uniqueness of vector components (Section 1.2.2), which in turn relies on the linear independence of the basis.

Matrices therefore form a vector space[34] – a fact that's not very useful, however. In analogy with vectors, vector spaces, and inner products of vectors, we need a rule for multiplying matrices. An $M \times N$ matrix $\boldsymbol{A}$ and an $N \times P$ matrix $\boldsymbol{B}$ determine as a product the matrix $\boldsymbol{C} \equiv \boldsymbol{AB}$, an $M \times P$ matrix, with entries

$$\sum_{j=1}^{P} (\boldsymbol{AB})_{ij} x_j = \sum_{k=1}^{N} A_{ik} (\boldsymbol{Bx})_k = \sum_{j=1}^{P} \sum_{k=1}^{N} A_{ik} B_{kj} x_j \Longrightarrow (\boldsymbol{AB})_{ij} = \sum_{k=1}^{N} A_{ik} B_{kj}. \tag{1.20}$$

The product of two matrices $(\boldsymbol{AB})$ requires that the number of columns of $\boldsymbol{A}$ equals the number of rows of $\boldsymbol{B}$, and the ij component of $(\boldsymbol{AB})$ can be interpreted as the scalar product of the ith row of $\boldsymbol{A}$ and the jth column of $\boldsymbol{B}$ (look ahead to Eq. (1.35)).

An N-dimensional vector $\boldsymbol{x}$ can be considered an $N \times 1$ matrix – a *column vector*. (A vector is a column because if $\boldsymbol{A}$ is $M \times N$ then $\boldsymbol{Ax}$ must be an $M \times 1$ vector and the number of columns of $\boldsymbol{A}$ must equal the number of rows of $\boldsymbol{x}$.) If we consider the jth column of $\boldsymbol{B}$ to be the vector $\boldsymbol{x}_j$ (whose ith component[35] is $(\boldsymbol{x}_j)_i = B_{ij}$) then $\boldsymbol{B}$ looks like

$$\boldsymbol{B} = \begin{pmatrix} | & | & | & | \\ \boldsymbol{x}_1 & \boldsymbol{x}_2 & \cdots & \boldsymbol{x}_P \\ | & | & | & | \end{pmatrix}.$$

According to Eq. (1.20) we have

$$\boldsymbol{AB} = \boldsymbol{A} \begin{pmatrix} | & | & | & | \\ \boldsymbol{x}_1 & \boldsymbol{x}_2 & \cdots & \boldsymbol{x}_P \\ | & | & | & | \end{pmatrix} = \begin{pmatrix} | & | & | & | \\ \boldsymbol{Ax}_1 & \boldsymbol{Ax}_2 & \cdots & \boldsymbol{Ax}_P \\ | & | & | & | \end{pmatrix}, \tag{1.21}$$

and so the matrix $\boldsymbol{C} = \boldsymbol{AB}$ is composed of column vectors $\boldsymbol{y}_j = \boldsymbol{Ax}_j$.

There are two basic operations on matrices that allow us to obtain new matrices: complex conjugation and transposition. The conjugate $\boldsymbol{A}^*$ of $\boldsymbol{A}$ is formed by replacing all the elements A_{ij} by their complex conjugates, A^*_{ij}. A matrix is real if $\boldsymbol{A}^* = \boldsymbol{A}$. Clearly, $(\boldsymbol{A}^*)^* = \boldsymbol{A}$. If $\boldsymbol{A}$ is represented as in Eq. (1.17), its transpose, denoted $\boldsymbol{A}^T$, is formed by switching rows with columns:

$$\boldsymbol{A}^T = \begin{pmatrix} A_{11} & A_{21} & \cdots & A_{M1} \\ A_{12} & A_{22} & \cdots & A_{M2} \\ \vdots & \vdots & \ddots & \vdots \\ A_{1N} & A_{2N} & \cdots & A_{MN} \end{pmatrix}.$$

[34] The *zero matrix* $\mathbf{0}$, an $M \times N$ matrix with all entries zero, is the zero element of a vector space of matrices.

[35] Don't be confused by the subscripts here: $\boldsymbol{x}_j$ is a *vector* and $(\boldsymbol{x}_j)_i$ is a component of that vector.

More succinctly, $(\boldsymbol{A}^T)_{ij} = (\boldsymbol{A})_{ji}$. If $\boldsymbol{A}$ is an $M \times N$ matrix, then $\boldsymbol{A}^T$ is an $N \times M$ matrix.

An important property of transposes is revealed when we consider the transpose of the product matrix. As we now show, $(\boldsymbol{AB})^T = \boldsymbol{B}^T\boldsymbol{A}^T$. For $\boldsymbol{A}$ an $M \times N$ matrix and $\boldsymbol{B}$ an $N \times P$ matrix, the components of $(\boldsymbol{AB})^T$ are

$$\begin{aligned}(\boldsymbol{AB})^T_{ij} = (\boldsymbol{AB})_{ji} &= \sum_{k=1}^{N} A_{jk}B_{ki} = \sum_{k=1}^{N} (\boldsymbol{A}^T)_{kj}(\boldsymbol{B}^T)_{ik} \\ &= \sum_{k=1}^{N} (\boldsymbol{B}^T)_{ik}(\boldsymbol{A}^T)_{kj} = (\boldsymbol{B}^T\boldsymbol{A}^T)_{ij}.\end{aligned}$$

1.4.4 Rank and nullity

Linear operators $\mathcal{A} : \mathbb{V} \to \mathbb{W}$ map $|0\rangle \in \mathbb{V}$ onto $|0\rangle \in \mathbb{W}$; $\mathcal{A}(|0\rangle) = |0\rangle$, see Eq. (1.2). Besides $|0\rangle$, however, there may be nonzero vectors of $\mathbb{V}$ that are mapped into $|0\rangle \in \mathbb{W}$.

Definition. *The set of vectors in* $\mathbb{V}$ *that are mapped onto* $|0\rangle \in \mathbb{W}$ *is a subspace of* $\mathbb{V}$ *called the* kernel *or the* nullspace *of* $\mathcal{A}$*, denoted* $\ker \mathcal{A}$*. The dimension of* $\ker \mathcal{A}$ *is called the* nullity.

Definition. *The* range *or the* image *of a linear operator* $\mathcal{A} : \mathbb{V} \to \mathbb{W}$ *is a subset of* $\mathbb{W}$ *consisting of all vectors* $\mathcal{A}|\psi\rangle \in \mathbb{W}$ *for* $|\psi\rangle \in \mathbb{V}$*. The dimension of the range is called the* rank *of* $\mathcal{A}$.

The rank and the nullity are connected by a fundamental relation, the *rank-nullity* theorem or the *dimension* theorem [8, p. 242].

Theorem 1.4.1. *For* $\mathcal{A} : \mathbb{V} \to \mathbb{W}$ *a linear operator,*

$$\dim \mathbb{V} = \dim \mathcal{A}(\mathbb{V}) + \dim \ker \mathcal{A}. \tag{1.22}$$

Rank plus nullity equals the dimension of the domain: Every vector in $\mathbb{V}$ has to go "somewhere" under the action of $\mathcal{A}$. Vectors in the nullspace of $\mathcal{A}$ are mapped into $|0\rangle \in \mathbb{W}$, which is a zero-dimensional subspace[36] of $\mathbb{W}$.

The rank of the matrix representing an operator is the dimension of the vector space spanned by its columns – the maximum number of linearly independent columns. There are many equivalent ways to define the rank of a matrix.

[36] What is the dimension of the subspace of $\mathbb{W}$ consisting solely of the zero vector $|0\rangle$? How many linearly independent combinations of $|0\rangle$ are there? Zero.

Example. The matrix

$$\begin{pmatrix} 1 & 0 & 1 \\ -2 & -3 & 1 \\ 3 & 3 & 0 \end{pmatrix}$$

has rank 2. The first two columns are linearly independent (check it!), but the third column is not linearly independent of the first two (subtract the second column from the first). See Exercise 1.15.

1.4.5 Bounded operators

Definition. *A linear operator $\mathcal{A}$ is* bounded *if there is a positive number b such that $\|\mathcal{A}\psi\| \leq b\|\psi\|$ for every vector ψ. The smallest number b is called the* norm *of $\mathcal{A}$ and is denoted $\|\mathcal{A}\|_{\text{op}}$ (to distinguish it from the vector norm). Thus, $\|\mathcal{A}\psi\| \leq \|\mathcal{A}\|_{\text{op}}\|\psi\|$.*

For $\mathcal{A}$, $\mathcal{B}$ bounded operators and for scalars a, operator norms have the following properties, which follow from the definition of operator norm:

$$\begin{aligned} \|\mathcal{A}+\mathcal{B}\|_{\text{op}} &\leq \|\mathcal{A}\|_{\text{op}} + \|\mathcal{B}\|_{\text{op}} \\ \|a\mathcal{A}\|_{\text{op}} &= |a| \quad \|\mathcal{A}\|_{\text{op}} \\ \|\mathcal{A}\mathcal{B}\|_{\text{op}} &\leq \|\mathcal{A}\|_{\text{op}} \quad \|\mathcal{B}\|_{\text{op}}. \end{aligned} \tag{1.23}$$

Bounded operators can't stretch the length of a vector by more than their norm. Operators on finite-dimensional spaces are automatically bounded – infinite-dimensional vector spaces require a separate treatment.

Theorem 1.4.2. *Every linear operator on a finite-dimensional space is bounded.*

Proof. Let $\mathcal{A}$ be a linear operator on an N-dimensional space which has orthonormal basis $\{\hat{\mathbf{e}}_i\}_{i=1}^N$, in terms of which $A_{jk} = (\hat{\mathbf{e}}_j, \mathcal{A}\hat{\mathbf{e}}_k)$. Let K be the largest of the numbers[37] $\left|\sum_{i=1}^N A_{ij}^* A_{ik}\right|$ for $j, k = 1, 2, \ldots, N$. For any vector ψ, $\psi = \sum_{i=1}^N a_i \hat{\mathbf{e}}_i$. Then,[38]

[37] For finite-dimensional vector spaces, there's no question of the existence of such an upper bound. For infinite dimensional spaces, an operator is bounded if there is a positive constant K such that $\sum_{k=1}^{\infty} |\sum_{j=1}^N A_{kj}a_j|^2 \leq K \sum_{j=1}^N |a_j|^2$ for all values of $\{a_j\}_{j=1}^N$ for $N = 1, 2, 3, \ldots$ [9, p. 93].

[38] How do we know which form of the inner product to use? Any finite dimensional vector space with an orthonormal basis can be identified with $\mathbb{E}^N$, the inner product for which is in Section 1.4.

$$\begin{aligned}\|\mathcal{A}\psi\|^2 &= (\mathcal{A}\psi,\mathcal{A}\psi) = \sum_{k=1}^{N}(\mathcal{A}\psi)_k^*(\mathcal{A}\psi)_k = \sum_{k=1}^{N}\left(\sum_{j=1}^{N}A_{kj}a_j\right)^*\left(\sum_{i=1}^{N}A_{ki}a_i\right)\\ &\le \sum_{i=1}^{N}\sum_{j=1}^{N}|a_i||a_j^*|\left|\sum_{k=1}^{N}A_{kj}^*A_{ki}\right| \le K\sum_{i=1}^{N}\sum_{j=1}^{N}|a_i||a_j| \le NK\sum_{i=1}^{N}|a_i|^2 = NK\|\psi\|^2,\end{aligned} \tag{1.24}$$

where we've used Eq. (1.19) for $(\mathcal{A}\psi)_k$, $\text{Re}z \le |z|$ in the first inequality, $|z^*| = |z|$, and the inequality[39] $\sum_{i=1}^N \sum_{j=1}^N |a_i||a_j| \le N\sum_{i=1}^N |a_i|^2$. Thus, $\mathcal{A}$ is bounded. Note that the bound is proportional to N. □

1.4.6 Inverses

An important concept in working with operators is that of the *inverse operator.*

Definition. *The operator $\mathcal{B}$ is the* inverse *of $\mathcal{A}$ if $\mathcal{AB} = \mathcal{BA} = \mathcal{I}$.*

The definition implies that if $\mathcal{A}$ maps ψ to $\mathcal{A}\psi$, the inverse $\mathcal{B}$ uniquely maps $\mathcal{A}\psi$ to ψ. Operators not possessing an inverse are said to be *singular.*[40] No operator has more than one inverse. To show uniqueness of the inverse (when it exists), assume linear operator $\mathcal{A}$ has two inverses, $\mathcal{B}$ and $\mathcal{C}$, in which case $\mathcal{BA} = \mathcal{I}$ and $\mathcal{AB} = \mathcal{I} = \mathcal{AC}$ so that $\mathcal{B} - \mathcal{C} = \mathcal{BA}(\mathcal{B} - \mathcal{C}) = \mathcal{B}(\mathcal{AB} - \mathcal{AC}) = 0$. Thus, *the* inverse of $\mathcal{A}$ is written $\mathcal{A}^{-1}$. What's required for $\mathcal{A}$ to possess an inverse is answered in the following theorem.

Theorem 1.4.3. *A linear operator $\mathcal{A}$ has an inverse if and only if for each nonzero vector ψ, there is one and only one vector ϕ such that $\psi = \mathcal{A}\phi$.*

Proof. If $\mathcal{A}^{-1}$ exists, then for each ψ, $\phi = \mathcal{A}^{-1}\psi$ is such that $\mathcal{A}\phi = \psi$. If χ is a vector such that $\mathcal{A}\chi = \psi$, then $\chi = \mathcal{A}^{-1}\psi = \mathcal{A}^{-1}\mathcal{A}\phi$, and hence $\chi = \phi$. That proves the "only if" part. For the converse, assume for each ψ, there is a unique ϕ such that $\psi = \mathcal{A}\phi$. We can define an operator $\mathcal{B}$ such that $\mathcal{B}\psi = \phi$. In that case, $\mathcal{BA}\phi = \phi$, and hence $\mathcal{B}$ is the inverse operator to $\mathcal{A}$: $\mathcal{BA} = \mathcal{I}$. We must show that $\mathcal{B}$ defined this way is linear. Assume for ψ_1 and ψ_2 that $\psi_1 = \mathcal{A}\phi_1$ and $\psi_2 = \mathcal{A}\phi_2$. Then, $\mathcal{B}(\psi_1 + \psi_2) = \mathcal{B}(\mathcal{A}\phi_1 + \mathcal{A}\phi_2) = \mathcal{BA}(\phi_1 + \phi_2) = \phi_1 + \phi_2 = \mathcal{B}\psi_1 + \mathcal{B}\psi_2$. For any scalar a, $\mathcal{B}a\psi = \mathcal{B}a\mathcal{A}\phi = \mathcal{BA}a\phi = a\phi = a\mathcal{B}\psi$. This completes the proof except for one point: What if ψ is the zero vector? If $\mathcal{A}$ has an inverse, there is no nonzero vector ϕ such that $\mathcal{A}\phi = 0$: If $\mathcal{A}\phi = 0$ and $\mathcal{A}^{-1}$ exists, then $\phi = \mathcal{A}^{-1}\mathcal{A}\phi = 0$, a contradiction. □

[39] This inequality follows from $|a||b| \le \frac{1}{2}(|a|^2 + |b|^2)$, which is implied by $(|a| - |b|)^2 \ge 0$.

[40] The projection operator in the example of Section 1.4.2 is singular – there is not a unique vector in three dimensions corresponding to its projection onto two dimensions. There are an unlimited number of vectors in three dimensions having the same projection onto two dimensions.

For the example in Section 1.4.2, $\mathcal{A}\hat{\mathbf{e}}_3 = 0$ and $\mathcal{A}$ is singular. There are several equivalent conditions for an operator to possess an inverse, which we summarize in the form of a theorem.

Theorem 1.4.4. *For linear operator $\mathcal{A}$ on vector space $\mathbb{V}_N$ with basis $\{\hat{\mathbf{e}}_i\}_{i=1}^N$, each of the following statements are necessary and sufficient conditions for $\mathcal{A}$ to possess an inverse:*

(i) There is no nonzero vector $\phi \in \mathbb{V}_N$ such that $\mathcal{A}\phi = 0$;

(ii) The set of vectors $\{\mathcal{A}\hat{\mathbf{e}}_1, \ldots, \mathcal{A}\hat{\mathbf{e}}_N\}$ is linearly independent;

(iii) There is a linear operator $\mathcal{B}$ such that $\mathcal{BA} = \mathcal{I}$;

(iv) The matrix representing $\mathcal{A}$ has nonzero determinant, $\det \boldsymbol{A} \neq 0$.

Comments. We've already shown many of these points; what we haven't shown (and won't) is that they're necessary and sufficient. For condition (ii), if $\mathcal{A}^{-1}$ exists, the vectors $\mathcal{A}\hat{\mathbf{e}}_1, \ldots, \mathcal{A}\hat{\mathbf{e}}_N$ are linearly independent if the set $\{\hat{\mathbf{e}}_i\}_{i=1}^N$ is linearly independent: If $\sum_{k=1}^N a_k \hat{\mathbf{e}}_k = 0$ and $\mathcal{A}^{-1}$ exists, then $\sum_{k=1}^N a_k \hat{\mathbf{e}}_k = \mathcal{A}^{-1} \sum_{k=1}^N a_k \mathcal{A}\hat{\mathbf{e}}_k = 0$, implying that the set $\{\mathcal{A}\hat{\mathbf{e}}_k\}$ is linearly independent. Condition (iv) invokes the determinant of a matrix, which we yet have to discuss (see Section 1.4.9). The determinant of the product of matrices $\boldsymbol{C} = \boldsymbol{AB}$ is the product of the determinants, $\det \boldsymbol{C} = \det \boldsymbol{A} \cdot \det \boldsymbol{B}$. If the determinant of the matrix representing $\mathcal{A}$ were zero, the determinant of the matrix corresponding to $\mathcal{A}^{-1}\mathcal{A} = \mathcal{I}$ would also be zero, which it is not ($\det \boldsymbol{I} = 1$). Note that the determinant of the matrix representation of the singular operator in the example in Section 1.4.2 is zero.

The matrix representation of the inverse operator $\mathcal{A}^{-1}$, denoted $\boldsymbol{A}^{-1}$, is related to the matrix representation of $\mathcal{A}$ though the relation $\boldsymbol{A}^{-1}\boldsymbol{A} = I$, which in terms of matrix elements satisfies $\sum_j A_{ij}^{-1} A_{jk} = \delta_{ik}$. An explicit expression for the elements of $\boldsymbol{A}^{-1}$ is given in Section 1.4.9.

1.4.7 Change of basis and the similarity transformation

Let $\{\hat{\mathbf{e}}_1, \hat{\mathbf{e}}_2, \ldots, \hat{\mathbf{e}}_N\}$ be an orthonormal basis for $\mathbb{V}_N$, in terms of which

$$\boldsymbol{x} = x_1\hat{\mathbf{e}}_1 + x_2\hat{\mathbf{e}}_2 + \cdots + x_N\hat{\mathbf{e}}_N. \tag{1.25}$$

Once the basis has been specified, $\boldsymbol{x}$ can be considered a $N \times 1$ matrix (a column vector)

$$\boldsymbol{x} = (x_1, x_2, \ldots, x_N)^T.$$

Suppose we have another set of orthonormal vectors $\{\hat{\mathbf{e}}'_1, \hat{\mathbf{e}}'_2, \ldots, \hat{\mathbf{e}}'_N\}$ that we want to use as a basis. In the new basis, the *same vector* $\boldsymbol{x}$ can be expressed[41]

[41] A vector $\boldsymbol{x}$ represents something physical and is independent of the basis that we use to represent it. Quantities that exist independently of the basis are referred to as *geometric objects*.

$$\boldsymbol{x} = x'_1\hat{\boldsymbol{e}}'_1 + x'_2\hat{\boldsymbol{e}}'_2 + \cdots + x'_N\hat{\boldsymbol{e}}'_N, \tag{1.26}$$

where the $\{x'_i\}$ are the components of $\boldsymbol{x}$ in the new basis. The new basis vectors $\hat{\boldsymbol{e}}'_j$, being elements of $\mathbb{V}_N$, can be represented as linear combinations of the old basis vectors:

$$\hat{\boldsymbol{e}}'_j = \sum_{i=1}^{N} S_{ij}\hat{\boldsymbol{e}}_i. \tag{1.27}$$

(Note the order of the indices; Eq. (1.27), like Eq. (1.14), is not matrix multiplication.) The expansion coefficients $\{S_{ij}\}$ in Eq. (1.27) cannot be arbitrary: For the new basis $\{\hat{\boldsymbol{e}}'_j\}$ to be orthonormal implies a constraint on the coefficients S_{ij}. We require

$$\delta_{kj} = (\hat{\boldsymbol{e}}'_k, \hat{\boldsymbol{e}}'_j) = \sum_{i=1}^{N}\sum_{m=1}^{N} S^*_{mk}S_{ij}(\hat{\boldsymbol{e}}_m, \hat{\boldsymbol{e}}_i) = \sum_{i=1}^{N}\sum_{m=1}^{N} S^*_{mk}S_{ij}\delta_{mi} = \sum_{i=1}^{N} S^*_{ik}S_{ij}. \tag{1.28}$$

Equation (1.28) informs us that the S_{ij} are elements of a *unitary* matrix; see Section 1.4.10. A unitary matrix $\boldsymbol{S}$ is such that $(S^{-1})_{ki} = S^*_{ik}$.

How are the components $\{x_i\}$ related to the $\{x'_i\}$? Equating the two basis expansions for $\boldsymbol{x}$, Eqs. (1.25) and (1.26), we have, using Eq. (1.27)

$$\boldsymbol{x} = \sum_{i=1}^{N} x_i\hat{\boldsymbol{e}}_i = \sum_{j=1}^{N} x'_j\hat{\boldsymbol{e}}'_j = \sum_{j=1}^{N} x'_j \sum_{i=1}^{N} S_{ij}\hat{\boldsymbol{e}}_i = \sum_{i=1}^{N}\sum_{j=1}^{N} S_{ij}x'_j\hat{\boldsymbol{e}}_i, \tag{1.29}$$

so, evidently,[42]

$$x_i = \sum_{j=1}^{N} S_{ij}x'_j. \tag{1.30}$$

Note that Eq. (1.30) (in contrast to Eq. (1.27)) *is* matrix multiplication. This relationship can be written in the form of a matrix equation

$$\boldsymbol{x} = \boldsymbol{S}\boldsymbol{x}' \quad \text{where} \quad \boldsymbol{x}' = (x'_1, x'_2, \ldots, x'_N)^T. \tag{1.31}$$

The matrix $\boldsymbol{S}$ must have an inverse because each basis consists of linearly independent vectors (see Section 1.4.9), and we can write

$$\boldsymbol{x}' = \boldsymbol{S}^{-1}\boldsymbol{x}. \tag{1.32}$$

Now suppose $\boldsymbol{y} = \boldsymbol{A}\boldsymbol{x}$. In the new basis, the equation is $\boldsymbol{y}' = \boldsymbol{A}'\boldsymbol{x}'$ where $\boldsymbol{y} = \boldsymbol{S}\boldsymbol{y}'$ and $\boldsymbol{x} = \boldsymbol{S}\boldsymbol{x}'$. How is $\boldsymbol{A}'$ related to $\boldsymbol{A}$? We have

$$\boldsymbol{y} = \boldsymbol{A}\boldsymbol{x} \implies \boldsymbol{S}\boldsymbol{y}' = \boldsymbol{A}\boldsymbol{S}\boldsymbol{x}' \implies \boldsymbol{y}' = \boldsymbol{S}^{-1}\boldsymbol{A}\boldsymbol{S}\boldsymbol{x}' \implies \boldsymbol{A}' = \boldsymbol{S}^{-1}\boldsymbol{A}\boldsymbol{S}.$$

---

[42] To pass from Eq. (1.29) to Eq. (1.30) relies on the uniqueness of the coefficients in a given basis expansion; the same point we noted in Section 1.4.2.

Two matrices related by $\boldsymbol{A}' = \boldsymbol{S}^{-1}\boldsymbol{A}\boldsymbol{S}$ are said to be *similar*; $\boldsymbol{A}' = \boldsymbol{S}^{-1}\boldsymbol{A}\boldsymbol{S}$ is known as a *similarity transformation*.

We're now in a position to address an issue raised in Section 1.2.5. A linear functional F, acting on a vector $\boldsymbol{x}$ having the basis expansion in Eq. (1.25), has the value (see Eq. (1.1)) $F(\boldsymbol{x}) = \sum_i x_i F(\hat{\mathbf{e}}_i)$. How do the quantities $F(\hat{\mathbf{e}}_i)$ transform under the change of basis described by Eq. (1.27)? Because F is linear,

$$F(\hat{\mathbf{e}}'_j) = F\left(\sum_i S_{ij}\hat{\mathbf{e}}_i\right) = \sum_i S_{ij} F(\hat{\mathbf{e}}_i). \tag{1.33}$$

Comparing Eq. (1.33) with Eq. (1.27), we see that the collection of terms $F(\hat{\mathbf{e}}_j)$ transforms under a change of basis in the same way that the basis vectors transform.[43] We can now verify the claim made in Section 1.2.5, that the value of $F(\boldsymbol{x})$ is independent of basis:

$$\begin{aligned} F(\boldsymbol{x}) &= \sum_j x'_j F(\hat{\mathbf{e}}'_j) = \sum_j \sum_i x'_j S_{ij} F(\hat{\mathbf{e}}_i) = \sum_j \sum_i \sum_k (S^{-1})_{jk} S_{ij} F(\hat{\mathbf{e}}_i) x_k \\ &= \sum_i \sum_k \delta_{ik} x_k F(\hat{\mathbf{e}}_i) = \sum_i x_i F(\hat{\mathbf{e}}_i) = F(\boldsymbol{x}), \end{aligned}$$

where we've used Eqs. (1.33) and (1.32) and that $\sum_j S_{ij}(S^{-1})_{jk} = \delta_{ik}$ is independent of basis (the identity operator has the representation $I_{jk} = \delta_{jk}$ in all orthonormal bases). Linear functionals map vectors onto scalars, $F : \mathbb{V} \to \mathbb{F}$, where the value $F(\psi \in \mathbb{V})$ is independent of basis.

1.4.8 Adjoints and Hermitian operators

Definition. *Associated with a linear operator $\mathcal{A}$ is another operator, its* adjoint *or* Hermitian conjugate, *$\mathcal{A}^\dagger$, such that $(\mathcal{A}^\dagger\phi, \psi) = (\phi, \mathcal{A}\psi)$ holds for all vectors ψ, ϕ.*[44]

The adjoint has the following properties, which hold as a consequence of its definition:

$$\begin{aligned} (\mathcal{A} + \mathcal{B})^\dagger &= \mathcal{A}^\dagger + \mathcal{B}^\dagger \\ (\mathcal{A}\mathcal{B})^\dagger &= \mathcal{B}^\dagger \mathcal{A}^\dagger \\ (a\mathcal{A})^\dagger &= a^* \mathcal{A}^\dagger \\ (\mathcal{A}^\dagger)^\dagger &= \mathcal{A}. \end{aligned} \tag{1.34}$$

[43] Quantities that transform in the same way under the same linear transformation are said to transform *cogrediently*. Quantities that transform in the same way as the basis vectors are said to transform *covariantly*.

[44] Technically, $\mathcal{A}^\dagger$ is a linear operator on elements of the dual space. For linear operator $\mathcal{A}$ on vector space $\mathbb{V}$, and for vectors $|\psi\rangle, |\phi\rangle \in \mathbb{V}$, $\langle \mathcal{A}^\dagger\phi|$ is a linear functional associated with $\langle\phi| \in \mathbb{V}^*$ such that $\langle \mathcal{A}^\dagger\phi|$ acting on $|\psi\rangle$ is the same as $\langle\phi|$ acting on $|\mathcal{A}\psi\rangle$, $\langle \mathcal{A}^\dagger\phi|\psi\rangle = \langle\phi|\mathcal{A}\psi\rangle$.

Theorem 1.4.5. *The adjoint $\mathcal{A}^\dagger$ associated with a bounded linear operator $\mathcal{A}$ is a bounded linear operator with $\|\mathcal{A}^\dagger\|_{\text{op}} = \|\mathcal{A}\|_{\text{op}}$.*

Proof. Consider that $\|\mathcal{A}^\dagger\psi\|^2 = (\mathcal{A}^\dagger\psi, \mathcal{A}^\dagger\psi) = (\psi, \mathcal{A}\mathcal{A}^\dagger\psi) \leq |(\psi, \mathcal{A}\mathcal{A}^\dagger\psi)| \leq \|\psi\| \ \ \|\mathcal{A}\mathcal{A}^\dagger\psi\| \leq \|\psi\| \ \ \|\mathcal{A}\|_{\text{op}}\|\mathcal{A}^\dagger\psi\|$, where we've used the Schwarz inequality. From this result, we infer that $\|\mathcal{A}^\dagger\psi\| \leq \|\mathcal{A}\|_{\text{op}}\|\psi\|$. Hence, $\mathcal{A}^\dagger$ is bounded with $\|\mathcal{A}^\dagger\|_{\text{op}} \leq \|\mathcal{A}\|_{\text{op}}$. If we were to reverse the roles of $\mathcal{A}$ and $\mathcal{A}^\dagger$ in this inequality, we would conclude that $\|\mathcal{A}\|_{\text{op}} \leq \|\mathcal{A}^\dagger\|_{\text{op}}$. Consistency requires that $\|\mathcal{A}^\dagger\|_{\text{op}} = \|\mathcal{A}\|_{\text{op}}$. □

Theorem 1.4.6. *If $\mathcal{A}$ has inverse $\mathcal{A}^{-1}$, then $\mathcal{A}^\dagger$ has an inverse, $(\mathcal{A}^\dagger)^{-1} = (\mathcal{A}^{-1})^\dagger$.*

Proof. Consider the two results: $(\phi, \psi) = (\mathcal{A}\mathcal{A}^{-1}\phi, \psi) = (\mathcal{A}^{-1}\phi, \mathcal{A}^\dagger\psi) = (\phi, (\mathcal{A}^{-1})^\dagger\mathcal{A}^\dagger\psi)$ and $(\phi, \psi) = (\mathcal{A}^{-1}\mathcal{A}\phi, \psi) = (\mathcal{A}\phi, (\mathcal{A}^{-1})^\dagger\psi) = (\phi, \mathcal{A}^\dagger (\mathcal{A}^{-1})^\dagger\psi)$. Thus, $(\mathcal{A}^{-1})^\dagger\mathcal{A}^\dagger = \mathcal{A}^\dagger(\mathcal{A}^{-1})^\dagger = \mathcal{I}$, which proves the theorem. □

For $\boldsymbol{A}$, the matrix representation of $\mathcal{A}$, the matrix representation of $\mathcal{A}^\dagger$, denoted $\boldsymbol{A}^\dagger$, is defined as the transpose of the complex conjugate of $\boldsymbol{A}$, $\boldsymbol{A}^\dagger \equiv (\boldsymbol{A}^*)^T = (\boldsymbol{A}^T)^*$. From the properties of the transpose, $(\boldsymbol{AB})^\dagger = \boldsymbol{B}^\dagger\boldsymbol{A}^\dagger$ and $(\lambda\boldsymbol{A})^\dagger = \lambda^*\boldsymbol{A}^\dagger$, the same as the rules for operators, Eq. (1.34).[45]

Since $\boldsymbol{x}$ is an $N \times 1$ matrix (a column vector) then, evidently, $\boldsymbol{x}^\dagger$ is a $1 \times N$ – a *row vector* – and if $\boldsymbol{y}$ is an $N \times 1$ vector, it's possible to operate on $\boldsymbol{y}$ by $\boldsymbol{x}^\dagger$ (since the dimensions match up correctly). We have[46]

$$\boldsymbol{x}^\dagger\boldsymbol{y} = \sum_{i=1}^{N} x_i^* y_i = \langle\boldsymbol{x}|\boldsymbol{y}\rangle. \tag{1.35}$$

As the product of a $1 \times N$ matrix and a $N \times 1$ matrix, the inner product is a 1×1 matrix (a scalar). We could also have created a new matrix $\boldsymbol{y}\boldsymbol{x}^\dagger$, which is the product of a $N \times 1$ matrix and a $1 \times N$ matrix. The result is an $N \times N$ matrix known as the "outer product" and has components $(\boldsymbol{y}\boldsymbol{x}^\dagger)_{ij} = y_i x_j^*$.

Definition. *A bounded operator $\mathcal{A}$ is* self-adjoint *or* Hermitian *if $\mathcal{A}^\dagger = \mathcal{A}$. Likewise for its matrix representation, a matrix for which $\boldsymbol{A}^\dagger = \boldsymbol{A}$ is said to be a* Hermitian matrix.

Self-adjoint operators are a particularly important class of operators that have several useful properties (see Section 1.5). In quantum mechanics,

---

[45]Only in an orthonormal basis is $\mathcal{A}^\dagger$ represented by the adjoint of the matrix representing $\mathcal{A}$.

[46]Kets are thus represented as column vectors and bras as row vectors. From this point of view, the i-j element C_{ij} of $\boldsymbol{C} = \boldsymbol{AB}$ is formed by the i-th row of $\boldsymbol{A}$ operating on the j-th column of $\boldsymbol{B}$. (An interpretation we will revisit in Chapter 11.)

all physical observables such as momentum, angular momentum, and spin are represented by self-adjoint operators.

Defining self-adjoint operators on infinite-dimensional spaces requires care – basically, are the operators defined for all vectors of the space and do the inner products exist? Consider $L^2[0,1]$. Let $\mathcal{A}$ be defined by $\mathcal{A}\psi(x) = x\psi(x)$ for every vector ψ in this space. For any ψ, $\|\mathcal{A}\psi\|^2 = \int_0^1 |\mathcal{A}\psi(x)|^2 \,\mathrm{d}x = \int_0^1 |x\psi(x)|^2 \,\mathrm{d}x \le \int_0^1 |\psi(x)|^2 \,\mathrm{d}x = \|\psi\|^2$. Thus, $\mathcal{A}$ is bounded. For any ψ and ϕ, $(\phi, \mathcal{A}\psi) = \int_0^1 \phi^*(x)\mathcal{A}\psi(x) \,\mathrm{d}x = \int_0^1 \phi^*(x)x\psi(x) \,\mathrm{d}x = \int_0^1 (\phi(x)x)^*\psi(x) \,\mathrm{d}x = \int_0^1 (\mathcal{A}\phi(x))^*\psi(x) \,\mathrm{d}x = (\mathcal{A}\phi, \psi)$. Thus, $\mathcal{A}$ is self-adjoint.

Now consider the same operator for the space $L^2(-\infty, \infty)$. $\mathcal{A}$ is not bounded for this space: $\|\mathcal{A}\psi\|^2 = \int_{-\infty}^{\infty} |x\psi(x)|^2 \,\mathrm{d}x$ can be any number larger than $\|\psi\|^2 = \int_{-\infty}^{\infty} |\psi(x)|^2 \,\mathrm{d}x$. Thus, $\mathcal{A}$ is not defined *for all vectors* in this space: There are vectors $\psi \in L^2(-\infty, \infty)$, i.e. functions $\psi(x)$ for which $\int_{-\infty}^{\infty} |\psi(x)|^2 \,\mathrm{d}x$ exists, such that $\int_{-\infty}^{\infty} |x\psi(x)|^2 \,\mathrm{d}x$ does not exist. In that case, $\mathcal{A}\psi$ is not an element of $L^2(-\infty, \infty)$.

For an operator to be Hermitian, it must be bounded, a result that holds for infinite-dimensional spaces [5, p. 296].

Theorem 1.4.7. *If a linear operator $\mathcal{A}$ is defined for all vectors of a space and if $(\phi, \mathcal{A}\psi) = (\mathcal{A}\phi, \psi)$ for all ϕ and ψ, then $\mathcal{A}$ is bounded.*

1.4.9 Determinants and the matrix inverse

$N \times N$ matrices contain N^2 independent pieces of information, and it can be a daunting task to understand what this information is trying to tell us. Of course, even in the case of vectors we still have N pieces of information, but interpreting vectors in terms of "magnitude" and "direction" allows us to get a pretty good intuitive feel for what's going on. It's natural, then, to ask if there are similar scalar quantities that can be associated with matrices which somehow "describe" the important behavior of the linear operator represented by the matrix.

One way to associate a number with a square matrix $\boldsymbol{A}$ is to form its *determinant* (denoted $|\boldsymbol{A}|$ or det $\boldsymbol{A}$), which, as we'll show, *is independent of basis.*[47] Finding the determinant of an $N \times N$ matrix for $N \gtrsim 4$ entails a lengthy calculation, one that's best done using a computer. For small matrices, it's not so difficult, and even if it is difficult, the determinant is an extremely important tool since it provides a closed-form expression for $\boldsymbol{A}^{-1}$ (among other things), and this is where we're eventually headed.

Before defining the determinant, we introduce another quantity, the *Levi-Civita permutation symbol* $\varepsilon_{i_1 i_2 \cdots i_N}$ for an N-dimensional space[48] – a

[47]One might wonder, what *is* the determinant, what does it represent? Determinants of real $N \times N$ matrices can be interpreted geometrically as volumes in N-dimensional Euclidean space, but to show that is beyond the intended scope of this book.

[48]The Levi-Civita symbol, used here in the definition of the determinant, is a useful quantity to know about in its own right – we return to it in Chapter 11.

symbol with N indices, where each index i_k can take on N values, $1 \leq i_k \leq N$, for $1 \leq k \leq N$. The symbol is defined, starting with the reference sequence of integers $12\cdots N$, to have the value $\varepsilon_{12\cdots N} = 1$. All other values of $\varepsilon_{i_1 i_2 \cdots i_N}$ are determined by the rule that under an interchange of indices i_j and i_k, the symbol changes sign (for any j and k):

$$\varepsilon_{i_1\cdots i_j \cdots i_k \cdots i_N} = -\varepsilon_{i_1\cdots i_k \cdots i_j \cdots i_N}. \tag{1.36}$$

For $N = 3$, with $\varepsilon_{123} = 1$, $\varepsilon_{132} = -1$. The Levi-Civita symbol is *totally antisymmetric* in its indices: under pairwise interchange of any two indices, it changes sign – Eq. (1.36). Because of this property, the symbol vanishes if two of its indices have the same value. By the rules, $\varepsilon_{113} = -\varepsilon_{113}$, implying $\varepsilon_{113} = 0$. There are $N!$ arrangements (*permutations*) of the integers $1\cdots N$. Hence, there are $N!$ nonzero values of $\varepsilon_{i_1\cdots i_N}$, either ± 1. The symbol has value $+1$ for even permutations of $1\cdots N$ (permutations attained by an even number of pairwise interchanges) and -1 for odd permutations.

Definition. *The* determinant *of an $N \times N$ matrix $\boldsymbol{A}$, whether real or complex, is the algebraic sum of $N!$, N-fold products of the matrix elements A_{ij}:*

$$\det \boldsymbol{A} \equiv \sum_{i_1=1}^{N} \cdots \sum_{i_N=1}^{N} \varepsilon_{i_1\cdots i_N} A_{1i_1} A_{2i_2} \cdots A_{Ni_N}. \tag{1.37}$$

There are $N!/2$ positively signed terms in the expansion of Eq. (1.37) and $N!/2$ negatively signed terms,[49] depending on whether the particular set of indices $i_1 \cdots i_N$ is an even or odd permutation of $12\cdots N$.

Example. Consider the 2×2 matrix $\boldsymbol{A} = \begin{pmatrix} a & b \\ c & d \end{pmatrix}$. Find it's determinant using Eq. (1.37).

$$\begin{aligned}\det \boldsymbol{A} &= \sum_{i=1}^{2}\sum_{j=1}^{2} \varepsilon_{ij} A_{1i} A_{2j} = \varepsilon_{11} A_{11} A_{21} + \varepsilon_{12} A_{11} A_{22} + \varepsilon_{21} A_{12} A_{21} + \varepsilon_{22} A_{12} A_{22} \\ &= \varepsilon_{12}(A_{11}A_{22} - A_{12}A_{21}) = ad - bc,\end{aligned}$$

a familiar result, where we've used $\varepsilon_{11} = \varepsilon_{22} = 0$ and $\varepsilon_{21} = -\varepsilon_{12} = -1$.

---

[49]The factorial of any integer $N > 1$ is always even.

Example. Use Eq. (1.37) to show that the determinant of the unit matrix is 1.

$$\det \boldsymbol{I} = \sum_{i_1=1}^{N} \cdots \sum_{i_N=1}^{N} \varepsilon_{i_1 i_2 \cdots i_N} I_{1i_1} I_{2i_2} \cdots I_{Ni_N}$$
$$= \sum_{i_1=1}^{N} \cdots \sum_{i_N=1}^{N} \varepsilon_{i_1 i_2 \cdots i_N} \delta_{1i_1} \delta_{2i_2} \cdots \delta_{Ni_N} = \varepsilon_{12\cdots N} = 1.$$

Equation (1.37) indicates a sum of N-fold products of matrix elements, where, for each term in the series, the matrix elements involved in the product are such that each factor is from a separate row of the matrix – the first entry from the first row, the second from the second row, etc. Determinants can also be defined as a sum over columns:

$$\det \boldsymbol{A} \equiv \sum_{i_1=1}^{N} \cdots \sum_{i_N=1}^{N} \varepsilon_{i_1 \cdots i_N} A_{i_1 1} A_{i_2 2} \cdots A_{i_N N}. \tag{1.38}$$

To show the equivalence of Eqs. (1.37) and (1.38) would require an argument lengthier than we have room for here; see [8, p. 320]. Accepting the equality of Eqs. (1.37) and (1.38), we have that $\det \boldsymbol{A} = \det \boldsymbol{A}^T$ (which is easily seen to be true for a 2×2 matrix).

From Eqs. (1.37) and (1.38), we can conclude, based on the antisymmetry of the Levi-Civita symbol, that the value of a determinant changes sign if either two rows of the matrix are interchanged, or if two columns are interchanged. By the same reasoning, the determinant *vanishes* if two rows or two columns are identical. That statement can be made stronger: The determinant vanishes if rows or columns are not linearly independent (the determinant of the matrix in the example in Section 1.4.4 vanishes). That fact was used in Section 1.4.7 to conclude that the transformation matrix $\boldsymbol{S}$ in Eq. (1.31) must have an inverse because bases are linearly independent.

Each nonzero term in Eq. (1.37) containing $A_{1\_}A_{2\_} \cdots A_{N\_}$ (where the blanks are filled in with the values of a permutation of $1 \cdots N$) has exactly one factor from each row and one factor from each column. For any row i therefore, we can express the determinant as a sum over the elements of the ith row, A_{ij}, for $j = 1, \cdots, N$, multiplied by all the other terms collected together that are associated with A_{ij}:

$$\det \boldsymbol{A} = A_{i1}C_{i1} + A_{i2}C_{i2} + \cdots + A_{iN}C_{iN} = \sum_{j=1}^{N} A_{ij}C_{ij}. \quad \text{(for any } i\text{)} \tag{1.39}$$

Each coefficient C_{ij} is called the *cofactor* of A_{ij}, and Eq. (1.39) is known as the *Laplace expansion in cofactors*. There are $N!$ nonzero terms in

Eq. (1.37), implying that each cofactor in Eq. (1.39) contains $(N-1)!$ terms. Perhaps the cofactors are themselves determinants? That's indeed what they are, up to a minus sign.

Definition. *The cofactor* $C_{ij} = (-1)^{i+j} M_{ij}$ *of an element* A_{ij} *of* $\boldsymbol{A}$*, where* M_{ij} *is the* (i, j) minor *of* $\boldsymbol{A}$*, the determinant of the matrix formed from* $\boldsymbol{A}$ *by removing the* i*th row and* j*th column.*

For example, the cofactor of element A_{23} of a 3×3 matrix $\boldsymbol{A}$ is

$$C_{23} = (-1)^{2+3} \begin{vmatrix} A_{11} & A_{12} \\ A_{31} & A_{32} \end{vmatrix} = -(A_{11}A_{32} - A_{12}A_{31}).$$

Theorem 1.4.8. *The determinant of a matrix product is the product of the determinants:* $\det(\boldsymbol{AB}) = \det \boldsymbol{A} \cdot \det \boldsymbol{B}$.

Comment. A proof of this important result is given Section 11.10. An immediate application is that, starting from $\boldsymbol{AA}^{-1} = I$, $\det \boldsymbol{A}^{-1} = (\det \boldsymbol{A})^{-1}$.

The determinant is independent of basis. Under a change of basis, matrices transform as $\boldsymbol{A}' = \boldsymbol{S}^{-1}\boldsymbol{AS}$ (Section 1.4.7), where $\boldsymbol{S}$ is the matrix effecting the basis transformation, Eq. (1.27). Using the property of determinants that $|\boldsymbol{AB}| = |\boldsymbol{A}||\boldsymbol{B}|$, the value of the determinant is independent of basis:

$$\begin{aligned}\det \boldsymbol{A}' &= \det(\boldsymbol{S}^{-1}\boldsymbol{AS}) = \det \boldsymbol{S}^{-1} \det \boldsymbol{A} \det \boldsymbol{S} \\ &= \det(\boldsymbol{S}^{-1}\boldsymbol{S}) \det \boldsymbol{A} = \det \boldsymbol{I} \det \boldsymbol{A} = \det \boldsymbol{A}.\end{aligned}$$

We summarize the properties of determinants:

1. $|\boldsymbol{A}^T| = |\boldsymbol{A}| \implies |\boldsymbol{A}^\dagger| = |\boldsymbol{A}|^*$.
2. If two rows (or columns) of $\boldsymbol{A}$ are interchanged, then $|\boldsymbol{A}|$ changes sign.
3. If any two rows (or columns) of $\boldsymbol{A}$ are the same, then $|\boldsymbol{A}| = 0$.
4. The determinant of a matrix is unchanged in value by adding a constant multiple of one row (column) to another.
5. If the elements of a row (column) are multiplied by a constant λ, then the determinant of the new matrix is λ times the determinant of the original matrix. If *all* elements of an $N \times N$ matrix are multiplied by λ, $\det(\lambda \boldsymbol{A}) = \lambda^N \det \boldsymbol{A}$.
6. $|\boldsymbol{AB}| = |\boldsymbol{A}||\boldsymbol{B}| = |\boldsymbol{BA}| \implies |\boldsymbol{A}^{-1}| = |\boldsymbol{A}|^{-1}$.

The inverse of matrix $\boldsymbol{A}$ exists when $\det \boldsymbol{A} \neq 0$ (noted in Section 1.4.6) and may be found using the cofactors of $\boldsymbol{A}$. If we were to replace row i of $\boldsymbol{A}$ by row k, the determinant of that modified matrix would vanish (equality of rows). That fact can be expressed through a modification of Eq. (1.39):

$$0 = \sum_{j=1}^{N} A_{kj} C_{ij}. \qquad (i \neq k) \tag{1.40}$$

Equations (1.39) and (1.40) can be combined into a single equation:

$$\sum_{j=1}^{N} A_{kj} C_{ij} = \delta_{ik} \det \boldsymbol{A}. \tag{1.41}$$

The inverse matrix $\boldsymbol{A}^{-1}$ is related to $\boldsymbol{A}$ through the relation $\boldsymbol{A}^{-1}\boldsymbol{A} = I$, or, in terms of matrix elements $\sum_j A_{ij}^{-1} A_{jk} = \delta_{ik}$. Comparing with Eq. (1.41), we have that the inverse matrix has elements

$$(A^{-1})_{ji} = \frac{1}{\det \boldsymbol{A}} C_{ij} = \frac{(-1)^{i+j}}{\det \boldsymbol{A}} M_{ij}. \tag{1.42}$$

Note the order of the indices. The inverse does not exist if $\det \boldsymbol{A} = 0$.

1.4.10 Unitary operators

Definition. *A linear operator $\mathcal{U}$ is* unitary *if $\|\mathcal{U}\phi\| = \|\phi\|$ for every vector ϕ.*

Unitary operators do not change the length of vectors (an important concept in quantum mechanics); obviously unitary operators are bounded. The condition $\|\mathcal{U}\phi\| = \|\phi\|$ for every ϕ implies that $\mathcal{U}$ has an inverse: There is no nonzero vector ϕ for which $\mathcal{U}\phi = 0$. An obvious example of a unitary operator is the identity operator $\mathcal{I}$.

Unitary operators have the property that for any two vectors ψ and ϕ:

$$(\mathcal{U}\psi, \mathcal{U}\phi) = (\psi, \phi). \tag{1.43}$$

To prove Eq. (1.43), let $\chi = \psi + a\phi$, where a is a complex scalar. It can be shown that

$$\begin{aligned} \|\chi\|^2 &= \|\psi\|^2 + |a|^2\|\phi\|^2 + 2\mathrm{Re}(a(\psi,\phi)) \\ \|\mathcal{U}\chi\|^2 &= \|\mathcal{U}\psi\|^2 + |a|^2\|\mathcal{U}\phi\|^2 + 2\mathrm{Re}(a(\mathcal{U}\psi,\mathcal{U}\phi)). \end{aligned} \tag{1.44}$$

By definition, $\|\mathcal{U}\chi\| = \|\chi\|$, etc., and the two results in Eq. (1.44) imply that

$$\mathrm{Re}(a(\mathcal{U}\psi,\mathcal{U}\phi)) = \mathrm{Re}(a(\psi,\phi)). \tag{1.45}$$

Equation (1.43) is implied by Eq. (1.45); see Exercise 1.18.

Example. For the space $L^2[a,b]$,

$$\mathcal{U}\psi(x) \equiv \mathrm{e}^{\mathrm{i}\omega x}\psi(x) \tag{1.46}$$

is unitary, where ω is a real number. We can see that $\mathcal{U}$ defined by Eq. (1.46) is unitary:

$$\|\mathcal{U}\psi\|^2 = (\mathcal{U}\psi, \mathcal{U}\psi) = \int_a^b |\mathrm{e}^{\mathrm{i}\omega x}\psi(x)|^2\,\mathrm{d}x = \int_a^b |\psi(x)|^2\,\mathrm{d}x = (\psi,\psi) = \|\psi\|^2.$$

Theorem 1.4.9. *A linear operator $\mathcal{U}$ is unitary if and only if $\mathcal{U}\mathcal{U}^\dagger = \mathcal{U}^\dagger \mathcal{U} = \mathcal{I}$.*

Proof. If $\mathcal{U}\mathcal{U}^\dagger = \mathcal{U}^\dagger\mathcal{U} = \mathcal{I}$, then $\mathcal{U}$ has an inverse, $\mathcal{U}^{-1} = \mathcal{U}^\dagger$. In this case, for any vector ψ, $\|\mathcal{U}\psi\|^2 = (\mathcal{U}\psi, \mathcal{U}\psi) = (\psi, \mathcal{U}^\dagger\mathcal{U}\psi) = \|\psi\|^2$. Thus, $\mathcal{U}$ is unitary. For the converse, assume $\mathcal{U}$ is unitary. Then, by Eq. (1.43), for any vectors ψ and ϕ, $(\phi, \mathcal{U}^\dagger\mathcal{U}\psi) = (\mathcal{U}\phi, \mathcal{U}\psi) = (\phi, \psi)$. Thus, $\mathcal{U}^\dagger\mathcal{U} = \mathcal{I}$. If $\mathcal{U}$ is unitary, then $\mathcal{U}^\dagger = \mathcal{U}^{-1}$ is unitary. □

The condition for unitarity, $\|\mathcal{U}\psi\| = \|\psi\|$, implies, because $\mathcal{U}$ is linear, that

$$\|\mathcal{U}\phi - \mathcal{U}\psi\| = \|\mathcal{U}(\phi - \psi)\| = \|\phi - \psi\|.$$

Thus, the "distance" between vectors ϕ and ψ, $\|\phi - \psi\|$, is preserved by a unitary operator, what's referred to in the advanced literature as an *isometric* operation.

Definition. *A* unitary matrix $\boldsymbol{U}$ *is one for which* $\boldsymbol{U}^\dagger = \boldsymbol{U}^{-1}$, *that* $(\boldsymbol{U}^{-1})_{jk} = (\boldsymbol{U})^*_{kj}$.

Unitary matrices have the properties that:

1. $|\boldsymbol{U}|^*|\boldsymbol{U}| = 1$, i.e. the determinant $|\boldsymbol{U}|$ has unit modulus, $|\boldsymbol{U}| = \mathrm{e}^{\mathrm{i}\theta}$ for some θ,
2. if $\boldsymbol{y} = \boldsymbol{U}\boldsymbol{x}$, then $\boldsymbol{y}^\dagger\boldsymbol{y} = \boldsymbol{x}^\dagger\boldsymbol{x}$. Unitary matrices preserve the length of vectors.

Definition. *A real $N \times N$ matrix having orthonormal columns is an* orthogonal matrix.

For an orthogonal matrix $\boldsymbol{A}$, the rules of matrix multiplication reveal that

$$\boldsymbol{A}^T\boldsymbol{A} = \begin{pmatrix} - & \boldsymbol{x}_1^T & - \\ - & \boldsymbol{x}_2^T & - \\ - & \cdots & - \\ - & \boldsymbol{x}_N^T & - \end{pmatrix} \begin{pmatrix} | & | & | & | \\ \boldsymbol{x}_1 & \boldsymbol{x}_2 & \cdots & \boldsymbol{x}_N \\ | & | & | & | \end{pmatrix} = \begin{pmatrix} 1 & 0 & & 0 \\ 0 & 1 & & 0 \\ 0 & 0 & \ddots & 0 \\ 0 & 0 & & 1 \end{pmatrix} = \boldsymbol{I}$$

and thus $\boldsymbol{A}^T = \boldsymbol{A}^{-1}$. Orthogonal matrices are unitary matrices with all elements real. For an orthogonal matrix $\boldsymbol{A}$, $\det \boldsymbol{A} = \pm 1$: $1 = |\boldsymbol{I}| = |\boldsymbol{A}^{-1}\boldsymbol{A}| = |\boldsymbol{A}^T\boldsymbol{A}| = |\boldsymbol{A}^T||\boldsymbol{A}| = |\boldsymbol{A}|^2$ (contrast with unitary matrices, $\det \boldsymbol{U} = e^{i\theta}$).

The importance of orthogonal matrices extends beyond the simple way by which $\boldsymbol{A}^{-1}$ can be calculated: Orthogonal matrices operate on real vectors without changing their magnitude. This behavior can be verified by writing $\boldsymbol{y} = \boldsymbol{A}\boldsymbol{x}$ so that

$$\langle \boldsymbol{y}|\boldsymbol{y}\rangle = \boldsymbol{y}^T\boldsymbol{y} = (\boldsymbol{A}\boldsymbol{x})^T\boldsymbol{A}\boldsymbol{x} = \boldsymbol{x}^T\boldsymbol{A}^T\boldsymbol{A}\boldsymbol{x} = \boldsymbol{x}^T\boldsymbol{A}^{-1}\boldsymbol{A}\boldsymbol{x} = \boldsymbol{x}^T\boldsymbol{I}\boldsymbol{x} = \boldsymbol{x}^T\boldsymbol{x} = \langle \boldsymbol{x}|\boldsymbol{x}\rangle.$$

Example. The classic example of an orthogonal matrix is one that rigidly rotates a two-dimensional coordinate system by the angle φ – see Figure 1.5. For (x', y') the coordinates of a fixed point in space in the rotated coordinate system, it's an exercise in trigonometry to show that they're related to the coordinates (x, y) of the same point in space in the unrotated coordinate system by a matrix equation:

$$\begin{pmatrix} x' \\ y' \end{pmatrix} = \begin{pmatrix} \cos\varphi & \sin\varphi \\ -\sin\varphi & \cos\varphi \end{pmatrix} \begin{pmatrix} x \\ y \end{pmatrix}. \tag{1.47}$$

The inverse of this matrix is especially easy to find since a rotation by $-\varphi$ will "undo" the original rotation. We have

$$\boldsymbol{A}^{-1} = \begin{pmatrix} \cos\varphi & -\sin\varphi \\ \sin\varphi & \cos\varphi \end{pmatrix} = \boldsymbol{A}^T$$

and so $\boldsymbol{A}$ is orthogonal. It's straightforward to show that $(x')^2 + (y')^2 = x^2 + y^2$, i.e. the length of the vector $\boldsymbol{r}$ is preserved under the rotation of coordinate axes.

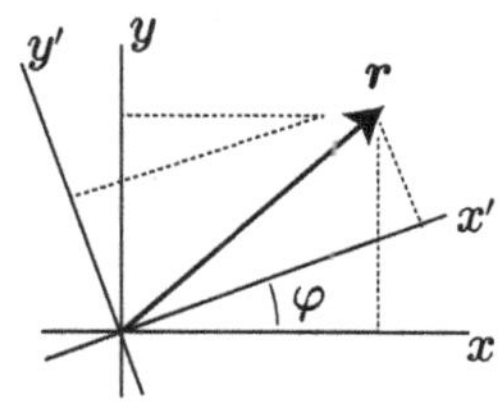

Figure 1.5 Coordinate systems having a common origin with axes rotated through a fixed angle φ.

1.4.11 The trace of a matrix

There is another basis-independent, numerical quantity associated with the elements of a square matrix (in addition to the determinant), the *trace*.

Definition. *For an $N \times N$ matrix $\boldsymbol{A}$, its* trace, *denoted* $\mathrm{Tr}\boldsymbol{A}$, *is the sum of its diagonal matrix elements,* $\mathrm{Tr}\boldsymbol{A} \equiv \sum_{i=1}^N A_{ii}$.

There are many applications of the trace operation in physics, e.g. statistical mechanics.

By definition, $\mathrm{Tr}\boldsymbol{A}^T = \mathrm{Tr}\boldsymbol{A}$ and $\mathrm{Tr}\boldsymbol{A}^* = (\mathrm{Tr}\boldsymbol{A})^*$. A more important property is that for $N \times N$ matrices $\boldsymbol{A}$ and $\boldsymbol{B}$, $\mathrm{Tr}(\boldsymbol{AB}) = \mathrm{Tr}(\boldsymbol{BA})$:

$$\mathrm{Tr}(\boldsymbol{AB}) = \sum_{i=1}^N (\boldsymbol{AB})_{ii} = \sum_{i=1}^N \sum_{j=1}^N A_{ij}B_{ji} = \sum_{j=1}^N \sum_{i=1}^N B_{ji}A_{ij} = \sum_{j=1}^N (\boldsymbol{BA})_{jj} = \mathrm{Tr}(\boldsymbol{BA}).$$

This property can be extended to the product of three or more matrices, the *cyclic invariance of the trace*: $\text{Tr}(\boldsymbol{ABC}) = \text{Tr}(\boldsymbol{CAB}) = \text{Tr}(\boldsymbol{BCA})$; see Exercise 1.24. We can immediately apply cyclic invariance to the trace of similar matrices (those connected by a similarity transformation). If $\boldsymbol{A}' = \boldsymbol{S}^{-1}\boldsymbol{AS}$, then

$$\text{Tr}\boldsymbol{A}' = \text{Tr}(\boldsymbol{S}^{-1}\boldsymbol{AS}) = \text{Tr}(\boldsymbol{SS}^{-1}\boldsymbol{A}) = \text{Tr}(\boldsymbol{IA}) = \text{Tr}\boldsymbol{A},$$

establishing the basis-independence of the trace.

1.5 Eigenvectors and their role in representing operators

In this section, we show, by making use of a special basis consisting of the *eigenvectors* of a linear operator, that working with operators is almost as easy as working with numbers. It isn't just for computational ease, however, that we study eigenvectors. Every eigenvector is associated with a number – its *eigenvalue* – which plays a fundamental role in quantum mechanics: The result of measurement can only be one of the eigenvalues of the operator representing a physical observable. The "eigenproblem" as defined in this section is quite fundamental to further developments in the theory and in applications; we return to it in Chapter 2.

1.5.1 Eigenvectors and eigenvalues

Definition. *For a linear operator $\mathcal{A} : \mathbb{V} \to \mathbb{V}$, if $\boldsymbol{x}$ is a nonzero vector of $\mathbb{V}$ such that $\mathcal{A}\boldsymbol{x} = \lambda\boldsymbol{x}$ for some scalar λ, $\boldsymbol{x}$ is said to be an* eigenvector *belonging to $\mathcal{A}$. An* eigenvalue *belonging to $\mathcal{A}$ is a scalar λ such that $\mathcal{A}\boldsymbol{x} = \lambda\boldsymbol{x}$ for some nonzero vector $\boldsymbol{x}$. The set of all eigenvalues of $\mathcal{A}$ is called its* spectrum.

The aforementioned statement has been carefully worded. Each eigenvector $\boldsymbol{x}$ of $\mathcal{A}$ determines an eigenvalue, but the association may not be one-to-one: There may be several linearly independent eigenvectors corresponding to the same eigenvalue – what are referred to as *degenerate* eigenvectors. We sidestep the complications associated with degenerate eigenvectors in this book, even though degeneracies are not an uncommon occurrence in physical applications; they play a fundamental role in statistical mechanics, for example. Note that while an eigenvector cannot be $|0\rangle$, eigenvalues equal to 0 *are* allowed.

Eigenvectors are a special class of vector. Every vector $\boldsymbol{x} \in \mathbb{V}$ is mapped into *some* vector $\mathcal{A}\boldsymbol{x} \in \mathbb{V}$, but only certain vectors are mapped into scalar multiples of themselves, $\mathcal{A}\boldsymbol{x} = \lambda\boldsymbol{x}$. We can label the eigenvalues and their corresponding eigenvectors[50]

[50]Non degenerate eigenvectors can be labeled with the same label as the eigenvalue. Otherwise, the eigenequation (any equation involving eigenvectors) is written $\mathcal{A}\boldsymbol{x}_{n,j} = \lambda_n \boldsymbol{x}_{n,j}$, with $j = 1, \ldots, p$, where p is the *multiplicity* of the eigenvalue.

with a *discrete* index n (an integer), $\mathcal{A}\boldsymbol{x}_n = \lambda_n \boldsymbol{x}_n$; continuous labels are possible with infinite-dimensional vector spaces. Operators $\mathcal{A}$ on finite-dimensional vector spaces $\mathbb{V}$ are bounded (Section 1.4.5) so that for all vectors $\boldsymbol{x} \in \mathbb{V}$, $\|\mathcal{A}\boldsymbol{x}\| \leq b\|\boldsymbol{x}\|$, where b is a finite number. For eigenvectors, $\mathcal{A}\boldsymbol{x}_n = \lambda_n \boldsymbol{x}_n$, where $|\lambda_n| < b$ for all n. Note that if $\boldsymbol{x}$ is an eigenvector, so is $\alpha\boldsymbol{x}$ for scalar α ($\mathcal{A}$ is linear: $\mathcal{A}(\alpha\boldsymbol{x}) = \lambda(\alpha\boldsymbol{x})$). To avoid this (mostly meaningless) ambiguity, we concentrate on normalized eigenvectors for which $\langle \boldsymbol{x}|\boldsymbol{x}\rangle = 1$.

What's required for a scalar to be an eigenvalue of a linear operator? We can write the eigenequation $\mathcal{A}\boldsymbol{x} = \lambda\boldsymbol{x}$ in the form $(\mathcal{A} - \lambda\mathcal{I})\boldsymbol{x} = \boldsymbol{0}$, where $\boldsymbol{0}$ is the zero vector, $|0\rangle$. An operator has an inverse if and only if no nonzero vector is mapped to $|0\rangle$ (Section 1.4.6). Yet nonzero vectors satisfying $(\mathcal{A} - \lambda\mathcal{I})\boldsymbol{x} = \boldsymbol{0}$ are what we seek. Our question is therefore answered, which we state as a theorem:

Theorem 1.5.1. *If λ is an eigenvalue of operator $\mathcal{A}$, the operator $\mathcal{A} - \lambda\mathcal{I}$ has no inverse.*

This theorem provides a way to calculate eigenvalues. For an operator to have an inverse, the determinant of its matrix representation must be nonzero (Section 1.4.6). For λ to be an eigenvalue of a linear operator, therefore, the determinant of the matrix representing $\mathcal{A} - \lambda\mathcal{I}$ must be zero,

$$|\boldsymbol{A} - \lambda\boldsymbol{I}| = 0. \tag{1.48}$$

Equation (1.48) is called the *characteristic* (or secular) equation of $\boldsymbol{A}$. For $\boldsymbol{A}$ an $N \times N$ matrix, the determinant $|\boldsymbol{A} - \lambda\boldsymbol{I}|$, when expanded out, is a polynomial[51] of degree N in λ, $P(\lambda) \equiv |\boldsymbol{A} - \lambda\boldsymbol{I}|$, the *characteristic polynomial* (or secular function). The roots of $P(\lambda)$ are the eigenvalues[52] of $\boldsymbol{A}$. Any polynomial of degree N can be written in the form[53]:

$$P(\lambda) = (-1)^N(\lambda - \lambda_1)(\lambda - \lambda_2)\cdots(\lambda - \lambda_N). \tag{1.49}$$

The quantities λ_i in Eq. (1.49) are the eigenvalues of $\boldsymbol{A}$.

---

[51]The determinant *is* a polynomial (in N-fold products of matrix elements) – see Eq. (1.37).

[52]That eigenvalues can be found from the roots of the characteristic polynomial is a prominent reason for studying determinants. Eigenvectors have geometric significance, vectors that are stretched into multiples of themselves. We can't find the eigenvalues, however, through geometry – hence the excursion into the algebraic properties of determinants.

[53]Equation (1.49) relies on a major result of mathematics, the Fundamental Theorem of Algebra [8, p. 116], which comes with some fine print. $P(\lambda)$ can be written in the form of Eq. (1.49) only if we allow for the possibility of complex-valued eigenvalues. Equation (1.49) assumes no degeneracies. A more careful statement is that $P(\lambda)$ has at least one root and never more than N distinct roots.

Example. Determine the eigenvalues and eigenvectors for the matrix

$$\boldsymbol{A} = \begin{pmatrix} 1 & 1 & 3 \\ 1 & 1 & -3 \\ 3 & -3 & -3 \end{pmatrix}.$$

Solution: The secular equation is

$$|\boldsymbol{A} - \lambda \boldsymbol{I}| = \begin{vmatrix} 1-\lambda & 1 & 3 \\ 1 & 1-\lambda & -3 \\ 3 & -3 & -3-\lambda \end{vmatrix} = 0.$$

Evaluating the determinant implies that $P(\lambda) = -\lambda^3 - \lambda^2 + 24\lambda - 36$, which can be factored, $P(\lambda) = -(\lambda - 2)(\lambda - 3)(\lambda + 6)$, implying that the eigenvalues are $\lambda_1 = 2, \lambda_2 = 3, \lambda_3 = -6$.

The eigenvectors are found by solving the eigenequation for known λ. For $\lambda_1 = 2$, the associated eigenvector $\boldsymbol{x} = (x_1, x_2, x_3)^T$ must satisfy

$$\begin{aligned} x_1 + x_2 + 3x_3 &= 2x_1 \\ x_1 + x_2 - 3x_3 &= 2x_2 \\ 3x_1 - 3x_2 - 3x_3 &= 2x_3. \end{aligned}$$

This is a set of three equations in three unknowns and is solved by $\boldsymbol{x} = (a, a, 0)^T$ for any a. Normalization requires $a = 1/\sqrt{2}$ (check it!). The other two eigenvectors are found in a similar way and we have

$$\boldsymbol{x}_1 = \frac{1}{\sqrt{2}}(1,1,0)^T, \quad \boldsymbol{x}_2 = \frac{1}{\sqrt{3}}(1,-1,1)^T, \quad \boldsymbol{x}_3 = \frac{1}{\sqrt{6}}(1,-1,-2)^T.$$

Theorem 1.5.2. *Let operator $\mathcal{A}$ have eigenvectors $\boldsymbol{x}_1, \boldsymbol{x}_2, \ldots, \boldsymbol{x}_m$ corresponding to eigenvalues $\lambda_1, \lambda_2, \ldots, \lambda_m$, with $\lambda_i \neq \lambda_j$ for $i \neq j$. Then the eigenvectors $\boldsymbol{x}_1, \boldsymbol{x}_2, \ldots, \boldsymbol{x}_m$ are linearly independent.*

For a proof, see [2, p. 109]. The theorem applies to any set of m nondegenerate eigenvectors. The eigenvectors found in the previous example are linearly independent (check it!). One infers from the theorem that *a linear operator acting on an N-dimensional space cannot have more than N eigenvectors having distinct eigenvalues.*

The eigenvalues of $\mathcal{A}$, found through the characteristic equation associated with its matrix representation $\boldsymbol{A}$, are independent of basis (determinants are basis independent; Section 1.4.9). Note from Eq. (1.49) that by setting $\lambda = 0$,

$$\det \boldsymbol{A} = \lambda_1 \lambda_2 \cdots \lambda_N \equiv \prod_{k=1}^{N} \lambda_k. \tag{1.50}$$

The determinant of any matrix is the product of its eigenvalues.[54] (The determinant of $\boldsymbol{A}$ in the previous example is -36, the same as the product of its eigenvalues.) If one of the eigenvalues is zero, therefore, the operator is not invertible.

That eigenvalues are independent of basis is reflected in the following statement for operators (one that's valid for infinite-dimensional vector spaces):

Theorem 1.5.3. *If λ is an eigenvalue of $\mathcal{A}$, then it's also an eigenvalue of $\tilde{\mathcal{A}} \equiv \mathcal{T}\mathcal{A}\mathcal{T}^{-1}$, where $\mathcal{T}$ is a linear operator possessing an inverse $\mathcal{T}^{-1}$.*

Proof. If λ is an eigenvalue of $\mathcal{A}$, there is a nonzero vector ψ such that $\mathcal{A}\psi = \lambda\psi$. If $\mathcal{T}$ has an inverse $\mathcal{T}^{-1}$, then $\mathcal{T}\psi$ is not zero. Then, $(\mathcal{T}\mathcal{A}\mathcal{T}^{-1})\mathcal{T}\psi = \lambda\mathcal{T}\psi$, proving the theorem. □

1.5.2 The eigenproblem for Hermitian and unitary operators

Prominent in physical applications are the eigenvalues and eigenvectors of Hermitian and unitary operators, the properties of which are summarized in the following theorems.

Theorem 1.5.4. *Eigenvalues of Hermitian operators are real numbers.*

Proof. If $\mathcal{A}$ is Hermitian and λ is an eigenvalue of $\mathcal{A}$ with eigenvector ψ, then

$$\lambda(\psi, \psi) = (\psi, \lambda\psi) = (\psi, \mathcal{A}\psi) = (\mathcal{A}\psi, \psi) = (\lambda\psi, \psi) = \lambda^*(\psi, \psi) \implies \lambda = \lambda^*.$$

□

Theorem 1.5.5. *Eigenvalues of unitary operators are complex numbers of unit magnitude.*

Proof. If $\mathcal{U}$ is a unitary operator and λ is an eigenvalue of $\mathcal{U}$ with eigenvector ψ, then

$$(\psi, \psi) = (\psi, \mathcal{U}^\dagger\mathcal{U}\psi) = (\mathcal{U}\psi, \mathcal{U}\psi) = (\lambda\psi, \lambda\psi) = |\lambda|^2(\psi, \psi) \implies \lambda = \mathrm{e}^{\mathrm{i}\theta}.$$

□

Theorem 1.5.6. *Eigenvectors of Hermitian or unitary operators corresponding to different eigenvalues are orthogonal.*

Proof. If $\mathcal{A}$ is Hermitian and ψ_1 and ψ_2 are eigenvectors of $\mathcal{A}$ corresponding to eigenvalues λ_1 and λ_2 with $\lambda_1 \neq \lambda_2$, then, because λ_1, λ_2 are real

$$\begin{aligned}(\lambda_1 - \lambda_2)(\psi_1, \psi_2) &= (\lambda_1\psi_1, \psi_2) - (\psi_1, \lambda_2\psi_2) \\ &= (\mathcal{A}\psi_1, \psi_2) - (\psi_1, \mathcal{A}\psi_2) = 0 \implies (\psi_1, \psi_2) = 0.\end{aligned}$$

If $\mathcal{U}$ is unitary and ψ_1 and ψ_2 are eigenvectors of $\mathcal{U}$ corresponding to eigenvalues λ_1 and λ_2 with $\lambda_1 \neq \lambda_2$, then

[54] For a matrix having d distinct eigenvalues, then $\det \boldsymbol{A} = \prod_{k=1}^{d} \lambda_k^{m_k}$, where m_k is the multiplicity of λ_k.

$$\lambda_1^*\lambda_2(\psi_1,\psi_2) = (\lambda_1\psi_1,\lambda_2\psi_2) = (\mathcal{U}\psi_1,\mathcal{U}\psi_2) = (\psi_1,\psi_2) \implies (\psi_1,\psi_2) = 0.$$

Because $\lambda_1^*\lambda_2 \neq 1$, $(\psi_1,\psi_2) = 0$. □

Example. Consider the matrix

$$\boldsymbol{A} = \begin{pmatrix} 1 & \mathrm{i} \\ -\mathrm{i} & 1 \end{pmatrix}.$$

As can be checked, $\boldsymbol{A}$ is Hermitian, it has two real eigenvalues, $\lambda = 0, 2$ which correspond to orthogonal eigenvectors $(\psi_{\lambda=0},\psi_{\lambda=2}) = 0$, where

$$\psi_{\lambda=0} = \frac{1}{\sqrt{2}}\begin{pmatrix} 1 \\ \mathrm{i} \end{pmatrix} \qquad \psi_{\lambda=2} = \frac{1}{\sqrt{2}}\begin{pmatrix} \mathrm{i} \\ 1 \end{pmatrix}.$$

Clearly, $\psi_{\lambda=0}$ and $\psi_{\lambda=2}$ are linearly independent. The fact that there is a zero eigenvalue implies that $\boldsymbol{A}$ is singular – it has zero determinant.

1.5.3 Diagonalizing matrices

Diagonal matrices $\boldsymbol{D}$ are particularly easy to work with. With nonzero entries only along the diagonal ($D_{ij} = D_{ii}\delta_{ij}$), the inverse of a diagonal matrix is itself diagonal with $(\boldsymbol{D}^{-1})_{ij} = (D_{ii})^{-1}\delta_{ij}$ (a property *not* shared by the inverse of general matrices). The nth power of a diagonal matrix (the effect of n copies of $\boldsymbol{D}$) $\boldsymbol{D}^n \equiv \boldsymbol{D}\cdots\boldsymbol{D}$ has elements $(\boldsymbol{D}^n)_{ij} = (D_{ii})^n\delta_{ij}$. If $\boldsymbol{B}$ is another diagonal matrix of the same size as $\boldsymbol{D}$, then $\boldsymbol{DB} = \boldsymbol{BD}$. The set of all $N \times N$ diagonal matrices is *commutative* under matrix multiplication (not true of general matrices).

Almost any operation we can do with matrices is simpler with diagonal matrices. A strategy for working with matrices is to seek a basis in which the representation of the operator is diagonal. Under a change of basis, matrices transform by $\boldsymbol{A} \to \boldsymbol{A}' = \boldsymbol{S}^{-1}\boldsymbol{AS}$ (Section 1.4.7). The goal is to find a matrix $\boldsymbol{S}$ such that the similarity transformation associated with a given matrix $\boldsymbol{A}$ yields a diagonal matrix $\boldsymbol{D}$:

$$\boldsymbol{S}^{-1}\boldsymbol{AS} = \boldsymbol{D}. \tag{1.51}$$

In this way, $\boldsymbol{A}$ and $\boldsymbol{D}$ have the same eigenvalues (Section 1.5.1). The process of finding a diagonal matrix $\boldsymbol{D}$ similar to a given matrix $\boldsymbol{A}$ is referred to as *diagonalizing* the matrix.

Equation (1.51) is equivalent to $\boldsymbol{AS} = \boldsymbol{SD}$ and, according to Eq. (1.21), $\boldsymbol{AS} = \boldsymbol{SD}$ can be interpreted as[55]

$$\boldsymbol{A}\boldsymbol{x}_i = D_{ii}\boldsymbol{x}_i, \tag{1.52}$$

---

[55]The jth element of the ith column of $\boldsymbol{S}$ is $(\boldsymbol{SD})_{ji} = \sum_{k=1}^N S_{jk}D_{ki} = \sum_{k=1}^N S_{jk}D_{ii}\delta_{ki} = D_{ii}S_{ji}$, where the last follows from $\boldsymbol{D}$ being diagonal so that $D_{ki} = D_{ii}\delta_{ki}$.

where $\boldsymbol{x}_i$ denotes the ith column of $\boldsymbol{S}$ (because $\boldsymbol{D}$ is diagonal). Equation (1.52) is in the form of an eigenequation. *Diagonalizing a matrix is tantamount to finding its eigenvalues.*

Definition. *A linear operator $\mathcal{A}$ on vector space $\mathbb{V}$ is said to be* diagonalizable *if there is a basis for $\mathbb{V}$ consisting entirely of the eigenvectors of $\mathcal{A}$.*

Any set of eigenvectors is linearly independent (Section 1.5.1); if for an N-dimensional space an operator has N distinct eigenvectors, they will span the space which implies we can find an orthonormal basis of eigenvectors. Eigenvectors of Hermitian and unitary operators naturally form orthogonal sets (Section 1.5.2).

Theorem 1.5.7. *If $\boldsymbol{A}$ is $N \times N$ with N linearly independent eigenvectors $\boldsymbol{x}_i$, $i = 1, 2, \ldots, N$ with associated eigenvalues λ_i, then the matrix $\boldsymbol{S}$ whose columns are formed from the eigenvectors so that*

$$\boldsymbol{S} = \begin{pmatrix} | & | & | & | \\ \boldsymbol{x}_1 & \boldsymbol{x}_2 & \cdots & \boldsymbol{x}_N \\ | & | & | & | \end{pmatrix}$$

will yield a similarity transformation of $\boldsymbol{A}$ that obeys

$$\boldsymbol{S}^{-1}\boldsymbol{A}\boldsymbol{S} = \begin{pmatrix} \lambda_1 & 0 & 0 & 0 \\ 0 & \lambda_2 & 0 & 0 \\ 0 & 0 & \ddots & 0 \\ 0 & 0 & 0 & \lambda_N \end{pmatrix} \equiv \boldsymbol{\Lambda}. \tag{1.53}$$

Proof. Since $\boldsymbol{A}\boldsymbol{x}_i = \lambda_i \boldsymbol{x}_i$, we have

$$\boldsymbol{A}\boldsymbol{S} = \boldsymbol{A}\begin{pmatrix} | & | & | & | \\ \boldsymbol{x}_1 & \boldsymbol{x}_2 & \cdots & \boldsymbol{x}_N \\ | & | & | & | \end{pmatrix} = \begin{pmatrix} | & | & | & | \\ \lambda_1\boldsymbol{x}_1 & \lambda_2\boldsymbol{x}_2 & \cdots & \lambda_N\boldsymbol{x}_N \\ | & | & | & | \end{pmatrix}$$

$$= \begin{pmatrix} | & | & | & | \\ \boldsymbol{x}_1 & \boldsymbol{x}_2 & \cdots & \boldsymbol{x}_N \\ | & | & | & | \end{pmatrix}\begin{pmatrix} \lambda_1 & 0 & 0 & 0 \\ 0 & \lambda_2 & 0 & 0 \\ 0 & 0 & \ddots & 0 \\ 0 & 0 & 0 & \lambda_N \end{pmatrix}$$

so $\boldsymbol{A}\boldsymbol{S} = \boldsymbol{S}\boldsymbol{\Lambda} \Longrightarrow \boldsymbol{S}^{-1}\boldsymbol{A}\boldsymbol{S} = \boldsymbol{\Lambda}$. (The matrix $\boldsymbol{S}$ is invertible because its columns are linearly independent.) □

Some points to note regarding this theorem: The result applies to matrices $\boldsymbol{A}$ having no repeated eigenvalues (i.e. the numbers $\lambda_1, \lambda_2, \ldots, \lambda_N$ are distinct). *Any matrix with distinct eigenvalues can be diagonalized*, but by implication, *not all matrices are diagonalizable.* The equation $\boldsymbol{AS} = \boldsymbol{S\Lambda}$ holds only if the columns of $\boldsymbol{S}$ are eigenvectors of $\boldsymbol{A}$. Eigenvectors, however, are not unique up to a multiplicative constant, and thus the diagonalizing matrix $\boldsymbol{S}$ is not unique. We can multiply the columns of $\boldsymbol{S}$ by any nonzero constants and produce a new diagonalizing $\boldsymbol{S}$. Note that real-valued symmetric matrices (for which $A_{ij} = A_{ji}$) are automatically Hermitian and so can always be diagonalized.

Example. We have shown that the eigenvalues and eigenvectors of

$$\boldsymbol{A} = \begin{pmatrix} 1 & 1 & 3 \\ 1 & 1 & -3 \\ 3 & -3 & -3 \end{pmatrix}$$

are $\lambda_1 = 2, \quad \lambda_2 = 3, \quad \lambda_3 = -6$, and

$$\boldsymbol{x}_1 = \frac{1}{\sqrt{2}} \begin{pmatrix} 1 \\ 1 \\ 0 \end{pmatrix}, \quad \boldsymbol{x}_2 = \frac{1}{\sqrt{3}} \begin{pmatrix} 1 \\ -1 \\ 1 \end{pmatrix}, \quad \boldsymbol{x}_3 = \frac{1}{\sqrt{6}} \begin{pmatrix} 1 \\ -1 \\ -2 \end{pmatrix}.$$

We use these eigenvectors to form $\boldsymbol{S}$ as

$$\boldsymbol{S} = \begin{pmatrix} 1/\sqrt{2} & 1/\sqrt{3} & 1/\sqrt{6} \\ 1/\sqrt{2} & -1/\sqrt{3} & -1/\sqrt{6} \\ 0 & 1/\sqrt{3} & -2/\sqrt{6} \end{pmatrix}.$$

Then it's easy to verify that

$$\begin{aligned} \boldsymbol{AS} &= \begin{pmatrix} 1 & 1 & 3 \\ 1 & 1 & -3 \\ 3 & -3 & -3 \end{pmatrix} \begin{pmatrix} 1/\sqrt{2} & 1/\sqrt{3} & 1/\sqrt{6} \\ 1/\sqrt{2} & -1/\sqrt{3} & -1/\sqrt{6} \\ 0 & 1/\sqrt{3} & -2/\sqrt{6} \end{pmatrix} \\ &= \begin{pmatrix} 1/\sqrt{2} & 1/\sqrt{3} & 1/\sqrt{6} \\ 1/\sqrt{2} & -1/\sqrt{3} & -1/\sqrt{6} \\ 0 & 1/\sqrt{3} & -2/\sqrt{6} \end{pmatrix} \begin{pmatrix} 2 & 0 & 0 \\ 0 & 3 & 0 \\ 0 & 0 & -6 \end{pmatrix} = \boldsymbol{S\Lambda}. \end{aligned}$$

Because the trace is invariant under similarity transformations, we have that

$$\text{Tr}\boldsymbol{A} = \text{Tr}(\boldsymbol{S\Lambda S}^{-1}) = \text{Tr}\boldsymbol{\Lambda} = \sum_{i=1}^{N} \lambda_i. \tag{1.54}$$

The trace of a matrix is the sum of its eigenvalues; the determinant of a matrix is the product of its eigenvalues, Eq. (1.50).

1.6 Hilbert space: Infinite-dimensional vector space

We've worked mostly with finite-dimensional vector spaces in this chapter. Not every vector space, however, is spanned by a finite number of vectors. Indeed, some of the most important applications of linear spaces involve *Hilbert space*, a generic term for infinite-dimensional vector spaces (made precise in the following). Function spaces, for example, require an infinite number of basis vectors to represent functions.[56] As an example, the functions $|e_n\rangle \equiv e^{2\pi inx}$, $n = 0, \pm 1, \pm 2, \ldots,$ comprise an infinite, orthonormal set in $L^2[0,1]$, $\langle e_m|e_n\rangle = \delta_{nm}$. According to Fourier's theorem (Chapter 4), any function $\psi(x)$, such that $\int_0^1 |\psi(x)|^2 \, dx$ is finite, may be represented by a Fourier series,[57]

$$|\psi\rangle = \sum_{n=-\infty}^{\infty} a_n|e_n\rangle, \tag{1.55}$$

where $a_n = \langle e_n|\psi\rangle$. Equation (1.55) generalizes Eq. (1.7) in that an infinite number of vectors are required to effect the equality in Eq. (1.55). *The basic concepts covered in this chapter can all be generalized to infinite dimensions.* The one fundamental difference between finite and infinite-dimensional spaces is the need to understand linear combinations of infinite numbers of vectors. Infinite summations make sense only when they're *convergent*. For finite-dimensional spaces, there's no question of the *existence* of finite sums of vectors, such as in Eq. (1.7). With infinite summations, we face two questions: (i) does a given infinite sum of vectors exist? and (ii) is the resultant sum of vectors (if it exists) an element of the vector space? Only in Hilbert space theory does the latter question arise. You would probably never ask whether a convergent, infinite sum of numbers is a number, or rather, is a number within your universe of numbers.

First consider the problem of summing an infinite sequence of complex numbers z_k, $S \equiv \sum_{k=1}^{\infty} z_k$. The classic way to associate a value with an infinite summation is to first define the *partial sums* $S_N \equiv \sum_{k=1}^{N} z_k$. If, as $N \to \infty$, the sequence of partial sums S_N tends to a limit, the infinite series *converges* to the limit, $\lim_{N\to\infty} S_N = S$. (Series that do not converge are said to *diverge*.) The convergence $S_N \to S$ means that the difference[58] $|S - S_N| \to 0$ as $N \to \infty$.

---

[56]There are an infinite number of points in any interval of the real line. To specify the values of a function $f(x)$ for $a \le x \le b$ requires in principle the specification of an infinite number of pieces of information.

[57]Fourier series are more general than Taylor series, although both are representations of functions involving infinite series (see Chapter 4). Note that the sum in Eq. (1.55) runs from $-\infty$ to $+\infty$.

[58]"Converging to zero" means tending to a number smaller than any positive number. A sequence $z_1, z_2, \ldots$ has limit l if, for a given positive number ϵ, a number N_0 can be found such that $|z_N - l| < \epsilon$ for all $N > N_0$.

In the same way, a sequence of *vectors* $\{\psi_k\}$ converges to a *limit vector* ψ if the norm of the difference vector converges to zero,[59] $\|\psi - \psi_k\| \to 0$ as $k \to \infty$. An infinite linear combination of vectors, $\psi \equiv \sum_{k=1}^{\infty} a_k\phi_k$ is defined if the partial sums $\psi_N \equiv \sum_{k=1}^{N} a_k\phi_k$ converge to a limit, $\psi_N \to \psi$. The limit of a convergent sequence is unique: If $\psi_N \to \psi$ and $\psi_N \to \chi$ as $N \to \infty$, then $\psi - \chi$ must be zero:

$$\|\psi - \chi\| = \|\psi - \psi_N + \psi_N - \chi\| \leq \|\psi - \psi_N\| + \|\psi_N - \chi\|,$$

where we've used the triangle inequality.

Infinite linear combinations can be added componentwise. For $\psi = \sum_{k=1}^{\infty} a_k\phi_k$ and $\chi = \sum_{k=1}^{\infty} b_k\phi_k$, then $\psi + \chi = \sum_{k=1}^{\infty}(a_k + b_k)\phi_k$; if the sequence of partial sums $\psi_N = \sum_{k=1}^{N} a_k\phi$ converges to ψ and $\chi_N = \sum_{k=1}^{N} b_k\phi_k$ converges to χ, the addition of partial sums $\psi_N + \chi_N = \sum_{k=1}^{N}(a_k + b_k)\phi_k$ converges to $\psi + \chi$ because $\|\psi + \chi - (\psi_N + \chi_N)\| \leq \|\psi - \psi_N\| + \|\chi - \chi_N\|$. Scalar multiplication of infinite linear combinations can also be done componentwise. If $\psi = \sum_{k=1}^{\infty} a_k\phi_k$, then $\alpha\psi = \sum_{k=1}^{\infty} \alpha a_k\phi_k$; if $\sum_{k=1}^{N} a_k\phi_k = \psi_N \to \psi$, then $\sum_{k=1}^{N} \alpha a_k\phi_k = \alpha\psi_N \to \alpha\psi$ because $\|\alpha\psi - \alpha\psi_N\| = |\alpha|\|\psi - \psi_N\|$.

Definition. *A set of complex numbers* $\{z_k\}_{k=1}^{\infty}$ *that converges to a limit* z *is* closed *if the limit point* z *is part of the set.*

The definition of closed set might ordinarily appear to us as rather "mathy," yet the same applies to vectors: A set of vectors $\{\psi_k\}_{k=1}^{\infty}$ that converges $\psi_k \to \psi$ is closed if the limit vector ψ is part of the set. *Infinite-dimensional subspaces are specified by closed sets of vectors*: Infinite-dimensional subspaces must contain all finite linear combinations $\sum_{k=1}^{N} a_k\psi_k$ as well as infinite, $\sum_{k=1}^{\infty} a_k\psi_k$, and hence they must contain the limits of sequences of partial sums.[60] A set of vectors $\{\psi_1, \psi_2, \ldots\}$ is a basis if they're linearly independent[61] and they span the space, i.e. if each vector ψ in the subspace can be expressed as a linear combination, $\psi = \sum_{k=1}^{\infty} a_k\psi_k$. A basis for an infinite-dimensional space must be a closed set of vectors. There is an additional criterion, however, that must be imposed. Infinity is a big "place": One can always add more to an infinite set and still have an infinite set, $\infty + \infty = \infty$. An infinite set of

[59] We're glossing over several distinctions concerning the convergence of a sequence of functions – pointwise convergence, uniform convergence, and convergence in the mean. Knopp [10] is a good resource.

[60] Finite-dimensional subspaces of infinite-dimensional spaces are certainly allowed – they are sometimes referred to as *linear manifolds* to distinguish them from infinite-dimensional subspaces. A subspace is a closed linear manifold, a needless distinction for finite-dimensional spaces.

[61] Mutually orthogonal vectors are linearly independent, so that box is checked if we make use of orthonormal basis sets. We employ orthonormal bases for another reason, however. The notion of completeness, required for Hilbert space theory, and defined momentarily, is framed in terms of orthogonality.

vectors could be a subset of another infinite set of vectors. How do we ensure there are not vectors in the space not expressible in terms of the basis for a given subspace, i.e. how do we ensure that the whole space is a subspace?

We reach for the concept of *completeness* (defined in Section 1.3.5). We've used (in this section) the traditional condition for convergence, but there's another sense in which convergence can be defined. A sequence of vectors converges if either of the conditions holds:

1. $\|\psi - \psi_n\| \to 0$ as $n \to \infty$,
2. $(\phi, \psi - \psi_n) \to 0$ as $n \to \infty$, for each ϕ in the space.

Condition (i) implies condition (ii). For every vector ϕ and for any n,

$$|(\phi, \psi - \psi_n)| \le \|\phi\| \cdot \|\psi - \psi_n\|,$$

by the Schwarz inequality. Thus, the truth of (i) implies the truth of (ii). The converse is true for finite-dimensional spaces. Let $\{\phi_1, \phi_2, \ldots, \phi_N\}$ be an orthonormal basis for finite-dimensional space $\mathbb{V}_N$. If we assume (ii), $(\phi_i, \psi - \psi_n) \to 0$ for each $i = 1, \ldots, N$ as $n \to \infty$, then because (see Section 1.3.5)

$$\|\psi - \psi_n\|^2 = \sum_{i=1}^{N} |(\phi_i, \psi - \psi_n)|^2, \tag{1.56}$$

it follows that $\|\psi - \psi_n\| \to 0$ as $n \to \infty$ – the truth of (ii) implies the truth of (i) for finite-dimensional spaces. Equation (1.56) holds because finite-dimensional spaces are *complete* – a basis for $\mathbb{V}_N$ is not contained in a larger set.

For infinite-dimensional spaces, it's possible that condition (ii) does not imply condition (i): It may be possible that a convergent sequence of vectors $\{\psi_n\}$ does not converge to a vector ψ *in the space*. The only way to preclude that is to assume the opposite.

Definition. *A complete inner-product space is called a* Hilbert space, *denoted* $\mathbb{H}$.

Hilbert space is *defined* so that condition (ii) applies.[62] There is no vector "outside of," i.e. orthogonal to, any vector in the space. Said differently, there is no vector not expressible in terms of a complete orthonormal basis,[63] i.e. *any* function can be expressed using a complete basis.[64]

---

[62] An equivalent definition is that $\mathbb{H}$ is complete if a sequence $\{\psi_n\}$ of elements of $\mathbb{H}$ satisfy the condition $\|\psi_n - \psi_m\| \to 0$ as $m, n \to \infty$ such that there exists an element ψ of $\mathbb{H}$ such that $\|\psi - \psi_n\| \to 0$ as $n \to \infty$.

[63] Note that we're assuming orthonormal bases for Hilbert space.

[64] All finite-dimensional spaces are complete, and hence, the term "finite-dimensional Hilbert space" is redundant; nevertheless, the term is used.

Equation (1.55) therefore implies the completeness relation for Hilbert space (the generalization of Eq. (1.11)),

$$\sum_{n=-\infty}^{\infty} |e_n\rangle\langle e_n| = \mathcal{I}. \tag{1.57}$$

Definition. *A Hilbert space is* separable *if it has an orthonormal basis consisting of a* countable[65] *number of vectors,* $\{\psi_1, \psi_2, \ldots\}$.

The definition portends the possibility of non separable Hilbert spaces – a topic outside the intended scope of this book. That $\mathbb{H}$ has a countably infinite set of basis vectors follows naturally from our definition of the dimension of a linear space (Section 1.2.2) – for every $N = 1, 2, 3 \ldots$, there is a set of N linearly independent elements of $\mathbb{H}$.

We've defined Hilbert space – a complete inner-product space – and we've provided the motivation for that definition. *Proving* that an infinite-dimensional space is complete is a task normally not undertaken by physicists. For most purposes, it suffices that l^2 and L^2 are separable Hilbert spaces.[66] It can be shown that, just as all finite-dimensional vector spaces are isomorphic to $\mathbb{E}^N$ (Section 1.3.4), all separable Hilbert spaces are isomorphic to $L^2[a, b]$ [12, p. 216]. In succeeding chapters, we'll encounter the so-called "special functions" of mathematical physics (spherical harmonics, Legendre polynomials, Bessel functions, etc.), which can all be treated within the framework of Hilbert space theory.

Definition. *A linear operator is continuous if* $A\psi_n \to A\psi$ *for any sequence of vectors* $\{\psi_n\}$ *that converges to the limit vector* ψ.

There are a few other topics we should discuss on how the results developed for finite-dimensional spaces generalize to the case of an infinite number of dimensions. Just as a function of a complex variable z is *continuous* if $f(z_n) \to f(z)$ as $z_n \to z$, the same property defines continuity of linear functionals: A linear functional F is continuous if $F(\psi_n)$ converges to $F(\psi)$ if the sequence of vectors $\{\psi_n\}$ converges to the limit vector ψ. A linear functional F_ψ defined by the inner product $F_\psi \equiv (\psi, \phi)$ is continuous if as $\phi_n \to \phi$, $(\psi, \phi_n) \to (\psi, \phi)$. For a finite-dimensional space, every linear functional is continuous. For infinite-dimensional spaces, if F is a continuous linear functional, and if $\psi = \sum_{k=1}^{\infty} a_k \phi_k$, then[67] $F(\psi) = \sum_{k=1}^{\infty} a_k F(\phi_k)$.

---

[65] A countable set is one that can be placed in one-to-one correspondence with some set of integers, perhaps all. A countable set is either a finite set or is *countably infinite.*

[66] A proof that l^2 is complete is given in [11, p. 70] and [9, p. 14] and that for L^2 in [5, sections 28, 32, 33].

[67] If F is continuous and the sequence of partial sums $\psi_N = \sum_{k=1}^{N} a_k \phi_k$ converges to ψ, the sequence of partial sums $F(\psi_N) = \sum_{k=1}^{N} a_k F(\phi_k)$ must converge to $F(\psi)$.

Theorem 1.6.1. *A linear operator is continuous if and only if it's bounded.*

Proof. Let $\mathcal{A}$ be a bounded linear operator with operator norm $\|\mathcal{A}\|_{\text{op}}$. If a sequence of vectors $\{\psi_n\}$ converges to a limit vector ψ, then $\|\mathcal{A}\psi - \mathcal{A}\psi_n\| = \|\mathcal{A}(\psi - \psi_n)\| \leq \|\mathcal{A}\|_{\text{op}} \cdot \|\psi - \psi_n\| \to 0$ as $n \to \infty$, so $\mathcal{A}\psi_n \to \mathcal{A}\psi$ as $n \to \infty$. Thus, $\mathcal{A}$ is continuous. Suppose now that $\mathcal{A}$ is not bounded. In that case, for each positive integer n, there must be a vector ψ_n such that $\|\mathcal{A}\psi_n\| > n\|\psi_n\|$. Let $\chi_n \equiv (n\|\psi_n\|)^{-1}\psi_n$ so that $\|\chi_n\| = 1/n$. Clearly, $\|\chi_n\| \to 0$ as $n \to \infty$. But $\|\mathcal{A}\chi_n\| > 1$, so that $\mathcal{A}\chi_n \not\to 0$ as $n \to \infty$. Thus, $\mathcal{A}$ is not continuous. □

Exercises

1.1. (a) Show that the set of all $N \times N$ matrices is a vector space under the usual rules of adding matrices and multiplying matrices by numbers.
(b) What is the dimension of the space of all $N \times N$ matrices? Hint: Find a basis for this space.

1.2. The Pauli spin matrices are defined as

$$\sigma_1 \equiv \begin{pmatrix} 0 & 1 \\ 1 & 0 \end{pmatrix} \qquad \sigma_2 \equiv \begin{pmatrix} 0 & -\mathrm{i} \\ \mathrm{i} & 0 \end{pmatrix} \qquad \sigma_3 \equiv \begin{pmatrix} 1 & 0 \\ 0 & -1 \end{pmatrix}.$$

Show that if we add the 2×2 unit matrix, call it $\sigma_0 \equiv \begin{pmatrix} 1 & 0 \\ 0 & 1 \end{pmatrix}$, the set $\{\sigma_0, \sigma_1, \sigma_2, \sigma_3\}$ is a basis for the space of complex 2×2 matrices. That is, show that any possible complex 2×2 matrix can be expressed as a linear combination of the basis matrices.

1.3. Show, using the axioms of inner products in Section 1.2, that $(a\phi, \psi) = a^*(\phi, \psi)$.

1.4. Fill in the steps from the inequality (1.5) to the Schwarz inequality, (1.4).

1.5. (a) Show that for square-integrable functions $f(x)$ and $g(x)$,

$$\left|\int f^*(x)g(x)\,\mathrm{d}x\right|^2 \leq \left(\int |f(x)|^2\,\mathrm{d}x\right)\left(\int |g(x)|^2\,\mathrm{d}x\right).$$

Hint: Use the inner product for L^2 and the Schwarz inequality.
(b) Now show, for square-integrable functions $f(x)$ and $g(x)$, that

$$\left(\int |f(x)+g(x)|^2\,\mathrm{d}x\right)^{1/2} \leq \left(\int |f(x)|^2\,\mathrm{d}x\right)^{1/2} + \left(\int |g(x)|^2\,\mathrm{d}x\right)^{1/2}.$$

Thus, the sum of square-integrable functions is square integrable. Hint: L^2 and the triangle inequality.

1.6. Show that $\|a\psi\| = |a| \cdot \|\psi\|$, the norm of $a\psi$ is the absolute magnitude of the scalar multiplied by the norm of ψ.

1.7. Derive the triangle inequality. Hint: Show that $\|\psi + \phi\|^2 = \|\psi\|^2 + \|\phi\|^2 + 2\text{Re}(\psi, \phi)$. Show, for any complex variable z, that $\text{Re}z \leq |z|$. Then apply the Schwarz inequality to arrive at the triangle inequality.

1.8. Show that a set of mutually orthogonal, nonzero vectors $\{\phi_1, \phi_2, \dots, \phi_N\}$ is linearly independent. Hint: If scalars $\{a_1, a_2, \dots, a_N\}$ are such that $\sum_{k=1}^{N} a_k \phi_k = 0$, then use orthogonality to show that $a_k = 0$ for all k. Linearly independent vectors need not be mutually orthogonal, but mutually orthogonal vectors are linearly independent.

1.9. Derive Bessel's inequality, (1.10). Hint: Start with $\|\psi'\|^2 \geq 0$, where ψ' is defined just before Eq. (1.9).

1.10. Derive Parseval's identity, Eq. (1.10). Note: This is a one-liner; it does not entail a lengthy calculation.

1.11. For $\phi(x)$, a smooth function defined on the real line (an element of a function space), which of the following is a linear functional for such functions?

(i) $F(\phi) = \int_1^2 \phi(t)\, \mathrm{d}t$,

(ii) $F(\phi) = \int_0^2 (\phi(t))^2\, \mathrm{d}t$,

(iii) $F(\phi) = \int_0^1 t^2 \phi(t)\, \mathrm{d}t$,

(iv) $F(\phi) = \int_0^1 \phi(t^2)\, \mathrm{d}t$,

(v) $F(\phi) = \frac{\mathrm{d}\phi}{\mathrm{d}t}$,

(vi) $F(\phi) = \left.\frac{\mathrm{d}^2\phi}{\mathrm{d}t^2}\right|_{t=1}$.

1.12. Which of the following operators $\mathcal{A}$ are linear operators on the space of continuous functions?

(i) $\mathcal{A}\psi(x) = x^3\psi(x)$,

(ii) $\mathcal{A}\psi(x) = \psi(x) + x^2$,

(iii) $\mathcal{A}\psi(x) = \psi(3x^2 + 1)$,

(iv) $\mathcal{A}\psi(x) = (\psi(x))^3$,

(v) $\mathcal{A}\psi(x) = x\frac{\mathrm{d}}{\mathrm{d}x}\psi(x)$,

(vi) $\mathcal{A}\psi(x) = \mathrm{e}^{\psi(x)}$,

(vii) $\mathcal{A}\psi(x) = \int_{-\infty}^{x} \mathrm{d}x'(x'\psi(x'))$.

1.13. In Section 1.5, we consider the "eigen" problem of operators acting on vectors such that $\mathcal{A}\boldsymbol{x} = \lambda\boldsymbol{x}$, where λ is a scalar. Show that the equivalent matrix equation is $\boldsymbol{A}\boldsymbol{x} = \lambda\boldsymbol{x}$, where $\boldsymbol{A}$ is the matrix representation of $\mathcal{A}$, and $\boldsymbol{x}$ here denotes the abstract vector $\boldsymbol{x}$ expressed in a basis. Hint: Here would be a great place to use the completeness relation, Eq. (1.11).

1.14. Explain why the matrix in the last example in Section 1.4.2 satisfies the rank-nullity theorem. Which vectors are mapped into $|0\rangle$?

1.15. The 3×3 matrix shown in the example of Section 1.4.4 has rank 2. By the rank-nullity theorem, there should be a one-dimensional space of vectors mapped to the zero vector. Find that class of vectors.

1.16. Show that the results in Eq. (1.23) follow from the definition of operator norm.

1.17. Show that

(a) If operator $\mathcal{B}$ is the inverse to operator $\mathcal{A}$, i.e. $\mathcal{A}\mathcal{B} = \mathcal{I}$, then $\mathcal{A}\mathcal{B} = \mathcal{B}\mathcal{A}$,

(b) If linear operators $\mathcal{A}$ and $\mathcal{B}$ have inverses $\mathcal{A}^{-1}$ and $\mathcal{B}^{-1}$, then the inverse of the compound operator $\mathcal{A}\mathcal{B}$ is such that $(\mathcal{A}\mathcal{B})^{-1} = \mathcal{B}^{-1}\mathcal{A}^{-1}$.

1.18. (a) Derive the results in Eq. (1.44). Show that they imply Eq. (1.45).

(b) For complex numbers z_1, z_2, and z_3, show that the requirement $\text{Re}(z_1 z_2) = \text{Re}(z_1 z_3)$ implies for arbitrary z_1, but fixed z_2 and z_3 (what we have in Eq. (1.45)), that $z_2 = z_3$. Hint: The real and imaginary parts of an arbitrary complex number can be varied independently of each other. You're asked to show that $\text{Re}(z_1(z_2 - z_3)) = 0$ implies, for arbitrary z_1, that $z_2 = z_3$.

1.19. Show that if $\mathcal{U}$ is a unitary operator and the set of vectors $\{\hat{\mathbf{e}}_i\}_{i=1}^N$ is an orthonormal basis, then the set $\{\mathcal{U}\hat{\mathbf{e}}_i\}_{i=1}^N$ is also an orthonormal basis. Hint: Use Eq. (1.43).

1.20. Show that the results in Eq. (1.34) follow from the definition of the adjoint operator.

1.21. Show that the inner product as specified by Eq. (1.35) is invariant under unitary transformations. Because orthonormal bases are connected by unitary transformations (Section 1.4.7), the inner product is independent of basis.

1.22. Derive the rotation matrix in Eq. (1.47).

1.23. Verify that Eq. (1.41) correctly produces the inverse matrix of a 2×2 matrix.

1.24. For $N \times N$ matrices $\boldsymbol{A}$, $\boldsymbol{B}$, and $\boldsymbol{C}$, prove the cyclic invariance of the trace: $\text{Tr}\ \boldsymbol{ABC} = \text{Tr}\ \boldsymbol{CAB} = \text{Tr}\ \boldsymbol{BCA}$.

1.25. The trace and the determinant are basis-independent characterizations of matrices. The two are connected in a simple way for matrices that differ infinitesimally from the unit matrix, $\boldsymbol{I} + \epsilon \boldsymbol{A}$, where ϵ is an infinitesimal. Show that $\det(\boldsymbol{I} + \epsilon \boldsymbol{A}) = 1 + \epsilon \text{Tr}\boldsymbol{A} + O(\epsilon^2)$, where $O(\epsilon^2)$ ("big O" notation) indicates that the order of the terms we're ignoring begin at second order in ϵ. This result is not as specialized as it might appear. Quantities such as $\boldsymbol{I} + \epsilon \boldsymbol{A}$ play a fundamental role in physics, e.g. infinitesimal Lorentz transformation (theory of relativity) or infinitesimal canonical transformations (classical mechanics). One way to study finite transformations is to consider the compound effect of many infinitesimal transformations.

1.26. Show that if operators $\mathcal{A}$ and $\mathcal{B}$ are connected by a similarity transformation, $\mathcal{A} = \mathcal{S}\mathcal{B}\mathcal{S}^{-1}$, then the characteristic polynomials associated with $\mathcal{A}$ and $\mathcal{B}$ are the same. Thus, the characteristic polynomial is independent of basis, and we may speak of the characteristic polynomial associated with an *operator* rather than its matrix representation. Hint: $\boldsymbol{A} - \lambda \boldsymbol{I} = \boldsymbol{S}(\boldsymbol{B} - \lambda \boldsymbol{I})\boldsymbol{S}^{-1}$.

2

Sturm–Liouville theory

In Chapter 1, we introduced linear operators on vector spaces, where our focus was on the general properties of operators and their matrix representations. We turn our attention now to the special, yet important, case of second-order linear differential operators.[1] We'll develop a body of mathematics known as *Sturm–Liouville theory*, which provides a unifying framework for discussing second-order differential equations.

A second-order linear differential equation with variable coefficients[2] specifies a linear differential operator:

$$\mathcal{L}f \equiv \left(p(x)\frac{\mathrm{d}^2}{\mathrm{d}x^2} + q(x)\frac{\mathrm{d}}{\mathrm{d}x} + r(x)\right) f(x) = h(x). \tag{2.1}$$

$\mathcal{L}$ acts on a function $f(x)$ and transforms it into another function, $\mathcal{L}f(x) = h(x)$. If $h(x) = 0$, Eq. (2.1) is said to be a *homogeneous* differential equation; otherwise, it's referred to as an *inhomogeneous* differential equation. For the most part, we consider the homogeneous case in this chapter; inhomogeneous problems are treated in Chapter 9.

What we know about linear operators comes from linear algebra, where the action of an operator $\mathcal{A}$ on elements of a vector space is represented by its matrix elements $A_{ij} \equiv \langle \boldsymbol{e}_i|\mathcal{A}\boldsymbol{e}_j\rangle$, where the $\{\boldsymbol{e}_i\}$ are an orthonormal set of basis vectors (Section 1.4.2). Do the same ideas apply to differential operators, which act on functions? They do, but it will take most of this chapter to show that. The key question that will concern us is, "what constitutes an orthonormal basis for a function space?"

[1] A large majority of the differential equations encountered in applications are second order in nature.

[2] The *coefficient functions* $p(x)$, $q(x)$, $r(x)$ are presumed continuous; $\mathcal{L}$ is a mapping from continuously twice-differentiable functions f on a given interval to another continuous function $h(x)$ on the same interval.

Mathematical Methods in Physics, Engineering, and Chemistry, First Edition.
Brett Borden and James Luscombe.

Every linear operator is associated with another linear operator, its adjoint.[3] Unitary and Hermitian operators, which possess orthogonal eigenvectors (Section 1.5.2), are defined in terms of their adjoints ($\mathcal{U}^{-1} = \mathcal{U}^\dagger$ and $\mathcal{A} = \mathcal{A}^\dagger$). Eigenvectors are linearly independent (Section 1.5.1), and bases for Hilbert space are orthonormal sets (Section 1.6). The eigenvectors of unitary and Hermitian *differential* operators can therefore serve as bases for function spaces. One of our first orders of business will be to find the adjoint of $\mathcal{L}$ in Eq. (2.1).

2.1 Second-order differential equations

2.1.1 Uniqueness and linear independence

We show in Chapter 5 how to construct solutions of the homogeneous differential equation $\mathcal{L}f = 0$ (with variable coefficient functions) in the form of power series, i.e. we show the *existence* of solutions in Chapter 5. In this section, we consider the *uniqueness* and the linear independence of solutions to Eq. (2.1). Uniqueness can be proven under very general circumstances [13, pp. 21–22].

Theorem 2.1.1. (Uniqueness) *If $p(x), q(x), r(x), h(x)$ in Eq. (2.1) are continuous functions over an open interval I of the real line, then at most one solution of Eq. (2.1) can pass through a given point such that $f(x_0) = c_0$ and $f'(x_0) = c_1$, where (c_0, c_1) are constants and $x_0 \in I$.*

Comments: The uniqueness theorem applies to the inhomogeneous version Eq. (2.1) – that's implicit in the statement of the theorem, but no harm in pointing it out explicitly. A similar theorem holds for the solution of nth-order differential equations, in which case there is only one solution passing through a point with $f(x_0) = c_0, f'(x_0) = c_1, \dots, f^{(n-1)}(x_0) = c_{n-1}$. The uniqueness theorem is familiar to students of physics: The solution of Newton's second law of motion is fully specified when the position and initial velocity of a particle are known. Only one solution can pass through a given point with specified initial conditions. This fact is used in advanced formulations of classical mechanics, that trajectories in *phase space* can never intersect.

The homogeneous version of Eq. (2.1) has two linearly independent solutions.[4]

Theorem 2.1.2. *Let f and g be two solutions of the homogeneous differential equation*

$$p(x)u'' + q(x)u' + r(x)u = 0, \tag{2.2}$$

[3]The fine print: Every *bounded* linear operator has a corresponding (bounded) adjoint operator.

[4]You may have seen this for differential equations involving constant coefficients; it applies for variable coefficients as well.

where p, q, and r are continuous functions over an open interval I. For some $x_0 \in I$, let $(f(x_0), f'(x_0))$ and $(g(x_0), g'(x_0))$ be linearly independent vectors. Then every solution ψ of Eq. (2.2) is in the form of a linear combination $\psi(x) = \alpha f(x) + \beta g(x)$ with constant coefficients α and β.

Proof. Clearly, for f and g solutions of Eq. (2.2), $\psi(x) = \alpha f(x) + \beta g(x)$ is also a solution. Conversely, assume that the function $\psi(x)$ satisfies Eq. (2.2). Then, at the given point x_0, constants α and β can be found such that

$$\psi(x_0) = \alpha f(x_0) + \beta g(x_0) \qquad \psi'(x_0) = \alpha f'(x_0) + \beta g'(x_0).$$

The constants α and β can be found using Cramer's rule:

$$\alpha = \frac{\psi_0 g_0' - g_0 \psi_0'}{f_0 g_0' - g_0 f_0'} \qquad \beta = \frac{f_0 \psi_0' - \psi_0 f_0'}{f_0 g_0' - g_0 f_0'},$$

where $f_0 \equiv f(x_0)$, $f_0' \equiv f'(x_0)$, and so forth. For coefficients α and β, the function $u(x) \equiv \psi(x) - \alpha f(x) - \beta g(x)$ satisfies Eq. (2.2) with the initial conditions $u(x_0) = u'(x_0) = 0$. By the uniqueness theorem, there is only one solution at $x = x_0$ with $u_0 = 0$ and $u_0' = 0$. By continuity, $u(x) = 0$ in a neighborhood of x_0, and in fact $u(x) = 0$ (the trivial solution) over the interval I. Thus, $\psi(x) = \alpha f(x) + \beta g(x)$ for all $x \in I$. □

The theorem relies on the linear independence of solutions at a point. The question of whether functions are linearly independent over an interval is fairly easily ascertained, as we now discuss. Linear independence of differentiable *functions* $y_1(x)$, $y_2(x)$ (not necessarily solutions of a differential equation), means that no linear combination

$$c_1 y_1(x) + c_2 y_2(x) = 0 \tag{2.3}$$

yields the identically zero function over some interval $[a, b]$ except for the trivial case $c_1 = c_2 = 0$. If Eq. (2.3) holds, then so does the equation that results by taking its derivative:

$$c_1 y_1'(x) + c_2 y_2'(x) = 0. \tag{2.4}$$

Equations (2.3) and (2.4) can be combined into a matrix equation

$$\begin{pmatrix} y_1(x) & y_2(x) \\ y_1'(x) & y_2'(x) \end{pmatrix} \begin{pmatrix} c_1 \\ c_2 \end{pmatrix} = 0. \tag{2.5}$$

An equation such as Eq. (2.5) has a nontrivial solution only if the determinant of the matrix is zero (see Section 1.5.1). In this case, we're *looking* for the trivial solution: y_1 and y_2 are linearly independent only if $c_1 = c_2 = 0$. Define the determinant[5]

[5] A determinant of functions is called (can you see it coming) a *functional determinant.*

Wronskian

$$W(y_1, y_2; x) \equiv \begin{vmatrix} y_1(x) & y_2(x) \\ y_1'(x) & y_2'(x) \end{vmatrix}, \tag{2.6}$$

the *Wronskian determinant*, or simply Wronskian. Because we *want* the trivial solution to Eq. (2.5), y_1 and y_2 will be linearly independent over an interval I if the Wronkisan is nonzero over the same interval.

Example. Use the Wronskian to test the linear independence of x^n and x^m.

$$W(x) = \begin{vmatrix} x^n & x^m \\ nx^{n-1} & mx^{m-1} \end{vmatrix} = (m-n)x^{n+m-1},$$

which is nonzero unless $m = n$. The functions x^n and x^m are linearly independent if $n \neq m$.

The Wronskian is a useful function in its own right. By taking its derivative, it follows from Eqs. (2.6) and (2.2) that

$$\frac{\mathrm{d}W(x)}{\mathrm{d}x} = -\frac{q(x)}{p(x)} W(x). \tag{2.7}$$

The Wronskian thus satisfies a first order differential equation that involves the coefficient functions from Eq. (2.2). Solutions of first order differential equations are unique – there's just one solution $W(x)$ of Eq. (2.7) having the value $W(a)$ at $x = a$. The solution of Eq. (2.7) can be written (if $W(x)$ is not identically zero)

$$W(x) = W(a) \exp\left(-\int_a^x \frac{q(x')}{p(x')}\,\mathrm{d}x'\right). \tag{2.8}$$

As we now show, the Wronskian of any solutions of Eq. (2.2) is identically positive, identically negative, or identically zero.

The Wronskian supplies a test of the linear independence of any two differentiable functions.[6] As specifically relates to the solutions of second-order linear differential equations, we have the following theorem.

Theorem 2.1.3. *If f and g are linearly independent solutions of Eq. (2.2), their Wronskian never vanishes.*

Proof. Suppose the Wronskian $W(f, g; x)$ vanishes at some point x_0. Then the vectors $(f(x_0), f'(x_0))$ and $(g(x_0), g'(x_0))$ would be linearly dependent, and we'd have the proportionality: $g(x_0) = kf(x_0)$ and $g'(x_0) = kf'(x_0)$ for some constant k (the columns of the Wronskian would be proportional). Form the function $\psi(x) \equiv g(x) - kf(x)$, a solution of Eq. (2.2)

[6]The Wronskian can be defined for n functions $f_1, \ldots, f_n$ that are $(n-1)$-times differentiable over an interval I.

by linearity. It also satisfies the initial condition $\psi(x_0) = 0$ and $\psi'(x_0) = 0$. By the uniqueness theorem (a result specifically for the solutions of differential equations), $\psi(x)$ must vanish identically. Therefore $g(x) = kf(x)$ for all x, contradicting the assumption of linear independence of f and g. □

Because $W(x) \neq 0$ for linearly independent solutions, it has one sign – it can't change sign over an interval for which the two solutions are linearly independent. We see this in Eq. (2.8), where the sign of $W(x)$ is governed by that of $W(a)$. The Wronskian can be used to derive an important property of the solutions to Eq. (2.2). The following theorem (the *Sturm separation theorem*) relates the relative positions of the *zeros* of solutions. A zero of a function is the point where its value is zero.

Theorem 2.1.4. *For $f(x)$ and $g(x)$ linearly independent solutions of Eq. (2.2), $f(x)$ vanishes at a point between two successive zeros of $g(x)$, i.e. the zeros of $f(x)$ and $g(x)$ alternate along the x-axis.*

Proof. If $g(x)$ vanishes at $x = x_i$, the Wronskian would have the value (from Eq. (2.6)) $W(f, g; x_i) = f(x_i)g'(x_i)$. Because f and g are linearly independent at x_i, $f(x_i) \neq 0$, and $g'(x_i) \neq 0$ (because by an argument that should now be familiar, if $g'(x_i) = 0$, then we would have $g(x) = 0$ by uniqueness). Suppose x_1 and x_2 are successive zeros of $g(x)$. Then $g'(x_1)$, $g'(x_2)$, $f(x_1)$, and $f(x_2)$ cannot be zero. Moreover, the nonzero numbers $g'(x_1)$ and $g'(x_2)$ cannot have the same sign. If $g(x)$ is increasing at $x = x_1$, then (by Rolle's theorem) $g(x)$ must be decreasing at $x = x_2$, and vice-versa. Because W has the same sign, it follows that $f(x_1)$ and $f(x_2)$ must also have opposite signs. Thus, $f(x)$ must vanish somewhere between x_1 and x_2. □

Example. Consider the differential equation $u'' + k^2u = 0$, which has trigonometric solutions, $\sin kx$ and $\cos kx$. The Sturm separation theorem implies that the zeros of $\sin kx$ (at locations $kx = n\pi$ for $n = 0, 1, 2, \ldots$) and $\cos kx$ (at $kx = \left(n + \frac{1}{2}\right)\pi$ for $n = 0, 1, \ldots$) must alternate, which they do!

2.1.2 The adjoint operator

We start with real-valued functions[7] $f(x)$ and $g(x)$ on the interval $[a, b]$, and consider the inner product $\langle g|\mathcal{L}f\rangle \equiv \int_a^b g(x)\mathcal{L}f(x)\, dx$. Referring to

[7]We work with real-valued functions for convenience because for many applications, real functions suffice. The all-important case of quantum mechanics, however, *requires* complex-valued functions – one can get away with real functions for bound states, but the more general case of scattering states requires that $\psi(x)$ be complex-valued. Nothing essential that we develop here relies on functions being real valued.

the definition of adjoint (Section 1.4.8), we want to define an operator $\mathcal{L}^\dagger$ such that $\langle g|\mathcal{L}f\rangle = \langle \mathcal{L}^\dagger g|f\rangle$. Through integration by parts, the derivatives of f as specified by $\mathcal{L}f$ in Eq. (2.1) can be transformed into derivatives of g. We find

$$\int_a^b g\mathcal{L}f \, \mathrm{d}x = [p(gf' - fg') + fg(q - p')]\Big|_a^b + \int_a^b [(pg)'' - (qg)' + rg]f \, \mathrm{d}x. \tag{2.9}$$

adjoint operator

The terms in square brackets in the integrand on the right side of Eq. (2.9) specify a linear operator, one that we'll provisionally denote $\mathcal{L}^\dagger$,

$$\mathcal{L}^\dagger g \equiv [p(x)g]'' - [q(x)g]' + r(x)g. \tag{2.10}$$

Equation (2.9) is then equivalent (by definition) to

$$\langle g|\mathcal{L}f\rangle - \langle \mathcal{L}^\dagger g|f\rangle = [p(gf' - fg') + fg(q - p')]\big|_a^b . \tag{2.11}$$

Equation (2.11) implies the following equation (known as *Lagrange's identity*),

$$g\mathcal{L}f - (\mathcal{L}^\dagger g)f = \frac{\mathrm{d}}{\mathrm{d}x}[p(gf' - g'f) + gf(q - p')]. \tag{2.12}$$

Evidently, Eq. (2.12) suggests that if a function $g(x)$ is a solution of $\mathcal{L}^\dagger g = 0$, then g is an *integrating factor* for $\mathcal{L}f$: For g such that $\mathcal{L}^\dagger g = 0$, then from Eq. (2.12), $g\mathcal{L}f =$ a total derivative (and so g is an integrating factor for the differential form $\mathcal{L}f$).

The operator $\mathcal{L}^\dagger$ in Eq. (2.10) is associated with $\mathcal{L}$ (it's based on the coefficient functions $p(x), q(x), r(x)$ of $\mathcal{L}$), but also (through Eq. (2.11)) with the values of the functions f, g and their derivatives on the boundaries of the system – the *boundary conditions*. Moral of the story: *Differential operators are not completely specified until we have information about the boundary conditions.* To accomplish the definition of adjoint ($\langle g|\mathcal{L}f\rangle = \langle \mathcal{L}^\dagger g|f\rangle$) requires that the boundary conditions be such that the right side of Eq. (2.11) vanishes. Before addressing boundary conditions, let's first see what is entailed for $\mathcal{L}$ to be self-adjoint, given the definition of $\mathcal{L}^\dagger$ in Eq. (2.10).

2.1.3 Self-adjoint operator

To have $\mathcal{L}$ self-adjoint, i.e. $\mathcal{L}f = \mathcal{L}^\dagger f$ for all f, places a restriction on the coefficient function $q(x)$, namely $q(x) = p'(x)$ (shown in Exercise 2.3). With $q = p'$ in Eq. (2.1), a self-adjoint, second-order linear differential operator has the form

$$\mathcal{L}f = \frac{\mathrm{d}}{\mathrm{d}x}\left(p(x)\frac{\mathrm{d}f}{\mathrm{d}x}\right) + r(x)f. \tag{2.13}$$

Consequently, Eq. (2.12) simplifies to

$$g\mathcal{L}f - (\mathcal{L}g)f = \frac{\mathrm{d}}{\mathrm{d}x}[p(gf' - g'f)]. \qquad \text{(self-adjoint operator)} \qquad (2.14)$$

We'll draw upon Eq. (2.14) in what follows.

The *test* for whether a differential operator $\mathcal{L}$ as specified in Eq. (2.1) is self-adjoint is if $q = p'$. This condition might seem overly restrictive, but that's not the case. Any second-order, linear differential operator can be converted into self-adjoint form by multiplying $\mathcal{L}f$ by the factor (Exercise 2.4)

$$\lambda(x) \equiv \frac{1}{p(x)} \mathrm{e}^{\int (q(x)/p(x))\mathrm{d}x}. \qquad (2.15)$$

Example. The Legendre differential operator (that we study later, Eq. (5.46)) is self-adjoint:

$$\mathcal{L}f \equiv (1 - x^2)f'' - 2xf' + l(l+1)f = \frac{\mathrm{d}}{\mathrm{d}x}((1 - x^2)f') + l(l+1)f.$$

Bessel's differential equation, however, as given by Eq. (5.25), is not self-adjoint:

$$x^2y'' + xy' + (x^2 - \nu^2)y = 0.$$

It can be placed in self-adjoint form by multiplying by the factor $\lambda(x)$, Eq. (2.15):

$$\lambda(x) = \frac{1}{x^2}\mathrm{e}^{\int (x/x^2)\mathrm{d}x} = \frac{1}{x^2}\mathrm{e}^{\ln x} = \frac{1}{x^2}x = \frac{1}{x}.$$

Bessel's equation in self-adjoint form is thus

$$\frac{\mathrm{d}}{\mathrm{d}x}(xy') + \frac{x^2 - \nu^2}{x}y = 0. \qquad (2.16)$$

2.2 Sturm–Liouville systems

Definition. *A* Sturm–Liouville equation *is a differential equation of the form:*

$$\frac{\mathrm{d}}{\mathrm{d}x}\left(p(x)\frac{\mathrm{d}y}{\mathrm{d}x}\right) + [\lambda w(x) - r(x)]y = 0. \qquad (2.17)$$

The new wrinkles here are the parameter λ and the function $w(x)$, the *weight function*.[8] For fixed λ, Eq. (2.17) can be written as an operator expression

---

[8]The parameter λ in the Sturm–Liouville equation often arises from the *separation constant* introduced in the method of separation of variables for reducing a PDE to an ODE (see Chapter 3). We've changed the factor of $r(x)$ in Eq. (2.1) to $-r(x)$ in Eq. (2.17). Nothing of consequence depends on this cosmetic change.

$$\mathcal{L}y = -\lambda w(x)y, \tag{2.18}$$

where $\mathcal{L}$ is self-adjoint. The Sturm–Liouville equation generalizes the eigenvalue problem to include the weight function, $w(x)$.

We'll refer to a Sturm–Liouville equation together with a specification of *boundary conditions* as a Sturm–Liouville *system.* A class of boundary conditions with wide applicability in the physical sciences consists of homogeneous linear relations between the value of y and that of its derivative y' at the boundaries:

$$\alpha_1 y(a) + \alpha_2 y'(a) = 0 \qquad \beta_1 y(b) + \beta_2 y'(b) = 0, \tag{2.19}$$

Dirichlet and Neumann boundary conditions

periodic boundary condition

for systems defined on $[a, b]$ and for $\alpha_1, \alpha_2, \beta_1, \beta_2$ real numbers. Boundary conditions in the form of Eq. (2.19) are referred to as *separated* boundary conditions; they each apply only at the separated boundary points of the system. When $\alpha_2 = \beta_2 = 0$, the boundary conditions are referred to as *Dirichlet conditions*, while those for $\alpha_1 = \beta_1 = 0$ are known as *Neumann conditions.* The case of α_1, α_2, β_1, β_2 nonzero is referred to as a *mixed* boundary condition. *Periodic boundary conditions* are another type of boundary condition, also with wide applicability to physical systems:

$$y(a) = y(b) \qquad y'(a) = y'(b). \tag{2.20}$$

nonhomogeneous boundary conditions

The boundary conditions specified by Eqs. (2.19) or (2.20) are termed *homogeneous* because solutions of the Sturm–Liouville equation (a linear equation) can be multiplied by nonzero constants and still satisfy the same boundary conditions. Boundary conditions not in the form of Eq. (2.19), where linear combinations of the function and its derivative are nonzero at the boundaries are termed *nonhomogeneous.* If f is a solution of a homogeneous differential equation satisfying homogeneous boundary conditions, the differential equation and the boundary conditions are satisfied by αf, where α is a constant. Suppose, however, f is the solution of a homogeneous differential equation $\mathcal{L}f = 0$ satisfying *nonhomogeneous* boundary conditions, e.g. $f(x) = g(x)$ for g a nonzero function of x at the boundary. Define a new function $v \equiv g - f$, implying that $\mathcal{L}v = \mathcal{L}g$ is an inhomogeneous differential equation satisfying *homogeneous* boundary conditions.[9] *Homogeneous differential equations with nonhomogeneous boundary conditions are equivalent to inhomogeneous differential equations with homogeneous boundary conditions* [14, p. 277]. Inhomogeneous differential equations are treated in Chapter 9. No loss of generality, therefore, is incurred by discussing homogeneous boundary conditions for homogeneous differential equations.

Self-adjoint operators satisfy Eq. (2.14). Integrating Eq. (2.14), we have a relation that holds for arbitrary functions f and g:

[9] It's assumed g can be extended continuously into the interior of the domain of f.

$$\int_a^b [g\mathcal{L}f - (\mathcal{L}g)f] \, \mathrm{d}x = p(gf' - g'f)\Big|_a^b . \tag{2.21}$$

The right side of Eq. (2.21) *vanishes* for f and g satisfying homogeneous boundary conditions (Eqs. (2.19) or (2.20)). For separated boundary conditions, $gf' - g'f = 0$ at each endpoint, while for periodic boundary conditions (and for $p(x)$ a periodic function[10]) $(gf' - g'f)|_a^b = 0$ (Exercise 2.7). Thus, for functions satisfying homogeneous boundary conditions, we have for self-adjoint operators the relation

Hermitian operator

$$\int_a^b f\mathcal{L}g \, \mathrm{d}x = \int_a^b g\mathcal{L}f \, \mathrm{d}x. \tag{2.22}$$

Equation (2.22) is frequently given as the defining property of a self-adjoint operator (as we did in Section 1.4.8). For Eq. (2.22) to agree with Eq. (2.21), however, the functions f and g must satisfy the boundary conditions. We'll refer to operators satisfying Eq. (2.22) as Hermitian. While it's common practice to refer to self-adjoint and Hermitian operators synonymously, there's a difference between Eq. (2.14) and its integrated form, Eq. (2.22).

Definition. *A* regular *Sturm–Liouville system has* $p(x) > 0$ *and* $w(x) > 0$ *for* $a \le x \le b$ *together with a specification of separated boundary conditions at the endpoints.*

regular SL system

Sturm–Liouville equations can be specified on finite, semi-infinite, or infinite intervals I. For finite intervals, I may include neither, one, or both endpoints. Excluding an endpoint may be necessary if for example $\lim_{x\to a} p(x) = 0$ or $\lim_{x\to a} w(x) = 0$. Only when I is a closed, finite interval with $p(x)$ and $w(x)$ positive on $[a, b]$ do we have a regular Sturm–Liouville system. Sturm–Liouville systems with $p(x) > 0$ and $w(x) > 0$ on the open interval (a, b) are technically classified as singular. Solutions to singular Sturm–Liouville systems can be obtained through the imposition of boundary conditions not always in the form of Eqs. (2.19) or (2.20), for example by the requirement that the solution be bounded at a singular endpoint.

Example. The Legendre differential equation, Eq. (5.46),

$$\frac{\mathrm{d}}{\mathrm{d}x}[(1 - x^2)y'] + l(l+1)y = 0 \qquad (-1 \le x \le 1)$$

is a singular Sturm–Liouville system because the coefficient $p(x) = (1 - x^2)$ vanishes at $x = \pm 1$. We show in Section 5.4 that requiring the solutions of the Legendre differential equation to be finite as $x \to \pm 1$ results in a solution in the form of polynomials.

---

[10] Periodic functions are treated in Chapter 4.

2.3 The Sturm–Liouville eigenproblem

A nontrivial solution of a Sturm–Liouville system (a solution of Eq. (2.17) satisfying the boundary conditions) is an eigenfunction, with the parameter λ the eigenvalue. The differential equations of mathematical physics are all in the Sturm–Liouville form – see Section 3.2. Sturm–Liouville theory is thus a unifying framework for understanding most if not all the differential equations you're likely to encounter in the physical sciences. It behooves us therefore to get to know the general properties of the solutions to the Sturm–Liouville eigenvalue problem. First, however, two examples of finding eigenvalues.

Example. Solve the Sturm–Liouville system

$$y'' + \lambda y = 0 \qquad (0 \le x \le b) \tag{2.23}$$

with Dirichlet boundary conditions $y(0) = 0$, $y(b) = 0$. There are no nontrivial solutions if $\lambda \le 0$ (Exercise 2.8), so take $\lambda > 0$. Equation (2.23) is easy to solve because it has constant coefficients: $y(x) = A\cos(\sqrt{\lambda}x) + B\sin(\sqrt{\lambda}x)$. Requiring $y(0) = 0$ implies that $A = 0$. The solution thus reduces to $y(x) = B\sin(\sqrt{\lambda}x)$. The other boundary condition $y(b) = 0$ places a requirement on the eigenvalues such that $\sin(\sqrt{\lambda}b) = 0$. The terms $\sqrt{\lambda}b$ must coincide with the zeros of the sine function, i.e. $\sqrt{\lambda}b = n\pi$, for integer n. The eigenvalues are $\lambda_n = n^2\pi^2/b^2$, with the eigenfunctions $y_n = A_n \sin(n\pi x/b)$, $n = 1, 2, 3, \ldots$. The case $n = 0$ is excluded because it results in a trivial solution.

Example. Work the same differential equation as Eq. (2.23), except with periodic boundary conditions $y(0) = y(b)$ and $y'(0) = y'(b)$. The general solution is the same as in the previous example. The boundary conditions lead to two simultaneous equations

$$\begin{pmatrix} 1 - \cos(\sqrt{\lambda}b) & -\sin(\sqrt{\lambda}b) \\ \sin(\sqrt{\lambda}b) & 1 - \cos(\sqrt{\lambda}b) \end{pmatrix} \begin{pmatrix} A \\ B \end{pmatrix} = 0. \tag{2.24}$$

Equation (2.24) possesses a nontrivial solution when the determinant of the matrix vanishes, implying that $\cos(\sqrt{\lambda}b) = 1$. The eigenvalues are thus $\lambda_n = 4n^2\pi^2/b^2$ corresponding to the eigenfunctions $y_n(x) = A_n\cos(2n\pi x/b) + B_n \sin(2n\pi x/b)$ for $n = 1, 2, 3, \ldots$. Additionally, $\lambda = 0$ is an eigenvalue corresponding to eigenfunction $y_0 = 1$.

We've just found the eigenvalues *for the same Sturm–Liouville equation* subject to two types of boundary conditions, in one case having the spectrum $\{\lambda_n = n^2\pi^2/b^2\}$, $n = 1, 2, 3, \ldots$, and in the other

$\{\lambda_n = 4n^2\pi^2/b^2\}$, $n = 0, 1, 2, \ldots$. These examples underscore that eigenvalues are *selected* (or determined) by the boundary conditions.[11] There is a qualitative difference in the eigenfunctions associated with the two types of boundary conditions. For the eigenvalue problem with separated boundary conditions, *there is at most one linearly independent eigenfunction corresponding to a single eigenvalue* λ. This follows because the Wronskian (of two possible linearly independent solutions) *vanishes at the endpoints for separated boundary conditions.* In the aforementioned examples, there are two linearly independent eigenfunctions associated with the same eigenvalue for the case of periodic boundary conditions, but only one for separated boundary conditions.

We now derive two of the most important properties of the solutions to Sturm–Liouville systems. These proofs mirror those given in Section 1.5.2.

Eigenvalues of regular Sturm–Liouville systems are real-valued

Remarkably, such a result follows just from the self-adjointness of $\mathcal{L}$ and the boundary conditions. Assume for the present that the eigenvalue λ can be a complex quantity, and, moreover, that the solutions $y(x)$ can be complex-valued as well.[12] Take the complex conjugate of Eq. (2.18),

$$\mathcal{L}y^* + \lambda^* w(x) y^* = 0. \tag{2.25}$$

Multiply Eq. (2.18) by y^* and Eq. (2.25) by y, and subtract

$$\underbrace{y^*\mathcal{L}g - y\mathcal{L}y^*}_{\frac{\mathrm{d}}{\mathrm{d}x}[p(x)(y'y^*-yy'^*)]} + (\lambda - \lambda^*) w(x) yy^* = 0, \tag{2.26}$$

where we've made use of Eq. (2.14) in Eq. (2.26) (let $f = y^*$ and $g = y$). Multiply Eq. (2.26) by $\mathrm{d}x$ and integrate between a and b:

$$(\lambda^* - \lambda)\int_a^b w(x)|y(x)|^2 \,\mathrm{d}x = [p(x)(y'y^* - yy'^*)]\Big|_a^b = 0. \tag{2.27}$$

The right side of Eq. (2.27) vanishes because of the boundary conditions (Exercise 2.7). Equation (2.27) implies that $\lambda^* = \lambda$, or that λ is real.[13]

---

[11]Schrödinger's 1926 paper on quantum mechanics was titled (in English translation) *Quantization as an Eigenvalue Problem.* This article, one of the hallmark achievements of twentieth-century physics, showed that quantized energy levels emerge "just" as an eigenvalue problem. Prior to Schrödinger's work, it was far from clear how quantized values of physical quantities occur.

[12]The functions $p(x)$, $w(x)$, and $r(x)$ in Eq. (2.17) are real-valued, but the solution $y(x)$ can be a complex-valued function of the real variable x, e.g. $y(x) = \mathrm{e}^{\mathrm{i}x}$.

[13]The integral $\int_a^b w|y|^2 \,\mathrm{d}x$ in Eq. (2.27) cannot vanish because w and $|y|^2$ are positive on $[a, b]$. A tenet of quantum theory is that the results of measurement (as in a laboratory) correspond to the eigenvalues of self-adjoint operators; as such, they must be real.

Eigenfunctions belonging to different eigenvalues are orthogonal

Let's play the same game for eigenfunctions belonging to distinct eigenvalues. With $\mathcal{L}y_n = -\lambda_n w(x) y_n$ and $\mathcal{L}y_m = -\lambda_m w(x) y_m$ $(m \neq n)$, form the quantity

$$\underbrace{y_m \mathcal{L} y_n - y_n \mathcal{L} y_m}_{\frac{\mathrm{d}}{\mathrm{d}x}[p(x)(y_n' y_m - y_n y_m')]} = (\lambda_m - \lambda_n) w(x) y_n(x) y_m(x), \tag{2.28}$$

where we've used Eq. (2.14). Integrate both sides of Eq. (2.28) and make use of the boundary conditions (as in Eq. (2.27)). That leaves us with the desired result:

$$(\lambda_m - \lambda_n) \int_a^b w(x) y_n(x) y_m(x)\, \mathrm{d}x = 0.$$

Because $m \neq n$ (by assumption), eigenfunctions belonging to different eigenvalues are orthogonal *with respect to the weight function* $w(x)$:

$$\int_a^b w(x) y_n(x) y_m(x)\, \mathrm{d}x = 0. \qquad (n \neq m) \tag{2.29}$$

normalized eigenfunction

Definition. *A normalized eigenfunction has unit norm with respect to* $w(x) > 0$*:*

$$\int_a^b w(x) |y_n(x)|^2\, \mathrm{d}x = 1. \tag{2.30}$$

Eigenfunctions can be normalized by letting $y_n \to \tilde{y}_n \equiv y_n / \sqrt{\int w(x)|y_n(x)|^2\, \mathrm{d}x}$.

Example. Consider the eigenfunctions obtained in the first example of Section 2.3, $y_n = \sin(n\pi x/b)$, $n = 1, 2, 3, \ldots$, defined on $[0, b]$. Show these are an orthogonal set. (Note that $w(x) = 1$ in this example.)

$$\begin{aligned}
\int_0^b \sin\left(\frac{n\pi x}{b}\right) \sin\left(\frac{m\pi x}{b}\right) \mathrm{d}x &= \frac{b}{\pi} \int_0^\pi \sin nu \ \sin mu \ \mathrm{d}u \\
&= \frac{b}{2\pi} \int_0^\pi [\cos(n-m)u - \cos(n+m)u]\, \mathrm{d}u \\
&= \frac{b}{2\pi} \left[\frac{1}{n-m} \int_0^{(n-m)\pi} \cos y\, \mathrm{d}y - \frac{1}{n+m} \int_0^{(n+m)\pi} \cos y\, \mathrm{d}y \right] \\
&= \frac{b}{2\pi} \left[\frac{1}{n-m} \sin y \Big|_0^{(n-m)\pi} - \frac{1}{n+m} \sin y \Big|_0^{(n+m)\pi} \right] \\
&= \frac{b}{2\pi} \left[\frac{\sin(n-m)\pi}{n-m} - \frac{\sin(n+m)\pi}{n+m} \right] = \frac{b}{2} \delta_{nm}.
\end{aligned}$$

Make sure you understand the final equality. We've worked through the details of this kind of integral; we won't always be so explicit. For $n \neq m$, the terms in square brackets vanish, and the functions are orthogonal, as per the result of Eq. (2.29). What if $n = m$? That limit requires some care, with the result shown.

We state an important theorem that helps us understand the orthogonality of eigenfunctions [13, p. 212].

Theorem 2.3.1. *A regular Sturm–Liouville system has an infinite sequence of real eigenvalues* $\lambda_0 < \lambda_1 < \lambda_2 < \cdots$ *with* $\lim_{n\to\infty} \lambda_n = \infty$. *The eigenfunction* $y_n(x)$ *belonging to* λ_n *has exactly* n *zeros on* (a, b) *and is uniquely determined up to a constant factor.*

The spectrum of regular Sturm–Liouville systems is thus a denumerably infinite set, $\{\lambda_n\}_{n=0}^{\infty}$, what's referred to as a *discrete spectrum.* discrete spectrum

---

Example. Equation (2.23) subject to $y(0) = y(b) = 0$ has the eigenvalue spectrum $\lambda_n = n^2\pi^2/b^2$ with eigenfunctions $y_n(x) = \sin(n\pi x/b)$, $n = 1, 2, 3, \ldots$. To illustrate the theorem, we should label the eigenvalues and eigenfunctions $\lambda_n = (n+1)^2\pi^2/b^2$ and $y_n(x) = \sin(n+1)\pi x/b$, $n = 0, 1, 2, \ldots$. The eigenfunction belonging to the smallest eigenvalue $\lambda_0 = \pi^2/b^2$, $y_0 = \sin \pi x/b$ has zero nodes in the open interval $(0, b)$, i.e. the vanishing of the eigenfunction at the endpoints don't "count" as nodes.

Example. Stationary states of a free quantum particle
A free particle of mass m and energy E, one subject to a uniform potential energy environment V_0, is described by the time-independent Schrödinger equation (where $\lambda = \sqrt{2m(E - V_0)}/\hbar$)

$$y'' + \lambda y = 0 \qquad (-\infty < x < \infty) \tag{2.31}$$

For $\lambda > 0$, Eq. (2.31) has two linearly independent solutions $\sin(\sqrt{\lambda}x)$ and $\cos(\sqrt{\lambda}x)$. For $\lambda = 0$, it has the one bounded solution, $y = 1$. For $\lambda < 0$, it has the linearly independent, unbounded solutions $\sinh(\sqrt{|\lambda|}x)$ and $\cosh(\sqrt{|\lambda|}x)$. The spectrum of the free particle is thus, on physical grounds, $\lambda \geq 0$. The spectrum in this case is *continuous* and does not conform to the aforementioned theorem. This system, however, is not a regular Sturm–Liouville system – there are no boundary conditions in the form of Eq. (2.19). continuous spectrum

---

Can eigenfunctions defined on $(-\infty, \infty)$ be normalized? Not necessarily. In the aforementioned example, $y(x) = \cos kx$ cannot be normalized over $(-\infty, \infty)$ for weight function $w(x) = 1$. Functions for which the normalization integral exists get a special name.

square-integrable function

Definition. *A function f is* square integrable with respect to weight function $w(x) > 0$ *over an interval I when*

$$\int_I w(x)|f(x)|^2 \, dx < \infty. \tag{2.32}$$

2.4 The Dirac delta function

Consider the Kronecker delta function, δ_{ij} (introduced in Section 1.3.4), a simple function of two indices. What does it do for us? By having the value 1, when $i = j$ and zero otherwise, δ_{ij} can select out of a summation (of quantities labeled by a discrete index) precisely one term, $A_i = \sum_j \delta_{ij} A_j$ (if the range of the summation index j includes the value i). Could there be a delta function with *continuous* indices, $\delta_{xx'}$, for x and x' real numbers?[14]

Instead of $\delta_{xx'}$, let's write[15] $\delta(x - x')$. How would we define such a quantity? We'd want, in emulating the Kronecker function, to define $\delta(x - x') = 0$ for $x \neq x'$. We'd also want an analogous "filtering" ability in an *integral*, $f(x) = \int \delta(x - x') f(x') \, dx'$, as long as the value of x occurs in the range of integration.[16] That ability, however, places a stringent requirement on $\delta(x - x')$ for $x' \to x$. Because $\delta(x - x') = 0$ for $x' \neq x$, it suffices to examine the behavior of $\delta(x - x')$ in an infinitesimal neighborhood of x. The filtering ability of $\delta(x - x')$ is captured by the following expression (for infinitesimal $\epsilon > 0$)

$$f(x) = \int_{x-\epsilon}^{x+\epsilon} \delta(x - x') f(x') \, dx' = f(x) \int_{x-\epsilon}^{x+\epsilon} \delta(x - x') \, dx'. \tag{2.33}$$

We can replace the value of $f(x')$ in the integral in Eq. (2.33) with its value at $x' = x$ because $\delta(x - x') = 0$ for $x' \neq x$. Equation (2.33), however, implies that

$$\int_{x-\epsilon}^{x+\epsilon} \delta(x - x') \, dx' = 1 \implies \int_{-\infty}^{\infty} \delta(x - x') \, dx' = 1, \tag{2.34}$$

where the range of integration can be extended in Eq. (2.34) because (again) $\delta(x - x') = 0$ for $x' \neq x$. We see that $\delta(x - x')$ has unit area under the curve, even though the function is zero whenever $x' \neq x$! How can that

[14] This is more than an idle question; we're going to need such a quantity, soon.

[15] We could just as well write the Kronecker function $\delta(i - j)$, with $\delta(0) = 1$ and $\delta((i - j) \neq 0) = 0$.

[16] A quantity labeled by a continuous index, such as f_x, is none other than a function of x, $f(x)$.

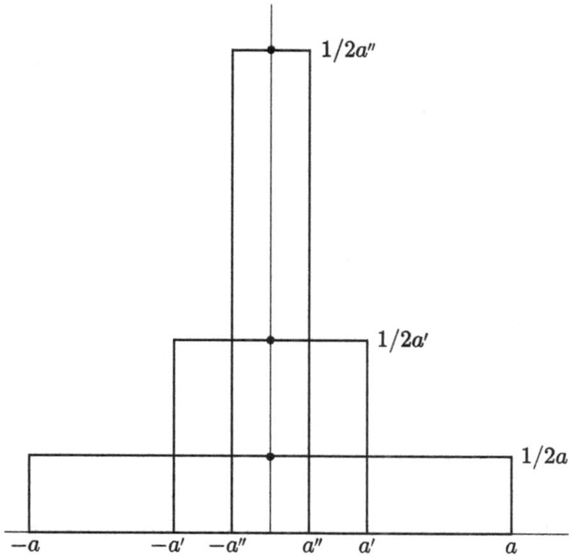

Figure 2.1 Sequence of unit-area rectangles.

be? By playing games with infinity. We want $\delta(0)$ to be *infinite*, but in such a way that $\delta(x)$ has unit area under the curve. We'll show how that can be achieved as the result of a certain limit process.

As a quick example, consider a sequence of rectangles, each of which encloses unit area, of width $2a$ and height $1/(2a)$ (see Figure 2.1). As $a \to 0$, we achieve a spike infinitely narrow and infinitely high, yet which encloses unit area.

The *Dirac delta function* is defined by the requirements

$$\delta(x - x') \equiv \begin{cases} \infty & x' = x, \quad \text{such that} \quad \int_{-\infty}^{\infty} \delta(x - x')\, \mathrm{d}x' = 1 \\ 0 & x' \neq x. \end{cases} \tag{2.35}$$

Is there a function with these properties? Yes and no. There is no *one* function in the traditional sense with these properties, yet it's possible to construct (in many ways) a *sequence of functions* that approach the Dirac function in a limiting process. Consider the sequence of normalized functions

Dirac delta function

$$\delta_n(x) \equiv \sqrt{\frac{n}{\pi}} \mathrm{e}^{-nx^2}. \qquad (n = 1, 2, 3, \ldots) \tag{2.36}$$

Such functions have unit area under the curve, $\int_{-\infty}^{\infty} \delta_n(x)\, \mathrm{d}x = 1$, for any[17] $n > 0$. For $n = 1, 2, 3, \ldots$, the width of these functions becomes smaller while the height grows taller; the width of $\delta_n(x)$ scales like $1/\sqrt{n}$, while the height $\delta_n(0) = \sqrt{n/\pi}$. The sequence $\{\delta_n(x)\}$ provides (as $n \to \infty$) support for what we require of the Dirac function: As $n \to \infty$, $\delta_n(x \neq 0) \to 0$ and $\delta_n(0) \to \infty$ with unit area under the curve. In the limit we achieve the desired filtering property:

[17] The integral $\int_{-\infty}^{\infty} \mathrm{e}^{-ax^2}\, \mathrm{d}x = \sqrt{\pi/a}$ for $a > 0$.

$$\lim_{n\to\infty}\int_{-\infty}^{\infty}\delta_n(x)f(x)\,\mathrm{d}x = f(0)\lim_{n\to\infty}\int_{-\infty}^{\infty}\delta_n(x)\,\mathrm{d}x = f(0)\lim_{n\to\infty}(1) = f(0).$$

It's sometimes written (loosely) that $\lim_{n\to\infty}\delta_n(x) = \delta(x)$, even though strictly speaking $\lim_{n\to\infty}\delta_n(x)$ *does not exist.* A more precise statement is

$$\lim_{n\to\infty}\int_{-\infty}^{\infty}\delta_n(x)f(x)\,\mathrm{d}x = \int_{-\infty}^{\infty}\delta(x)f(x)\,\mathrm{d}x.$$

The Dirac function, known as a *generalized function* [15], has meaning only "inside" an integral, but that doesn't stop people from writing $\delta(x)$ in formulas *as if* it were an ordinary function, knowing it will show up eventually inside an integral. The sequence of normalized functions in Eq. (2.36) is not unique; there are many ways to realize the Dirac function (see Exercise 2.10). We'll encounter several different examples (*instantiations*) of the Dirac function in the forthcoming chapters.

2.5 Completeness

We now come to perhaps the most important property of the eigenfunctions of Sturm–Liouville systems, that *as orthogonal sets of functions* they form *complete sets.*

completeness

Definition. *Let* $\{\phi_k(x)\}$, $k = 1, 2, 3, \ldots$ *be a set of bounded, square-integrable functions on an interval* I*, orthogonal with respect to a positive weight function* $w(x)$ *so that*

$$\int_I w(x)\phi_k(x)\phi_l(x)\,\mathrm{d}x = 0 \quad (k \neq l)\ .$$

The set of functions $\{\phi_k\}$ *is* complete *if any square-integrable function* $f(x)$ *can be expressed as an infinite linear combination of the elements of the set:*

$$f(x) = \sum_{n=1}^{\infty} c_n\phi_n(x). \tag{2.37}$$

eigenfunction expansion

Equation (2.37) is referred to as an *eigenfunction expansion.* Two questions must be addressed before accepting Eq. (2.37): How are the coefficients c_n determined for a given function f, and in what sense does the infinite series converge to the function $f(x)$? The first question is easy to answer, the second less so.

orthonormal set

In answering the first question, it's usually easier to work with *orthonormal* functions, those that are orthogonal over an interval I and are normalized to unity:

$$\int_I w(x)\phi_n(x)\phi_m(x)\,\mathrm{d}x = \delta_{nm}. \tag{2.38}$$

We'll assume in what follows an orthonormal set of functions $\{\phi_k\}$. Multiply Eq. (2.37) by $w(x)\phi_m(x)$, integrate term by term over the interval, and use Eq. (2.38):

$$\int_I w(x)\phi_m(x)f(x)\,\mathrm{d}x = \sum_{n=1}^{\infty} c_n \int_I w(x)\phi_m(x)\phi_n(x)\,\mathrm{d}x = \sum_{n=1}^{\infty} c_n\delta_{nm} = c_m.$$

Fourier coefficients

The expansion coefficients (also known as the *Fourier coefficients*) are thus given by[18]

$$c_m = \int_I w(x)\phi_m(x)f(x)\,\mathrm{d}x. \tag{2.39}$$

Substitute Eq. (2.39) into Eq. (2.37):

$$f(x) = \int_I \left(\sum_{n=1}^{\infty} w(x')\phi_n(x')\phi_n(x)\right) f(x')\,\mathrm{d}x'. \tag{2.40}$$

Comparing Eq. (2.40) with Eq. (2.33), we recognize that the terms in parentheses constitute a Dirac delta function (because $f(x)$ is an arbitrary function):

completeness relation

$$\sum_{n=1}^{\infty} w(x')\phi_n(x')\phi_n(x) = \delta(x-x'). \tag{2.41}$$

Equation (2.41) is called the *completeness relation.* A set of functions $\{\phi_k\}$ satisfying Eqs. (2.38) and (2.41) is referred to as a *complete, orthonormal set.* We'll see how different functions satisfy the completeness relation in forthcoming chapters.

complete, orthonormal set

The second issue, convergence, involves, as a question of mathematics: If the functions $\phi_k(x)$ are continuous, or differentiable, or integrable, is the same true of the function $f(x)$ obtained from the infinite series, Eq. (2.37)? We're going to sidestep such topics and content ourselves by simply quoting some of the major theorems in this subject.

As discussed in Section 1.6, convergence of infinite series is addressed through analysis of the partial sums, $s_n(x) \equiv \sum_{k=1}^{n} c_k\phi_k(x)$. Consider the mean-square "error" between $f(x)$ and the partial sum *at a fixed order* of approximation n, generated through *some* choice of coefficients $\sum_{k=1}^{n} \gamma_k\phi_k(x)$,

$$E_n(\gamma_1,\ldots,\gamma_n) \equiv \int_I \left[f(x) - \sum_{k=1}^{n}\gamma_k\phi_k(x)\right]^2 w(x)\,\mathrm{d}x.$$

What coefficients $\{\gamma_k\}$ minimize the mean-square error? It can be shown that E_n is minimized by the coefficients given by Eq. (2.39) [14, p. 424].

[18] Equation (2.39) is identical in form to Eq. (1.6), the coefficients for a vector expressed as a linear combination of orthonormal basis vectors. All that has changed is the nature of the inner product – which in the context of a Sturm–Liouville system could include a weight function.

The coefficients c_k $(k \leq n)$ provide the "best" approximation at order[19] n. Convergence is achieved when $E_n \to 0$ as $n \to \infty$.

piecewise continuous

Definition. *A* piecewise continuous *function is continuous except at a finite number of points of discontinuity.*

We're guaranteed that Eq. (2.37) converges by the following theorem, known as the *expansion theorem* [14, p. 427]:

expansion theorem

Theorem 2.5.1. *Every piecewise continuous function $f(x)$ with a square-integrable first derivative may be expanded in an infinite series of orthogonal functions $\{\phi_k\}$ which converges in all subdomains free of points of discontinuity; at the points of discontinuity it represents the arithmetic mean of the right- and left-hand limits.*

The eigenfunctions of a Sturm–Liouville system can thus be used to represent functions $f(x)$ that are not continuous! In regions where f is continuous, Eq. (2.37) converges to f. At points of discontinuity, the series converges to the arithmetic mean of the values of the function on the two sides of the discontinuity. We'll see this at work when we come to Fourier series, Chapter 4

2.6 Recap

Functions can be treated as vectors, as elements of a vector space. Of relevance to this chapter is the space L^2 of all square-integrable functions. Eigenfunctions of Sturm–Liouville systems are complete, orthonormal sets. A complete set of functions *spans the space* of functions through Eq. (2.37), and, because the eigenfunctions are orthogonal, they're linearly independent. The eigenfunctions of Sturm–Liouville systems thus constitute *an orthonormal basis* for L^2. It was one of the triumphs of late 19th-century and early twentieth-century mathematics to show that square-integrable functions can be expressed as infinite series of the eigenfunctions of Sturm–Liouville systems. This realization provides the mathematical foundation for much of what we'll do in forthcoming chapters.

Summary

- Any differential equation of the form of Eq. (2.1) can be put into self-adjoint form, Eq. (2.13). A self-adjoint differential operator is one that's identical to its adjoint.
- Self-adjoint linear differential equations have two important properties: Their eigenvalues are real, and eigenfunctions corresponding to

[19]The kth coefficient γ_k in the list $(\gamma_1, \cdots, \gamma_k, \cdots, \gamma_n)$ which gives the best approximation to f at order n is, remarkably, the same for all $n \geq k$. Such a property does not hold for least-squares approximations to a function using nonorthogonal functions.

different eigenvalues are orthogonal. For separated boundary conditions, there is only one linearly independent eigenfunction for a given eigenvalue (the Wronskian vanishes at the boundaries). For periodic boundary conditions, there are two linearly independent eigenfunctions for a given eigenvalue.

- The Dirac delta function has the filtering property $f(x) = \int \delta(x - x')f(x')\,\mathrm{d}x'$. There are many ways to realize the Dirac function.
- The eigenfunctions of self-adjoint differential equations form complete, orthonormal sets that can be used to represent square-integrable functions, $f(x) = \sum_{n=1}^{\infty} c_n \phi_n(x)$, where the coefficients are given by Eq. (2.39).
- A complete orthonormal set of functions $\{\phi_k\}$ satisfies the completeness relation, Eq. (2.41).

Exercises

2.1. Derive Eq. (2.9).

2.2. The *adjoint equation* is defined as the differential equation $\mathcal{L}^\dagger f = 0$. Show, for $\mathcal{L}$ defined in Eq. (2.1) and for $\mathcal{L}^\dagger$ defined in Eq. (2.10), that the adjoint of the adjoint equation is the original differential equation, $(\mathcal{L}^\dagger f)^\dagger = \mathcal{L}f = 0$. Expand out $\mathcal{L}^\dagger f$ so that it has the form of $\mathcal{L}f$. Now find the adjoint of that equation; it should agree precisely with $\mathcal{L}f$.

2.3. Show that in order for $\mathcal{L}$ in Eq. (2.1) to be such that $\mathcal{L} = \mathcal{L}^\dagger$ with $\mathcal{L}^\dagger$ defined in Eq. (2.10), it must be true that $p' = q$.

2.4. Show that Eq. (2.1) can be placed in self-adjoint form by multiplying by $\lambda(x)$ defined in Eq. (2.15). Start by multiplying Eq. (2.1) by an unknown function $\lambda(x)$. The requirement that the modified differential equation (after multiplying by $\lambda(x)$) be in self-adjoint *form* $((\tilde{p}f')' + \tilde{r}f = 0)$ is that $(\lambda(x)p(x))' = \lambda(x)q(x)$. Show that $\lambda(x)$ must be as given in Eq. (2.15), and thus $\tilde{p}(x) = \lambda(x)p(x)$ and $\tilde{r}(x) = \lambda(x)r(x)$. Show that if $q = p'$, then $\lambda(x)$ is a constant, i.e. the differential equation is already in self-adjoint form.

2.5. The functions $T_n(x)$ satisfy the differential equation

$$\sqrt{1-x^2}\frac{\mathrm{d}^2 T_n}{\mathrm{d}x^2} - \frac{x}{\sqrt{1-x^2}}\frac{\mathrm{d}T_n}{\mathrm{d}x} + \frac{n^2}{\sqrt{1-x^2}}T_n = 0$$

for each n. Determine an orthogonality condition for $T_n(x)$.

2.6. Reduce the following differential equations to self-adjoint form:

(a) $(1-x^2)y'' - xy' + \lambda y = 0$. (The Chebyshev differential equation.) Hint: Show that $\lambda(x) = 1/\sqrt{1-x^2}$.

(b) $x^2y'' + xy' + y = 0$. A: $(\mathrm{d}/\,\mathrm{d}x)(xy') + y/x = 0$.

(c) $y'' + \tan xy' = 0$.

2.7. (a) Show that the boundary conditions in Eq. (2.19) imply that the group of terms $y'y^* - yy'^*$ vanish at either of the endpoints. The coefficients $\alpha_1, \alpha_2, \beta_1$, and β_2 are real numbers. Show this first by assuming α_2 (or β_2) to be zero, then consider the general case where $\alpha_2 \neq 0$.

(b) Show that $[y'y^* - yy'^*]|_a^b$ vanishes for periodic boundary conditions.

(c) Suppose two functions f and g each satisfy at a point the boundary condition in Eq. (2.19). We thus have the pair of equations, at $x = a$:

$$\alpha_1 f(a) + \alpha_2 f'(a) = 0 \qquad \alpha_1 g(a) + \alpha_2 g'(a) = 0,$$

or that

$$\begin{pmatrix} f(a) & f'(a) \\ g(a) & g'(a) \end{pmatrix} \begin{pmatrix} \alpha_1 \\ \alpha_2 \end{pmatrix} = 0.$$

Argue that the Wronskian of two functions f and g each satisfying separated boundary conditions vanishes at an endpoint.

2.8. Consider the differential equation, Eq. (2.23), $y'' + \lambda y = 0$, together with the boundary conditions $y(0) = y(b) = 0$. Show there are no nontrivial solutions meeting the boundary conditions if $\lambda \leq 0$.

2.9. Derive Eq. (2.24) given in the latter example of Section 2.3. Refer to the previous example for details.

2.10. Show that the functions

$$\delta_n(x) = \frac{n}{\pi}\frac{1}{1+n^2x^2} \qquad (n = 1, 2, 3, \ldots)$$

form a sequence suitable for realizing the Dirac delta function.

(a) Show that for any n, $\delta_n(x)$ has unit area under the curve. Hint: $\frac{\mathrm{d}}{\mathrm{d}x}\tan^{-1}x = \frac{1}{1+x^2}$.

(b) Show that as $n \to \infty$, $\delta_n(x \neq 0) \to 0$ and $\delta_n(0) \to \infty$.

2.11. Suppose the normalized eigenfunctions of a Sturm–Liouville system $\{\phi_n(x)\}$ are complex valued. Show that the completeness relation, Eq. (2.41), is then

$$\sum_{n=1}^{\infty} w(x')\phi_n^*(x')\phi_n(x) = \delta(x - x'). \tag{2.42}$$

It's helpful to use Dirac notation (but not necessary), which underscores that a function $f(x) \equiv |f\rangle$ can be treated as a vector (an element of a vector space, L^2). In Dirac notation, Eq. (2.37) is written $|f\rangle = \sum_n a_n|\phi_n\rangle$. The inner product for complex vector spaces is given as the integral $\int_I w(x)f^*(x)g(x)\,\mathrm{d}x \equiv \langle f|g\rangle$, where we've included the weight function w. The orthonormality of the basis functions is written $\langle \phi_k|\phi_l\rangle = \delta_{kl}$ in Dirac notation.

2.12. Find the eigenfunctions and eigenvalues for the differential equation $y_n''(x) = -\lambda_n y_n(x)$ on the interval $0 \leq x \leq a$ subject to the boundary conditions:

(a) $y(0) = 0 \qquad y(a) = 0,$

(b) $y(0) = 0 \qquad y'(a) = 0,$

(c) $y'(0) = 0 \qquad y'(a) = 0,$

(d) $y(0) + ay'(0) = 0 \qquad y(a) - ay'(a) = 0.$

Each case is separate. Is $\lambda = 0$ allowed in any of these cases? For case (d), find the equation that determines the eigenvalues, and give an argument that there is an infinite set of eigenfunctions and eigenvalues. Hint: Present a graphical solution of the transcendental equation

$$\tan x = \frac{2x}{1-x^2},$$

where $x \equiv \sqrt{\lambda}a$.

3

Partial differential equations

How do electrical charges interact through space? How do masses interact gravitationally? The paradigm in physics is that interactions between objects are mediated by *fields* that propagate in space and time.[1] There are scalar fields, vector fields, and tensor fields – assignments to points of space and time of scalar, vector, or tensor quantities. Fields are governed, in the manner by which they change in time and space, by partial differential equations, the form of which depend on the physical application at hand.[2] The goal of this chapter is to introduce: (i) the simplest types of partial differential equations (PDEs) for scalar fields; and (ii) a method for solving them – the method of separation of variables.

3.1 A survey of partial differential equations

3.1.1 The continuity equation

Consider an amount $Q(t)$ of a conserved quantity contained in a fixed volume V at time t (the precise nature of the substance is immaterial; it could be charge, matter, or energy). To say that Q is *conserved* means that it can't just "disappear" from V, the only way Q can decrease is if the substance "leaves the building," i.e. *flows* through the surface S bounding V. The conservation of Q is modeled by the equation

[1]The use of the word *field* here is not the same as the field $\mathbb{F}$ introduced in the definition of vector space.

[2]Many PDEs can be developed through reasoning based on direct experience with physical systems. In other cases, PDEs governing fields are found by first developing a *Lagrangian* of the field, with the PDE following from an appropriate *action principle* – a topic not covered in this book. Still other PDEs have been found by *inspired guesswork*, the truth of which is ascertained by comparison with experiment.

Mathematical Methods in Physics, Engineering, and Chemistry, First Edition.
Brett Borden and James Luscombe.

$$\frac{\mathrm{d}}{\mathrm{d}t}Q(t) = -\oint_S \boldsymbol{J}(\boldsymbol{r},t)\cdot \mathrm{d}\boldsymbol{s}, \tag{3.1}$$

where $\boldsymbol{J}(\boldsymbol{r},t)$ is the local current density vector (dimensions of substance per unit surface area per unit time) and $\mathrm{d}\boldsymbol{s} = \hat{\boldsymbol{n}}\mathrm{d}s$ is the infinitesimal element of surface area, considered as a vector quantity in the direction of the outward unit normal $\hat{\boldsymbol{n}}$ of S at location $\boldsymbol{r}$. Equation (3.1) is known as a *balance equation*: The rate of change of Q is accounted for by the net flow through S. If the net flow is positive (negative), Q decreases (increases) in time. For fixed V, Eq. (3.1) is equivalent to

$$\begin{aligned}\frac{\mathrm{d}}{\mathrm{d}t}Q(t) &= \frac{\mathrm{d}}{\mathrm{d}t}\int_V \rho(\boldsymbol{r},t)\,\mathrm{d}^3r = \int_V \frac{\partial\rho(\boldsymbol{r},t)}{\partial t}\,\mathrm{d}^3r \\ &= -\oint_S \boldsymbol{J}(\boldsymbol{r},t)\cdot \mathrm{d}\boldsymbol{s} = -\int_V \boldsymbol{\nabla}\cdot\boldsymbol{J}(\boldsymbol{r},t)\,\mathrm{d}^3r,\end{aligned} \tag{3.2}$$

where $\rho(\boldsymbol{r},t)$ is the local density function (substance per volume), and we've invoked the divergence theorem (Appendix A). We can take the time derivative inside the integral in the second equality in Eq. (3.2) because V is fixed. Because Eq. (3.2) holds for any V, we have the differential form of Eq. (3.1) known as the *continuity equation*:

continuity equation

$$\frac{\partial\rho}{\partial t} + \boldsymbol{\nabla}\cdot\boldsymbol{J} = 0. \tag{3.3}$$

The continuity equation applies to any conserved quantity. Equation (3.3) relates the time rate of change of the *density field* $\rho(\boldsymbol{r},t)$ at location $\boldsymbol{r}$ to the divergence (at $\boldsymbol{r}$) of the *flow field* $\boldsymbol{J}(\boldsymbol{r},t)$, a vector field. The divergence of $\boldsymbol{J}$ at a point is nonzero only if the density (at $\boldsymbol{r}$) is changing in time: $\boldsymbol{\nabla}\cdot\boldsymbol{J} = -\partial\rho/\partial t$. A positive divergence of $\boldsymbol{J}$ is locally accounted for by a decrease in the density – the missing density at a point is what flows away at that point.

3.1.2 The diffusion equation

The *diffusion equation* follows from the continuity equation by using a phenomenological model of the current density vector, $\boldsymbol{J} = -D\,\boldsymbol{\nabla}\,\rho$. This relation (*Fick's law*) models flow as occurring against the direction of gradients in ρ, which is what is observed macroscopically – substances naturally flow in such a way as to remove density inhomogeneities (diffusive motion). The *diffusion coefficient* D is a material-specific proportionality factor between the flux and the density gradient. When Fick's law is combined with the continuity equation, we obtain the PDE for diffusive motion of the density field[3]:

[3]Drift motion of charges in response to an external field is described by a far-simpler PDE. Ohm's law $\boldsymbol{J} = \sigma\boldsymbol{E}$ combined with the continuity equation and Gauss's law ($\boldsymbol{\nabla}\cdot\boldsymbol{E} = \rho/\epsilon_0$) leads to $\partial\rho/\partial t + (\sigma/\epsilon_0)\rho = 0$, where σ is the electrical conductivity of a material and ϵ_0 is a physical constant, the permittivity of free space.

$$\frac{\partial \rho}{\partial t} - D\nabla^2 \rho = 0. \tag{3.4}$$

homogeneous diffusion equation

Note the characteristic feature of the diffusion equation: First order derivatives in time dependence, second order derivatives in space dependence. You should be able to see from Eq. (3.4) that D has dimensions of length squared per time.

3.1.3 The free-particle Schrödinger equation

The time-dependent Schrödinger equation for a free particle of mass m is[4]:

$$i\hbar \frac{\partial \psi}{\partial t} + \frac{\hbar^2}{2m} \nabla^2 \psi = 0, \tag{3.5}$$

where $\hbar$ is Planck's constant and ψ (a complex-valued quantity) is such that $\psi^*(\boldsymbol{r}, t)\psi(\boldsymbol{r}, t)\, \mathrm{d}^3 r$ is the probability of finding the particle within a volume $\mathrm{d}^3 r$ at location $\boldsymbol{r}$ at time t. The free-particle Schrödinger equation has the form of the diffusion equation with a *complex* diffusion coefficient, $D = i\hbar/(2m)$. The mathematical method of solving Eq. (3.5) is the same as that for Eq. (3.4) – the same math has the same solution. The Schrödinger equation gives rise to a continuity equation with $\rho \equiv \psi^* \psi$ and $\boldsymbol{J} \equiv i\hbar/(2m)(\psi\, \boldsymbol{\nabla}\, \psi^* - \psi^*\, \boldsymbol{\nabla}\, \psi)$.

3.1.4 The heat equation

The *heat conduction equation*, or simply the *heat equation*, is a variant of the diffusion equation. Start from the continuity equation with the conserved quantity being thermal energy. Heat transfers $\mathrm{d}Q$ at constant volume (so that no energy of work is transferred to the system) are related to temperature changes $\mathrm{d}T$ through the relation $\mathrm{d}Q = C_V\, \mathrm{d}T$, where C_V is the heat capacity of the substance at constant volume. The *specific heat* c_v is the heat capacity C_V per mass. The change in thermal energy *density* $\mathrm{d}u$ associated with $\mathrm{d}T$ is then $\mathrm{d}u = c_v \rho\, \mathrm{d}T$ (first law of thermodynamics), where ρ is the mass density. *Fourier's law*, $\boldsymbol{J} = -\kappa\, \boldsymbol{\nabla}\, T$, captures the natural tendency of heat to flow against temperature gradients (second law of thermodynamics), where the *thermal conductivity* κ is a material-specific quantity. Combining Fourier's law with the continuity equation, we obtain the PDE governing the *temperature field* $T(\boldsymbol{r}, t)$,

$$\frac{\partial T}{\partial t} - \alpha \nabla^2 T = 0, \tag{3.6}$$

heat equation

where $\alpha \equiv \kappa/(\rho c_v)$ is the *thermal diffusivity.*

[4]The Schrödinger equation can't be *derived* from something more fundamental. It can be motivated, but not derived; it's something entirely new. We wouldn't teach it, however, if it hadn't been successfully tested against the results of experimental findings. Nature is the arbiter of physical truth.

3.1.5 The inhomogeneous diffusion equation

Suppose we have a quantity that's *not* conserved. In that case, we can modify the balance equation (3.1) by introducing an additional *source* term:

$$\frac{\mathrm{d}}{\mathrm{d}t}Q(t) = -\oint_S \boldsymbol{J}(\boldsymbol{r},t)\cdot \mathrm{d}\boldsymbol{s} + \int_V s(\boldsymbol{r},t)\,\mathrm{d}^3r. \tag{3.7}$$

Here $s(\boldsymbol{r},t)$ is the local *source density* function (or simply *source function*) representing the rate at which the substance comprising Q in V is changing by a means other than flowing through the surface bounding V. Such a situation occurs in semiconductors where charge carriers are locally produced by exposure to electromagnetic radiation. Invoking Fick's law as aforementioned, we obtain the *inhomogeneous diffusion equation*

inhomogeneous diffusion equation

$$\frac{\partial \rho(\boldsymbol{r},t)}{\partial t} - D\nabla^2\rho(\boldsymbol{r},t) = s(\boldsymbol{r},t). \tag{3.8}$$

The mathematics involved in solving inhomogeneous PDEs like Eq. (3.8) is sufficiently different from that involved in the solution of the corresponding homogeneous PDE that we postpone the treatment of such equations until Chapter 9.

3.1.6 Schrödinger equation for a particle in a potential field

The Schrödinger equation for a particle in a potential energy environment characterized by the function $V(\boldsymbol{r})$ has the form of the inhomogeneous diffusion equation:

$$\mathrm{i}\hbar\frac{\partial \psi}{\partial t} + \frac{\hbar^2}{2m}\nabla^2\psi = V(\boldsymbol{r})\psi(\boldsymbol{r},t). \tag{3.9}$$

The inhomogeneous term in Eq. (3.9) involves the product of ψ (what we're trying to solve for) and the potential energy function V. This is a special type of equation, treated in Chapter 10.

3.1.7 The Poisson equation

In electromagnetism, the static (time-independent) electric field vector $\boldsymbol{E}(\boldsymbol{r})$ is related to the charge density function $\rho(\boldsymbol{r})$ through Gauss's law: $\boldsymbol{\nabla}\cdot\boldsymbol{E} = \rho/\epsilon_0$. Charges are the *source* of the static electric field. The static $\boldsymbol{E}$ field is obtained from the gradient of a scalar field $\phi(\boldsymbol{r})$, the *electrostatic potential function*, with $\boldsymbol{E} = -\boldsymbol{\nabla}\phi$. By combining $\boldsymbol{E} = -\boldsymbol{\nabla}\phi$ with Gauss's law, we obtain the *Poisson equation*, an inhomogeneous, second order PDE describing the electrostatic potential field:

Poisson equation

$$\nabla^2\phi = -\rho(\boldsymbol{r})/\epsilon_0. \tag{3.10}$$

Solving the Poisson equation for a given charge density function $\rho(\boldsymbol{r})$ is one of the major tasks of the theory of electrostatics. PDEs in the form of the Poisson equation also follow from the inhomogeneous diffusion equation under *steady-state conditions*, where all time derivatives are zero.

3.1.8 The Laplace equation

The *Laplace equation* is the version of the Poisson equation when $\rho(\boldsymbol{r}) = 0$,

$$\nabla^2 \phi = 0. \tag{3.11}$$

Laplace equation

The Laplace equation holds a special place in the pantheon of PDEs. Under steady-state conditions, the temperature field satisfies the Laplace equation, $\nabla^2 T = 0$.

3.1.9 The wave equation

Consider a string of mass density ρ (mass/length), under tension T stretched along the x-axis. At time t, let the string have transverse displacement $\psi(x,t)$ (see Figure 3.1). From Newton's second law,

$$(T\sin\theta)_{x+\mathrm{d}x} - (T\sin\theta)_x = \rho\,\mathrm{d}x\frac{\partial^2\psi}{\partial t^2}.$$

For a displacement sufficiently small that $\partial\psi/\partial x \ll 1$, we can approximate $\sin\theta \approx \tan\theta \approx \partial\psi/\partial x$ so that $T\sin\theta \approx T\partial\psi/\partial x$, in which case the equation of motion becomes in the limit $\mathrm{d}x \to 0$,

$$\frac{\partial^2\psi}{\partial x^2} - \frac{1}{c^2}\frac{\partial^2\psi}{\partial t^2} = 0, \tag{3.12}$$

where $c^2 \equiv T/\rho$. Equation (3.12) is the *one-dimensional wave equation* where c is the speed of wave propogation on the string. By extending the derivation to a stretched membrane and then to a three-dimensional elastic medium, the homogeneous wave equation in any number of dimensions is

$$\nabla^2\psi - \frac{1}{c^2}\frac{\partial^2\psi}{\partial t^2} = 0. \tag{3.13}$$

wave equation

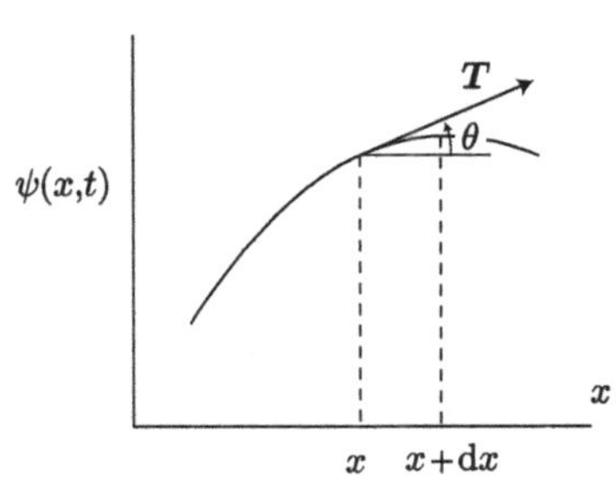

Figure 3.1 Displacement field $\psi(x,t)$ of a string under tension T.

Table 3.1 Generic form of PDEs for scalar fields $\psi(\boldsymbol{r},t)$ most commonly encountered.

$\nabla^2\psi(\boldsymbol{r},t) = s(\boldsymbol{r})$	Poisson equation (Laplace if $s(\boldsymbol{r}) = 0$)
$\nabla^2\psi(\boldsymbol{r},t) - \frac{1}{\alpha}\frac{\partial\psi(\boldsymbol{r},t)}{\partial t} = s(\boldsymbol{r},t)$	Diffusion equation
$\nabla^2\psi(\boldsymbol{r},t) - \frac{1}{c^2}\frac{\partial^2\psi(\boldsymbol{r},t)}{\partial t^2} = s(\boldsymbol{r},t)$	Wave equation

3.1.10 Inhomogeneous wave equation

In electrodynamics, it's shown that the time-dependent scalar potential $\phi(\boldsymbol{r},t)$ satisfies the *inhomogeneous wave equation*, which we cite without derivation[5]:

inhomogeneous wave equation

$$\nabla^2\phi(\boldsymbol{r},t) - \frac{1}{c^2}\frac{\partial^2\phi(\boldsymbol{r},t)}{\partial t^2} = -\rho(\boldsymbol{r},t)/\epsilon_0. \tag{3.14}$$

3.1.11 Summary of PDEs

We see the pattern now. Table 3.1 shows the generic form of the most commonly encountered[6] PDEs for scalar fields, which all involve second-order spatial derivatives (in the form of the Laplacian operator ∇^2; see Appendix A) and up to second-order time derivatives, *with or without source functions*. In the next few chapters, we concentrate on the *homogeneous* versions of these PDEs (no source terms), saving inhomogeneous differential equations for Chapter 9.

3.2 Separation of variables and the Helmholtz equation

We now introduce the method of *separation of variables*, a procedure for constructing solutions of PDEs. Consider the time-dependent heat equation, Eq. (3.6), which we reproduce here:

$$\nabla^2\psi - \frac{1}{\alpha}\frac{\partial\psi}{\partial t} = 0. \tag{3.6}$$

We use ψ to denote temperature – we're about to use T for another purpose. We're looking for a temperature distribution that's space and time dependent, $\psi = \psi(\boldsymbol{r},t)$. The separation of variables method *guesses* the solution of Eq. (3.6) to be in the form of a *product* of functions, where each of these functions depends on only a single variable:

separation of variables

$$\psi(\boldsymbol{r},t) = R(\boldsymbol{r})T(t). \tag{3.15}$$

[5]See any book on classical electrodynamics. Equation (3.14) holds in the Lorenz gauge.

[6]It's not implied that Table 3.1 is exhaustive of the types of PDEs encountered in physical applications.

Does such a guess work? Only one way to find out. Substituting Eq. (3.15) into Eq. (3.6) yields

$$T\nabla^2 R = \frac{1}{\alpha} R \frac{dT}{dt}. \tag{3.16}$$

The partial derivative with respect to time in Eq. (3.6) has been replaced with an ordinary derivative in Eq. (3.16) (T depends only on t and so, as far as $T(t)$ is concerned, there's no difference between $\partial/\partial t$ and d/dt). Now, divide[7] Eq. (3.15) by $\psi = RT$:

$$\underbrace{\frac{1}{R(\boldsymbol{r})}\nabla^2 R(\boldsymbol{r})}_{\text{depends only on spatial variables}} = \underbrace{\frac{1}{\alpha}\frac{1}{T(t)}\frac{dT(t)}{dt}}_{\text{depends only on time}}. \tag{3.17}$$

We now come to a key step in this method, one that we'll frequently invoke. The terms on the left of Eq. (3.17) depend only on spatial variables, while the terms on the right depend only on time. *Such an equality is possible only if both sides are equal to a constant.* (The term on the left is independent of t, the term on the right is independent of $\boldsymbol{r}$, yet they are equal for all values of $\boldsymbol{r}$ and t.) Let's rewrite Eq. (3.17) to make explicit the equality with a constant:

$$\frac{1}{R(\boldsymbol{r})}\nabla^2 R(\boldsymbol{r}) = -k^2 = \frac{1}{\alpha}\frac{1}{T(t)}\frac{dT(t)}{dt}, \tag{3.18}$$

where $-k^2$ is the *separation constant*, written so that no matter what k turns out to be, the separation constant is manifestly negative. How did we know to make it negative? Experience. If that makes you uneasy, try it with a positive separation constant and see how your answer turns out. If it doesn't make sense physically, reconsider.

separation constant

The equality on the right side of Eq. (3.18) implies an ordinary differential equation (ODE) that's readily solved: (check it!)

$$T(t) = Ae^{-\alpha k^2 t}, \tag{3.19}$$

where A is a constant. A negative separation constant thus leads to a temperature that *decays* in time. Unless something was forcing the temperature to indefinitely increase in time, an exponentially decaying solution in time conforms to our physical expectations.

That leaves the left side of Eq. (3.18):

$$\nabla^2 R(\boldsymbol{r}) + k^2 R(\boldsymbol{r}) = 0. \tag{3.20}$$

Helmholtz equation

Equation (3.20) is known as the *Helmholtz equation*, a PDE involving the spatial variables only. To make progress with Eq. (3.20), we must declare

[7] If your inner mathematician is asking about $RT = 0$, we're explicitly *not* looking for the trivial solution.

a coordinate system (which we'll do shortly). Note what's happened, however. By guessing a product solution $\psi(\boldsymbol{r},t) = R(\boldsymbol{r})T(t)$, the original PDE, Eq. (3.19), a differential equation in four variables, has been reduced to a PDE involving three variables plus an ODE, the solution of which is readily obtained to produce the time variation, Eq. (3.19). That's been accomplished at the expense of introducing an unknown quantity (k^2) into the analysis. The quantity k is determined by the boundary conditions.

A Helmholtz equation can always be obtained from second-order PDEs in which the space and time derivatives occur separately.[8] Consider the wave equation, Eq. (3.13). By separating variables with $\psi(\boldsymbol{r},t) = R(\boldsymbol{r})T(t)$, we obtain the analog of Eq. (3.18):

$$\frac{1}{R}\nabla^2 R = -k^2 = \frac{1}{c^2}\frac{1}{T}\frac{\mathrm{d}^2 T}{\mathrm{d}t^2}. \tag{3.21}$$

The time-dependent part $T(t)$ is readily obtained: (check it!)

$$T(t) = A\mathrm{e}^{\mathrm{i}kct} + B\mathrm{e}^{-\mathrm{i}kct}. \tag{3.22}$$

The spatial part $R(\boldsymbol{r})$ is obtained as the solution of the Helmholtz equation, Eq. (3.20).

Each of the (homogeneous) PDEs in Table 3.1 can therefore be reduced to the Helmholtz equation using the separation of variables method.[9] It behooves us then to study the Helmholtz equation, which we do now in the three most widely used coordinate systems – rectangular, cylindrical, and spherical coordinates. (Refer to Section A.5 for a review of these coordinate systems and the associated Laplacian operator representations.)

3.2.1 Rectangular coordinates

The Helmholtz equation in rectangular coordinates is, from Eq. (3.20),

$$\left(\frac{\partial^2}{\partial x^2} + \frac{\partial^2}{\partial y^2} + \frac{\partial^2}{\partial z^2} + k^2\right) R(x,y,z) = 0. \tag{3.23}$$

To solve Eq. (3.23) try as a solution a product of three unknown functions,

$$R(x,y,z) = X(x)Y(y)Z(z). \tag{3.24}$$

Substitute Eq. (3.24) into Eq. (3.23) and divide by $R = XYZ$ to obtain

$$\frac{1}{X}X'' + \frac{1}{Y}Y'' + \frac{1}{Z}Z'' + k^2 = 0, \tag{3.25}$$

---

[8]There is only one time direction (right?), but several spatial directions.

[9]The Laplace equation is a special case, corresponding to $k^2 = 0$.

where the notation X'' indicates a second derivative of X with respect to its argument, $X'' \equiv \mathrm{d}^2X/\mathrm{d}x^2$. Once more we make the separation-constant argument; rewrite Eq. (3.25) in the form

$$\underbrace{X^{-1}X''}_{\text{depends only on } x} = \underbrace{-(k^2 + Y^{-1}Y'' + Z^{-1}Z'')}_{\text{depends only on } y \text{ and } z}.$$

The term on the left is independent of y and z, the terms on the right are independent of x; the only way we can have an equality for all x, y, z is if each side is equal to a constant. By repeating the argument for each term in Eq. (3.25), each is equal to a constant:

$$\frac{1}{X}X'' = -k_1^2 \qquad \frac{1}{Y}Y'' = -k_2^2 \qquad \frac{1}{Z}Z'' = -k_3^2, \tag{3.26}$$

where $k_1^2 + k_2^2 + k_3^2 = k^2$. We've taken the separation constants in Eq. (3.26) to each be negative; that might need to be revised depending on the particulars of the system.

Thus, the Helmholtz equation (a PDE in three variables) has been reduced to three ordinary differential equations (ODEs), a step that comes at the expense of introducing three unknown constants. Actually, only two are independent because of the constraint $k_1^2 + k_2^2 + k_3^2 = k^2$ (if k^2 is known); alternatively, k_1, k_2, and k_3 could be found through the imposition of boundary conditions, with k^2 then defined as $k^2 = k_1^2 + k_2^2 + k_3^2$. The ODEs in Eq. (3.26) have simple solutions,

$$X(x) = A_1\mathrm{e}^{\mathrm{i}k_1x} + B_1\mathrm{e}^{-\mathrm{i}k_1x} \quad Y(y) = A_2\mathrm{e}^{\mathrm{i}k_2y} + B_2\mathrm{e}^{-\mathrm{i}k_2y} \quad Z(z) = A_3\mathrm{e}^{\mathrm{i}k_3z} + B_3\mathrm{e}^{-\mathrm{i}k_3z}, \tag{3.27}$$

and thus we've found a family of solutions to Eq. (3.23).

Example. Laplace equation in two dimensions

Consider the two-dimensional steady state heat equation, $\nabla^2 T = 0$, or

$$\frac{\partial^2 T}{\partial x^2} = -\frac{\partial^2 T}{\partial y^2}.$$

Writing $T(x, y) = X(x)Y(y)$ implies, from the PDE,

$$\frac{1}{X}X'' = -\frac{1}{Y}Y''.$$

From now on we'll skip the argument that such an equality can be achieved only if each side is equal to a constant (did you get that?). Introducing a separation constant $-\alpha^2$, X and Y must satisfy

$$X'' + \alpha^2 X = 0 \qquad Y'' - \alpha^2 Y = 0.$$

In this case, the solutions for X and Y can be written

$$X(x) = A\cos\alpha x + B\sin\alpha x \qquad Y(y) = C\cosh\alpha y + D\sinh\alpha y. \tag{3.28}$$

The solution to the PDE ($\nabla^2 T = 0$) is incomplete until the constants α and A, B, C, D are known; these will be determined once boundary conditions have been specified.

3.2.2 Cylindrical coordinates

The Helmholtz equation in cylindrical coordinates is:

$$\left(\frac{1}{\rho}\frac{\partial}{\partial\rho}\left(\rho\frac{\partial}{\partial\rho}\right) + \frac{1}{\rho^2}\frac{\partial^2}{\partial\phi^2} + \frac{\partial^2}{\partial z^2} + k^2\right)\psi(\rho,\phi,z) = 0, \tag{3.29}$$

where we've changed notation again ($R \leftrightarrow \psi$); we'll use R for another purpose. Try as a solution, a product of three unknown functions:

$$\psi(\rho,\phi,z) = R(\rho)\Phi(\phi)Z(z). \tag{3.30}$$

Substitute Eq. (3.30) into Eq. (3.29) and divide by ψ:

$$\frac{1}{R\rho}\frac{\mathrm{d}}{\mathrm{d}\rho}\left(\rho\frac{\mathrm{d}R}{\mathrm{d}\rho}\right) + \frac{1}{\rho^2}\frac{1}{\Phi}\Phi'' + \frac{1}{Z}Z'' + k^2 = 0. \tag{3.31}$$

In the usual way (i.e. proceeding by an argument that we hope is familiar by now), let's introduce a separation constant. Let

$$\frac{1}{\Phi}\Phi'' = -m^2, \tag{3.32}$$

the solution of which is

$$\Phi(\phi) = \mathrm{e}^{im\phi}. \tag{3.33}$$

We won't bother with integration constants here; that will come later when we assemble the solution $\psi(\rho,\phi,z) = R(\rho)\Phi(\phi)Z(z)$. We also don't bother to split out separate solutions for $\pm m$; that's taken care of by letting m have positive and negative values in Eq. (3.33). The quantity m is particularly simple to ascertain. The solution for Φ must be *single valued*: If the azimuthal coordinate ϕ is permitted to "wrap around" the z-axis, we demand that

single-valued solution

$$\Phi(\phi + 2\pi) = \Phi(\phi). \tag{3.34}$$

The only way Eq. (3.34) can be satisfied by $\Phi(\phi) = \mathrm{e}^{im\phi}$ is if *m is an integer*. Equation (3.34) can be considered a *type* of boundary condition, one imposed by the internal consistency of the solution. With m determined, Eq. (3.31) reduces to

$$\frac{1}{R\rho}\frac{\mathrm{d}}{\mathrm{d}\rho}\left(\rho\frac{\mathrm{d}R}{\mathrm{d}\rho}\right) - \frac{m^2}{\rho^2} + \frac{1}{Z}Z'' + k^2 = 0. \qquad (m = \text{integer}) \tag{3.35}$$

Time for another separation constant. Let

$$\frac{1}{Z}Z'' = \alpha^2 - k^2. \tag{3.36}$$

The right side of Eq. (3.36) could be positive or negative depending on the system, i.e. depending on boundary conditions. If $\alpha^2 > k^2$, the solution of Eq. (3.36) is

$$Z(z) = A\exp\left(\sqrt{\alpha^2 - k^2}z\right) + B\exp\left(-\sqrt{\alpha^2 - k^2}z\right). \tag{3.37}$$

For $\alpha^2 < k^2$, the form of $Z(z)$ would be in terms of complex exponentials. The important point is that the form of $Z(z)$ is known.

Having introduced the separation constants m and α, Eq. (3.35) reduces to

radial equation

$$R'' + \frac{1}{\rho}R' + \left(\alpha^2 - \frac{m^2}{\rho^2}\right)R = 0. \tag{3.38}$$

The differential equation for $R(\rho)$ is called the *radial equation.* Equation (3.38) can be simplified through a change of variables. Let $x \equiv \alpha\rho$, where we assume that $\alpha \neq 0$. It can then be shown (do this) that Eq. (3.38) is the same as

Bessel differential equation

$$R''_m(x) + \frac{1}{x}R'_m(x) + \left(1 - \frac{m^2}{x^2}\right)R_m(x) = 0, \tag{3.39}$$

where we've labeled the solution of Eq. (3.39) by the value of m. Because for every m there's a new ODE to solve; it's convenient to label the solutions by m.

In some sense we're done, or at least we *will* be done when we know how to solve Eq. (3.39). We started out with the Helmholtz equation in cylindrical coordinates, Eq. (3.29), for $\psi(\rho, \phi, z)$, a function of three variables. By "guessing" that $\psi = R\Phi Z$ can be written as the product of three unknown functions, we've been able, by introducing two separation constants, to obtain the form of $\Phi(\phi)$, Eq. (3.33), and that for $Z(z)$, Eq. (3.37). It remains to solve Eq. (3.39), which is likely unfamiliar because it's an ODE with nonconstant coefficients. Equation (3.39) is *Bessel's differential equation,* and its solutions are *Bessel functions.* We'll learn all about Bessel functions in Chapter 7. There are two linearly-independent solutions of Eq. (3.39): $J_m(x)$, *the* Bessel function, and $N_m(x)$, the *Neumann function.* The solution of Eq. (3.39) can therefore be written (without knowing the form of J_m and N_m):

Bessel and Neumann function

$$R_m(x) = A_m J_m(x) + B_m N_m(x), \tag{3.40}$$

where A_m and B_m are constants.

For the special case of $\alpha = 0$, Eq. (3.38) reduces to

$$R'' + \frac{1}{\rho}R' - \frac{m^2}{\rho^2}R = 0. \tag{3.41}$$

The solution to Eq. (3.41) has the form (check it!)

$$R(\rho) = \begin{cases} a\rho^m + b\rho^{-m} & m \neq 0 \\ a + b\ln\rho. & m = 0 \end{cases} \tag{3.42}$$

3.2.3 Spherical coordinates

Let's do this procedure in spherical coordinates. From Eq. (3.20),

$$\left(\frac{1}{r^2}\frac{\partial}{\partial r}\left(r^2\frac{\partial}{\partial r}\right) + \frac{1}{r^2\sin\theta}\frac{\partial}{\partial\theta}\left(\sin\theta\frac{\partial}{\partial\theta}\right) + \frac{1}{r^2\sin^2\theta}\frac{\partial^2}{\partial\phi^2} + k^2\right)\psi(r,\theta,\phi) = 0. \tag{3.43}$$

We're going to implement separation of variables here in two stages. Write

$$\psi(r,\theta,\phi) = R(r)Y(\theta,\phi), \tag{3.44}$$

(i.e. separate the radial coordinate from the angular coordinates) in which case we have, substituting Eq. (3.44) into Eq. (3.43):

$$\frac{1}{Rr^2}\frac{\mathrm{d}}{\mathrm{d}r}(r^2R') + \frac{1}{r^2Y}\left[\frac{1}{\sin\theta}\frac{\partial}{\partial\theta}\left(\sin\theta\frac{\partial Y}{\partial\theta}\right) + \frac{1}{\sin^2\theta}\frac{\partial^2 Y}{\partial\phi^2}\right] + k^2 = 0. \tag{3.45}$$

Equation (3.45) suggests that we introduce a separation constant, call it[10] $-\lambda$:

$$\frac{1}{Y}\left[\frac{1}{\sin\theta}\frac{\partial}{\partial\theta}\left(\sin\theta\frac{\partial Y}{\partial\theta}\right) + \frac{1}{\sin^2\theta}\frac{\partial^2 Y}{\partial\phi^2}\right] = -\lambda, \tag{3.46}$$

and thus the radial part satisfies the differential equation

$$R'' + \frac{2}{r}R' + \left(k^2 - \frac{\lambda}{r^2}\right)R = 0. \tag{3.47}$$

Except for a pesky factor of two, Eq. (3.47) looks like Bessel's equation, Eq. (3.38). Equation (3.47) can be put into the form of Eq. (3.38) through a change of variables. Let $R(r) = F(r)/\sqrt{r}$. As can be shown, the function $F(r)$ satisfies

$$F'' + \frac{1}{r}F' + \left(k^2 - \frac{\nu^2}{r^2}\right)F = 0, \qquad \nu \equiv \sqrt{\lambda + \frac{1}{4}} \tag{3.48}$$

which is in the form of Bessel's equation.[11] *The radial function $R(r)$ is then known in terms of Bessel functions.* Assuming $k \neq 0$,

$$R_\nu(r) = A\frac{J_\nu(kr)}{\sqrt{kr}} + B\frac{N_\nu(kr)}{\sqrt{kr}}. \tag{3.49}$$

[10] The separation constant λ is not in the form of the square of a quantity, as have all separation constants introduced up to this point. We'll discuss in Chapter 5 how the value of λ is determined.

[11] Bessel functions are defined for arbitrary values of the parameter ν, and are not restricted to integers as in Eq. (3.38).

What we *don't* know, therefore, is the function $Y(\theta, \phi)$, what's called the *spherical harmonic function.* To solve Eq. (3.46), separate variables again and write

spherical harmonic function

$$Y(\theta, \phi) = P(\theta)\Phi(\phi). \tag{3.50}$$

Substituting Eq. (3.50) into Eq. (3.46),

$$\frac{1}{\sin\theta}\frac{1}{P(\theta)}\frac{\mathrm{d}}{\mathrm{d}\theta}(\sin\theta P') + \frac{1}{\sin^2\theta}\frac{1}{\Phi}\Phi'' = -\lambda. \tag{3.51}$$

Introduce a separation constant. Let $\Phi''/\Phi = -m^2$, which has as a solution (just as with Eq. (3.32)) $\Phi(\phi) = \mathrm{e}^{\mathrm{i}m\phi}$ with m an integer (if Φ has unrestricted domain). Equation (3.51) is therefore equivalent to

$$P''(\theta) + \cot(\theta)P'(\theta) + \left(\lambda - \frac{m^2}{\sin^2\theta}\right)P(\theta) = 0. \qquad (m = \text{integer}) \tag{3.52}$$

associated Legendre differential equation

Legendre differential equation

Equation (3.52) is the *associated Legendre differential equation.* For $m = 0$, it's called simply the Legendre differential equation. It turns out that the solutions to the associated Legendre equation, Eq. (3.52) can be obtained from the solutions to the Legendre equation (Chapter 6). Equation (3.52) is likely an unfamiliar differential equation because of the variable coefficients. We show in Chapter 6 that Eq. (3.52) has well-behaved solutions (finite for $0 \le \theta \le \pi$) only for special values of λ – those generated by the formula $\lambda = l(l+1)$ for $l = 0, 1, 2, 3, \ldots$. The solutions to Eq. (3.52) are therefore labeled by two integers, $P_l^m(\theta)$.

With $\lambda = l(l+1)$, the index ν in Eq. (3.49) (the *order* of the Bessel function) simplifies: $\nu = \sqrt{\lambda + \frac{1}{4}} = l + \frac{1}{2}$. Bessel functions of half-integer order are related to another type of Bessel function, known as *spherical Bessel functions* (Chapter 7).

3.3 The paraxial approximation

An approximation of the Helmholtz equation finding wide use in diffraction optics is known as the *paraxial approximation* [16, p. 145]. It's a separation of variables technique *of sorts.* Write the function $R(\boldsymbol{r})$ in Eq. (3.20) in the following form:

$$R(x, y, z) = u(x, y, z)\mathrm{e}^{\mathrm{i}kz}, \tag{3.53}$$

where $u(x, y, z)$ is an unknown function. Substituting Eq. (3.53) in Eq. (3.20), we find (check it!)

$$\frac{\partial^2}{\partial z^2}u + \nabla_\perp^2 u + 2\mathrm{i}k\frac{\partial}{\partial z}u = 0, \tag{3.54}$$

where $\nabla_\perp^2 \equiv \partial^2/\partial x^2 + \partial^2/\partial y^2$. Equation (3.54) is equivalent to the Helmholtz equation, Eq. (3.20); no approximations have been made yet. The approximation consists of the following assumptions:

1. $\left|\frac{\partial^2 u}{\partial z^2}\right| \ll \left|2k\frac{\partial u}{\partial z}\right| = \frac{4\pi}{\lambda}\left|\frac{\partial u}{\partial z}\right|,$

   where $k = 2\pi/\lambda$ with λ the wavelength of the spatial part of a wave that's modeled by the Helmholtz equation;

2. and $\left|\frac{\partial^2 u}{\partial z^2}\right| \ll \left|\frac{\partial^2 u}{\partial x^2}\right|, \left|\frac{\partial^2 u}{\partial y^2}\right|.$

The first approximation is valid if the spatial variation of u, as measured by $\partial u/\partial z$, is small over the nominal wavelength of the field. The second approximation is justified in applications involving optical beams. Assuming the validity of these assumptions, Eq. (3.54) can be replaced with

$$\left(\nabla_\perp^2 + 2ik\frac{\partial}{\partial z}\right) u(x, y, z) = 0, \tag{3.55}$$

the paraxial approximation of the Helmholtz equation.

3.4 The three types of linear PDEs

The material in this section is optional, and could be skipped on a first reading – in other words, read it when you need it. Table 3.1 lists the most commonly encountered PDEs in applications. These equations, however, are not in the most general form a second-order PDE can have. Restricting ourselves to two independent variables, second-order PDEs can have the form:

$$\left[R(x,y)\frac{\partial^2}{\partial x^2} + S(x,y)\frac{\partial^2}{\partial x \partial y} + T(x,y)\frac{\partial^2}{\partial y^2}\right]\Phi(x,y) = F(x, y, \Phi, \Phi_x, \Phi_y), \tag{3.56}$$

where the function F is known, and which possibly contains first-order partial derivatives, denoted $\Phi_x \equiv \partial\Phi/\partial x$ and $\Phi_y \equiv \partial\Phi/\partial y$ (a fairly standard notation). If confronted with Eq. (3.56), with its variable coefficient functions $R(x, y)$, $S(x, y)$, $T(x, y)$, a question one might reasonably ask (to one's advantage): Can coordinate systems be found in which Eq. (3.56) simplifies to a more manageable form (such as those in Table 3.1)? The answer is in the affirmative, as we now show.

A transformation from (x, y) as the independent variables to a new set of variables (ξ, η) is specified by functions[12] $\xi = \xi(x, y)$ and $\eta = \eta(x, y)$. It can be shown (through systematic use of the chain rule, Exercise 3.6) that under the change of coordinates $\xi = \xi(x, y)$ and $\eta = \eta(x, y)$, Eq. (3.56) implies

$$\left[A(\xi_x, \xi_y)\frac{\partial^2}{\partial \xi^2} + B(\xi_x, \xi_y; \eta_x, \eta_y)\frac{\partial^2}{\partial \xi \partial \eta} + A(\eta_x, \eta_y)\frac{\partial^2}{\partial \eta^2}\right]\phi(\xi, \eta) = f(\xi, \eta, \phi, \phi_\xi, \phi_\eta), \tag{3.57}$$

[12] Coordinate transformations are discussed in Section 11.2.2.

where $\phi(\xi, \eta) = \Phi(x, y)$, and A and B are the functions

$$A(u, v) \equiv Ru^2 + Suv + Tv^2$$
$$B(u_1, v_1; u_2, v_2) \equiv 2Ru_1u_2 + S(u_1v_2 + u_2v_1) + 2Tv_1v_2, \tag{3.58}$$

and where the quantities u_1, u_2, v_1, v_2 are "placeholders" for the variables in Eq. (3.57). How can ξ and η can be chosen so that Eq. (3.57) takes the simplest form possible? We'll be guided by an analogy: It's known from the theory of conic sections that the quadratic equation $Ax^2 + Bxy + Cy^2 = 0$ describes a hyperbola, a parabola, or an ellipse depending on whether the *discriminant* $B^2 - 4AC$, is respectively positive, zero, or negative. The discriminant of the *quadratic form*[13] $A(u, v)$ in Eq. (3.58), $S^2 - 4RT$, plays a similar role in classifying PDEs. It can be shown (Exercise 3.7) that the discriminant of the coefficient functions from Eq. (3.57) is related to that from Eq. (3.56) through the relation

$$B^2(u_1, v_1; u_2, v_2) - 4A(u_1, v_1)A(u_2, v_2) = (S^2 - 4RT)(u_1v_2 - v_1u_2)^2. \tag{3.59}$$

Equation (3.59) implies that the *sign* of the discriminant is the same in all coordinate systems connected by invertible coordinate transformations (Exercise 3.8). Said differently, the sign of the discriminant is an *intrinsic* property of the PDE, independent of the coordinate system used. The three possible values of the discriminant: positive; zero; and negative, imply there are three (and only three) possible types of linear PDEs.

three types of linear PDEs

3.4.1 Hyperbolic PDEs

If $S^2 - 4RT > 0$, the quadratic equation

$$R\lambda^2 + S\lambda + T = 0 \tag{3.60}$$

has two real roots, $\lambda_\pm(x, y) = (-S \pm \sqrt{S^2 - 4RT})/(2R)$. We can make the coefficients of $\partial^2/\partial\xi^2$ and $\partial^2/\partial\eta^2$ *vanish* if ξ and η can be found such that

$$\frac{\partial \xi}{\partial x} = \lambda_\pm(x, y)\frac{\partial \xi}{\partial y} \qquad \frac{\partial \eta}{\partial x} = \lambda_\mp(x, y)\frac{\partial \eta}{\partial y}. \tag{3.61}$$

To see this, note that, from Eq. (3.58),

$$A(\xi_x, \xi_y) = R(\xi_x)^2 + S\xi_x\xi_y + T(\xi_y)^2 = (\xi_y)^2\left[R\left(\frac{\xi_x}{\xi_y}\right)^2 + S\left(\frac{\xi_x}{\xi_y}\right) + T\right]. \tag{3.62}$$

Substituting $\xi_x/\xi_y = \lambda_\pm$ into Eq. (3.62), we have $A(\xi_x, \xi_y) = 0$ from Eq. (3.60). By the same reasoning, $A(\eta_x, \eta_y) = 0$ for $\eta_x/\eta_y = \lambda_\mp$.

---

[13] A quadratic *form* is a polynomial of degree two in several variables that's not set equal to something (in which case it becomes a quadratic equation).

Equation (3.61) involves *first-order* PDEs, which are equivalent to certain first-order ODEs [17, p. 62]. Quantities ξ and η satisfying Eq. (3.61) can be any functions f and g of x and y (for α and β constants),[14] such that

$$\xi = f(x, y) = \alpha \qquad \eta = g(x, y) = \beta, \tag{3.63}$$

where y and x are connected through the ODEs

$$\frac{\mathrm{d}y}{\mathrm{d}x} + \lambda_{\pm}(x, y) = 0. \tag{3.64}$$

To see how this comes about, consider that for $\xi = $ constant,

$$\mathrm{d}\xi = \xi_x \,\mathrm{d}x + \xi_y \,\mathrm{d}y = 0 \implies \frac{\mathrm{d}y}{\mathrm{d}x} = -\frac{\xi_x}{\xi_y} = -\lambda_{\pm}(x, y), \tag{3.65}$$

characterisic curves

where we've used Eq. (3.61). The ODEs in Eq. (3.64) are called the *characteristic equations* of the PDEs in Eq. (3.61), the solutions of which, when written in the form of Eq. (3.63), are called the *characteristic curves.*

With Eq. (3.61) satisfied, $A(\xi_x, \xi_y) = A(\eta_x, \eta_y) = 0$, implying from Eq. (3.59) that $B^2 > 0$, i.e. $B \neq 0$. We can thus divide Eq. (3.57) by B to conclude that Eq. (3.56) can be reduced to the form

$$\frac{\partial^2 \phi}{\partial \xi \partial \eta} = \Phi(\xi, \eta, \phi, \phi_\xi, \phi_\eta) \tag{3.66}$$

hyperbolic PDE

when the discriminant is positive. PDEs in the form of Eq. (3.66) are said to be *hyperbolic.*

Example. The wave equation is a hyperbolic PDE,

$$\left(\frac{\partial^2}{\partial x^2} - \frac{1}{c^2} \frac{\partial^2}{\partial t^2} \right) \psi(x, t) = 0. \tag{3.67}$$

Comparing with Eq. (3.56), $R = 1$, $S = 0$, $T = -1/c^2$. Thus, $S^2 - 4RT > 0$. The roots of Eq. (3.60) are $\lambda_{\pm} = \pm 1/c$, implying that the solutions to Eq. (3.64) are $x = \pm ct$. We can take $\xi = x - ct$ and $\eta = x + ct$ as the characteristic curves. The wave equation thus reduces to an equation of the form of Eq. (3.66),

$$\frac{\partial^2 \phi}{\partial \xi \partial \eta} = 0,$$

the solution of which is simply $\phi(\xi, \eta) = f(\xi) + g(\eta)$, where f and g are differentiable functions. The solution to the wave equation as originally stated is then

$$\psi(x, t) = f(x - ct) + g(x + ct). \tag{3.68}$$

The functions f and g are determined from boundary conditions.

---

[14] Whereas the solutions of ODEs involve arbitrary constants, the solutions of PDEs involve arbitrary functions – functions that are found by imposing boundary conditions.

3.4.2 Parabolic PDEs

If $S^2 - 4RT = 0$, the roots of Eq. (3.60) are equal, implying that we can force *either* of the coefficients of $\partial^2/\partial\xi^2$ and $\partial^2/\partial\eta^2$ to vanish, but not both. We can choose ξ, for example, so that $A(\xi_x, \xi_y) = 0$, implying from Eq. (3.59) that $B(\xi_x, \xi_y; \eta_x, \eta_y) = 0$. The quantity η can be any function of x and y that's independent of ξ. In that way, $A(\eta_x, \eta_y) \neq 0$ because otherwise η would be a function of ξ. Dividing by $A(\eta_x, \eta_y)$, Eq. (3.57) reduces to

$$\frac{\partial^2\phi}{\partial\eta^2} = \Phi(\xi, \eta, \phi, \phi_\xi, \phi_\eta). \tag{3.69}$$

PDEs in the form of Eq. (3.69) are said to be *parabolic*. parabolic PDE

Example. The diffusion equation naturally occurs in the form of a parabolic PDE:

$$\frac{\partial^2\phi}{\partial x^2} = \frac{1}{D}\frac{\partial\phi}{\partial t}, \tag{3.70}$$

where D is the diffusion constant. Here, the function F is simply $\frac{1}{D}\partial\phi/\partial t$.

3.4.3 Elliptic PDEs

If $S^2 - 4RT < 0$, the roots of Eq. (3.60) are a complex-conjugate pair. Thus, Eq. (3.57) can be reduced to a PDE in the form of Eq. (3.66), with the exception that ξ and η are complex conjugates, $\eta = \xi^*$. Define the independent real variables α and β,

$$\alpha \equiv \frac{1}{2}(\eta + \xi) \qquad \beta \equiv \frac{\mathrm{i}}{2}(\eta - \xi).$$

As is straightforward to show,

$$\frac{\partial^2}{\partial\xi\partial\eta} = \frac{1}{4}\left(\frac{\partial^2}{\partial\alpha^2} + \frac{\partial^2}{\partial\beta^2}\right), \tag{3.71}$$

implying that Eq. (3.56) can be reduced to the form

$$\left(\frac{\partial^2}{\partial\alpha^2} + \frac{\partial^2}{\partial\beta^2}\right)\phi(\alpha, \beta) = \Phi(\alpha, \beta, \phi, \phi_\alpha, \phi_\beta). \tag{3.72}$$

PDEs in the form of Eq. (3.72) are said to be *elliptic*. elliptic PDE

Example. The Laplace equation naturally occurs in the form of an elliptic PDE:

$$\left(\frac{\partial^2}{\partial x^2} + \frac{\partial^2}{\partial y^2}\right)\phi(x, y) = 0.$$

3.5 Outlook

Separation of variables is a great technique – it reduces a difficult problem, a PDE, to a set of more manageable problems, ODEs. It's not a general method, however. A PDE expressed in a particular coordinate system is said to be *separable* if it can be reduced to separate ODEs in those coordinates; otherwise, it's *nonseparable*. We've just seen that the Helmholtz equation is separable in rectangular, cylindrical, and spherical coordinates. In general, the Helmholtz equation is separable in eleven different coordinate systems [18, pp. 494–523, 655–666] – only three of which are used with any regularity.

The most commonly encountered PDEs can be reduced to the Helmholtz equation, Eq. (3.20). The Helmholtz equation in turn can be reduced, in separable coordinate systems, to three ODEs. In doing so, we encountered (in Cartesian, cylindrical, and spherical coordinates) four types of functions, one familiar and three likely unfamiliar.

- Complex exponentials, $e^{im\phi}$, where m is often an integer. That's the familiar.
- Bessel functions, of which there are two kinds, $J_k(x)$, *the* Bessel function, and the Neumann function $N_k(x)$, a second, linearly independent solution of the Bessel differential equation. The order k is most often an integer or half-integer.
- Spherical harmonics, functions of the spherical angles, $Y(\theta, \phi) = P(\theta)e^{im\phi}$, where $P(\theta)$ is the solution of the associated Legendre differential equation, Eq. (3.52).
- Spherical Bessel functions, which are related to Bessel functions of half-integer order, $J_{l+\frac{1}{2}}(x)$. It's true that if we understood Bessel functions of arbitrary order, we'd understand Bessel functions of half-integer order, but their properties are sufficiently specialized that it's convenient to give them a separate name.

A major task of this book is dedicated to fleshing out the solutions of the Bessel and Legendre differential equations. It's an arduous undertaking, but one well worth the effort. We'll take it a step at a time, but in the end, dear aspiring student of advanced science, you'll be expected to know these functions, as well as you know the other transcendental functions.

Summary

- The method of separation of variables reduces a PDE in n variables to n ODEs. Such a "separation" occurs at the expense of introducing $n - 1$ separation constants, which are determined through the imposition of boundary conditions.

- The homogeneous PDEs in Table 3.1 reduce to the Helmholtz equation by writing $\psi(\boldsymbol{r},t) = R(\boldsymbol{r})T(t)$. The function $T(t)$ satisfies a simple ODE with solutions given either by Eqs. (3.19) or (3.22). The hard work is in solving the Helmholtz equation, the solutions of which depend on the coordinate system employed.
- The new kids on the block are (in all likelihood) the Bessel and Legendre differential equations, Eq. (3.38) and Eq. (3.52). Upcoming chapters are devoted to solving these differential equations and to the properties of their solutions, the Bessel functions and the Legendre polynomials. The goal has been to provide an early introduction to these functions to acclimate you; Bessel functions and Legendre polynomials are part of the "lexicon" of physical applications.

Exercises

3.1. The following equation is a solution to the one-dimensional homogeneous diffusion equation:

$$\psi(x,t) = \frac{1}{\sqrt{4\pi Dt}} \int_{-\infty}^{\infty} \mathrm{d}x' \exp\left(-\frac{(x-x')^2}{4Dt}\right) \psi(x',0). \tag{3.73}$$

(a) Show this. Verify that Eq. (3.73) solves the differential equation

$$\frac{\partial^2 \psi}{\partial x^2} - \frac{1}{D}\frac{\partial \psi}{\partial t} = 0. \tag{3.74}$$

You should find that all parts of Eq. (3.73) are required to satisfy the diffusion equation, except apparently the factor of 4π.

(b) Use the Dirac delta function to show that $\psi(x,t)$ in Eq. (3.73) properly reduces to the correct initial value $\psi(x,0)$ as $t \to 0$ (and hence the factor of 4π is necessary). Use the sequence of functions in Eq. (2.36), but instead of $\delta_n(x) = (\sqrt{n/\pi})\mathrm{e}^{-nx^2}$ as $n \to \infty$, rewrite it as $\delta_\epsilon(x) = (1/\sqrt{\pi\epsilon})\exp(-x^2/\epsilon)$ as $\epsilon \to 0$.

3.2. Show that Eq. (3.48) follow from Eq. (3.47) with the substitution $R(r) = F(r)/\sqrt{r}$.

3.3. Fill in the steps leading to Eq. (3.49).

3.4. Derive Eq. (3.52) from Eq. (3.51).

3.5. Show that Eq. (3.52) can be written, under the change of variables $x = \cos\theta$,

$$\frac{\mathrm{d}}{\mathrm{d}x}\left[(1-x^2)P'(x)\right] + \left(\lambda - \frac{m^2}{1-x^2}\right)P(x) = 0. \tag{3.75}$$

Hint: Use the chain rule to transform the derivatives:

$$\frac{\mathrm{d}}{\mathrm{d}\theta} = \frac{\mathrm{d}x}{\mathrm{d}\theta}\frac{\mathrm{d}}{\mathrm{d}x} = -\sin\theta\frac{\mathrm{d}}{\mathrm{d}x} = -\sqrt{1-x^2}\frac{\mathrm{d}}{\mathrm{d}x}.$$

3.6. Consider an invertible coordinate transformation $(x, y) \leftrightarrow (x', y')$, with transformation equations $x' = x'(x, y)$, $y' = y'(x, y)$, and $x = x(x', y')$, $y = y(x', y')$. Show that under the transformation, the second partial derivative transforms as:

$$\frac{\partial^2}{\partial x^2} = \left(\frac{\partial x'}{\partial x}\right)^2 \frac{\partial^2}{\partial x'^2} + \left(\frac{\partial y'}{\partial x}\right)^2 \frac{\partial^2}{\partial y'^2} + 2\frac{\partial x'}{\partial x}\frac{\partial y'}{\partial x}\frac{\partial^2}{\partial y' \partial x'}$$
$$+ \frac{\partial^2 x'}{\partial x^2}\frac{\partial}{\partial x'} + \frac{\partial^2 y'}{\partial x^2}\frac{\partial}{\partial y'}.$$

Hint: By the chain rule,

$$\frac{\partial}{\partial x} = \frac{\partial x'}{\partial x}\frac{\partial}{\partial x'} + \frac{\partial y'}{\partial x}\frac{\partial}{\partial y'} \qquad \frac{\partial}{\partial y} = \frac{\partial x'}{\partial y}\frac{\partial}{\partial x'} + \frac{\partial y'}{\partial y}\frac{\partial}{\partial y'}.$$

The second derivative therefore transforms under the chain rule as

$$\frac{\partial^2}{\partial x^2} = \frac{\partial}{\partial x}\frac{\partial}{\partial x} = \frac{\partial}{\partial x}\left(\frac{\partial x'}{\partial x}\frac{\partial}{\partial x'} + \frac{\partial y'}{\partial x}\frac{\partial}{\partial y'}\right)$$
$$= \frac{\partial x'}{\partial x}\frac{\partial}{\partial x}\frac{\partial}{\partial x'} + \frac{\partial^2 x'}{\partial x^2}\frac{\partial}{\partial x'} + \frac{\partial y'}{\partial x}\frac{\partial}{\partial x}\frac{\partial}{\partial y'} + \frac{\partial^2 y'}{\partial x^2}\frac{\partial}{\partial y'}$$

Now substitute the relations for $\partial/\partial x$ and $\partial/\partial y$ from the chain rule.

3.7. Derive Eq. (3.59) starting from the definitions in Eq. (3.58).

3.8. (a) Verify the following equation:

$$\begin{pmatrix} A(\xi_x, \xi_y) & \frac{1}{2}B \\ \frac{1}{2}B & A(\eta_x, \eta_y) \end{pmatrix} = \begin{pmatrix} \xi_x & \xi_y \\ \eta_x & \eta_y \end{pmatrix} \begin{pmatrix} R & \frac{1}{2}S \\ \frac{1}{2}S & T \end{pmatrix} \begin{pmatrix} \xi_x & \eta_x \\ \xi_y & \eta_y \end{pmatrix},$$

where the functions A and B are given in Eq. (3.58).

(b) By taking the determinant of both sides of this matrix equation, show that

$$B^2 - 4A(\xi_x, \xi_y)A(\eta_x, \eta_y) = (S^2 - 4RT)J^2, \qquad (3.76)$$

where $J \equiv \begin{vmatrix} \xi_x & \xi_y \\ \eta_x & \eta_y \end{vmatrix}$ is the Jacobian determinant. Equation (3.76) is the same as Eq. (3.59). The sign of the discriminant is therefore invariant under coordinate transformations with $J \neq 0$.

3.9. Derive Eq. (3.71).

3.10. Show that the characteristics of hyperbolic PDEs with constant coefficients are linear functions, $\xi = ax + by$.

4

Fourier analysis

In Section 3.2.1, we saw that elementary solutions to the Helmholtz equation in rectangular coordinates are in the form of complex exponentials $e^{i\alpha x}$, where α is often an integer. Moreover, a principal result of Chapter 2 is that the eigensolutions to Sturm–Liouville systems form a complete orthonormal set of basis functions that can be used to represent square-integrable functions. Evidently, since the Helmholtz equation is linear, *general* solutions can be expressed as a superposition of eigensolutions. The question is, then, under what conditions are the elementary solutions also eigensolutions? (And the answer to this question, of course, depends on boundary conditions.)

Section 2.2 calls out three types of boundary conditions applicable to the Sturm–Liouville eigenproblem: Dirichlet; Neumann; and periodic. We begin in this chapter by considering periodic boundary conditions (and, more generally, *periodic functions*). Periodic functions *repeat* their values over regular intervals, or *periods* (see Figure 4.1).

Many phenomena in nature exhibit periodic behavior, either in space or time, or both, and are described by periodic functions. Periodic functions are a special class of functions; most functions are *aperiodic*. We develop a powerful method for analyzing periodic functions known as *Fourier series*. Then, we show how Fourier series can be extended to *Fourier integrals*, which apply to aperiodic functions.

4.1 Fourier series

Definition. *A function $f(x)$ is* periodic, *with period P, if $f(x+P) = f(x)$ for all values of x in the domain of the function.* periodic function

Mathematical Methods in Physics, Engineering, and Chemistry, First Edition.
Brett Borden and James Luscombe.

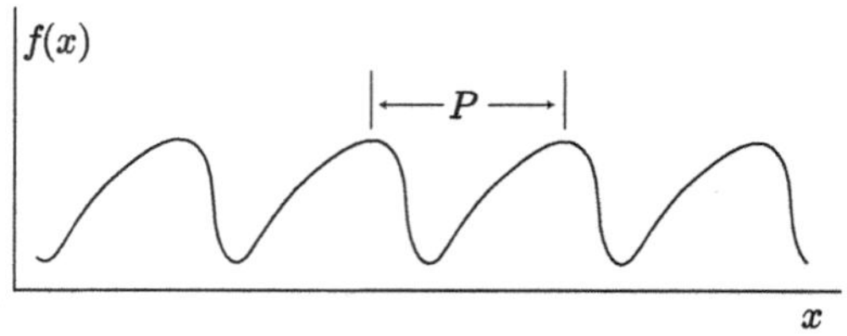

Figure 4.1 Periodic function with period P.

Example. Trigonometric functions are periodic, with period 2π:

$$e^{i(x+2\pi)} = e^{ix}e^{2\pi i} = e^{ix}.$$

We've used $e^{2\pi i} = \cos 2\pi + i \sin 2\pi = 1$. Clearly, $\cos x$ and $\sin x$ are periodic functions with period 2π. The functions $\cos nx$ and $\sin nx$ are periodic with period 2π when n is an integer: $\cos n(x + 2\pi) = \cos(nx + 2\pi n) = \cos nx \cos 2\pi n - \sin nx \sin 2\pi n = \cos nx$. Likewise, $\sin(nx + 2\pi n) = \cos nx \sin 2\pi n + \sin nx \cos 2\pi n = \sin nx$.

The Sturm–Liouville system $y'' + n^2 y = 0$ $(n = 0, 1, 2, \ldots)$ subject to *periodic boundary conditions* specified at $x = 0$ and $x = 2\pi$ has linearly independent solutions $\cos nx$ and $\sin nx$. We therefore expect (based on Eq. (2.37)) that any function y of period 2π can be represented as an expansion in terms of these functions:

Fourier series

$$y(x) = \frac{1}{2}a_0 + \sum_{n=1}^{\infty}[a_n \cos nx + b_n \sin nx]. \tag{4.1}$$

Equation (4.1) is known as a *Fourier series.*[1] This equation clearly represents a function of period 2π because the basis functions have that periodicity; we also know (from Section 2.5), that such a series converges to a piecewise continuous function $y(x)$ in regions where y is continuous, and to the arithmetic mean of the function at points of discontinuity. We obtain the coefficients $\{a_n\}$ and $\{b_n\}$ from the orthogonality of the eigenfunctions. Such orthogonality relations are a consequence of the integral, for integer n and m:

$$\int_0^{2\pi} e^{i(n-m)x}\, dx = 2\pi\delta_{nm}. \tag{4.2}$$

Using Eq. (4.2), we have

$$\int_0^{2\pi} \cos mx \cos nx\, dx = \pi\delta_{nm} \qquad \int_0^{2\pi} \sin nx \sin mx\, dx = \pi\delta_{nm}$$
$$\int_0^{2\pi} \sin nx \cos mx\, dx = 0. \tag{4.3}$$

Multiplying Eq. (4.1) first by $\cos mx$ and then by $\sin mx$, integrating term by term over $(0, 2\pi)$, and using the relations in Eq. (4.3), we obtain

[1] Also known as a trigonometric series.

expressions for the *Fourier coefficients* (the same as what would be obtained from Eq. (2.39)):

Fourier coefficients

$$a_n = \frac{1}{\pi}\int_0^{2\pi} \cos nx\ y(x)\ dx \qquad n = 0, 1, 2, \ldots$$

$$b_n = \frac{1}{\pi}\int_0^{2\pi} \sin nx\ y(x)\ dx. \qquad n = 1, 2, 3, \ldots \tag{4.4}$$

Each of the quantities $\{a_n\}$ and $\{b_n\}$ is a "correlation function" – a measure of the extent to which $y(x)$ resembles either $\cos nx$ or $\sin nx$; the Fourier coefficients are the *projection* of $y(x)$ (considered as a vector) onto the basis "vectors" (functions). Note that the constant term $a_0/2$ in Eq. (4.1) is the *mean value* of the function over a period: From Eq. (4.4), $a_0/2 = (1/(2\pi))\int_0^{2\pi} y(x)\ dx$.

Example. Periodic step function

The periodic step function $f(x)$ is a piecewise continuous function of period 2π (see Figure 4.2), $f(x) = f(x + 2\pi)$, which over one period (specified on $[0, 2\pi]$) has the values

$$f(x) = \begin{cases} 1 & 0 \le x < \pi \\ 0 & \pi \le x < 2\pi. \end{cases} \tag{4.5}$$

The Fourier coefficients are obtained using Eq. (4.4):

$$a_n = \frac{1}{\pi}\int_0^{2\pi} f(x)\cos nx\ dx = \frac{1}{\pi}\int_0^{\pi} \cos nx\ dx = \delta_{n,0},$$

$$b_n = \frac{1}{\pi}\int_0^{2\pi} f(x)\sin nx\ dx = \frac{1}{\pi}\int_0^{\pi} \sin nx\ dx = -\frac{1}{\pi}\frac{\cos nx}{n}\bigg|_0^{\pi}$$

$$= -\frac{1}{n\pi}[(-1)^n - 1] = \begin{cases} 0 & n \text{ even} \\ \frac{2}{n\pi} & n \text{ odd}. \end{cases}$$

The Fourier representation of the periodic step function is then, from Eq. (4.1):

$$f(x) = \frac{1}{2} + \frac{2}{\pi}\left(\sin x + \frac{1}{3}\sin 3x + \frac{1}{5}\sin 5x + \cdots\right). \tag{4.6}$$

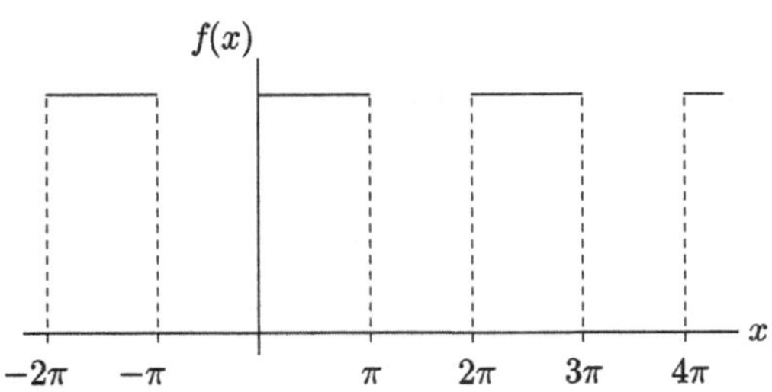

Figure 4.2 Periodic step function.

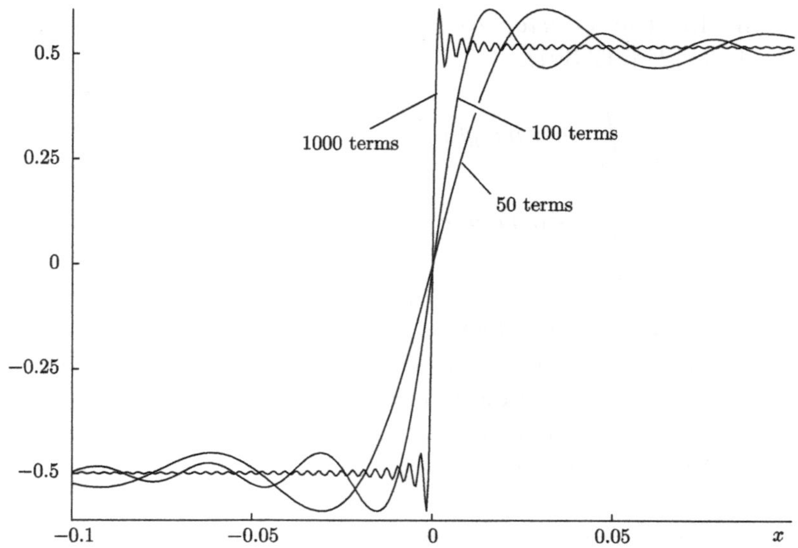

Figure 4.3 Partial sums of the infinite series in Eq. (4.6) in the neighborhood of $x = 0$.

Gibbs phenomenon

The preceding example illustrates the convergence issues mentioned in Section 2.5. Figure 4.3 displays three partial sums of the infinite series in Eq. (4.6) (including the factor of $2/\pi$) containing $N = 50$, 100, and 1000 terms, for values of x near the discontinuity at $x = 0$. The partial sum for any finite N constitutes the best approximation to the function for that number of terms in the series (Section 2.5). In Figure 4.3, we see that the partial sums have relatively large oscillations near $x = 0$, where they produce values both larger than, and less than, the value of the function. This "overshoot" phenomenon (also called the *Gibbs phenomenon*) does not go away with increasing N; every partial sum overshoots the function it is approximating at a discontinuity. We're guaranteed, however, (by the theorem in Section 2.5) that the partial sums converge as $N \to \infty$ to the function in regions where the function is continuous. There's no contradiction here (partial sums having an overshoot, yet as $N \to \infty$ partial sums having no overshoot): The value of x where the maximum overshoot occurs simply moves closer to the discontinuity with increasing N. At the discontinuity, the function converges to the arithmetic mean of the values of the function on the two sides of the discontinuity, zero in this case. Numerically, it is found that the overshoot is about 9% of the correct value of the function.

We can have some fun with Eq. (4.6). For $x = \pi/2$, the step function has the value 1 (Eq. (4.5)). Equation (4.6) implies, for $x = \pi/2$, that

$$\frac{\pi}{4} = 1 - \frac{1}{3} + \frac{1}{5} - \frac{1}{7} + \cdots = \sum_{n=0}^{\infty} \frac{(-1)^n}{2n+1}.$$

The partial sums associated with this series progressively approach the value $\pi/4$.

Fourier cosine and sine series

Fourier cosine and sine series

Suppose the period of our function (of period 2π) was specified not over the interval $[0, 2\pi]$ (as with the periodic step function), but over $[-\pi, \pi]$. The integrals in equations (4.2) and (4.3) are unaltered by changing the limits of integration from 0 and 2π to $-\pi$ and π. The expressions for the Fourier coefficients in Eq. (4.4) are therefore unchanged, save for the limits of integration:

$$a_m = \frac{1}{\pi}\int_{-\pi}^{\pi} f(x)\cos mx \quad dx \quad \text{and} \quad b_m = \frac{1}{\pi}\int_{-\pi}^{\pi} f(x)\sin mx \quad dx. \tag{4.7}$$

even, odd functions

Expressing the Fourier coefficients in terms of integrals with limits of integration symmetric about the origin allows us to exploit *symmetries* of the function. If $f(x)$ is *even* ($f(-x) = f(x)$), then from Eq. (4.7) $b_m = 0$ for all m, and the resulting Fourier series will contain only the terms in $\cos mx$; this is known as a *Fourier cosine series.* Similarly, if $f(x)$ is *odd* ($f(-x) = -f(x)$), all the coefficients a_m vanish, leaving us with a *Fourier sine series.* Thus, when $f(x)$ has such symmetries, we can immediately invoke a Fourier cosine or sine series. Most functions, however, have no particular symmetry, and we must use the full Fourier series, Eq. (4.1).

Example. Full-wave rectifier

Rectifiers are electrical devices that convert alternating currents (AC) (alternately positive and negative) into direct current (DC). The first stage of a *full-wave rectifier* converts the whole of a waveform to one of constant polarity (either positive or negative), as illustrated in Figure 4.4. At this point, the rectified current is not DC (which would be constant and independent of x). Filter circuits are employed to remove the frequency components of the rectified signal. Let's use Fourier theory to find the frequencies that must be removed from the signal to produce DC.

Write the rectified signal in the form

$$f(x) = \begin{cases} \sin \omega x & 0 < \omega x < \pi \\ -\sin \omega x & -\pi < \omega x < 0 \end{cases}$$

where ω denotes the frequency of the current (say, $60 \times 2\pi$ rad/s). Because $f(x)$ is even, no terms of the form $\sin(n\omega x)$ will appear in the series expansion. We have from Eq. (4.7),

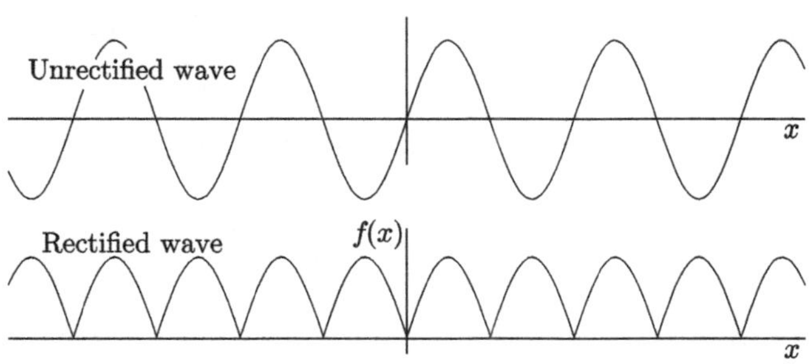

Figure 4.4 Full-wave rectification.

$$a_0 = \frac{1}{\pi}\int_{-\pi}^{0} -\sin\omega x'\,\mathrm{d}(\omega x') + \frac{1}{\pi}\int_{0}^{\pi}\sin\omega x'\,\mathrm{d}(\omega x')$$
$$= \frac{2}{\pi}\int_{0}^{\pi}\sin\omega x'\,\mathrm{d}(\omega x') = \frac{4}{\pi}$$
$$a_n = \frac{2}{\pi}\int_{0}^{\pi}\sin\omega x'\cos(n\omega x)\,\mathrm{d}(\omega x') = \begin{cases} -\frac{2}{\pi}\frac{2}{n^2-1} & n \text{ even} \\ 0 & n \text{ odd} \end{cases}$$

The resulting Fourier series is

$$f(x) = \frac{2}{\pi} - \frac{4}{\pi}\sum_{n=2,4,6,\ldots}^{\infty}\frac{\cos(n\omega x)}{n^2-1}. \tag{4.8}$$

The original frequency ω has been removed by the rectification process, with the lowest frequency now in the current 2ω. The direct current component has magnitude $2/\pi$ and the higher frequencies fall off as n^{-2} (which is pretty fast).

4.2 The exponential form of Fourier series

complex Fourier series

Fourier series for periodic functions of period 2π can also be written in terms of complex exponentials,

$$y(x) = \sum_{n=-\infty}^{\infty} c_n \mathrm{e}^{\mathrm{i}nx}. \tag{4.9}$$

Equation (4.9) can be derived from Eq. (4.1) through the substitutions $\cos nx = \frac{1}{2}(\mathrm{e}^{\mathrm{i}nx} + \mathrm{e}^{-\mathrm{i}nx})$ and $\sin nx = \frac{1}{2\mathrm{i}}(\mathrm{e}^{\mathrm{i}nx} - \mathrm{e}^{-\mathrm{i}nx})$, where $c_n = \frac{1}{2}(a_n - \mathrm{i}b_n)$ (Exercise 4.5). Equation (4.9) can be written directly, however, (without the intermediary of Eq. (4.1)) as an instance of Eq. (2.37), because the functions $\{\mathrm{e}^{\mathrm{i}nx}/\sqrt{2\pi}\}$ are a complete orthonormal set of solutions of the Sturm–Liouville system $y'' + n^2 y = 0$ for integers $n = 0, \pm 1, \pm 2, \ldots$. The coefficients c_n are found in the usual way, from orthogonality. In order to make use of Eq. (4.2), multiply Eq. (4.9) by $\mathrm{e}^{-\mathrm{i}mx}$ and integrate from $-\pi$ to π:

$$\int_{-\pi}^{\pi}\mathrm{e}^{-\mathrm{i}mx}y(x)\,\mathrm{d}x = \sum_{n=-\infty}^{\infty} c_n\int_{-\pi}^{\pi}\mathrm{e}^{\mathrm{i}(n-m)x}\,\mathrm{d}x = \sum_{n=-\infty}^{\infty} 2\pi c_n\delta_{nm} = 2\pi c_m.$$

The Fourier coefficients are thus given by the expression

$$c_n = \frac{1}{2\pi}\int_{-\pi}^{\pi}\mathrm{e}^{-\mathrm{i}nx}y(x)\,\mathrm{d}x. \tag{4.10}$$

Equation (4.10) can also be derived from $c_n = \frac{1}{2}(a_n - \mathrm{i}b_n)$.

The appearance of "negative i" in Eq. (4.10) underscores that the eigenfunctions e^{inx} are elements of a *complex vector space*. Let the normalized eigenfunctions be denoted $|e_n\rangle \equiv e^{inx}/\sqrt{2\pi}$ (Dirac notation), so that, using the inner product for complex spaces (Chapter 1), Eq. (4.2) is simply $\langle e_n|e_m\rangle = \delta_{nm}$. Equation (4.10) can then be seen as the projection of $|y\rangle$ onto $|e_n\rangle$, with $c_n = \langle e_n|y\rangle/\sqrt{2\pi}$.

Completeness of the functions e^{inx}

The completeness relation for the set of orthonormal functions $\{e^{inx}/\sqrt{2\pi}\}$ is completeness

$$\frac{1}{2\pi}\sum_{n=-\infty}^{\infty} e^{in(x-x')} = \delta(x-x'). \tag{4.11}$$

Note that we must sum over positive and negative values of n, i.e. over all eigenfunctions of the Sturm–Liouville equation[2] $y'' = -n^2 y$. Completeness relations, which emerge from the internal consistency of equations like Eq. (2.40), are difficult to show for any given set of functions. It's possible in the case of complex exponentials, however, to demonstrate Eq. (4.11).

To show that the sum in Eq. (4.11), which we can write as

$$\frac{1}{2\pi}\sum_{n=-\infty}^{\infty} e^{in\alpha} = \delta(\alpha), \tag{4.12}$$

represents the Dirac delta function, it must be shown to have two properties: We must show that (i) for $\alpha \neq 0$, it sums to zero,

$$\sum_{n=-\infty}^{\infty} e^{in\alpha} = 0 \qquad (\alpha \neq 0) \tag{4.13}$$

and (ii) it has unit area under the curve (Section 2.4), which we can write (for $\epsilon > 0$)

$$\frac{1}{2\pi}\int_{-\epsilon}^{\epsilon}\sum_{n=-\infty}^{\infty} e^{in\alpha} \quad d\alpha = 1. \tag{4.14}$$

The first property is the more difficult to show. We start by breaking up the infinite sum in Eq. (4.13) into two other infinite summations:

$$\sum_{n=-\infty}^{\infty} e^{in\alpha} = 1 + \sum_{n=1}^{\infty} e^{in\alpha} + \sum_{n=1}^{\infty} e^{-in\alpha} = 1 + 2\sum_{n=1}^{\infty}\cos n\alpha, \tag{4.15}$$

so that the sum in Eq. (4.13) produces a real number.

[2]The functions e^{inx} and e^{-inx} are linearly independent (for $n \neq 0$).

To sum $\sum_{n=1}^{\infty} e^{in\alpha}$, we resort to a trick: Define another sum

$$S_a \equiv \sum_{n=1}^{\infty} (a e^{i\alpha})^n, \tag{4.16}$$

where $0 < a < 1$. The sum we want is $\lim_{a \to 1} S_a$. The quantity S_a can be evaluated for any $a < 1$; it's the geometric series. We find

$$S_a = \frac{a e^{i\alpha}}{1 - a e^{i\alpha}}. \tag{4.17}$$

Using Eq. (4.17), we have, mirroring the terms in Eq. (4.15):

$$1 + S_a + S_a^* = \frac{1 - a^2}{1 - 2a\cos\alpha + a^2}. \tag{4.18}$$

The quantity on the right of Eq. (4.18) has the properties we want from a delta function: For $\alpha \neq 0$, the limit $a \to 1$ has the value zero. For $\alpha = 0$, the limit $a \to 1$ does not exist (as noted in Section 2.4). Thus, we have shown Eq. (4.13).

To show the second property, integrate Eq. (4.15):

$$\begin{aligned}\int_{-\epsilon}^{\epsilon} \sum_{n=-\infty}^{\infty} e^{in\alpha} \, d\alpha &= \int_{-\epsilon}^{\epsilon} \left(1 + 2\sum_{n=1}^{\infty} \cos n\alpha\right) d\alpha \\ &= 2\epsilon + 2\sum_{n=1}^{\infty} \int_{-\epsilon}^{\epsilon} \cos n\alpha \, d\alpha = 2\epsilon + 4\sum_{n=1}^{\infty} \frac{\sin n\epsilon}{n}.\end{aligned} \tag{4.19}$$

The infinite series on the right of Eq. (4.19) is known (Exercise 4.7):

$$\sum_{n=1}^{\infty} \frac{\sin nx}{n} = \frac{1}{2}(\pi - x). \qquad (0 < x < 2\pi) \tag{4.20}$$

Apply Eq. (4.20) to Eq. (4.19) for finite $\epsilon > 0$ and we're done; Eq. (4.14) holds. We've therefore shown the completeness relation in Eq. (4.11).

4.3 General intervals

It goes without saying that if all this theory applies only to functions with period $p = 2\pi$, then this Fourier series stuff will have limited utility. Fortunately, everything we've done has been defined in terms of integrals, and integrals allow us to change variables.

Suppose $f(x)$ is a periodic function with period $p = 2l$ so that $f(x + 2nl) = f(x)$. Then, under the change of variables $x = x'l/\pi$, we see that

$$f\frac{(x' + 2n\pi)l}{\pi} = f(x'l/\pi + 2nl) = f(x + 2nl) = f(x) = f(x'l/\pi)$$

and so $f(x'l/\pi)$ is periodic with period 2π. Consequently, we can simply make a change of variables in all of our previous results and obtain Fourier series representations for *any* periodic function. In particular, for the complex Fourier series we obtain: If $f(x)$ is periodic with period $2l$, then $\mathrm{e}^{\mathrm{i}n\pi x/l}$ is also periodic with period $2l$ and

$$f(x) = \sum_{n=-\infty}^{\infty} c_n \mathrm{e}^{\mathrm{i}n\pi x/l} \tag{4.21}$$

where

Fourier series for general intervals

$$c_n = \frac{1}{2\pi}\int_{-\pi}^{\pi} f\left(\frac{x'l}{\pi}\right)\mathrm{e}^{-\mathrm{i}nx'}\,\mathrm{d}x' = \frac{1}{2l}\int_{-l}^{l} f(x)\mathrm{e}^{-\mathrm{i}n\pi x/l}\,\mathrm{d}x. \tag{4.22}$$

In the last step, we used the change of variables $x = x'l/\pi$. It should be clear that the same argument can be applied to sine and cosine series of functions periodic on a general interval to yield

$$f(x) = \frac{a_0}{2} + \sum_{n=1}^{\infty} a_n \cos\left(\frac{n\pi x}{l}\right) + \sum_{n=1}^{\infty} b_n \sin\left(\frac{n\pi x}{l}\right)$$

where

$$a_n = \frac{1}{l}\int_{-l}^{l} f(x)\cos\left(\frac{n\pi x}{l}\right)\mathrm{d}x \quad \text{and} \quad b_n = \frac{1}{l}\int_{-l}^{l} f(x)\sin\left(\frac{n\pi x}{l}\right)\mathrm{d}x. \tag{4.23}$$

Similarly, Eq. (4.2) becomes

$$\frac{1}{2l}\int_{-l}^{l} \mathrm{e}^{\mathrm{i}(n-m)\pi x/l}\,\mathrm{d}x = \delta_{nm}. \tag{4.24}$$

Note that when calculating the coefficients c_n for the complex series, we integrate over one period p of $f(x)$ and divide the integral by p. But when calculating the coefficients for the sine and cosine series, we integrate over one period and divide the integral by $p/2$. This detail is a consequence of the complex series involving a sum over both positive and negative values of n while the sine and cosine series sum only over positive n.

Example. Periodic ramp

Consider the periodic function whose values on the principle interval are $f(x) = x$, $-1 < x < 1$ (see Figure 4.5). This function has period $p = 2$, and its complex Fourier series has coefficients (from Eq. (4.22))

$$c_n = \frac{1}{2}\int_{-1}^{1} x\mathrm{e}^{-\mathrm{i}n\pi x}\,\mathrm{d}x.$$

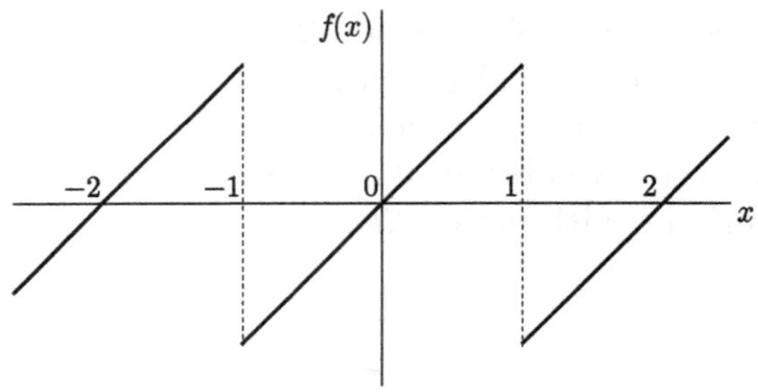

Figure 4.5 Periodic ramp function of period $p = 2$.

When $n = 0$, we see that $c_0 = 0$. For $n \neq 0$, we integrate by parts to obtain

$$c_n = \frac{1}{2}\int_{-1}^{1} x\mathrm{e}^{-\mathrm{i}n\pi x}\,\mathrm{d}x = \frac{1}{2}\left[\frac{x\mathrm{e}^{-\mathrm{i}n\pi x}}{-\mathrm{i}n\pi}\bigg|_{-1}^{1} - \frac{1}{-\mathrm{i}n\pi}\int_{-1}^{1}\mathrm{e}^{-\mathrm{i}n\pi x}\,\mathrm{d}x\right]$$
$$= \frac{\mathrm{i}\cos(n\pi)}{n\pi} = \begin{cases} -\mathrm{i}/n\pi & n \text{ odd} \\ +\mathrm{i}/n\pi & n \text{ even} \end{cases} = (-1)^n\frac{\mathrm{i}}{n\pi}.$$

Consequently,

$$f(x) = \frac{\mathrm{i}}{\pi}\sum_{\substack{n=-\infty \\ n\neq 0}}^{\infty}(-1)^n\frac{\mathrm{e}^{\mathrm{i}n\pi x}}{n}. \tag{4.25}$$

The series in Eq. (4.25) can be split into two series

$$f(x) = \sum_{n=1}^{\infty}(-1)^n\frac{\mathrm{i}\mathrm{e}^{\mathrm{i}n\pi x}}{n\pi} - \sum_{n=1}^{\infty}(-1)^n\frac{\mathrm{i}\mathrm{e}^{-\mathrm{i}n\pi x}}{n\pi}$$

(the second series accounts for all the $-n$ terms). These series can be recombined as follows:

$$f(x) = \sum_{n=1}^{\infty}(-1)^n\frac{\mathrm{i}}{n\pi}(\mathrm{e}^{\mathrm{i}n\pi x} - \mathrm{e}^{-\mathrm{i}n\pi x}) = -2\sum_{n=1}^{\infty}\frac{(-1)^n}{n\pi}\sin(n\pi x)$$

which is a Fourier sine series (as it should be, since $f(x)$ is an odd function).

Example. Temperature distribution in a plate

Let $T(x, y)$ denote the temperature at position (x, y) on a rectangular plate (see Figure 4.6). We know from Eq. (3.6) that the steady-state temperature distribution satisfies the Laplace equation

$$\nabla^2 T = 0 \quad \Rightarrow \quad \frac{\partial^2 T}{\partial x^2} + \frac{\partial^2 T}{\partial y^2} = 0.$$

Suppose that the edges of the plate lie at $x = 0$, $x = L$, $y = 0$, and $y = H$. Suppose also that three of the edges are maintained at zero temperature $T(0, y) = T(L, y) = T(x, 0) = 0$, while the fourth edge is

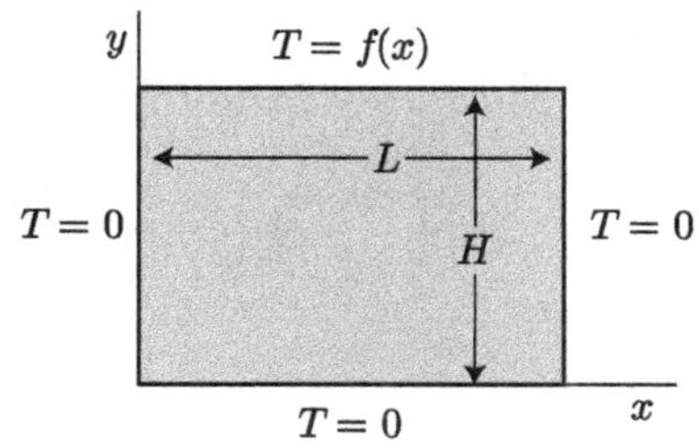

Figure 4.6 Temperature distribution in a rectangular plate.

maintained at the temperature distribution $T(x, H) = f(x)$, where we leave $f(x)$ as an unspecified function. Given these boundary conditions, what is $T(x, y)$ at an arbitrary point?

Separation of variables for the two-dimensional Laplace equation in rectangular coordinates was considered in the last example in Section 3.2.1. We found the general solution

$$T(x, y) = [A\cos(kx) + B\sin(kx)][Ce^{ky} + De^{-ky}]$$

where A, B, C, and D and k are unknown constants.

Now enforce the boundary conditions. At $x = 0$, we want

$$T(0, y) = A[Ce^{ky} + De^{-ky}] = 0. \tag{4.26}$$

Equation (4.26) can be satisfied only if $A = 0$. The general solution is thus reduced to

$$T(x, y) = B\sin(kx)[Ce^{ky} + De^{-ky}].$$

At $x = L$, we want that

$$T(L, x) = B\sin(kL)[Ce^{ky} + De^{-ky}] = 0,$$

implying either $B = 0$ or $\sin(kL) = 0$. Since $B = 0$ results in $T(x, y) = 0$ everywhere, we choose

$$\sin(kL) = 0 \quad \Rightarrow \quad k = \frac{n\pi}{L} \quad n = 1, 2, 3, \ldots .$$

Thus, a boundary condition has been used to determine the separation constant. The question is now, what value of n should we choose? For a given value of n, the function

$$T_n(x, y) \equiv B_n \sin\left(\frac{n\pi x}{L}\right)\left[C_n \exp\left(\frac{n\pi y}{L}\right) + D_n \exp\left(-\frac{n\pi y}{L}\right)\right]$$

satisfies two of the four boundary conditions.

Enforcing the boundary condition at $y = 0$ yields (for all values of n)

$$T_n(x, 0) = B_n \sin\left(\frac{n\pi x}{L}\right)[C_n + D_n] = 0,$$

which will vanish if $C_n = D_n = 0$ (not useful) or $C_n = -D_n$. Consequently, any solution must be of the form:

$$T_n(x,y) = \underbrace{B_n C_n}_{\text{rename this as } a_n/2} \sin\left(\frac{n\pi x}{L}\right)\left[\exp\left(\frac{n\pi y}{L}\right) - \exp\left(-\frac{n\pi y}{L}\right)\right]$$
$$= a_n \sin\left(\frac{n\pi x}{L}\right)\sinh\left(\frac{n\pi y}{L}\right). \tag{4.27}$$

Equation (4.27) satisfies three of the four boundary conditions for each value of n. A superposition of the solutions $T_n(x,y)$, for all values of n, therefore also satisfies three of the four boundary conditions:

$$T(x,y) = \sum_{n=1}^{\infty} a_n \sin\left(\frac{n\pi x}{L}\right)\sinh\left(\frac{n\pi y}{L}\right). \tag{4.28}$$

We have one more boundary condition to try to enforce: $T(x,H) = f(x)$. With $y = H$ in Eq. (4.28), we have

$$T(x,H) = \sum_{n=1}^{\infty}\left\{a_n \sinh\left(\frac{n\pi H}{L}\right)\right\}\sin\left(\frac{n\pi x}{L}\right), \qquad 0 < x < L$$

which we recognize as a Fourier sine series. It follows that the fourth boundary condition will be satisfied if the constant coefficients $b_n \equiv a_n \sinh\left(\frac{n\pi H}{L}\right)$ can be found for which

$$f(x) = \sum_{n=1}^{\infty} b_n \sin\left(\frac{n\pi x}{L}\right).$$

But we know from Eq. (4.23) that

$$b_n = a_n \sinh\left(\frac{n\pi H}{L}\right) = \frac{2}{L}\int_0^L f(x)\sin\left(\frac{n\pi x}{L}\right)\,dx$$

makes this happen and so, solving for the a_n, we conclude

$$T(x,y) = \sum_{n=1}^{\infty}\frac{b_n}{\sinh(n\pi H/L)}\sin\left(\frac{n\pi x}{L}\right)\sinh\left(\frac{n\pi y}{L}\right). \tag{4.29}$$

Equation (4.29) is as far as we can take the solution, until $f(x)$ is specified; in that way, the coefficients b_n would be determined.

Example. Vibrating string

Consider a vibrating string of length l with amplitude $\psi(x,t)$ described by the one-dimensional wave equation, Eq. (3.12), where c is the speed of wave propagation on the string. Assume Dirichlet ("clamped") boundary conditions at $x = 0$ and $x = l$, i.e. $\psi(0, t) = 0$ and $\psi(l, t) = 0$. Separating variables in the usual way, with $\psi(x, t) = X(x)T(t)$, we know that the time-varying part is, from Eq. (3.22), given in terms of oscillatory solutions with frequency $\omega = kc$, where k is the separation constant in the Helmholtz equation, $X''(x) + k^2 X(x) = 0$. The spatial part of this example is similar to that in the previous example (in the x-direction), so we can borrow what we found there. The solution to the wave equation can be written as

$$\psi(x, t) = \sum_{n=1}^{\infty} \sin(k_n x)[A_n \cos(\omega_n t) + B_n \sin(\omega_n t)], \tag{4.30}$$

where $k_n \equiv n\pi/l$ and $\omega_n \equiv k_n c$. Equation (4.30) satisfies the wave equation and the prescribed spatial boundary conditions. What's new in this problem is the role of *initial conditions*. The initial *shape* of the string is, from Eq. (4.30) with $t = 0$,

$$\psi(x, 0) = \sum_{n=1}^{\infty} A_n \sin(k_n x).$$

The initial velocity of the string is, differentiating Eq. (4.30) with respect to time,

$$\left.\frac{\partial \psi}{\partial t}(x, t)\right|_{t=0} = \sum_{n=1}^{\infty} \omega_n B_n \sin(k_n x).$$

The coefficients A_n and B_n are then obtained using the orthogonality of the sine functions:

$$A_n = \frac{2}{l} \int_0^l \sin(n\pi x/l) \psi(x, 0) \, \mathrm{d}x$$

$$B_n = \frac{2}{n\pi c} \int_0^l \sin(n\pi x/l) \left.\frac{\partial \psi}{\partial t}\right|_{t=0} \mathrm{d}x.$$

The solution of the wave equation thus requires initial conditions, in addition to spatial boundary conditions, before a complete solution is known.

4.4 Parseval's theorem

Periodic functions of arbitrary period, $f(x) = f(x + 2l)$, can thus be represented by Fourier series, Eq. (4.21). We know from Section 2.5 that arbitrary square-integrable functions can be represented by eigenfunction expansions, of which the Fourier series is a particular instance (using the basis set $\{e^{in\pi x/l}\}$). The Fourier coefficients, Eq. (4.22), are a measure of the extent to which $f(x)$ has components along each of the basis vectors.

If the basis is *complete*, i.e. the eigenfunction expansion completely constructs the function from a linear combination of the basis set – the set of coefficients $\{c_n\}$ in their entirety should in some sense be equivalent to the function $f(x)$.

This equivalence between a function and its Fourier coefficients is established by integrating the square of the function over one period:

$$\int_{-l}^{l} |f(x)|^2 \,\mathrm{d}x = \int_{-l}^{l} f^*(x) f(x) \,\mathrm{d}x = \sum_{n=-\infty}^{\infty} \sum_{m=-\infty}^{\infty} c_n^* c_m \int_{-l}^{l} \mathrm{e}^{\mathrm{i}(n-m)\pi x/l} \,\mathrm{d}x$$

$$= 2l \sum_{n=-\infty}^{\infty} \sum_{m=-\infty}^{\infty} c_n^* c_m \delta_{nm} = 2l \sum_{n=-\infty}^{\infty} |c_n|^2, \tag{4.31}$$

Parseval's theorem

where we've used Eqs. (4.21) and (4.24). Equation (4.31) is known as *Parseval's theorem* . It's often expressed in the form

$$\sum_{n=-\infty}^{\infty} |c_n|^2 = \frac{1}{2l} \int_{-l}^{l} |f(x)|^2 \,\mathrm{d}x \equiv \overline{|f|^2}, \tag{4.32}$$

showing that the sum of the squares of the Fourier coefficients is equal to the *average value* of the square-magnitude of the function over one period,[3] $\overline{|f|^2}$.

Parseval's theorem is yet another reason for thinking of functions as infinite-dimensional vectors: If $f(x)$ is a vector with component representation $f(x) = \sum_{n=-\infty}^{\infty} c_n \mathrm{e}^{\mathrm{i}n\pi x/l}$ in the orthonormal basis $\{\mathrm{e}^{\mathrm{i}n\pi x/l}\}$, then the magnitude-squared of f should be equal to the sum of the squares of its components.[4] Parseval's theorem is actually another version of completeness[5] – the idea being that the Fourier series expansion in terms of $\mathrm{e}^{\mathrm{i}n\pi x/l}$ correctly represents any expandable function (the basis set spans the Hilbert space of periodic functions with period $p = 2l$). Parseval's theorem is very important in physics and engineering since, in many cases, the *energy* contained in a physical quantity described by the function f is equal to $\overline{|f|^2}$. This energy will be the same no matter which representation is used to compute it.

Example. Amaze your friends using Parseval's theorem. Consider the periodic ramp. The left side of Parseval's theorem in this case has the value $\overline{f^2} = \frac{1}{2} \int_{-1}^{1} x^2 \,\mathrm{d}x = \frac{1}{3}$. Using the coefficients for its Fourier

---

[3] The corresponding relation in terms of the cosine and sine coefficients is $\overline{|f|^2} = (a_0/2)^2 + \frac{1}{2} \sum_{n=1}^{\infty} (|a_n|^2 + |b_n|^2)$ (Exercise 4.5).

[4] By our definition of the inner product of two functions and the fact that $\overline{|f|^2} = \langle f|f \rangle$ for any function f.

[5] A necessary and sufficient condition for a set of orthonormal functions to be complete is that Parseval's equality be obeyed [19, p. 301].

representation in Eq. (4.25), Parseval's theorem leads to the unexpected result

$$\frac{\pi^2}{6} = \sum_{n=1}^{\infty} \frac{1}{n^2}.$$

Who knew?

4.5 Back to the delta function

The Dirac delta function was introduced in Section 2.4 with its properties defined by Eq. (2.35). Here, we develop another representation of the delta function. Consider the function D_t defined by the integral

$$D_t(x - x') \equiv \frac{1}{2\pi} \int_{-t}^{t} \mathrm{e}^{\mathrm{i}(x-x')u} \, \mathrm{d}u = \frac{1}{\pi} \frac{\sin(x - x')t}{x - x'}.$$

We'll show that $\lim_{t\to\infty} D_t(x - x') = \delta(x - x')$. The function D_t has unit area under the curve for all t:

$$\int_{-\infty}^{\infty} D_t(x - x') \, \mathrm{d}x = \frac{1}{\pi} \int_{-\infty}^{\infty} \frac{\sin(x - x')t}{x - x'} \, \mathrm{d}x = \frac{1}{\pi} \int_{-\infty}^{\infty} \frac{\sin y}{y} \, \mathrm{d}y = 1,$$

where we've substituted $y = (x - x')t$, implying $\mathrm{d}y = t \, \mathrm{d}x$, and we've made use of the integral $\int_{-\infty}^{\infty} (\sin x/x) \, \mathrm{d}x = \pi$ (Exercise 4.7).

sinc function

Definition. *The* sinc *function is defined* [6] *as* $\mathrm{sinc}(x) \equiv \sin x/x$.

The sinc function is plotted in Figure 4.7. Note that it has the value 1 at the origin.

The function $D_t(x - x')$ has the "right stuff" for the delta function: $D_t(0) = t/\pi$, while for $x \neq x'$, its width scales like π/t. For large values of t, $D_t(0)$ grows indefinitely, and its width shrinks while maintaining unit area under the curve. Thus, we have a representation of the delta function in terms of an *integral*, a result that we'll use often:

integral representation of delta function

$$\delta(x - x') = \frac{1}{2\pi} \int_{-\infty}^{\infty} \mathrm{e}^{\mathrm{i}(x-x')u} \, \mathrm{d}u. \tag{4.33}$$

Of course, delta functions have meaning only when they show up inside integrals (Section 2.4). Compare Eq. (4.33) with Eq. (4.11); Eq. (4.33) is the continuum version of Eq. (4.11).

[6] Another definition is $\mathrm{sinc}(x) \equiv \sin(\pi x)/(\pi x)$. Its zeros occur at integer values of x.

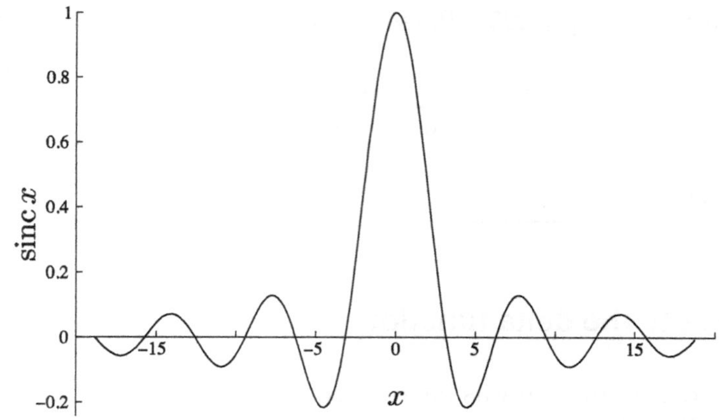

Figure 4.7 Sinc(x) versus x.

Properties of the delta function

properties of $\delta(x)$

We note several useful properties of the delta function:

- *Scaling property:* For constant a, $\int_{-\infty}^{\infty} \delta(ax)f(x) = f(0)/|a|$. Consider separately positive and negative values of a and don't forget to change the limits of integration for negative a. This property is usually written $\delta(ax) = \delta(x)/|a|$.
- *Evenness:* $\delta(-x) = \delta(x)$. Use the scaling property and set $a = -1$.
- *Derivative* δ': Integrate by parts

$$\int_{-\infty}^{\infty} \delta'(x)f(x)\,\mathrm{d}x = \delta(x)f(x)\big|_{-\infty}^{\infty} - \int_{-\infty}^{\infty} \delta(x)f'(x)\,\mathrm{d}x = -f'(0).$$

- *Action on functions:* $\delta(f(x)) = \sum_i \frac{1}{|f'(x_i)|}\delta(x - x_i)$ where the sum is over the simple zeros of the function, $f(x_i) = 0$. This follows because in the vicinity of a zero, $f(x) \approx (x - x_i)f'(x_i)$, and it's only when the delta function encounters the value zero that it "wakes up." In particular, $\delta(x^2 - a^2) = \frac{1}{2|a|}[\delta(x + a) + \delta(x - a)]$.

Relation to the step function

The unit *step function*[7] $\theta(x)$ is defined such that

$$\theta(x) \equiv \begin{cases} 1 & x > 0 \\ 0 & x < 0. \end{cases} \tag{4.34}$$

The step function finds use in many applications as an "on–off" switch. For our purposes, we note that the derivative of $\theta(x)$ is the same as the Dirac delta function,

[7] Also known as the Heaviside function.

$$\theta'(x) = \delta(x). \tag{4.35}$$

To prove this statement, integrate by parts, for any function $f(x)$:

$$\begin{aligned}\int_{-\infty}^{\infty} f(x)\theta'(x)\,\mathrm{d}x &= [\theta(x)f(x)]\Big|_{-\infty}^{\infty} - \int_{-\infty}^{\infty} \theta(x)f'(x)\,\mathrm{d}x\\ &= f(\infty) - \int_{0}^{\infty} f'(x)\,\mathrm{d}x\\ &= f(\infty) - (f(\infty) - f(0)) = f(0).\end{aligned}$$

Comparing with the filtering property of $\delta(x)$ (i.e. $\int_{-\infty}^{\infty} \delta(x)f(x)\,\mathrm{d}x = f(0)$), we have that $\theta'(x) = \delta(x)$.

4.6 Fourier transform

The Fourier series representation of a periodic function is a pretty useful tool, but it's one that applies only to periodic functions. A situation that happens quite often, however, is one in which the values of a function are known only over a finite domain – say, time-varying data measured during a one-minute interval. We can still use the tools of Fourier analysis here if we simply *assume* that the data are periodic outside of the interval (i.e. we consider a *periodic extension* of our measurements[8]).

Another situation for which the Fourier series is an important analysis tool occurs when a function $f(x)$ is of the form

$$f(x) = g(x) + h(x)$$

where $g(x)$ is periodic with period p and $h(x)$ is periodic with period q. It is sometimes useful to expand such functions in a Fourier series, but what period is appropriate?

In the simple case that q is an integer multiple of p, this problem can be examined by Fourier series since when $q = Np$, *both* $g(x)$ and $h(x)$ will repeat with periodicity q and so we can construct a Fourier series using Eq. (4.21) with $2l = q = Np$. Clearly, if p/q a rational number, then we can use the same scheme since it will always be possible to find an l for which $Np = 2l$ and $Mq = 2l$ and thereby construct a Fourier series with components associated with $g(x)$ and $h(x)$.

If we don't know p and q up-front, however, we have a problem with constructing the Fourier series. We could simply choose l to be very large and hope for the best. If, somehow, we could set $2l = \infty$ (which is a horrible abuse of notation, but you get the idea), then we could be assured that if $g(x)$ and $h(x)$ are periodic for *some* p and q, they will be periodic in the interval $2l$.

[8]Note that there are many situations in which this assumption is quite justified, but that discussion would take us too far afield.

Mathematically, the idea works as follows. For the Fourier series of a function of nominal period $2l$, write Eq. (4.21) as

$$f(x) = \sum_{n=-\infty}^{\infty} c_n e^{i\alpha_n x}, \tag{4.36}$$

where $\alpha_n \equiv n\pi/l$. The Fourier coefficients c_n are, from Eq. (4.22),

$$c_n = \frac{1}{2l}\int_{-l}^{l} f(x')e^{-i\alpha_n x'}\,dx' = \frac{\Delta\alpha}{2\pi}\int_{-l}^{l} f(x')e^{-i\alpha_n x'}\,dx',$$

where $\Delta\alpha \equiv \alpha_{n+1} - \alpha_n = \pi/l$. Substituting c_n back into Eq. (4.36), we have

$$\begin{aligned} f(x) &= \sum_{n=-\infty}^{\infty} \frac{\Delta\alpha}{2\pi}\int_{-l}^{l} f(x')e^{-i\alpha_n x'}\,dx' e^{i\alpha_n x} \\ &= \frac{1}{2\pi}\sum_{n=-\infty}^{\infty}\left[\int_{-l}^{l} f(x')e^{-i\alpha_n x'}\,dx'\right] e^{i\alpha_n x}\Delta\alpha \\ &\equiv \frac{1}{2\pi}\sum_{n=-\infty}^{\infty} A(\alpha_n)e^{i\alpha_n x}\Delta\alpha. \end{aligned} \tag{4.37}$$

Now, what happens if l becomes large? The quantities α_n "pack down" as $l \to \infty$ to become a continuous variable,[9] $\alpha_n \to \alpha$. In the limit $l \to \infty$, the sum in Eq. (4.37) is expressed as an integral:

$$f(x) = \lim_{l\to\infty}\frac{1}{2\pi}\sum_{n=-\infty}^{\infty} A(\alpha_n)e^{i\alpha_n x}\Delta\alpha \equiv \frac{1}{2\pi}\int_{-\infty}^{\infty} F(\alpha)e^{i\alpha x}\,d\alpha \tag{4.38}$$

where the function $F(\alpha)$ is the integral,

$$F(\alpha) = \int_{-\infty}^{\infty} f(x')e^{-i\alpha x'}\,dx'. \tag{4.39}$$

Fourier transform

The function $F(\alpha)$ in Eq. (4.39) is known as the *Fourier transform* of $f(x)$. Equations (4.38) and (4.39) are the *continuum generalizations* of Eqs. (4.9) and (4.10), the extension of a Fourier series (that represents functions with finite periods) to a Fourier *integral* that represents any function. Arbitrary functions can be considered periodic functions of infinite period.

[9]It's a subtle mathematical point of how a countably infinite collection of numbers can rigorously be treated as a continuous variable, beyond the scope of the present discussion.

Example. Fourier transform of a cosine: When $f(x) = \cos(\alpha_0 x)$, a straightforward calculation reveals

$$\begin{aligned}F(\alpha) &= \int_{-\infty}^{\infty} \cos(\alpha_0 x)\mathrm{e}^{-\mathrm{i}\alpha x'}\,\mathrm{d}x' = \int_{-\infty}^{\infty} \frac{\mathrm{e}^{\mathrm{i}\alpha_0 x'} + \mathrm{e}^{-\mathrm{i}\alpha_0 x'}}{2}\mathrm{e}^{-\mathrm{i}\alpha x'}\,\mathrm{d}x' \\ &= \frac{2\pi}{2}\frac{1}{2\pi}\int_{-\infty}^{\infty} (\mathrm{e}^{-\mathrm{i}(\alpha-\alpha_0)x'} + \mathrm{e}^{-\mathrm{i}(\alpha+\alpha_0)x'})\,\mathrm{d}x' \\ &= \pi\left(\frac{1}{2\pi}\int_{-\infty}^{\infty} \mathrm{e}^{-\mathrm{i}(\alpha-\alpha_0)x'}\,\mathrm{d}x' + \frac{1}{2\pi}\int_{-\infty}^{\infty} \mathrm{e}^{-\mathrm{i}(\alpha+\alpha_0)x'}\,\mathrm{d}x'\right) \\ &= \pi(\delta(\alpha-\alpha_0) + \delta(\alpha+\alpha_0)),\end{aligned} \tag{4.40}$$

where we've used Eq. (4.33).

So the Fourier transform of a circular function picks out the frequency α_0 of the function. (The transform actually picks out the *two* frequencies α_0 and $-\alpha_0$: since $\cos(\alpha_0 x) = \cos(-\alpha_0 x)$ is an even function then $F(\alpha)$ must also be even.)

Example. An important example is the Fourier transform of the *top hat function*,

$$f(x) = \begin{cases} 1 & -s < x < s \\ 0 & |x| > s, \end{cases}$$

see Figure 4.8. This function is clearly not periodic. Its Fourier transform is, from Eq. (4.39),

$$F(\alpha) = \int_{-\infty}^{\infty} f(x)\mathrm{e}^{-\mathrm{i}\alpha x}\,\mathrm{d}x = \int_{-s}^{s} \mathrm{e}^{-\mathrm{i}\alpha x}\,\mathrm{d}x = \frac{2\sin\alpha s}{\alpha} = 2s\,\mathrm{sinc}(\alpha s). \tag{4.41}$$

Thus, the Fourier transform of the top hat function is the sinc function, Figure 4.7.

This example brings out an important property of *waves*. The parameter α in $\mathrm{e}^{\mathrm{i}\alpha x}$ is (for x a spatial variable) the *wavenumber*, or *spatial frequency*; it's related to the wavelength λ of a wave through $\alpha = 2\pi/\lambda$. Small values of α correspond to waves with long wavelength; large values of α describe waves with short wavelength. The function $F(\alpha)$ in

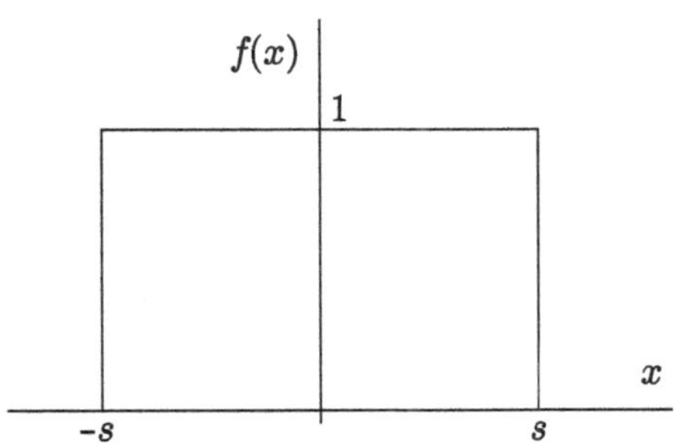

Figure 4.8 The "top hat" function.

Eq. (4.41) represents the spatial *superposition* of waves having different values of α. Waves with $\alpha \approx 0$ add coherently to produce a relatively large value of $F(\alpha)$, but waves for $|\alpha| \gtrsim \pi/s$ add destructively to produce a relatively small value of $F(\alpha)$, as we see in Figure 4.7.

The spatial width of the top hat function is clearly $\Delta x = 2s$. A convenient measure of the width of its Fourier transform is $\Delta\alpha = 2\pi/s$. The product of the two widths, $\Delta x\Delta\alpha = 4\pi$. What's important here is not the precise value of $\Delta x\Delta\alpha$, but rather that it's a *constant.* If Δx is made smaller (smaller values of s), the width of the Fourier transform must become larger to preserve the value of the product $\Delta x\Delta\alpha$. Conversely, if Δx is made larger, the width of the Fourier transform must become smaller. We see this effect in the previous example: A wave of a single frequency, $\cos\alpha_0 x$, of unlimited spatial extent ($\Delta x \to \infty$), has a Fourier transform $\delta(\alpha - \alpha_0)$ of zero width.

This feature ($\Delta x\Delta\alpha =$ constant) is a property of waves, independent of the physical manifestation of the wave (e.g, electromagnetic or acoustic). In quantum mechanics, the de Broglie hypothesis is that a material particle of momentum p is "associated" with a wave of wavelength $\lambda = h/p$, where h is Planck's constant.[10] Turning that relation around, a particle of wavelength λ has momentum $p = \hbar\alpha$, where $\hbar \equiv h/(2\pi)$. Multiplying the general result for waves $\Delta x\Delta\alpha =$ constant by $\hbar$, we have the Heisenberg uncertainty principle $\Delta p\Delta x \sim \hbar$. The uncertainty principle is then nothing new; it's a wave phenomenon. Once we sign-off on the concept of de Broglie waves, the uncertainty principle follows as a consequence.

uncertainty principle

Fourier transform as operator

It's useful to represent the Fourier transform as the result of an *operator.* Equation (4.39) can be written

$$\mathcal{F}f(\alpha) = \int_{-\infty}^{\infty} f(x)\mathrm{e}^{-\mathrm{i}\alpha x}\,\mathrm{d}x = F(\alpha). \tag{4.42}$$

inverse Fourier transform

The integral in Eq. (4.38) that transforms F back into f is known as the *inverse* Fourier transform, which we can write in operator form:

$$\mathcal{F}^{-1}F(x) = \frac{1}{2\pi}\int_{-\infty}^{\infty} F(\alpha)\mathrm{e}^{\mathrm{i}\alpha x}\,\mathrm{d}\alpha = f(x). \tag{4.43}$$

[10]We put "associated" in quotation marks because one can't visualize a de Broglie wave. Particles having an associated wavelength is a feature of the microscopic world supported by countless experiments, but one can't visualize how the association works. The classical concepts of particle (definite location) and wave (spread out) are seemingly incompatible. To quote Dirac [20, p. 10], it's up to us to find ways "of looking at the fundamental laws which makes their self-consistency obvious."

That $\mathcal{F}^{-1}$ is the inverse of $\mathcal{F}$ is readily shown:

$$\begin{aligned}\mathcal{F}(\mathcal{F}^{-1}F)(\alpha) &= \frac{1}{2\pi}\int_{-\infty}^{\infty} \mathrm{e}^{-\mathrm{i}\alpha x}\,\mathrm{d}x \int_{-\infty}^{\infty} F(\alpha')\mathrm{e}^{\mathrm{i}\alpha' x}\,\mathrm{d}\alpha' \\ &= \int_{-\infty}^{\infty} \mathrm{d}\alpha' \left(\frac{1}{2\pi}\int_{-\infty}^{\infty} \mathrm{d}x\mathrm{e}^{\mathrm{i}x(\alpha'-\alpha)}\right) F(\alpha') \\ &= \int_{-\infty}^{\infty} \mathrm{d}\alpha'\delta(\alpha-\alpha')F(\alpha') = F(\alpha), \qquad (4.44)\end{aligned}$$

where we've used Eq. (4.33) once again.[11] The functions $f(x)$ and $F(\alpha)$ related through Eqs. (4.38) and (4.39) are referred to as a *Fourier transform pair*. The relation between them can be symbolized $f \Longleftrightarrow F$.

Fourier transform pair

4.7 Convolution integral

One of the important properties of the Fourier transform is the way it deals with linear operators, and one of the more important of these operators is known as the *convolution*. The convolution integral appears often in physics and engineering. For functions $f(x)$ and $k(x)$, the convolution is defined by

convolution integral

$$m(x) \equiv \int_{-\infty}^{\infty} k(x-s)f(s)\,\mathrm{d}s. \qquad (4.45)$$

The convolution integral is often written symbolically as $m(x) = k(x) * f(x)$. (The convolution is also known as the "resultant" or the "Faltung.") All linear, shift-invariant systems are described by convolutions.

Example. A motivational example.

(Linear) measurement systems are often described by linear integral operators $\mathcal{A}$. The operator $\mathcal{A} : f(x) \mapsto m(x)$ describes the fidelity by which the measurements $m(x)$ capture the true values of an "object" $f(x)$ and (see Eq. (1.12))

$$m(x) = \int_a^b A(x,s)f(s)\,\mathrm{d}s\ , \qquad (4.46)$$

where the function $A(x, s)$ characterizes the measurement system.[12]

---

[11] Some people find the unsymmetrical role of the factor of 2π in the Fourier transform and the inverse operation to be an esthetic irritant and find ways to democratically share it between them. It's only necessary that a factor of $(2\pi)^{-1}$ appear *somewhere* in the two equations. Other equally valid definitions exist.

[12] An "ideal measurement system" will be one for which

$$f(x) = \int_a^b A(x,s)f(s)\,\mathrm{d}s$$

When the system response to a shifted object $f(x-y)$ is itself of the form $m(x-y)$ for all y, the system is said to be *shift-invariant.* Shift invariance guarantees that the system will measure objects the same no matter when they start (or, alternatively, where they're located). When the system is also linear, then we have the important case of a *linear, shift-invariant system.* If Eq. (4.46) represents a linear, shift-invariant system, then

$$m(x-y) = \int_a^b A(x,s)f(s-y)\,\mathrm{d}s = \int_{a-y}^{b-y} A(x,s'+y)f(s')\,\mathrm{d}s'$$

(where we have substituted $s' = s - y$). But straightforward substitution into Eq. (4.46) yields

$$m(x-y) = \int_a^b A(x-y,s)f(s)\,\mathrm{d}s$$

(assuming a and b are independent of x).

Evidently, linear and shift-invariant systems require

$$\int_a^b A(x-y,s)f(s)\,\mathrm{d}s = \int_{a-y}^{b-y} A(x,s+y)f(s)\,\mathrm{d}s.$$

This will be true when $a \to -\infty$, $b \to \infty$, *and* the function A is of the special form

$$A(x,s) = k(x-s)$$

for some function k. Evidently, linear, shift-invariant systems are described by convolutions

$$m(x) = \int_{-\infty}^{\infty} k(x-s)f(s)\,\mathrm{d}s.$$

The Fourier transform of the convolution $m(x)$ is

$$\begin{aligned} M(\alpha) = \int_{-\infty}^{\infty} m(x)\mathrm{e}^{-\mathrm{i}\alpha x}\,\mathrm{d}x &= \int_{-\infty}^{\infty}\left[\int_{-\infty}^{\infty} k(x-s)f(s)\,\mathrm{d}s\right]\mathrm{e}^{-\mathrm{i}\alpha x}\,\mathrm{d}x \\ &= \int_{-\infty}^{\infty} f(s)\left[\int_{-\infty}^{\infty} k(x-s)\mathrm{e}^{-\mathrm{i}\alpha x}\,\mathrm{d}x\right]\mathrm{d}s. \end{aligned}$$

---

for any object $f(x)$, since such a system will tell us everything we want to know about f. Such systems are not generally realizable and, typically, the problem of "measurement" is reduced to determining *something about* the object $f(s)$ from measurements $m(x)$.

Under the substitution $y = x - s$, this equation becomes

$$M(\alpha) = \int_{-\infty}^{\infty} f(s) \left[\int_{-\infty}^{\infty} k(y) \mathrm{e}^{-\mathrm{i}\alpha(y+s)} \, \mathrm{d}y \right] \mathrm{d}s$$

$$= \int_{-\infty}^{\infty} f(s) \mathrm{e}^{-\mathrm{i}\alpha s} \left[\int_{-\infty}^{\infty} k(y) \mathrm{e}^{-\mathrm{i}\alpha y} \, \mathrm{d}y \right] \mathrm{d}s$$

$$= \left[\int_{-\infty}^{\infty} k(y) \mathrm{e}^{-\mathrm{i}\alpha y} \, \mathrm{d}y \right] \left[\int_{-\infty}^{\infty} f(s) \mathrm{e}^{-\mathrm{i}\alpha s} \, \mathrm{d}s \right] = K(\alpha) F(\alpha). \quad (4.47)$$

Equation (4.47) is the *Fourier convolution theorem* and says that a convolution in the spatial domain becomes a simple product in the Fourier domain:

Fourier convolution theorem

$$k(x) * f(x) \Longleftrightarrow K(\alpha) F(\alpha).$$

The convolution theorem works both ways, and it can be shown that a convolution between two functions in the Fourier domain becomes (under inverse Fourier transformation) a simple product in the spatial domain:

$$K(\alpha) * F(\alpha) \Longleftrightarrow k(x) f(x).$$

Example. "Radio" transmission of information usually requires that a time-varying signal $f(t)$ be *modulated* (multiplied) by a carrier signal with a particular frequency ω_0 (the carrier frequency). The transmitted signal is a product of two functions in the time-domain, $f_{\text{trans}}(t) = f(t)\cos(\omega_0 t)$.

The frequency dependence of the transmitted signal will be different from that of $f(t)$ (owing to the modulation). Using the convolution theorem and Eq. (4.40), we have

$$F_{\text{trans}}(\omega) = \int_{-\infty}^{\infty} f(t) \cos(\omega_0 t) \mathrm{e}^{-\mathrm{i}\omega t} \, \mathrm{d}t = F(\omega) * \left\{ \int_{-\infty}^{\infty} \cos(\omega_0 x) \mathrm{e}^{-\mathrm{i}\alpha x} \, \mathrm{d}x \right\}$$

$$= F(\omega) * \{\pi(\delta(\alpha - \omega_0) + \delta(\alpha + \omega_0))\}$$

$$= \frac{1}{2\pi} \int_{-\infty}^{\infty} F(\omega - \beta) \{\pi(\delta(\beta - \alpha_0) + \delta(\beta + \alpha_0))\} \, \mathrm{d}\beta$$

$$= \frac{1}{2} [F(\omega - \omega_0) + F(\omega + \omega_0)]$$

and so the effect of modulation is to shift the frequency dependence of $f(t)$ by $\pm\omega_0$.

Example. The *Discrete Fourier Transform* (DFT) can be developed by considering the situation in which a periodic function is known only at discrete points $x \in \{x_n\}$, $n = 0, \pm 1, \pm 2, \ldots$. (This situation occurs, for example, when the function has been measured at discrete times and is known as a "discrete sampling" of the function.) When the discrete

discrete Fourier transform

points are of the form $x_n = n\Delta x$, then the function $f(x)$ can only be *approximated* by the function[13]

$$f_{\text{sampled}}(x) = f(x) \sum_{n=-\infty}^{\infty} \delta(n\Delta x - x).$$

The Fourier Transform F_{sampled} of f_{sampled} can be defined by this last relation and is related to the Fourier transform $F(\alpha)$ of $f(x)$ by the convolution theorem. To apply the convolution theorem, we must first determine the Fourier transform of

$$k(x) = \sum_{n=-\infty}^{\infty} \delta(n\Delta x - x).$$

We start by recognizing that since $k(x)$ is periodic (with period $p = \Delta x$), it can be represented as a series. From Eqs. (4.21) and (4.22), we obtain

$$c_n = \frac{1}{\Delta x} \int_{-\Delta x/2}^{\Delta x/2} \delta(x) \mathrm{e}^{-\mathrm{i}2\pi n x/\Delta x} \, \mathrm{d}x = \frac{1}{\Delta x}$$

and

$$k(x) = \sum_{n=-\infty}^{\infty} c_k \mathrm{e}^{\mathrm{i}2\pi n x/\Delta x} = \frac{1}{\Delta x} \sum_{n=-\infty}^{\infty} \mathrm{e}^{\mathrm{i}2\pi n x/\Delta x}.$$

Then, the Fourier transform of $k(x)$ follows as

$$\begin{aligned} K(\alpha) &= \int_{-\infty}^{\infty} k(x') \mathrm{e}^{-\mathrm{i}\alpha x} \, \mathrm{d}x = \int_{-\infty}^{\infty} \left\{ \frac{1}{\Delta x} \sum_{n=-\infty}^{\infty} \mathrm{e}^{\mathrm{i}2\pi n x/\Delta x} \right\} \mathrm{e}^{-\mathrm{i}\alpha x} \, \mathrm{d}x \\ &= \frac{2\pi}{\Delta x} \sum_{n=-\infty}^{\infty} \frac{1}{2\pi} \int_{-\infty}^{\infty} \mathrm{e}^{-\mathrm{i}x(\alpha - 2\pi n/\Delta x)} \, \mathrm{d}x \\ &= \frac{2\pi}{\Delta x} \sum_{n=-\infty}^{\infty} \delta\left(\alpha - \frac{2\pi n}{\Delta x}\right), \end{aligned}$$

where we've used Eq. (4.33).

[13]The function $f_{\text{sampled}}(x)$ is defined for *all* x and so Fourier transform analysis is appropriate. $f_{\text{sampled}}(x)$ is not generally equal to $f(x)$, however (it's equal to 0 when $x \neq n\Delta x$) and so may not be a particularly good approximation to $f(x)$ in practical circumstances. Such questions are answered by so-called *sampling theorems*, and this topic is beyond the scope of our discussion.

Finally, we can now apply the convolution theorem (using this last result) to obtain

$$
\begin{aligned}
F_{\text{sampled}} &= F(\alpha) * K(\alpha) = \frac{1}{2\pi}\int_{-\infty}^{\infty} F(\alpha-\beta)K(\beta)\,\mathrm{d}\beta \\
&= \frac{1}{2\pi}\int_{-\infty}^{\infty} F(\alpha-\beta)\left\{\frac{2\pi}{\Delta x}\sum_{n=-\infty}^{\infty}\delta\left(\beta-\frac{2\pi n}{\Delta x}\right)\right\}\,\mathrm{d}\beta \\
&= \frac{1}{\Delta x}\sum_{n=-\infty}^{\infty}\int_{-\infty}^{\infty} F(\alpha-\beta)\delta\left(\beta-\frac{2\pi n}{\Delta x}\right)\,\mathrm{d}\beta \\
&= \frac{1}{\Delta x}\sum_{n=-\infty}^{\infty} F\left(\alpha-\frac{2\pi n}{\Delta x}\right)
\end{aligned}
$$

and, so, the Fourier transform of a discrete sampling of the function $f(x)$ is a superposition of discrete shifts of its Fourier transform $F(\alpha)$.

Summary

- Periodic functions $f(x) = f(x + 2\pi)$ can be represented by Fourier series, in either of the equivalent forms Eq. (4.1) or Eq. (4.9). A Fourier series is "just" an eigenfunction expansion (Eq. (2.37)) in the complete orthonormal basis $\{\mathrm{e}^{inx}/\sqrt{2\pi}\}$. Fourier series can represent periodic functions of arbitrary periodicity, $f(x) = f(x + 2l)$ using a modified basis set, $\{\mathrm{e}^{in\pi x/l}/\sqrt{2l}\}$. The function "inherits" the periodicity of the basis functions.
- Periodic functions even (odd) about the origin are represented by Fourier cosine (sine) series.
- Parseval's theorem is an alternate version of the completeness idea. Each Fourier coefficient c_n is the projection of $f(x)$ onto the basis functions, e^{inx}. If the basis is complete, i.e. the Fourier series completely constructs a function from a linear combination of the basis set, the set of coefficients $\{c_n\}$, taken in their entirety are equivalent to the function. Equation (4.31) (Parseval's theorem) equates the sum of the squares of the coefficients c_n to the average value of the square of the function over one period.
- The Dirac delta function has an important integral representation, Eq. (4.33).
- The Fourier transform is the generalization of Fourier series to arbitrary functions, which can be seen as periodic functions with infinite period.
- The convolution integral, Eq. (4.45), is an operation on two functions to produce a third function that is in some sense a modified version of one of the original functions. The convolution theorem, Eq. (4.47), states that the Fourier transform of the convolution integral is equal to the product of the Fourier transforms of the original functions.

Exercises

4.1. Is $\tan x$ a periodic function? What's the period? Hint:

$$\tan(A+B) = \frac{\tan A + \tan B}{1 - \tan A \tan B}.$$

4.2. Do the integrals in Eqs. (4.2) and (4.3).

4.3. Derive the expressions for the Fourier coefficients, Eq. (4.4).

4.4. Show that the integrals in Eqs. (4.2) and (4.3) are unaltered if the limits of integration are changed to $-\pi$ and π.

4.5. (a) Show that Eq. (4.1) can be written in terms of complex exponentials

$$y(x) = \sum_{n=-\infty}^{\infty} c_n \mathrm{e}^{inx}$$

where $c_n = \frac{1}{2}(a_n - \mathrm{i}b_n)$ and $c_{-n} = \frac{1}{2}(a_n + \mathrm{i}b_n)$.

(b) Show, using the results of Eq. (4.7), together with $c_n = \frac{1}{2}(a_n - \mathrm{i}b_n)$, that

$$c_n = \frac{1}{2\pi}\int_{-\pi}^{\pi} \mathrm{e}^{-inx} f(x)\, \mathrm{d}x,$$

the same as Eq. (4.10).

(c) Show that

$$\sum_{n=-\infty}^{\infty} |c_n|^2 = \left(\frac{a_0}{2}\right)^2 + \frac{1}{2}\sum_{n=1}^{\infty}(|a_n|^2 + |b_n|^2).$$

4.6. Show that the sum in Eq. (4.15) is real by showing that it's equal to its complex conjugate. Let $S \equiv \sum_{n=-\infty}^{\infty} \mathrm{e}^{in\alpha}$. Show that $S^* = S$.

4.7. Show that the periodic function $f(x) = \pi - x$ defined over one period $0 < x < 2\pi$ has the Fourier series representation

$$\pi - x = 2\sum_{n=1}^{\infty} \frac{\sin nx}{n}.$$

Hint: Use the complex Fourier series, and show first that $c_0 = 0$. You should find that $c_n = -\mathrm{i}/n$ for $n \neq 0$.

4.8. Apply Parseval's theorem to the case of the full-wave rectifier, end of Section 4.1. You should find another way to compute the value of π. Hint: $\sin^2 x = \frac{1}{2}(1 - \cos 2x)$. A: $\pi^2/8 = 1 + 2\sum_{n=1}^{\infty}(4n^2 - 1)^{-2}$.

4.9. Find the Fourier series of period $p = 2$ for $f(x) = (x-1)^2$ on $0 < x < 2$ and use your result to evaluate the series

$$\sum_{n=1}^{\infty} \frac{1}{n^4}$$

(Hint: This problem works out easiest if you find the exponential Fourier series.)

4.10. Consider the function $f(x) = \cos kx$ for $-\pi < x < \pi$, where k is a real number. Derive the Fourier series for this function. Before starting this problem, is $f(x)$ even or odd? Which type of Fourier series should you be thinking about?

(a) Show that the Fourier coefficients

$$a_n = \frac{2}{\pi}\int_0^{\pi} \cos kx \cos nx \; \mathrm{d}x = \frac{(-1)^n}{\pi}\frac{2k \sin k\pi}{(k^2 - n^2)},$$

and hence

$$\cos kx = \frac{2k \sin k\pi}{\pi}\left(\frac{1}{2k^2} - \frac{\cos x}{k^2 - 1} + \frac{\cos 2x}{k^2 - 4} - \cdots\right). \tag{4.48}$$

Equation (4.48) is an example of a *partial fraction expansion.* Equation (4.8) is a partial fraction expansion of the rectified wave in Figure 4.4.

(b) What if $k = 1$? Is this series well defined? Show that the series reproduces $\cos x$ for $k = 1$.

(c) Set $x = \pi$ in Eq. (4.48).

$$k\pi \cot k\pi = 1 + 2k^2\left(\frac{1}{k^2 - 1} + \frac{1}{k^2 - 4} + \frac{1}{k^2 - 9} + \cdots\right) \tag{4.49}$$

is the partial fraction expansion of the cotangent function. We will derive Eq. (4.49) in Chapter 8 using a different method; see Exercise 8.19.

4.11. Show that $\int_{-\infty}^{\infty} (\sin x/x)\; \mathrm{d}x = \pi$ (used in Section 4.5). Because the integrand is even, the problem reduces to showing that $\int_0^{\infty} (\sin x/x)\; \mathrm{d}x = \pi/2$. Approach this integral in stages:

(a) Consider first the integral $I \equiv \int_0^{\infty} \mathrm{e}^{-xy} \sin y \; \mathrm{d}y$ where $x > 0$. By integrating by parts twice, you should find that

$$I = \frac{1}{1 + x^2}.$$

(b) Now make use of a trick:

$$\int_0^{\infty} \frac{\sin x}{x}\; \mathrm{d}x = \int_0^{\infty} \mathrm{d}x \sin x \left(\int_0^{\infty} \mathrm{e}^{-xy} \mathrm{d}y\right) = \int_0^{\infty} \mathrm{d}y \left(\int_0^{\infty} \mathrm{d}x \sin x \; \mathrm{e}^{-xy}\right).$$

What you're asked to show follows from this result, when combined with the result of part (a). Hint:

$$\mathrm{d}(\tan^{-1} y) = \frac{\mathrm{d}y}{1 + y^2}.$$

4.12. Evaluate $\int_0^5 \delta(x^2 - 9) g(x)\; \mathrm{d}x$, where $g(x)$ is a smooth function. A: $\frac{1}{6} g(3)$.

4.13. Use Eq. (4.33) to show that

$$\int_{-\infty}^{\infty} \cos \alpha x \cos \beta x \, \mathrm{d}x = \pi(\delta(\alpha - \beta) + \delta(\alpha + \beta))$$
$$\int_{-\infty}^{\infty} \cos \alpha x \sin \beta x \, \mathrm{d}x = 0$$
$$\int_{-\infty}^{\infty} \sin \alpha x \sin \beta x \, \mathrm{d}x = \pi(\delta(\alpha - \beta) - \delta(\alpha + \beta)). \tag{4.50}$$

Hint: The Dirac delta function is an even function.

4.14. We derived Parseval's theorem for Fourier series in Section 4.4, but a similar result holds for the Fourier transform. Show that

$$\frac{1}{2\pi}\int_{-\infty}^{\infty} F^*(\omega)F(\omega) \, \mathrm{d}\omega = \int_{-\infty}^{\infty} f^*(x)f(x) \, \mathrm{d}x. \tag{4.51}$$

Take two copies of Eq. (4.39), multiply them together appropriately, and make use of Eq. (4.33).

4.15. We have in Eq. (4.44) that $\mathcal{F}[\mathcal{F}^{-1}] = I$, the identity operator. What is the Fourier transform of the Fourier transform? Is it the original function? Show that $\mathcal{F}[\mathcal{F}[f(x)]] = 2\pi f(-x)$. Hint: $\mathcal{F}[\mathcal{F}[f]] = \int_{-\infty}^{\infty} \mathrm{d}\alpha \;\; \mathrm{e}^{-\mathrm{i}\alpha x} \int_{-\infty}^{\infty} \mathrm{d}x' f(x')\mathrm{e}^{-\mathrm{i}\alpha x'}$. The Fourier transform is not the same as its inverse; they are different animals.

4.16. Show that if $f(x)$ is an even function, it's Fourier transform can be written

$$F(\alpha) = 2\int_{0}^{\infty} f(x)\cos(\alpha x) \, \mathrm{d}x$$

what's termed a *Fourier cosine transform*, where

$$f(x) = \frac{1}{\pi}\int_{0}^{\infty} F(\alpha)\cos(\alpha x) \, \mathrm{d}\alpha.$$

Repeat with the *Fourier sine transform*, that if $f(x)$ is odd, then

$$F(\alpha) = 2\int_{0}^{\infty} f(x)\sin(x\alpha) \, \mathrm{d}x \qquad f(x) = \frac{1}{\pi}\int_{0}^{\infty} F(\alpha)\sin(\alpha x) \, \mathrm{d}\alpha.$$

4.17. Consider the function

$$f(t) = \begin{cases} 0 & t < 0 \\ \mathrm{e}^{-t/T}\sin \omega_0 t & t > 0. \end{cases}$$

Such a function might represent the displacement of a damped harmonic oscillator, or the electric field in a radiated wave, or the current in an antenna.

(a) Show that the Fourier transform of $f(t)$ is

$$F(\omega) = \frac{1}{2}\left(\frac{1}{\omega + \omega_0 - \frac{\mathrm{i}}{T}} - \frac{1}{\omega - \omega_0 - \frac{\mathrm{i}}{T}}\right).$$

(b) We can interpret the physical meaning of $F(\omega)$ with the help of Parseval's theorem, Eq. (4.51). If $f(t)$ represents a radiated electric field, the radiated power is proportional to $|f(t)|^2$, with the total energy radiated proportional to $\int_0^\infty |f(t)|^2\,dt$, which, by Parseval's theorem, is equal to $(1/(2\pi))\int_{-\infty}^{\infty} |F(\omega)|^2\,d\omega$. Thus, $|F(\omega)|^2$ is proportional to the energy radiated per unit frequency interval. Assume that T is sufficiently large so that $\omega_0 T \gg 1$, i.e. there are many oscillations in the "damping time," T. Then, $F(\omega)$ is sharply peaked about $\omega = \pm\omega_0$. Show that for frequencies near $\omega = \omega_0$ that

$$|F(\omega)| \approx \frac{1}{2}\frac{1}{\sqrt{(\omega - \omega_0)^2 + 1T^2}}.$$

It's easily shown that

$$|F(\omega_0 \pm 1/T)| = \frac{1}{\sqrt{2}}|F(\omega_0)|,$$

so that the frequencies $\omega = \omega_0 \pm 1/T$ are the "half-power" points. The characteristic frequency width of $|F(\omega)|$ in the vicinity of $\omega = \omega_0$ is therefore $2/T$ (when $\omega_0 T \gg 1$). This result is another "uncertainty principle," quite close to the Heisenberg uncertainty principle of quantum mechanics mentioned in the last example in Section 4.6. The length of time T during which a system oscillates is inversely proportional to the width $2/T$, which is a measure of the "uncertainty" of the frequency.

4.18. Equation (4.47) can be interpreted as saying that the convolution operator is *diagonalized* when represented in the Fourier basis

$$\phi_\alpha(x) = \frac{1}{\sqrt{2\pi}}\exp(i\alpha x).$$

Use Eq. (1.16) to formally demonstrate this property for both the Fourier series and the Fourier transform.

5

Series solutions of ordinary differential equations

We showed in Chapter 3 that the partial differential equations (PDEs) most commonly encountered in applications reduce to the Helmholtz equation[1] ($\nabla^2 R(\boldsymbol{r}) + k^2 R(\boldsymbol{r}) = 0$) when the space and time variables are separated, $\psi(\boldsymbol{r}, t) = R(\boldsymbol{r})T(t)$. By separating the spatial variables, $R(\boldsymbol{r}) = X(x)Y(y)Z(z)$, the problem of solving the Helmholtz equation reduces to solving a set of ordinary differential equations (ODEs). In some cases, the ODEs obtained in this manner are sufficiently simple that solutions can immediately be written down, as in Eqs. (3.26) or (3.32). In two cases, however, we encountered ODEs sufficiently nontrivial that they bear the names of their inventors: the Bessel differential equation, Eq. (3.38), and the Legendre differential equation, Eq. (3.52), or it's equivalent, Eq. (3.75). What makes these equations difficult is that they involve *variable* coefficients of the derivatives in the differential equation.

Consider the generic linear, homogeneous ODE

$$f(x)\frac{\mathrm{d}^2 y}{\mathrm{d}x^2} + g(x)\frac{\mathrm{d}y}{\mathrm{d}x} + h(x)y = 0. \tag{5.1}$$

The coefficient functions $f(x)$, $g(x)$, and $h(x)$ are real-valued functions of the independent variable x. Equation (5.1) can be put in *standard form* by dividing through by $f(x)$: standard form

$$y'' + P(x)y' + Q(x)y = 0, \tag{5.2}$$

where $P(x) \equiv g(x)/f(x)$ and $Q(x) \equiv h(x)/f(x)$. Your inner mathematician may be asking: What if $f(x)$ vanishes at certain points? The functions $P(x)$ and $Q(x)$ would then not be defined at those points. That situation can and does occur and is something we'll have to address.

[1] The Laplace equation is a special case of the homogeneous Helmholtz equation.

Mathematical Methods in Physics, Engineering, and Chemistry, First Edition.
Brett Borden and James Luscombe.

We introduce a method for obtaining solutions to linear ODEs having variable coefficients: the *method of Frobenius* or the *method of undetermined coefficients.* The method seeks solutions in the form of *analytic functions*, a wide class of functions. A complete exposition of the subject of analytic functions requires the theory of functions of a complex variable. Because knowledge of that theory is not required just yet (it is, however, treated in Chapter 8), we present some properties of analytic functions.

5.1 The Frobenius method

5.1.1 Power series

power series

A function y of the real variable x is *analytic* at x_0 if there are coefficients a_n and a positive number r such that the *power series*

$$y(x) = \sum_{n=0}^{\infty} a_n (x - x_0)^n \tag{5.3}$$

is convergent for $|x - x_0| < r$, the *interval of convergence*, with r the *radius of convergence.* The radius of convergence can be determined from an analysis of the coefficients a_n at large n and could be infinite (Appendix B). The point x_0 is called the *center* of the series; by a translation of coordinates, the series can be taken to be centered about the origin. Analytic functions are infinitely differentiable, or *smooth*, at every point of the interval of convergence. Not all smooth functions are analytic, however, as an example in Appendix B demonstrates. *A function is analytic at x_0 if its Taylor series about x_0 converges to the function in a neighborhood of x_0.*

Power series have the following properties [10, chapter 5].

- If the series in Eq. (5.3) converges for $|x - x_0| < r$, $y(x)$ may be expanded about any other point x_1 interior to the interval of convergence. The radius of convergence of the new series is at least equal to the positive number $r - |x_1 - x_0|$.
- Functions $y(x)$ represented by power series such as Eq. (5.3) are continuous and differentiable at every interior point x_1 of the interval of convergence. The derivative $y'(x)$ is represented by the series obtained by differentiating Eq. (5.3) term by term and has the same interval of convergence. Derivatives of all orders $y^{(n)}$ may be represented by series in the same way.
- If x and x_0 in Eq. (5.3) are replaced with complex numbers z and z_0, the power series $\sum_n a_n (z - z_0)^n$ converges for $|z - z_0| < r$, i.e. for z interior to the *circle of convergence* centered at z_0 of radius r. In this way, we obtain an *extension* of the function y to the complex plane: a function analytic for $|z - z_0| < r$. Although the radius of convergence for

$y(x)$ considered as a real-valued function can be established from various convergence tests, such procedures are often impractical. The good news is, the radius of convergence for complex functions is often easier to determine. The circle of convergence must contain on its periphery at least one *singularity* where the function is not analytic, and such singularities are readily recognizable (see Chapter 8). For example, the real function $f(x) = (1+x^2)^{-1}$ has the power series about $x=0$: $f(x) = \sum_{n=0}^{\infty}(-1)^n x^{2n}$, which is convergent for $|x|<1$. The complex function $f(z) = (1+z^2)^{-1}$ is analytic everywhere except at $z = \pm \mathrm{i}$; the radius of convergence is therefore $r=1$.

- Two power series may be added and subtracted term by term for x interior to the common interval of convergence of both series. Likewise, two power series can be multiplied together for x in the interior of the common intervals of convergence. If f and g are analytic at x_0, so is f/g if $g(x_0) \neq 0$. It may happen that f/g is analytic at x_0 even if $g(x_0)=0$, provided $\lim_{x\to x_0} f(x)/g(x)$ exists.

5.1.2 Introductory example

By way of introduction to the Frobenius method, consider the differential equation:

$$\frac{\mathrm{d}y}{\mathrm{d}x} - 2xy = 0. \tag{5.4}$$

Equation (5.4) is a first order differential equation; we'll get to second order differential equations shortly. Equation (5.4) can be integrated to obtain its solution:

$$\int \frac{\mathrm{d}y}{y} = \int 2x\,\mathrm{d}x \quad \Longrightarrow \quad y = A\mathrm{e}^{x^2}.$$

Not all differential equations can be reduced to quadratures (as in this example), which underscores the value of the Frobenius method (that we're about to introduce). We illustrate the method with this relatively simple example, for which we already know the answer, before turning to Eq. (5.2) with its possible nonanalytic coefficient functions $P(x)$ and $Q(x)$.

We seek a solution to Eq. (5.4) in the form of a power series:

$$y = \sum_{n=0}^{\infty} a_n x^n. \tag{5.5}$$

(The interval of convergence will emerge when we're done.) The goal at first is to determine the coefficients $\{a_n\}$ such that the power series solves the differential equation. The derivative of y (which appears in Eq. (5.4), what we're trying to solve) is represented by the power series obtained by differentiating Eq. (5.5):

$$\frac{\mathrm{d}y}{\mathrm{d}x} = \sum_{n=0}^{\infty} n a_n x^{n-1} = \sum_{n=1}^{\infty} n a_n x^{n-1}$$

(the first term corresponding to $n = 0$ is zero). The differential equation, Eq. (5.4), is then equivalent to the difference in power series[2]:

$$\frac{dy}{dx} - 2xy = 0 \implies \sum_{n=1}^{\infty} n a_n x^{n-1} - \sum_{n=0}^{\infty} 2a_n x^{n+1} = 0. \tag{5.6}$$

To make progress, we must combine the two series in Eq. (5.6) into one. This takes a few steps because the powers of x in the two series are different. Change the *dummy* summation index in the first series to $k = n - 1$. In the second series, change the dummy index to $k = n + 1$. Then[3]

$$\sum_{n=1}^{\infty} n a_n x^{n-1} - \sum_{n=0}^{\infty} 2a_n x^{n+1} = \sum_{k=0}^{\infty} (k+1)a_{k+1}x^k - \sum_{k=1}^{\infty} 2a_{k-1}x^k.$$

We can't combine the series quite yet: The first starts at $k = 0$ while the second starts at $k = 1$. "Split off" the $k = 0$ term from the first series, *then* combine the series:

$$\sum_{k=0}^{\infty}(k+1)a_{k+1}x^k - \sum_{k=1}^{\infty} 2a_{k-1}x^k = a_1x^0 + \sum_{k=1}^{\infty}(k+1)a_{k+1}x^k - \sum_{k=1}^{\infty} 2a_{k-1}x^k$$
$$= a_1x^0 + \sum_{k=1}^{\infty}[(k+1)a_{k+1} - 2a_{k-1}]x^k.$$

The differential equation, Eq. (5.4), is thus equivalent to the power-series equation

$$a_1x^0 + \sum_{k=1}^{\infty}[(k+1)a_{k+1} - 2a_{k-1}]x^k = 0. \tag{5.7}$$

uniqueness of power series

We now come to a key point of the method – one that we'll use repeatedly – the *uniqueness of power series.*[4] *The only way a power series in x can equal zero is if the coefficient of every power of x is zero*, i.e. $0 = \sum_{n=0}^{\infty} b_n x^n \implies b_n = 0$ for all n. Powers of x are linearly independent (see Section 2.1), and the only way a combination of linearly independent terms can sum to zero is if the coefficients are zero. In Eq. (5.7), therefore, the coefficient of each power of x must independently vanish, i.e.

$$a_1 = 0$$
$$(k+1)a_{k+1} - 2a_{k-1} = 0. \qquad k = 1, 2, 3, \ldots$$

[2] We can combine these series because they have a common interval of convergence.

[3] This substitution will put the two series "in phase."

[4] Suppose one has two power series for the same function, with the same interval of convergence, $f(x) = \sum_{n=0}^{\infty} a_n x^n$ and $f(x) = \sum_{n=0}^{\infty} b_n x^n$. Subtract the two expressions: $0 = \sum_{n=0}^{\infty}(a_n - b_n)x^n$. The only way this can hold, because the functions $\{x^n\}$ are linearly independent, is if $b_n = a_n$. There is only one power-series representation of a given function.

This requirement is in the form of a *recursion relation.* Because $k+1 \neq 0$,

$$a_{k+1} = \frac{2}{k+1}a_{k-1}. \qquad k = 1, 2, 3, \ldots$$

To determine explicit values for the $\{a_k\}$, we *iterate* the recursion relation:

$$\begin{aligned}
k &= 1\,, \quad a_2 = \frac{2}{2}a_0 = a_0 \\
k &= 2\,, \quad a_3 = \frac{2}{3}a_1 = 0 \\
k &= 3\,, \quad a_4 = \frac{2}{4}a_2 = \frac{1}{2}a_0 = \frac{1}{2!}a_0 \\
k &= 4\,, \quad a_5 = \frac{2}{5}a_3 = 0 \\
k &= 5\,, \quad a_6 = \frac{2}{6}a_4 = \frac{1}{3\cdot 2!}a_0 = \frac{1}{3!}a_0 \\
k &= 6\,, \quad a_7 = \frac{2}{7}a_5 = 0 \\
k &= 7\,, \quad a_8 = \frac{2}{8}a_6 = \frac{1}{4\cdot 3!}a_0 = \frac{1}{4!}a_0
\end{aligned}$$

and so forth. The terms a_n for n odd are therefore all zero, while the even terms are in the form $a_{2n} = a_0/n!$. There are *no restrictions* on the coefficient a_0; once its value is set, all of the other nonzero coefficients are uniquely determined. Thus

$$\begin{aligned}
y &= \sum_{n=0}^{\infty} a_n x^n = a_0 + 0 + a_2x^2 + 0 + a_4x^4 + 0 + a_6x^6 + \cdots \\
&= a_0\left[1 + x^2 + \frac{x^4}{2!} + \frac{x^6}{3!} + \cdots\right] = a_0\sum_{n=0}^{\infty}\frac{x^{2n}}{n!}
\end{aligned}$$

is the solution to Eq. (5.4) (which we recognize as the answer: $y = a_0 e^{x^2}$).

5.1.3 Ordinary points

Having illustrated the method by finding the solution to Eq. (5.4) (a first-order differential equation), let's apply it to second-order differential equations.

Definition. *A point $x = x_0$ is an* ordinary point *of Eq. (5.2) if $P(x)$ and $Q(x)$ are analytic at x_0 – that is, $P(x)$ and $Q(x)$ have power series representations in $(x - x_0)$. A point that's not an ordinary point is said to be a* singular point *of the equation.*[5] ordinary point

[5]Singular points need not be real numbers.

Example. The differential equation $y'' + (\ln x)y = 0$ has a singular point at $x = 0$; $Q(x) = \ln x$ does not possess a power series about $x = 0$.

We now state, without proof, a theorem on the solutions of Eq. (5.2).[6]

Theorem 5.1.1. *For $x = x_0$ an ordinary point of Eq. (5.2), there exists a unique power series solution, Eq. (5.3), for each choice of the coefficients a_0 and a_1, where the radius of convergence is at least as large as the smaller of the radii of convergence of the power series representing the coefficient functions.*

Comments:

- From Eq. (5.3), $y(x_0) = a_0$ and $y'(x_0) = a_1$. The method therefore builds in these data in determining the solution in the form of a power series. The values $y(x)$ at $x \neq x_0$ (of an analytic function) are determined by $y(x_0)$ and $y'(x_0)$.
- A consequence of this theorem is that if $P(x)$ and $Q(x)$ are *polynomials*, then power-series solutions in the form of Eq. (5.3) are *valid for all values of x* (because polynomials have an infinite radius of convergence). And the good news is that we will mostly be interested in cases when Eq. (5.2) has polynomial coefficients.

Example. The second-order differential equation

$$y'' - 2xy = 0 \tag{5.8}$$

has an ordinary point at $x = 0$. Because $Q(x) = -2x$ is a polynomial, Eq. (5.8) has power series solutions $y = \sum_{n=0}^{\infty} a_n x^n$ convergent for $|x| < \infty$. The second derivative has the power series representation $y'' = \sum_{n=2}^{\infty} n(n-1)a_n x^{n-2}$. The differential equation is thus equivalent to the combination of power series:

$$y'' - 2xy = 0 \implies \sum_{n=2}^{\infty} n(n-1)a_n x^{n-2} - \sum_{n=0}^{\infty} 2a_n x^{n+1} = 0.$$

In the first series, let $k = n - 2$, while in the second let $k = n + 1$:

$$\begin{aligned}
0 &= \sum_{k=0}^{\infty} (k+2)(k+1)a_{k+2} x^k - \sum_{k=1}^{\infty} 2a_{k-1} x^k \\
&= 2a_2 x^0 + \sum_{k=1}^{\infty} (k+2)(k+1)a_{k+2} x^k - \sum_{k=1}^{\infty} 2a_{k-1} x^k \\
&= 2a_2 x^0 + \sum_{k=1}^{\infty} [(k+2)(k+1)a_{k+2} - 2a_{k-1}]x^k.
\end{aligned}$$

[6]This theorem is a combination of theorems 1 and 2 in chapter 4 of [19].

Invoking the uniqueness of power series, we have

$$a_2 = 0$$

$$a_{k+2} = \frac{2}{(k+2)(k+1)} a_{k-1}. \qquad k = 1, 2, \ldots$$

Iterating the recursion relation,

$$\begin{aligned}
k &= 1\,, \quad a_3 = \frac{2a_0}{3 \cdot 2} \\
k &= 2\,, \quad a_4 = \frac{2a_1}{4 \cdot 3} \\
k &= 3\,, \quad a_5 = \frac{2a_2}{5 \cdot 4} = 0 \\
k &= 4\,, \quad a_6 = \frac{2a_3}{6 \cdot 5} = \frac{2^2}{6 \cdot 5 \cdot 3 \cdot 2} a_0 \\
k &= 5\,, \quad a_7 = \frac{2a_4}{7 \cdot 6} = \frac{2^2}{7 \cdot 6 \cdot 4 \cdot 3} a_1 \\
k &= 6\,, \quad a_8 = \frac{2a_5}{8 \cdot 7} = 0 \\
k &= 7\,, \quad a_9 = \frac{2a_6}{9 \cdot 8} = \frac{2^3}{9 \cdot 8 \cdot 6 \cdot 5 \cdot 3 \cdot 2} a_0 \\
k &= 8\,, \quad a_{10} = \frac{2a_7}{10 \cdot 9} = \frac{2^3}{10 \cdot 9 \cdot 8 \cdot 7 \cdot 6 \cdot 4 \cdot 3} a_1
\end{aligned}$$

and so on. It follows that

$$\begin{aligned}
y(x) = a_0 &\left[1 + \frac{2}{3 \cdot 2} x^3 + \frac{2^2}{6 \cdot 5 \cdot 3 \cdot 2} x^6 + \frac{2^3}{9 \cdot 8 \cdot 6 \cdot 5 \cdot 3 \cdot 2} x^9 + \cdots \right] \\
&+ a_1 \left[x + \frac{2}{4 \cdot 3} x^4 + \frac{2^2}{7 \cdot 6 \cdot 4 \cdot 3} x^7 + \frac{2^3}{10 \cdot 9 \cdot 8 \cdot 7 \cdot 6 \cdot 4 \cdot 3} x^{10} + \cdots \right].
\end{aligned}$$

There are no restrictions on a_0 or a_1. Thus, we have the two solutions:

$$y_1(x) = a_0 \left[1 + \sum_{k=1}^{\infty} \frac{2^k [1 \cdot 4 \cdot 7 \cdots (3k-2)]}{(3k)!} x^{3k}\right]$$

$$y_2(x) = a_1 \left[x + \sum_{k=1}^{\infty} \frac{2^k [2 \cdot 5 \cdot 8 \cdots (3k-1)]}{(3k+1)!} x^{3k+1}\right].$$

It can be verified (ratio test, see Appendix B) that each series converges for $|x| < \infty$.

Example. The differential equation

$$(x^2+1)y''+xy'-y=0 \tag{5.9}$$

has singular points at $x=\pm\mathrm{i}$ and so a power series solution will converge for $|x|<1$ (because $|\mathrm{i}|=1$). Let $y=\sum_{n=0}^{\infty}a_nx^n$ and substitute into Eq. (5.9):

$$\begin{aligned}
(x^2+1)&\sum_{n=2}^{\infty}n(n-1)a_nx^{n-2}+x\sum_{n=1}^{\infty}na_nx^{n-1}-\sum_{n=0}^{\infty}a_nx^n\\
&=\sum_{n=2}^{\infty}n(n-1)a_nx^n+\sum_{n=2}^{\infty}n(n-1)a_nx^{n-2}+\sum_{n=1}^{\infty}na_nx^n-\sum_{n=0}^{\infty}a_nx^n\\
&=(2a_2-a_0)x^0+6a_3x+\sum_{k=2}^{\infty}[(k+2)(k+1)a_{k+2}+(k^2-1)a_k]x^k.
\end{aligned}$$

Thus,

$$\begin{aligned}
2a_2-a_0=0 \quad&\Longrightarrow\quad a_2=\frac{1}{2}a_0\\
a_3=0 \quad&\Longrightarrow\quad a_3=0\\
(k+2)(k+1)a_{k+2}+(k^2-1)a_k=0 \quad&\Longrightarrow\quad a_{k+2}=-\frac{k-1}{k+2}a_k. \quad k=2,3,4,\dots
\end{aligned}$$

Iterating the recursion relation yields

$$\begin{aligned}
k=2,\quad a_4&=-\frac{1}{4}a_2=-\frac{1}{2\cdot 4}a_0=-\frac{1}{2^2 2!}a_0\\
k=3,\quad a_5&=-\frac{2}{5}a_3=0\\
k=4,\quad a_6&=-\frac{3}{6}a_4=\frac{3}{2\cdot 4\cdot 6}a_0=\frac{1\cdot 3}{2^3 3!}a_0\\
k=5,\quad a_7&=-\frac{4}{7}a_5=0\\
k=6,\quad a_8&=-\frac{5}{8}a_6=-\frac{3\cdot 5}{2\cdot 4\cdot 6\cdot 8}a_0=\frac{1\cdot 3\cdot 5}{2^4 4!}a_0\\
k=7,\quad a_9&=-\frac{6}{9}a_7=0\\
k=8,\quad a_{10}&=-\frac{7}{10}a_8=\frac{3\cdot 5\cdot 7}{2\cdot 4\cdot 6\cdot 8\cdot 10}a_0=\frac{1\cdot 3\cdot 5\cdot 7}{2^5 5!}a_0
\end{aligned}$$

and so on. It follows that the two solutions to Eq. (5.9) are

$$y(x)=a_1x+a_0\left[1+\frac{1}{2}x^2+\sum_{k=2}^{\infty}(-1)^{k-1}\frac{1\cdot 3\cdot 5\cdots(2k-3)}{2^kk!}x^{2k}\right]$$

for $|x|<1$.

Example. A power series solution to

$$y'' - (1 + x)y = 0$$

yields $a_2 = \frac{1}{2}a_0$ and the *three-term* recursion relation (show this)

$$a_{k+2} = \frac{a_k + a_{k-1}}{(k+1)(k+2)} \qquad k = 1, 2, 3, \ldots$$

Because this second-order equation will have two solutions (with coefficients a_0 and a_1), we can simplify the iteration by first choosing $a_0 \neq 0$ and $a_1 = 0$ to obtain one of the solutions, and then choosing $a_0 = 0$ and $a_1 \neq 0$ to obtain the other. If we set $a_1 = 0$, we find

$$\begin{aligned}
a_2 &= \frac{1}{2}a_0 \\
a_3 &= \frac{a_1 + a_0}{2 \cdot 3} = \frac{a_0}{2 \cdot 3} = \frac{1}{6}a_0 \\
a_4 &= \frac{a_2 + a_1}{3 \cdot 4} = \frac{a_0}{2 \cdot 3 \cdot 4} = \frac{1}{24}a_0 \\
a_5 &= \frac{a_3 + a_2}{4 \cdot 5} = \frac{a_0}{4 \cdot 5}\left(\frac{1}{2 \cdot 3} + \frac{1}{2}\right) = \frac{1}{30}a_0
\end{aligned}$$

and so on. Then

$$y_1(x) = a_0\left[1 + \frac{1}{2}x^2 + \frac{1}{6}x^3 + \frac{1}{24}x^4 + \frac{1}{30}x^5 + \cdots\right].$$

Similarly, if we choose $a_0 = 0$ then

$$\begin{aligned}
a_2 &= 0 \\
a_3 &= \frac{a_1 + a_0}{2 \cdot 3} = \frac{a_1}{2 \cdot 3} = \frac{1}{6}a_1 \\
a_4 &= \frac{a_2 + a_1}{3 \cdot 4} = \frac{a_1}{3 \cdot 4} = \frac{1}{12}a_1 \\
a_5 &= \frac{a_3 + a_2}{4 \cdot 5} = \frac{a_1}{2 \cdot 3 \cdot 4 \cdot 5} = \frac{1}{120}a_1
\end{aligned}$$

and so forth. Hence, another solution is

$$y_2(x) = a_1\left[x + \frac{1}{6}x^3 + \frac{1}{12}x^4 + \frac{1}{120}x^5 + \cdots\right].$$

Each series converges for all finite values of x.

5.1.4 Regular singular points

For $x = x_0$ a singular point, we may not be able to find a power-series solution in the form of Eq. (5.3). We may, however, be able to find a solution of the form

$$y = \sum_{n=0}^{\infty} a_n (x - x_0)^{n+r}, \tag{5.10}$$

where r is a constant to be determined.[7]

Singular points are classified as *regular* or *irregular*:

regular singular point

Definition. *A singular point $x = x_0$ of Eq. (5.2) is said to be a* regular singular point *if both $(x - x_0)P(x)$ and $(x - x_0)^2 Q(x)$ are analytic at x_0; that is, both $(x - x_0)P(x)$ and $(x - x_0)^2 Q(x)$ have a power series in $(x - x_0)$ with a positive radius of convergence. A singular point that is not regular is said to be an* irregular singular point *of the equation.*

When $P(x)$ and $Q(x)$ can be expressed as the ratio of polynomials with no common factors and the factor $(x - x_0)$ appears *at most* to the first power in the denominator of $P(x)$ and *at most* to the second power in the denominator of $Q(x)$, then $x = x_0$ is a regular singular point.

Example. Both $x = 0$ and $x = -1$ are singular points of the differential equation

$$x^2(x+1)^2 y'' + (x^2 - 1)y' + 2y = 0$$

For this equation, we form

$$P(x) = \frac{(x^2 - 1)}{x^2(x+1)^2} = \frac{x - 1}{x^2(x+1)} \quad \text{and} \quad Q(x) = \frac{2}{x^2(x+1)^2}$$

Since $(x - 0) = x$ appears to the second power in the denominator of $P(x)$, we see that 0 is an irregular singular point. The singular point $x = -1$, however, is regular.

---

method of Frobenius

At a regular singular point, we use the following theorem [19, p. 234]:

Theorem 5.1.2. *If $x = x_0$ is a regular singular point of Eq. (5.2), then at least one series solution exists of the form of Eq. (5.10).*

Note carefully what the theorem says: There is *at least one* power-series solution in this case, but not necessarily two.

---

[7]The index r here is *not* the radius of convergence.

Example. Consider the following differential equation:

$$3xy'' + y' - y = 0, \tag{5.11}$$

for which $x = 0$ is a regular singular point (check it!). Try a solution of the form

$$y = \sum_{n=0}^{\infty} a_n x^{n+r}. \tag{5.12}$$

Differentiating Eq. (5.12), we obtain

$$y' = \sum_{n=0}^{\infty} (n+r) a_n x^{n+r-1} \quad \text{and} \quad y'' = \sum_{n=0}^{\infty} (n+r)(n+r-1) a_n x^{n+r-2}$$

(note that the $n = 0$ and $n = 1$ terms *do not* vanish in this case). Substitution into Eq. (5.11) yields

$$\begin{aligned} &3xy'' + y' - y \\ &= 3\sum_{n=0}^{\infty}(n+r)(n+r-1)a_n x^{n+r-1} + \sum_{n=0}^{\infty}(n+r)a_n x^{n+r-1} - \sum_{n=0}^{\infty} a_n x^{n+r} \\ &= x^r \left(r(3r-2)a_0 x^{-1} + \sum_{k=0}^{\infty}[(k+r+1)(3k+3r+1)a_{k+1} - a_k]x^k \right) = 0, \end{aligned}$$

which implies

$$\begin{aligned} r(3r-2)a_0 &= 0 \\ (k+r+1)(3k+3r+1)a_{k+1} - a_k &= 0. \quad k = 0, 1, 2, \ldots \end{aligned} \tag{5.13}$$

We might be tempted to set $a_0 = 0$ to satisfy Eq. (5.13). Nothing is gained by doing that, however – it leads to the solution $y = 0$. Thus, we can't conclude that $a_0 = 0$. Instead, we choose r so that Eq. (5.13) is satisfied with $a_0 \neq 0$:

$$r(3r-2) = 0 \quad \Longrightarrow \quad r = \frac{2}{3} \quad \text{or} \quad r = 0. \tag{5.14}$$

indicial equation

The recursion relation is

$$a_{k+1} = \frac{a_k}{(k+r+1)(3k+3r+1)}. \quad k = 0, 1, 2, \ldots \quad r = 0, \frac{2}{3} \tag{5.15}$$

We thus obtain *two* recursion relations corresponding to the choices of r in Eq. (5.14). For $r = \frac{2}{3}$, Eq. (5.15) reduces to

$$r = \frac{2}{3} \qquad a_{k+1} = \frac{a_k}{(k+1)(3k+5)}, \qquad k = 0, 1, 2, \ldots \tag{5.16}$$

while for $r = 0$ we obtain

$$r = 0 \qquad a_{k+1} = \frac{a_k}{(k+1)(3k+1)}. \qquad k = 0, 1, 2, \ldots \tag{5.17}$$

Iterating Eq. (5.16) yields

$$\begin{aligned}
r = \frac{2}{3} \quad a_1 &= \frac{a_0}{5 \cdot 1} \\
a_2 &= \frac{a_1}{8 \cdot 2} = \frac{a_0}{2!5 \cdot 8} \\
a_3 &= \frac{a_2}{11 \cdot 3} = \frac{a_0}{3!5 \cdot 8 \cdot 11} \\
a_4 &= \frac{a_3}{14 \cdot 4} = \frac{a_0}{4!5 \cdot 8 \cdot 11 \cdot 14} \\
&\vdots \\
a_n &= \frac{a_0}{n!5 \cdot 8 \cdot 11 \cdots (3n+2)}, \qquad n = 1, 2, 3, \ldots
\end{aligned}$$

while iterating Eq. (5.17) yields

$$\begin{aligned}
r = 0 \quad a_1 &= \frac{a_0}{1 \cdot 1} \\
a_2 &= \frac{a_1}{2 \cdot 4} = \frac{a_0}{2!1 \cdot 4} \\
a_3 &= \frac{a_2}{3 \cdot 7} = \frac{a_0}{3!1 \cdot 4 \cdot 7} \\
a_4 &= \frac{a_3}{4 \cdot 10} = \frac{a_0}{4!1 \cdot 4 \cdot 7 \cdot 10} \\
&\vdots \\
a_n &= \frac{a_0}{n!1 \cdot 4 \cdot 7 \cdots (3n-2)}. \qquad n = 1, 2, 3, \ldots
\end{aligned}$$

Thus, we obtain two series solutions:

$$y_1(x) = a_0 x^{2/3} \left[1 + \sum_{n=1}^{\infty} \frac{1}{n!5 \cdot 8 \cdot 11 \cdots (3n+2)} x^n \right]$$

and

$$y_2(x) = a_0 x^0 \left[1 + \sum_{n=1}^{\infty} \frac{1}{n!1 \cdot 4 \cdot 7 \cdots (3n-2)} x^n \right].$$

Neither solution is a constant multiple of the other, and so these solutions are linearly independent. The general solution of Eq. (5.11) is

$$y(x) = Ay_1(x) + By_2(x)$$

and the ratio test can be used to show that both series converge for all finite x.

Equation (5.14) is an example of the *indicial equation*, and the values $r = \frac{2}{3}$ and $r = 0$ are called the *indicial roots* or *exponents* of the singularity. The Frobenius method only guarantees that we can find at least one solution of the assumed series form. In the aforementioned example, however, we found two. Now consider the following.

Example. The method of Frobenius applied to

$$xy'' + 3y' - y = 0 \tag{5.18}$$

yields the indicial equation $r(r+2) = 0$ with the roots $r_1 = 0$ and $r_2 = -2$ (show this). The series solution for $r = 0$ is

$$y_1(x) = a_0 x^0 \left[1 + \sum_{n=1}^{\infty} \frac{2}{n!(n+2)!} x^n\right] = a_0 \sum_{n=0}^{\infty} \frac{2}{n!(n+2)!} x^n. \qquad |x| < \infty$$

The solution for $r_2 = -2$ is

$$y_2(x) = a_0 x^{-2} \sum_{n=2}^{\infty} \frac{2}{(n-2)!n!} x^n = a_0 \sum_{n=2}^{\infty} \frac{2}{(n-2)!n!} x^{n-2}.$$

Substituting the dummy index $k = n - 2$, however, yields

$$y_2(x) = a_0 \sum_{k=0}^{\infty} \frac{2}{k!(k+2)!} x^k = y_1(x).$$

Thus, in this example, the method of Frobenius does not produce two independent solutions. The other solution must be obtained in some other way (see in the following).

It's useful to distinguish three possible cases associated with the indicial roots. Let the roots be denoted r_1 and r_2 with $r_1 \geq r_2$.

Case I

If r_1 and r_2 are distinct and do not differ by an integer, there exist two linearly independent solutions of Eq. (5.2) of the form

$$y_1 = \sum_{n=0}^{\infty} a_n x^{n+r_1} \quad \text{and} \quad y_2 = \sum_{n=0}^{\infty} b_n x^{n+r_2}$$

where $a_0 \neq 0$ and $b_0 \neq 0$.

Case II

If $r_1 - r_2 = N$, where N is a positive integer, then there exist two linearly independent solutions of Eq. (5.2) of the form

$$y_1(x) = \sum_{n=0}^{\infty} a_n x^{n+r_1} \qquad a_0 \neq 0$$

$$y_2(x) = C\ y_1(x)\ \ln x + \sum_{n=0}^{\infty} b_n x^{n+r_2} \qquad b_0 \neq 0$$

where the constant C *could* be zero. (Recall that we've assumed $r_1 > r_2$.)

Case III

When $r_1 = r_2$, the second solution will *always* contain a logarithm:

$$y_1(x) = \sum_{n=0}^{\infty} a_n x^{n+r_1} \qquad a_0 \neq 0$$

$$y_2(x) = y_1(x)\ \ln x + \sum_{n=1}^{\infty} b_n x^{n+r_2}.$$

Note that the series component of the solution for $y_2(x)$ starts at $n = 1$.

Example. In the previous example, we found that the method of Frobenius provides only one solution to Eq. (5.18) ($xy'' + 3y' - y = 0$):

$$y_1(x) = a_0 \sum_{n=0}^{\infty} \frac{2}{n!(n+2)!} x^n. \tag{5.19}$$

Equation (5.19) is the solution associated with the root $r_1 = 0$. Because the other root is $r_2 = -2$, we see that Eq. (5.18) falls under the category of a differential equation with indicial roots differing by an integer (case II). Thus, we try a solution of the form

$$y_2(x) = C\ y_1(x)\ \ln x + \sum_{n=0}^{\infty} b_n x^{n-2}. \tag{5.20}$$

Differentiating Eq. (5.20) yields

$$y_2' = C\frac{y_1}{x} + Cy_1' \ln x + \sum_{n=0}^{\infty} (n-2) b_n x^{n-3}$$

$$y_2'' = -C\frac{y_1}{x^2} + \frac{2Cy_1'}{x} + Cy_1'' \ln x + \sum_{n=0}^{\infty} (n-2)(n-3) b_n x^{n-4}.$$

Substituting into Eq. (5.18) and rearranging in the usual way, we have

$$xy_2'' + 3y_2' - y_2 = C\underbrace{[xy_1'' + 3y_1' - y_1]}_{\text{zero}}\ln x + 2C\left(y_1' + \frac{y_1}{x}\right)$$
$$+ \sum_{n=0}^{\infty}(n-2)nb_n x^{n-3} - \sum_{n=0}^{\infty} b_n x^{n-2} = 0.$$

Differentiating y_1 from Eq. (5.19) and rearranging, we find

$$-(b_0 + b_1)x^{-2} + \sum_{k=0}^{\infty}\left[\frac{4(k+1)a_0C}{k!(k+2)!} + k(k+2)b_{k+2} - b_{k+1}\right]x^{k-1} = 0.$$

Thus, $b_1 = -b_0$ and

$$\frac{4(k+1)a_0C}{k!(k+2)!} + k(k+2)b_{k+2} - b_{k+1} = 0. \quad k = 0, 1, 2, \ldots$$

When $k = 0$, the recursion relation yields $b_1 = 2a_0C$. For $k \geq 1$, we can write

$$b_{k+2} = \frac{b_{k+1}}{k(k+2)} + \frac{2(k+1)}{k!(k+2)!k(k+2)}b_0. \quad k = 1, 2, 3, \ldots$$

The coefficient b_2 is arbitrary. Iteration of the recursion relations results in

$$b_3 = \frac{b_2}{3} + \frac{2}{9}b_0$$
$$b_4 = \frac{b_3}{8} + \frac{1}{64}b_0 = \frac{b_2}{24} + \frac{25}{576}b_0$$
$$\vdots$$

From Eq. (5.20),

$$y_2(x) = Cy_1(x)\ln x + b_0x^{-2} + b_1x^{-1} + b_2 + b_3x + \cdots$$
$$= Cy_1(x)\ln x + b_0x^{-2} - b_0x^{-1} + b_2 + \left(\frac{b_2}{3} + \frac{2}{9}b_0\right)x + \cdots$$

In this case, the constants C and b_2 are *not* independent.

Example. Sometimes it's possible to generate two solutions in the "usual" way even if $r_1 - r_2 = N$. Consider the differential equation

$$xy'' + (x-6)y' - 3y = 0$$

If we try $y = \sum_{n=0}^{\infty} a_n x^{n+r}$, we obtain

$$x^r \left[r(r-7)a_0 x^{-1} + \sum_{k=0}^{\infty} [(k+r+1)(k+r-6)a_{k+1} + (k+r-3)a_k]x^k \right] = 0$$

$$\implies r_1 = 7 \; , r_2 = 0 \quad \text{and} \quad (k+r+1)(k+r-6)a_{k+1} + (k-3)a_k = 0$$

For the smaller root, the recursion relation becomes

$$(k+1)(k-6)a_{k+1} + (k-3)a_k = 0 \qquad k = 0, 1, 2, \ldots \tag{5.21}$$

and we cannot divide by $(k+1)(k-6)$ until $k > 6$ (since $k = 6$ would have us dividing by zero). Iterating Eq. (5.21), we obtain

$$\begin{aligned}
&1 \cdot (-6)a_1 + (-3)a_0 = 0 \\
&2 \cdot (-5)a_2 + (-2)a_1 = 0 \\
&3 \cdot (-4)a_3 + (-1)a_2 = 0 \\
&4 \cdot (-3)a_4 + 0a_3 = 0 \\
&5 \cdot (-2)a_5 + 1a_4 = 0 \\
&6 \cdot (-1)a_6 + 2a_5 = 0 \\
&7 \cdot 0a_7 + 3a_6 = 0,
\end{aligned}$$

the last four of which imply that $a_4 = a_5 = a_6 = 0$ while a_0 and a_7 can be chosen arbitrarily. Thus

$$a_1 = -\frac{1}{2}a_0 \quad a_2 = -\frac{1}{5}a_1 = \frac{1}{10}a_0 \quad a_3 = -\frac{1}{12}a_2 = -\frac{1}{120}a_0$$

and for $k \geq 7$

$$a_{k+1} = \frac{-(k-3)a_k}{(k+1)(k-6)} \implies a_8 = \frac{-4}{8 \cdot 1}a_7 \; , \; a_9 = \frac{-5}{9 \cdot 2}a_8 = \frac{4 \cdot 5}{2 \cdot 8 \cdot 9}a_7 \quad \cdots$$

If we choose $a_7 = 0$ and $a_0 \neq 0$, we obtain the solution

$$y_1 = a_0 \left[1 - \frac{1}{2}x + \frac{1}{10}x^2 - \frac{1}{120}x^3 \right]$$

If, however, we choose $a_7 \neq 0$ and $a_0 = 0$, then we obtain the series solution

$$y_2 = a_7 \left[x^7 + \sum_{k=8}^{\infty} \frac{(-1)^{k+1} 4 \cdot 5 \cdot 6 \cdots (k-4)}{(k-7)! 8 \cdot 9 \cdot 10 \cdots n} x^k \right]$$

The general solution is $y = y_1 + y_2$. (And we didn't need to use the second root $r_1 = 7$.)

5.2 Wronskian method for obtaining a second solution

The Frobenius method produces at least one solution to Eq. (5.2), but in case of only one, obtaining the "second solution" by the means discussed in Section 5.1.4 can be rather laborious. In this section, we develop another method for finding a second, linearly independent solution if one solution is already known.

If we know one solution $y_1(x)$ to Eq. (5.2), we can use the Wronskian to find a second, linearly independent, solution $y_2(x)$. From the definition of W, Eq. (2.6), we note that

$$\frac{W(x)}{y_1^2(x)} = \frac{y_1y_2' - y_2y_1'}{y_1^2} = \frac{\mathrm{d}}{\mathrm{d}x}\left(\frac{y_2(x)}{y_1(x)}\right). \tag{5.22}$$

Integrating Eq. (5.22), we have

$$y_2(x) = y_1(x)\left(A + \int^x \left(\frac{W(x')}{y_1^2(x')}\right) \mathrm{d}x'\right), \tag{5.23}$$

where A is a constant of integration. Of course, if y_1 and y_2 are solutions to Eq. (5.2), so is the combination $y_2 + Cy_1$, where C is a constant. We are thus free to discard the constant A in Eq. (5.23). Note that, for any C, we have (using the properties of determinants)

$$\begin{vmatrix} y_1 & y_2 + Cy_1 \\ y_1' & y_2' + Cy_1' \end{vmatrix} = \begin{vmatrix} y_1 & y_2 \\ y_1' & y_2' \end{vmatrix}.$$

Substituting for $W(x)$ from Eq. (2.8), we have a second linearly independent solution y_2 in terms of y_1:

$$y_2(x) = Cy_1(x) \int \frac{1}{y_1^2(x)} \exp\left(-\int^x P(x')\,\mathrm{d}x'\right) \mathrm{d}x. \tag{5.24}$$

Example. Consider the differential equation $y'' + y = 0$. This equation has linearly independent solutions $y_1(x) = \cos x$ and $y_2(x) = \sin x$. Use Eq. (5.24) to infer the second solution given $y_1 = \cos x$. In this case, the coefficient function $P(x) = 0$, implying from Eq. (2.8) that the Wronskian is a constant. Using Eq. (5.24),

$$y_2(x) = A\cos x \int \frac{\mathrm{d}x}{\cos^2 x} = A\cos x \tan x = A\sin x.$$

5.3 Bessel and Neumann functions

We now have the machinery to solve the Bessel differential equation, Eq. (3.39). We reproduce Eq. (3.39) here, generalizing the index m to an arbitrary constant, ν, which we take to be nonnegative, $\nu \geq 0$: Bessel's equation

$$x^2y''(x) + xy'(x) + (x^2 - \nu^2)y(x) = 0. \tag{5.25}$$

In physical applications, ν is often an integer or half-integer.

The point $x = 0$ is a regular singular point of Eq. (5.25), and thus, we try as a solution $y(x) = \sum_{n=0}^{\infty} a_n x^{n+r}$. We find that Eq. (5.25) is equivalent to

$$a_0x^r(r^2 - \nu^2) + a_1x^{r+1}((r+1)^2 - \nu^2) + \sum_{k=2}^{\infty}[a_k((k+r)^2 - \nu^2) + a_{k-2}]x^{k+r} = 0.$$

We eliminate the first term through the indicial equation $r^2 - \nu^2 = 0$, with roots $r_1 = \nu$ and $r_2 = -\nu$, and we set $a_1 = 0$ implying $a_1 = a_3 = \cdots = a_{2n+1} = \cdots = 0$. The recursion relation is, utilizing $r^2 = \nu^2$,

$$a_k = -\frac{1}{k(k+2r)}a_{k-2}. \qquad k = 2, 3, \ldots \quad \text{with} \quad a_1 = 0 \tag{5.26}$$

Iterating Eq. (5.26), the solution to Eq. (5.25) is, for either value of r,

$$y(x) = a_0 x^r \sum_{n=0}^{\infty} \frac{(-1)^n (x/2)^{2n}}{n!(1+r)(2+r)\cdots(n+r)}. \tag{5.27}$$

The denominator in Eq. (5.27) can be tidied up using the gamma function (Appendix C): $\Gamma(n+r+1) = (n+r)(n+r-1)\cdots(r+1)\Gamma(r+1)$. It's traditional to take $a_0 = (2^r\Gamma(r+1))^{-1}$ – a convention that we adopt. For $r = \nu \geq 0$, we have from Eq. (5.27) the Bessel function of order ν:

Bessel function

$$J_\nu(x) \equiv \left(\frac{x}{2}\right)^\nu \sum_{n=0}^{\infty} \frac{(-1)^n}{n!\Gamma(n+\nu+1)} \left(\frac{x}{2}\right)^{2n}. \tag{5.28}$$

The series in Eq. (5.28) converges for all $|x| < \infty$. The first few Bessel functions $J_n(x)$ for integer n are shown in Figure 5.1. These curves are obtained by summing Eq. (5.28) for $\nu = 0, 1, 2, 3$.

The other solution follows from Eq. (5.28) by letting $\nu \to -\nu$:

$$J_{-\nu}(x) = \left(\frac{x}{2}\right)^{-\nu} \sum_{n=0}^{\infty} \frac{(-1)^n}{n!\Gamma(n-\nu+1)} \left(\frac{x}{2}\right)^{2n}. \tag{5.29}$$

$J_\nu(x)$ and $J_{-\nu}(x)$ are the *Bessel functions of the first kind* of order ν and $-\nu$. General solutions to Bessel's equation are written

$$y(x) = AJ_\nu(x) + BJ_{-\nu}(x) \qquad (\nu \neq \text{integer})$$

for arbitrary constants A and B.

The Frobenius method tells us that if $r_1 - r_2 = 2\nu \neq$ integer, Eqs. (5.28) and (5.29) are the two linearly independent solutions of Eq. (5.25). It turns out, however, that for $2\nu =$ odd integer, J_ν and $J_{-\nu}$ are well

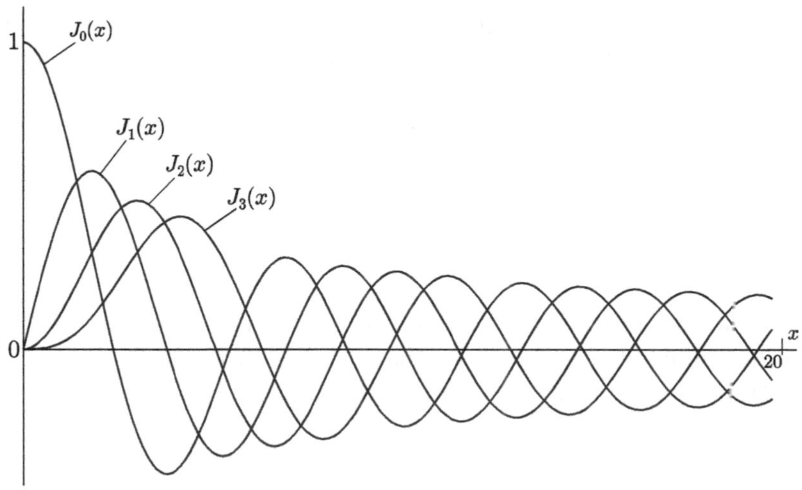

Figure 5.1 The first few Bessel functions $J_n(x)$.

defined and independent. It's only for $\nu =$ integer that we have the difficult problem of obtaining the second solution. The Wronskian of J_ν and $J_{-\nu}$ is given by[8]

$$W(J_\nu(x), J_{-\nu}(x)) = -\frac{2 \sin \pi\nu}{\pi x}. \tag{5.30}$$

For $\nu \neq n =$ integer, J_ν and $J_{-\nu}$ are linearly independent.

We can show directly that J_{-m} is not independent of J_m for integer m, in which case $J_{-m}(x) = (-1)^m J_m(x)$. With $\nu = m$ in Eq. (5.29), $\Gamma(n - m + 1)$ diverges for $n \leq m - 1$ (Appendix C). The series in Eq. (5.29) therefore starts at $n = m$:

$$\begin{aligned} J_{-m}(x) &= \left(\frac{x}{2}\right)^{-m} \sum_{n=m}^{\infty} \frac{(-1)^n}{n!\Gamma(n-m+1)} \left(\frac{x}{2}\right)^{2n} = \left(\frac{x}{2}\right)^{-m} \sum_{k=0}^{\infty} \frac{(-1)^{m+k}}{(k+m)!\Gamma(k+1)} \left(\frac{x}{2}\right)^{2(k+m)} \\ &= (-1)^m \left(\frac{x}{2}\right)^{m} \sum_{k=0}^{\infty} \frac{(-1)^k}{k!\Gamma(m+k+1)} \left(\frac{x}{2}\right)^{2k} = (-1)^m J_m(x), \end{aligned} \tag{5.31}$$

where we changed indices $k = n - m$ in the second equality and we used Eq. (5.28).

Neumann functions

For $\nu = m =$ integer, we do not have a second, linearly independent solution to Bessel's Eq. (5.25). That's too bad because many (if not most) applications of Bessel functions in physics involve just this case! From

[8]We don't have the tools yet to efficiently evaluate the Wronskian – for that we need the recursion relations satisfied by Bessel functions, Chapter 7.

the analysis of Section 5.1.4 (cases II and III), we know that the second solution will be of the form

$$\nu = m \quad y_2(x) = CJ_m(x)\ln x + \sum_{n=0}^{\infty} b_n x^{n-m} \quad b_0 \neq 0\,, C \neq 0\,, x > 0$$

$$\nu = 0 \quad y_2(x) = J_0(x)\ln x + \sum_{n=1}^{\infty} b_n x^n .$$

The plan would be to substitute these expressions into Eq. (5.25) and determine the coefficients $\{b_n\}$ so that a solution is obtained.

It's traditional, however, to proceed differently. Define the *Neumann function*[9]

Neumann function

$$N_\nu(x) \equiv \frac{\cos(\pi\nu)J_\nu(x) - J_{-\nu}(x)}{\sin(\pi\nu)} . \tag{5.32}$$

For general index ν, the Neumann functions, being a linear combination of J_ν and $J_{-\nu}$, are solutions of Bessel's equation. They are linearly independent of the Bessel functions $J_\nu(x)$, as can be seen from the Wronskian:

$$\begin{vmatrix} J_\nu & N_\nu \\ J'_\nu & N'_\nu \end{vmatrix} = -\frac{1}{\sin\pi\nu}\begin{vmatrix} J_\nu & J_{-\nu} \\ J'_\nu & J'_{-\nu} \end{vmatrix} = -\frac{1}{\sin\pi\nu}\left(\frac{-2\sin\pi\nu}{\pi x}\right) = \frac{2}{\pi x}, \tag{5.33}$$

where we've used Eq. (5.30). For arbitrary ν, N_ν is independent of J_ν. If we can show that the Neumann functions are well defined for $\nu = n$, we're done; we will have found a general second solution to the Bessel differential equation.

From Eq. (5.32),

$$N_n(x) = \lim_{\nu\to n}\left[\frac{\cos(\pi\nu)J_\nu(x) - J_{-\nu}(x)}{\sin(\pi\nu)}\right] = \frac{(-1)^n J_n - J_{-n}}{\sin(n\pi)} . \tag{5.34}$$

Using $J_{-n} = (-1)^n J_n$, we see that Eq. (5.34) yields the indeterminate form 0/0. In such cases we invoke l'Hopital's rule – take the limit of the ratio of derivatives. The derivatives we need, however, are *with respect to the order* ν of the Bessel functions:

$$\begin{aligned} N_n(x) &= \lim_{\nu\to n}\left[\frac{\cos(\pi\nu)\frac{\mathrm{d}J_\nu(x)}{\mathrm{d}\nu} - \pi\sin(\pi\nu)J_\nu(x) - \frac{\mathrm{d}J_{-\nu}(x)}{\mathrm{d}\nu}}{\pi\cos\pi\nu}\right] \\ &= \frac{1}{\pi}\left[\frac{\mathrm{d}J_n(x)}{\mathrm{d}n} - (-1)^n\frac{\mathrm{d}J_{-n}(x)}{\mathrm{d}n}\right] . \end{aligned} \tag{5.35}$$

Assuming for the present that the derivatives with respect to order are well defined (they are), we should verify that $N_n(x)$ as formally specified by Eq. (5.35) solves Bessel's equation. Equation (5.25) (it does). This is done in Exercise 5.7.

[9]Mathematicians call this the *Weber* function and denote it by $Y_\nu(x)$.

From Eq. (5.28),

$$\frac{\mathrm{d}J_n}{\mathrm{d}n} = \ln\left(\frac{x}{2}\right) J_n(x) + \sum_{k=0}^{\infty} \frac{(-1)^k (x/2)^{2k+n}}{k!} \frac{\mathrm{d}}{\mathrm{d}n}\left(\frac{1}{\Gamma(k+n+1)}\right), \tag{5.36}$$

while from Eq. (5.29),

$$\frac{\mathrm{d}J_{-n}}{\mathrm{d}n} = -\ln\left(\frac{x}{2}\right) J_{-n}(x) + \sum_{k=0}^{\infty} \frac{(-1)^k (x/2)^{2k-n}}{k!} \frac{\mathrm{d}}{\mathrm{d}n}\left(\frac{1}{\Gamma(k-n+1)}\right). \tag{5.37}$$

To make progress, we need the derivative of the gamma function. Using the chain rule,

$$\begin{aligned}
\frac{\mathrm{d}}{\mathrm{d}n}\left(\frac{1}{\Gamma(k+n+1)}\right) &= \frac{\mathrm{d}}{\mathrm{d}t}\left(\frac{1}{\Gamma(t)}\right)\bigg|_{t=k+n+1} \\
\frac{\mathrm{d}}{\mathrm{d}n}\left(\frac{1}{\Gamma(k-n+1)}\right) &= -\frac{\mathrm{d}}{\mathrm{d}t}\left(\frac{1}{\Gamma(t)}\right)\bigg|_{t=k-n+1}.
\end{aligned} \tag{5.38}$$

Derivatives of $\Gamma^{-1}(t)$ are evaluated in Appendix C; see Eqs. (C.15) and (C.16).

Combining Eq. (5.38) with Eqs. (5.36) and (5.37) (using the results in Appendix C), and then combining with Eq. (5.35), we find

$$\begin{aligned}
N_n(x) =& \frac{2}{\pi} \ln\left(\frac{x}{2}\right) J_n(x) - \frac{1}{\pi} \sum_{k=0}^{n-1} \left(\frac{x}{2}\right)^{2k-n} \frac{(n-k+1)!}{k!} \\
&- \frac{1}{\pi} \sum_{k=0}^{\infty} \frac{(-1)^k (x/2)^{2k+n}}{k!(k+n)!} (\psi(k+n+1) + \psi(k+1)),
\end{aligned} \tag{5.39}$$

where the function $\psi(k)$ is defined in Eq. (C.14).

The Neumann functions $N_n(x)$ for $n = 0, 1, 2, 3, 4$ are shown in Figure 5.2. These curves are obtained by summing Eq. (5.39). Neumann functions are also known as *Bessel functions of the second kind.* The general solution to Bessel's equation is then

$$y(x) = AJ_\nu(x) + BN_\nu(x) \qquad \text{(no restriction on } \nu) \tag{5.40}$$

where A and B are arbitrary constants. The values of $J_\nu(x)$ and $N_\nu(x)$ are extensively tabulated. We'll develop the properties of Bessel functions in Chapter 7. The goal here has been to develop the power series representations of J_ν and N_ν. We note that in practice, if the domain of the problem includes the origin, Neumann functions must be excluded because they diverge as $x \to 0$, i.e. set $B = 0$ in Eq. (5.40).

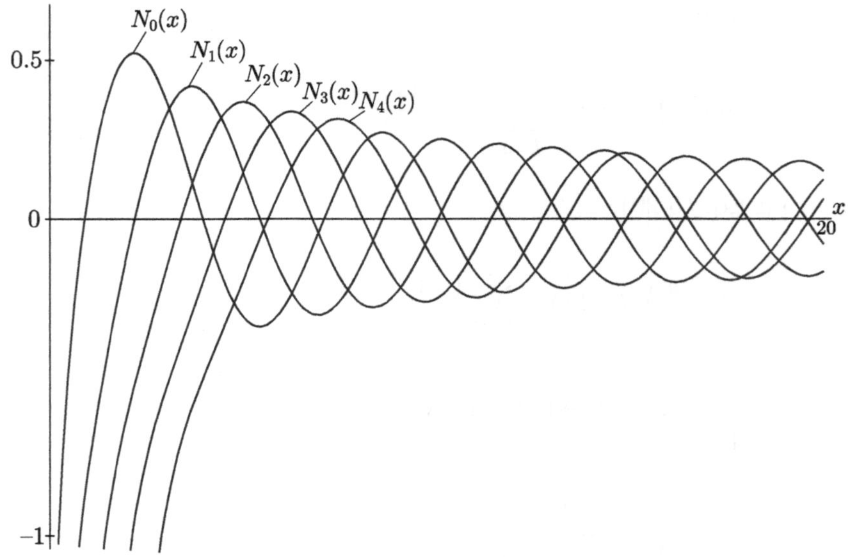

Figure 5.2 The first few Neumann functions $N_n(x)$.

5.4 Legendre polynomials

The associated Legendre differential equation, Eq. (3.75) has two parameters: λ, presently of unknown value, and the integer m. In this section, we consider its solutions for $m = 0$. We return in Chapter 6 to the case of $m \neq 0$. Thus, Eq. (3.75) with $m = 0$ is

$$(1 - x^2)y'' - 2xy' + \lambda y = 0. \tag{5.41}$$

Equation (5.41) has regular singular points at $x = \pm 1$, while $x = 0$ is an ordinary point. We expect to find power series solutions centered at $x = 0$ that converge at least for $|x| < 1$. With $y = \sum_{n=0}^{\infty} a_n x^n$, we find in the usual way the recursion relation for the coefficients

$$a_{k+2} = \frac{k(k+1) - \lambda}{(k+2)(k+1)} a_k. \qquad k = 0, 1, 2, \ldots \tag{5.42}$$

There are two linearly independent solutions:

$$y_1(x) = a_0 \left(1 - \frac{\lambda}{2} x^2 - \frac{\lambda(6-\lambda)}{4!} x^4 - \frac{\lambda(6-\lambda)(20-\lambda)}{6!} x^6 - \cdots \right), \tag{5.43}$$

and

$$y_2(x) = a_1 x \left(1 + \frac{2-\lambda}{3!} x^2 + \frac{(2-\lambda)(12-\lambda)}{5!} x^4 + \frac{(2-\lambda)(12-\lambda)(30-\lambda)}{7!} x^6 + \cdots \right). \tag{5.44}$$

Let's analyze the terms in the series for large k (for either Eqs. (5.43) or (5.44)). Using Eq. (5.42) we see that, for fixed λ,

$$\lim_{k \to \infty} \frac{a_{k+2}}{a_k} = 1. \tag{5.45}$$

By the ratio test, Eq. (5.45) tells us that the series (Eqs. (5.43) or (5.44)) are convergent for $|x| < 1$. The ratio test, however, is inconclusive for $x = \pm 1$; further analysis is required. It can be shown these series are divergent for $x = \pm 1$ [14, p. 326]. Equations (5.43) and (5.44) thus produce (for general values of λ) arbitrarily large values as $|x| \to 1$. Yet, there's nothing special in physical applications associated with $x = \pm 1$ (or $\theta = 0, \pi$). We'd "like" solutions of Eq. (5.41) that are *finite* for $|x| \leq 1$.

Divergent solutions can be prevented by choosing λ in such a way that the recursion relation Eq. (5.42) *terminates*. From Eq. (5.42), the series terminate when λ has any of the values given by $\lambda = l(l+1)$, $l = 0, 1, 2, \ldots$. For a given value of l, one of the series in Eqs. (5.43) or (5.44) reduces to a polynomial, while the other is discarded as it does not meet the requirement that the solution be finite for $|x| \leq 1$. Thus, for the values of λ generated by the formula $\lambda = l(l+1)$, we have a progression

Table 5.1 Legendre polynomals $P_n(x)$ for $0 \leq n \leq 5$.

n	$P_n(x)$
0	1
1	x
2	$\frac{1}{2}(3x^2 - 1)$
3	$\frac{1}{2}(5x^3 - 3x)$
4	$\frac{1}{8}(35x^4 - 30x^2 + 3)$
5	$\frac{1}{8}(63x^5 - 70x^3 + 15x)$

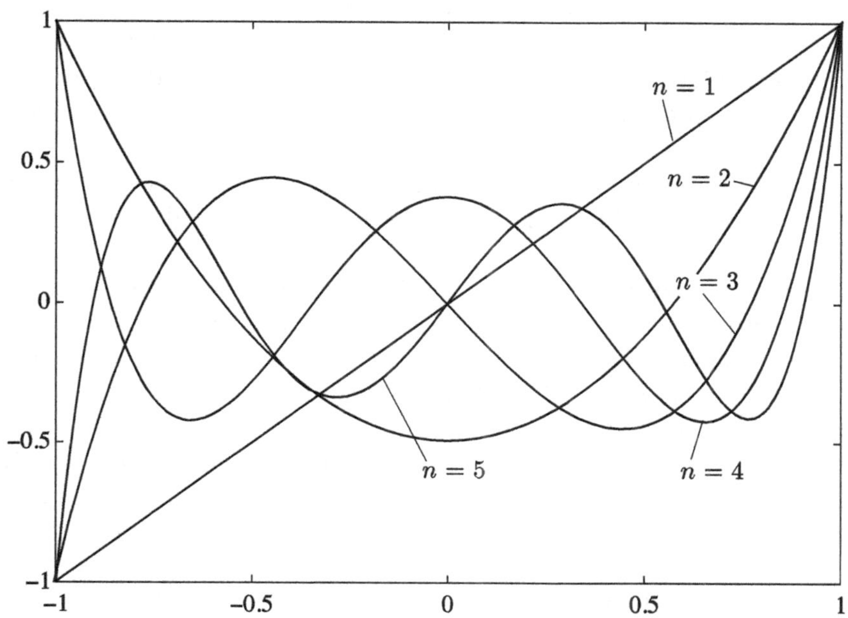

Figure 5.3 Legendre polynomials $P_n(x)$ for $n = 0, 1, 2, 3, 4, 5$.

of solutions to Eq. (5.41) that are alternately even and odd about the origin.

Legendre polynomials

With $\lambda = l(l+1)$ in Eq. (5.41), we have the differential equation for the *Legendre polynomials*:

$$(1-x^2)P_l'' - 2xP_l' + l(l+1)P_l = 0.$$
$$l = 0, 1, 2, \ldots \tag{5.46}$$

Expressions for $P_n(x)$ are listed in Table 5.1 for $n = 0, 1, 2, 3, 4, 5$. By convention, the Legendre polynomials are crafted so that $P_n(1) = 1$ for all n. Graphs of $P_n(x)$ are shown in Figure 5.3. We'll consider the properties of the Legendre polynomials in Chapter 6.

Summary

This chapter introduced the Frobenius method for solving linear homogeneous ODEs with variable coefficients, $y'' + P(x)y' + Q(x)y = 0$.

- An ordinary point x_0 is one for which the functions $P(x)$ and $Q(x)$ are analytic. At such points, try a solution of the form $y(x) = \sum_{n=0}^{\infty} a_n (x - x_0)^n$. If the functions P and Q are polynomials, the power series converges for all x.
- A regular singular point x_0 is one for which $(x - x_0)P(x)$ and $(x - x_0)^2 Q(x)$ are analytic. At such points, try a solution of the form $y(x) = \sum_{n=0}^{\infty} a_n (x - x_0)^{n+r}$. The exponent r must be obtained from the solution of the indicial equation.
- Several possibilities exist depending on the roots of the indicial equation. If the roots are distinct and do not differ by an integer, the Frobenius method produces two distinct solutions. If the roots are equal or differ by an integer, the method is guaranteed to produce at least one solution. In case of only one solution, rules were given for finding the second solution.
- An alternative method for finding the second linearly independent solution, if one solution is already known, is through the use of the Wronskian.
- Convergent power series solutions for the Bessel differential equation were derived, those for $J_\nu(x)$ and $J_{-\nu}(x)$, which are the two linearly independent solutions if $\nu \neq$ integer. For $\nu =$ integer n, $J_{-n}(x) = (-1)^n J_n(x)$. The Neumann function $N_\nu(x)$ is a linearly independent second solution valid for all ν.
- Power series solutions for the Legendre differential Eq. (5.41) (involving the parameter λ) are not well behaved for $|x| \to 1$ for arbitrary λ. To arrive at physically acceptable solutions, λ must be restricted to the special values generated by the formula $\lambda = l(l+1)$ for $l = 0, 1, 2, \ldots$. For these special values, the power series reduce to polynomials.

Exercises

5.1. Show that if a function $y(x) = \sum_{n=0}^{\infty} a_n (x - x_0)^n$ converges for $|x - x_0| < r$, then the power series representing its derivative, $y'(x) = \sum_{n=1}^{\infty} n a_n (x - x_0)^{n-1}$, is convergent within the same interval of convergence. Hint:

$$\lim_{n\to\infty} \left(\frac{n+1}{n} \frac{a_{n+1}}{a_n} \right) = \lim_{n\to\infty} \left(\frac{a_{n+1}}{a_n} \right).$$

5.2. Find the two linearly independent solutions of $x^2 y'' - 2xy' + 2y = 0$. This is kind of a trick problem. Applying the Frobenius method, you should find that there is no recursion relation for the coefficients of a power series. That implies the solution of the differential equation is not in the form of an infinite summation.

5.3. Solve the differential equation $xy'' + \left(\frac{1}{2} - x\right) y' - \frac{1}{2} y = 0$. You should find that the roots of the indicial equation are $r = 0$ and $r = \frac{1}{2}$. A: $y_1(x) = \mathrm{e}^x$,

$$y_2(x) = \sqrt{x} \sum_{n=0}^{\infty} \frac{(2x)^n}{(2n+1)(2n-1)\cdots 5 \cdot 3 \cdot 1}.$$

5.4. Find the solutions to the differential equation

$$y'' + \left(1 - \frac{1}{x}\right) y' - \frac{1}{x} y = 0.$$

You should find for the roots of the indicial equation $r = 0$ and $r = 2$. The solution for $r = 0$ is easy to recognize (A: $y_1(x) = \mathrm{e}^{-x}$). Use the Wronskian method to find the second solution. (A: $y_2(x) = x - 1$)

5.5. Show that the differential equation

$$4x^2 y''(x) + (1 - x^2) y(x) = 0$$

has two solutions in the form

$$y_1(x) = \sqrt{x} \left(1 + \frac{x^2}{16} + \frac{x^4}{1024} + \cdots \right)$$

$$y_2(x) = y_1(x) \ln x - \frac{x^{5/2}}{16} + \cdots$$

for x near zero.

5.6. Work out the steps leading to Eq. (5.27) from Eq. (5.26).

5.7. Verify that the group of terms in Eq. (5.35) is a solution to Eq. (5.25). Formally differentiate equation (5.25) with respect to n:

$$x^2 \frac{\mathrm{d}^2}{\mathrm{d}x^2} \left(\frac{\mathrm{d}J_n}{\mathrm{d}n} \right) + x \frac{\mathrm{d}}{\mathrm{d}x} \left(\frac{\mathrm{d}J_n}{\mathrm{d}n} \right) + (x^2 - n^2) \left(\frac{\mathrm{d}J_n}{\mathrm{d}n} \right) = 2nJ_n(x).$$

Show that when we do the same for $-n$, we obtain:

$$x^2 \frac{\mathrm{d}^2}{\mathrm{d}x^2} \left(\frac{\mathrm{d}J_{-n}}{\mathrm{d}n} \right) + x \frac{\mathrm{d}}{\mathrm{d}x} \left(\frac{\mathrm{d}J_{-n}}{\mathrm{d}n} \right) + (x^2 - n^2) \left(\frac{\mathrm{d}J_{-n}}{\mathrm{d}n} \right) = 2nJ_{-n}(x).$$

Use $J_{-n}(x) = (-1)^n J_n(x)$ and the formulas derived here to show that $N_n(x)$ as specified by Eq. (5.35) is a solution to Eq. (5.25).

5.8. Derive Eq. (5.36) and Eq. (5.37). Hint: $(x/2)^n = \mathrm{e}^{n\ln(x/2)}$.

5.9. For what values of the constant k does the differential equation

$$xy'' - 2xy' + (k - 3x)y = 0 \tag{5.47}$$

have a nontrivial, bounded solution for $0 \le x < \infty$?

(a) Clearly you're going to try a series solution to Eq. (5.47). Instead of a direct attack on Eq. (5.47), however, let's build in from the outset that we want a bounded solution. How do we solve differential equations? Guess. Try as a solution, $y(x) = \mathrm{e}^{-\alpha x} f(x)$. Show that the form of Eq. (5.47) simplifies for $\alpha = 1$ and $\alpha = -3$. Keep $\alpha = 1$, as it leads to a bounded solution – *if* $f(x)$ is bounded over $[0, \infty)$. For $\alpha = 1$, show that $f(x)$ satisfies the equation

$$xf'' - 4xf' + kf = 0. \tag{5.48}$$

(b) As $x = 0$ is a regular singular point of Eq. (5.48), try a series solution in the form of Eq. (5.10). Show that the roots of the indicial equation are $r = 0$ and $r = 1$. We anticipate that finding a second, linearly independent solution falls under the conditions specified in Section 5.1.4. Show in fact that one cannot develop a power series solution of Eq. (5.48) in the case of $r = 0$.

(c) Show that the recursion relation for the expansion coefficients a_m (associated with $r = 1$) is satisfied by

$$a_m = \frac{(4m - k)(4(m-1) - k)\cdots(4 - k)}{(m+1)!m!} a_0 \qquad m = 1, 2, \ldots$$

(d) For these expansion coefficients, is the infinite series $\sum_{n=0}^{\infty} a_n x^n$ convergent? Perform the ratio test of the coefficients (see Appendix B). Show that

$$\frac{a_{n+1}}{a_n} = \frac{4(n+1) - k}{(n+2)(n+1)}.$$

Is there any restriction on k that prevents one from having a bounded solution? What is the radius of convergence of the series?

(e) Suppose k has the values $k = 4m$, $m = 1, 2, 3, \ldots$. For such values of k, show that the solution to Eq. (5.47) is in the form

$$y(x) = x\mathrm{e}^{-x} p_{m-1}(x)$$

where $p_{m-1}(x)$ is a polynomial of degree $m - 1$. Give expressions for the first few polynomials. Show that $y = x\mathrm{e}^{-x}$ solves Eq. (5.47) for $k = 4$. Show that $f = x$ solves Eq. (5.48) for $k = 4$.

6

Spherical harmonics

The spherical harmonic functions $Y(\theta, \phi)$ were introduced in Chapter 3 as satisfying Eq. (3.46), which we reproduce here:

$$\left[\frac{1}{\sin\theta}\frac{\partial}{\partial\theta}\left(\sin\theta\frac{\partial}{\partial\theta}\right)+\frac{1}{\sin^2\theta}\frac{\partial^2}{\partial\phi^2}\right]Y(\theta,\phi)\equiv L^2Y(\theta,\phi)=-\lambda Y(\theta,\phi), \tag{3.46}$$

where L^2 denotes the *angular part* of the Laplacian operator – what remains of ∇^2 if the problem is restricted to the surface of the unit sphere.[1] The spherical harmonics are thus eigenfunctions of L^2; they occur in applications when a system possesses *spherical symmetry*.[2] In this chapter, we develop the properties of the spherical harmonics.

We found in Chapter 3 that we could write $Y(\theta, \phi)$ in product form,

$$Y(\theta,\phi)=P(\theta)\mathrm{e}^{\mathrm{i}m\phi}, \tag{6.1}$$

where m is an integer and where the function $P(\theta)$ satisfies Eq. (3.75), the associated Legendre differential equation (with $x = \cos\theta$):

$$(1-x^2)\frac{\mathrm{d}^2P}{\mathrm{d}x^2}-2x\frac{\mathrm{d}P}{\mathrm{d}x}+\left(\lambda-\frac{m^2}{1-x^2}\right)P=0. \tag{3.75}$$

Equation (3.75) has regular singular points at $x = \pm 1$ (Exercise 6.1), and thus we can try power-series solutions about those points,

[1]The notation L^2 is motivated by quantum theory, where what we're calling L^2 is, up to a multiplicative factor of $-\hbar^2$, the operator representing the square of the orbital angular momentum. The "square" notation shouldn't cause confusion: L^2 is a linear operator, the same as ∇^2.

[2]Electrons in isolated atoms move in a central potential field $V(r)$ depending only on the radial coordinate r and not on the angular coordinates θ, ϕ; the wave function is then in the product form $\psi(r, \theta, \phi) = R(r)Y(\theta, \phi)$.

Mathematical Methods in Physics, Engineering, and Chemistry, First Edition.
Brett Borden and James Luscombe.

$P(x) = \sum_{n=0}^{\infty} a_n (x \mp 1)^{n+r}$. The indicial equation is the same for either expansion about $x = 1$ or $x = -1$ (Exercise 6.2),

$$r^2 - \frac{m^2}{4} = 0. \tag{6.2}$$

Therefore, for expansions about either $x = \pm 1$, there are two solutions corresponding to the roots of the indicial equation, $\pm|m|/2$. For the expansion about $x = 1$, one solution is finite for $x \to 1$, varying like $(1-x)^{|m|/2}$, while the other solution diverges, varying like $(1-x)^{-|m|/2}$. For the expansion about $x = -1$, we find the same: one solution finite for $x \to -1$, with the other divergent. If we try to extend the solution finite about $x = 1$ to $x = -1$, it diverges there. We can't find, for *general* values of λ, solutions finite at both $x = \pm 1$.

We found in Section 5.4 that for $m = 0$, the only solutions finite at $x = \pm 1$, the Legendre polynomials, $P_l(x)$, occur for *special* values of λ, those generated by the formula $\lambda = l(l+1)$, $l = 0, 1, 2, \cdots$. We show in Section 6.2 that the same applies for $m \neq 0$ (solutions finite at $x = \pm 1$ when $\lambda = l(l+1)$). First, however, we look at the Legendre polynomials in greater detail.

6.1 Properties of the Legendre polynomials, $P_l(x)$

6.1.1 Rodrigues formula

parity property

Examining Eqs. (5.43) and (5.44), the power-series solutions of the Legendre differential equation for general λ (Eq. (5.41)), we see that for *even* values of l, the polynomial solutions $P_l(x)$ (that the power series reduce to for $\lambda = l(l+1)$) contain only even powers of x, while for *odd* values of l, $P_l(x)$ contains only odd powers of x. Thus, we have the *parity property* $P_l(-x) = (-1)^l P_l(x)$. As polynomials of degree l, the highest power of x in $P_l(x)$ is x^l. With these observations, we can write $P_l(x)$ in the form

$$P_l(x) = \sum_{k=0}^{[l/2]} a_k x^{l-2k}, \tag{6.3}$$

where $[l/2]$ denotes the greatest integer less than or equal to $\frac{1}{2}l$, i.e. $\frac{1}{2}l$ for even l and $\frac{1}{2}(l-1)$ for l odd.[3]

In the usual way, we obtain a recursion relation for the coefficients a_k (substitute Eq. (6.3) into the differential equation for $P_l(x)$, Eq. (5.46)):

$$a_k = -\frac{(l-2k+2)(l-2k+1)}{2k(2l+1-2k)} a_{k-1}. \qquad \left(k = 1, \ldots, \left[\frac{l}{2}\right]\right) \tag{6.4}$$

[3]For example, $[5/2] = 2$. Note that a_0 in Eq. (6.3) is the coefficient of x^l. Equation (6.3) "starts at the top" and works down to the lowest power of x in decrements of two, either x^0 for even l or x for odd l. By writing Eq. (6.3) in terms of $[l/2]$, even and odd values of l are treated on equal footing.

By iterating Eq. (6.4), we obtain an explicit expression for a_k (Exercise 6.4),

$$a_k = (-1)^k \frac{(l!)^2(2l-2k)!}{k!(2l)!(l-2k)!(l-k)!}a_0. \qquad \left(k = 1, \ldots, \left[\frac{l}{2}\right]\right) \tag{6.5}$$

By convention, a_0 is taken to have the value[4] $a_0 = (2l)!/(2^l(l!)^2)$. With this choice of a_0 in Eq. (6.5), the values of a_k are given by

$$a_k = (-1)^k \frac{(2l-2k)!}{2^l k!(l-2k)!(l-k)!}. \qquad \left(k = 0, 1, \ldots, \left[\frac{l}{2}\right]\right) \tag{6.6}$$

Combining Eq. (6.6) with Eq. (6.3), we have an explicit formula for $P_l(x)$:

$$P_l(x) = \frac{1}{2^l}\sum_{k=0}^{[l/2]} \frac{(-1)^k}{k!}\frac{(2l-2k)!}{(l-2k)!(l-k)!}x^{l-2k}. \tag{6.7}$$

Equation (6.7) can be simplified (considerably) by invoking the binomial expansion:

$$(x^2-1)^l = \sum_{k=0}^{l}(-1)^k\binom{l}{k}x^{2l-2k} = \left(\sum_{k=0}^{[l/2]} + \sum_{k=[l/2]+1}^{l}\right)(-1)^k\binom{l}{k}x^{2l-2k}, \tag{6.8}$$

where $\binom{l}{k} \equiv l!/(k!(l-k)!)$ is the *binomial coefficient*.[5] For even values of l, the sum from $k=0$ to $[l/2]$ in Eq. (6.8) contains the powers of x, $(x^{2l}, x^{2l-2}, \ldots, x^l)$, while the sum from $[l/2]+1$ to l has powers $(x^{l-2}, \ldots, x^0)$. For odd values of l, the sums have the powers $(x^{2l}, x^{2l-2}, \ldots, x^{l+1})$ and $(x^{l-1}, \ldots, x^0)$. As can be verified (Exercise 6.7),

binomial coefficient

$$\frac{\mathrm{d}^l}{\mathrm{d}x^l}x^{2l-2k} = \frac{(2l-2k)!}{(l-2k)!}x^{l-2k}, \tag{6.9}$$

just the term that appears in Eq. (6.7). By differentiating Eq. (6.8), we have, using Eq. (6.9) and that the lth derivative of the second sum in Eq. (6.8) vanishes,

$$\begin{aligned}\frac{\mathrm{d}^l}{\mathrm{d}x^l}(x^2-1)^l &= \sum_{k=0}^{[l/2]}(-1)^k\binom{l}{k}\frac{\mathrm{d}^l}{\mathrm{d}x^l}x^{2l-2k}\\ &= \sum_{k=0}^{[l/2]}(-1)^k\binom{l}{k}\frac{(2l-2k)!}{(l-2k)!}x^{l-2k} = 2^l l!\, P_l(x),\end{aligned}$$

[4]Recall that for Bessel functions (Section 5.3), there's also a conventional choice of a_0.

[5]The binomial theorem is $(x+y)^n = \sum_{k=0}^{n}\binom{n}{k}x^{n-k}y^k$ for n a positive integer.

where the last equality follows from Eq. (6.7). For all l, therefore,

Rodrigues formula

$$P_l(x) = \frac{1}{2^l l!}\frac{\mathrm{d}^l}{\mathrm{d}x^l}(x^2-1)^l. \tag{6.10}$$

Equation (6.10) is known as the *Rodrigues formula*[6]; it (or Eq. (6.7)) can be used to generate the expressions listed in Table 5.1. We should verify that $P_l(x)$ expressed in the form of Eq. (6.10) satisfies the differential Eq. (5.46) (it does); see Exercise 6.8. We infer from Eq. (6.10) that $P_l(1) = 1$ (Exercise 6.9).[7] We see the parity property $P_l(-1) = (-1)^l$ at work in Figure 5.3.

6.1.2 Orthogonality

The differential equation for $P_l(x)$, Eq. (5.46), is a Sturm–Liouville equation subject to the boundary condition that $P_l(x)$ is finite at $x = \pm 1$. We're therefore guaranteed that $\{P_l(x)\}_{l=0}^{\infty}$ is a complete orthogonal set, $\int_{-1}^{1} P_n(x)P_m(x)\,\mathrm{d}x = 0$, $n \neq m$ (see Eq. (2.29)).

For $n = m$, use the Rodrigues formula to calculate the normalization integral:

$$\int_{-1}^{1} P_l^2(x)\,\mathrm{d}x = \frac{1}{(2^l l!)^2}\int_{-1}^{1}\frac{\mathrm{d}^l}{\mathrm{d}x^l}(x^2-1)^l\left(\frac{\mathrm{d}^l}{\mathrm{d}x^l}(x^2-1)^l\right)\,\mathrm{d}x. \tag{6.11}$$

Equation (6.11) presents us with a challenging integral; we're going to evaluate it through repeated integration by parts. Integrate the right side of Eq. (6.11) by parts:

$$\begin{aligned}\int_{-1}^{1}\frac{\mathrm{d}^l}{\mathrm{d}x^l}(x^2-1)^l\left(\frac{\mathrm{d}^l}{\mathrm{d}x^l}(x^2-1)^l\right)\,\mathrm{d}x &= \frac{\mathrm{d}^{l-1}}{\mathrm{d}x^{l-1}}(x^2-1)^l\left(\frac{\mathrm{d}^l}{\mathrm{d}x^l}(x^2-1)^l\right)\Bigg|_{-1}^{1}\\ &- \int_{-1}^{1}\frac{\mathrm{d}^{l-1}}{\mathrm{d}x^{l-1}}(x^2-1)^l\left(\frac{\mathrm{d}^{l+1}}{\mathrm{d}x^{l+1}}(x^2-1)^l\right)\,\mathrm{d}x.\end{aligned}$$

The integrated part vanishes (Exercise 6.10). Repeat l times:

$$\begin{aligned}\int_{-1}^{1} P_l^2(x)\,\mathrm{d}x &= \frac{(-1)^l}{(2^l l!)^2}\int_{-1}^{1}(x^2-1)^l\frac{\mathrm{d}^{2l}}{\mathrm{d}x^{2l}}(x^2-1)^l\,\mathrm{d}x\\ &= \frac{(-1)^l(2l)!}{(2^l l!)^2}\int_{-1}^{1}(x^2-1)^l\,\mathrm{d}x,\end{aligned} \tag{6.12}$$

where we've used the result of Exercise 6.11. Noting that $(x^2-1)^l = (x-1)^l(x+1)^l$, integrate by parts the integral in Eq. (6.12):

[6]The Rodrigues formula is also written $P_l(x) = [(-1)^l/(2^l l!)](\,\mathrm{d}^l/\,\mathrm{d}x^l)(1-x^2)^l$.

[7]That $P_l(1) = 1$ can be shown from Eq. (6.7), but to do so requires dexterity in the use of binomial coefficients.

$$\int_{-1}^{1} (x-1)^l (x+1)^l \, dx = \frac{1}{l+1}(x-1)^{l+1}(x+1)^l \Big|_{-1}^{1} - \frac{l}{l+1}\int_{-1}^{1} (x-1)^{l+1}(x+1)^{l-1} \, dx,$$

where the integrated part vanishes. Repeating l times,

orthogonality relation

$$\int_{-1}^{1} (x-1)^l (x+1)^l \, dx = \frac{(-1)^l (l!)^2}{(2l)!}\int_{-1}^{1} (x-1)^{2l} dx = \frac{(-1)^l (l!)^2}{(2l)!}\frac{2^{2l+1}}{2l+1}. \tag{6.13}$$

Combining Eq. (6.13) with Eq. (6.12), we have the desired result

$$\int_{-1}^{1} P_l(x) P_m(x) \, dx = \frac{2}{2l+1}\delta_{lm}. \tag{6.14}$$

6.1.3 Completeness

Any square-integrable function can be expanded in a complete set of square-integrable orthogonal functions (Section 2.5). In terms of the Legendre polynomials, we have for any function $f(x)$ defined on $-1 \le x \le 1$:

Legendre series

$$f(x) = \sum_{l=0}^{\infty} a_l P_l(x). \tag{6.15}$$

Equation (6.15) is known as a *Legendre series*. The expansion coefficients are obtained from the orthogonality property, Eq. (6.14). We find (Exercise 6.12)

$$a_l = \frac{2l+1}{2}\int_{-1}^{1} P_l(x) f(x) \, dx. \tag{6.16}$$

Note that a_0 is the average of the function over the interval $[-1, 1]$. It's straightforward to show that Parseval's theorem in this case is given by[8] (Exercise 6.13)

$$\frac{1}{2}\int_{-1}^{1} |f(x)|^2 \, dx = \sum_{n=0}^{\infty} \frac{|a_n|^2}{2n+1}. \tag{6.17}$$

The completeness relation follows by substituting Eq. (6.16) into Eq. (6.15):

$$\frac{1}{2}\sum_{l=0}^{\infty} (2l+1) P_l(x) P_l(x') = \delta(x - x'). \tag{6.18}$$

Equation (6.18) provides another representation of the Dirac delta function.

[8]Compare with Eq. (6.13).

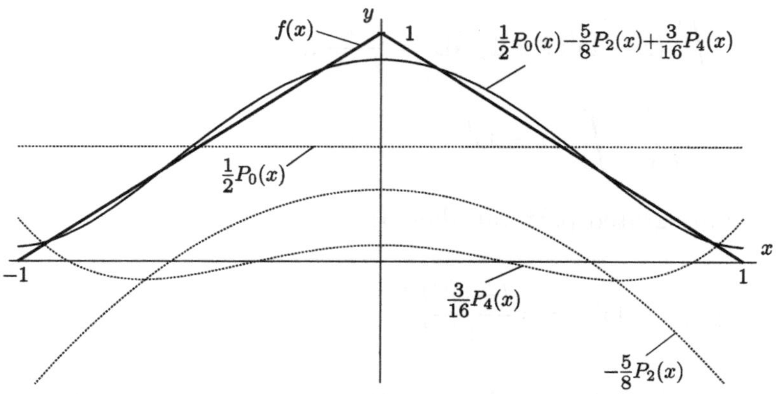

Figure 6.1 First few terms of the Legendre series for the function $f(x)$.

Example. Find the first few terms of the Legendre series for the following function:

$$f(x) = \begin{cases} 1+x & -1 \le x < 0 \\ 1-x & 0 \le x \le 1 \end{cases}.$$

The coefficients are found using Eq. (6.16). Noting that f is even, all coefficients for odd values of n vanish, $a_{2n+1} = 0$. For even values of n,

$$\begin{aligned} a_{2n} &= \frac{4n+1}{2}\left[\int_{-1}^{0} P_{2n}(x)f(x)\,\mathrm{d}x + \int_{0}^{1} P_{2n}(x)f(x)\,\mathrm{d}x\right] \\ &= (4n+1)\int_0^1 P_{2n}(x)f(x)\,\mathrm{d}x. \end{aligned}$$

Thus, using the expressions for $P_{2n}(x)$ in Table 3.1,

$$a_0 = \int_0^1 (1-x)\,\mathrm{d}x = \frac{1}{2} \qquad a_2 = \frac{5}{2}\int_0^1 (3x^2-1)(1-x)\,\mathrm{d}x = -\frac{5}{8}$$

$$a_4 = \frac{9}{8}\int_0^1 (35x^4 - 30x^2 + 3)(1-x)\,\mathrm{d}x = \frac{3}{16}.$$

From Eq. (6.15),

$$f(x) = \frac{1}{2}P_0(x) - \frac{5}{8}P_2(x) + \frac{3}{16}P_4(x) + \cdots.$$

The result of summing the first three terms in the series is shown in Figure 6.1.

6.1.4 Generating function

There is a special function, $g(x, y)$, a function of two variables, the *generating function*, having the property

$$g(x,y) \equiv \frac{1}{\sqrt{1-2xy+y^2}} = \sum_{n=0}^{\infty} P_n(x) y^n, \qquad (|y|<1) \tag{6.19}$$

generating function

where[9] $|y| < 1$. Thus, $g(x,y)$ can be seen either as a Legendre series with coefficients y^n, or equivalently as a power series in y with $P_n(x)$ as the coefficients.[10] The validity of Eq. (6.19) can be demonstrated by showing that it satisfies Parseval's theorem, Eq. (6.17).[11] Square $g(x,y)$ in Eq. (6.19) and integrate:

$$\int_{-1}^{1} \frac{dx}{1-2yx+y^2} = \sum_{n=0}^{\infty}\sum_{m=0}^{\infty} y^{n+m} \int_{-1}^{1} P_n(x)P_m(x)\, dx = 2\sum_{n=0}^{\infty} \frac{y^{2n}}{2n+1}, \tag{6.20}$$

where we've used Eq. (6.14). The integral on the left of Eq. (6.20) can be evaluated by elementary means (Exercise 6.14),

$$\int_{-1}^{1} \frac{dx}{1-2yx+y^2} = \frac{1}{y}\ln\left(\frac{1+y}{1-y}\right) = 2\sum_{n=0}^{\infty}\frac{y^{2n}}{2n+1}, \tag{6.21}$$

where the second equality in Eq. (6.21) is the Taylor series of the function obtained from the integral. The series in Eqs. (6.21) and (6.20) are identical, showing the identity of the two sides of Eq. (6.19).

Example. Electrostatic potential

The electrostatic potential φ due to a point charge q is (in SI units) given by the expression

$$\varphi = \frac{1}{4\pi\epsilon_0}\frac{q}{r}, \tag{6.22}$$

where r denotes the distance from the observer to the charge. This distance is most easily evaluated when the charge is located at the origin of coordinates. What if, however, the origin of coordinates is not so ideally situated – if, for example, q is located at a point $z = a$ on the z-axis?[12] It turns out that the generating function is well suited for just such a situation.

---

[9] With $x = \cos\theta$, $\sqrt{1-2xy+y^2} = \sqrt{(e^{i\theta}-y)(e^{-i\theta}-y)}$. With y considered a complex variable, $g(x,y)$ has singularities at $y = e^{\mp i\theta}$. Real values of y are therefore restricted to $|y| < 1$.

[10] By the way, $|P_l(x)| \leq 1$; Exercise 6.17.

[11] One can also expand $[1-2xy+y^2]^{-1/2}$ in a Taylor series, treating $|2xy - y^2|$ as a small quantity. The quantity multiplying y^n for any given n is the Legendre polynomial $P_n(x)$. See Exercise 6.15.

[12] Such a situation occurs whenever you have multiple charges at different locations (and there is more than one charge in the universe).

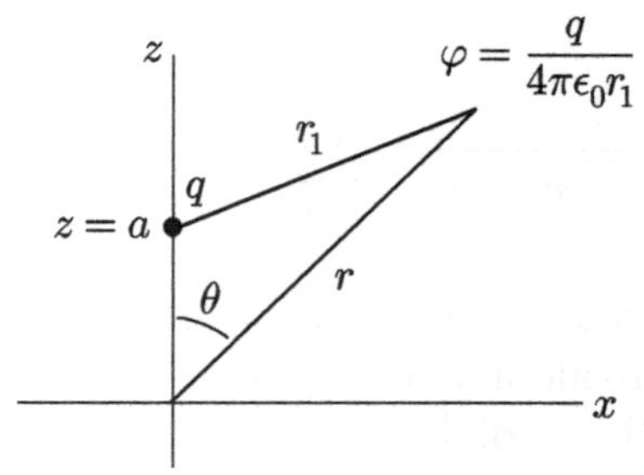

Figure 6.2 Electrostatic potential φ for a charge q displaced from the origin.

The electrostatic potential at a distance r_1 from a charge q located at $(0, 0, a)$ is

$$\varphi = \frac{q}{4\pi\epsilon_0 r_1}.$$

How do we express this result using a coordinate system with the origin displaced? Referring to Figure 6.2, apply the law of cosines $r_1^2 = r^2 + a^2 - 2ar\cos\theta$, so that

$$\frac{1}{r_1} = \frac{1}{(r^2 - 2ar\cos\theta + a^2)^{1/2}} = \frac{1}{r}\left(1 - \frac{2a\cos\theta}{r} + \frac{a^2}{r^2}\right)^{-1/2}. \tag{6.23}$$

The angle θ in Eq. (6.23) is (from Figure 6.2) the spherical polar angle, but that's an artifact of the way we've set up the problem. The angle θ in general is the angle between the vector location of the source element q (call it $\boldsymbol{a}$ in Figure 6.2) and the field point location $\boldsymbol{r}$, i.e. $\boldsymbol{r} \cdot \boldsymbol{a} = ar\cos\theta$.

We consider the case where[13] $r > a$. Equation (6.23) is then precisely in the form of the generating function, Eq. (6.19), with $y = a/r$ and $x = \cos\theta$. Thus,

$$\varphi = \frac{q}{4\pi\epsilon_0}\frac{1}{r}\sum_{n=0}^{\infty} P_n(\cos\theta)\left(\frac{a}{r}\right)^n. \tag{6.24}$$

Compare Eq. (6.24) with Eq. (6.22); as r becomes progressively larger compared to a, the predictions of Eq. (6.24) rapidly approach those of Eq. (6.22). Equation (6.24) has been obtained through the "luxury" of invoking the generating function; it's instructive to go through the process of explicitly expanding out the terms in Eq. (6.23) for $r > a$ to show that one obtains an expansion involving the Legendre polynomials – you're asked to do that in Exercise 6.15.

Example. Electric dipole

Suppose we have two charges: one with strength $+q$ located at $z = a$ and the other with strength $-q$ located at $z = -a$ (an electric dipole). As previously noted, the angle θ in Eq. (6.23) denotes the angle between

[13] The argument can easily be modified to handle the case of $r < a$. The result will be slightly different, but still a Legendre series. See Exercise 6.16.

the vector $\boldsymbol{r}$ and the vector from the origin to the charge q. Referring to Figure 6.3, the two angles are θ and $\pi - \theta$. The potential can be found by superposition. Using Eq. (6.24),

$$\varphi = \frac{q}{4\pi\epsilon_0 r_1} + \frac{-q}{4\pi\epsilon_0 r_2}$$
$$= \frac{q}{4\pi\epsilon_0 r}\sum_{n=0}^{\infty}[P_n(\cos\theta) - P_n(\cos(\pi-\theta))]\left(\frac{a}{r}\right)^n.$$

But $P_n(\cos(\pi-\theta)) = P_n(-\cos\theta) = (-1)^n P_n(\cos\theta)$, so the even terms in the series cancel, while the odd terms double up. We are left with

$$\varphi = \frac{2q}{4\pi\epsilon_0 r}\sum_{n \text{ odd}} P_n(\cos\theta)\left(\frac{a}{r}\right)^n.$$

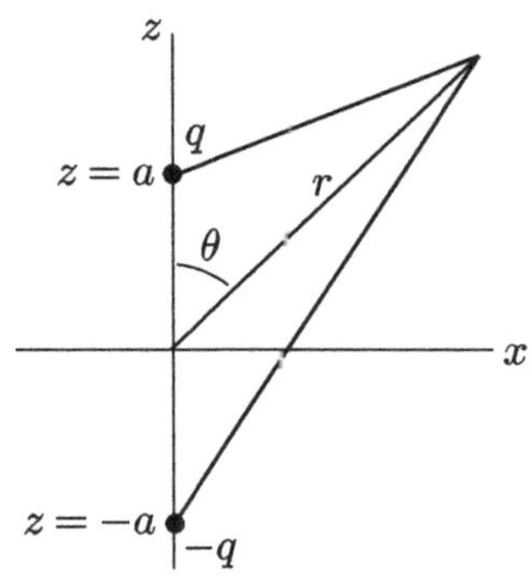

Figure 6.3 Electric dipole.

6.1.5 Recursion relations

The generating function can be used to develop additional properties of the Legendre polynomials. Differentiate $g(x,y)$ in Eq. (6.19) with respect to y:

$$\frac{\partial g(x,y)}{\partial y} = \frac{x-y}{(1-2xy+y^2)^{3/2}} = \sum_{n=0}^{\infty} nP_n(x)y^{n-1}.$$

Multiply by $1 - 2xy + y^2$ to obtain

$$(x-y)\frac{1}{(1-2xy+y^2)^{1/2}} = (1-2xy+y^2)\sum_{n=0}^{\infty} nP_n(x)y^{n-1}.$$

Substituting from Eq. (6.19) for $(1-2xy+y^2)^{-1/2}$ and rearranging, we have

$$(x-y)\sum_{n=0}^{\infty} P_n(x)y^n - (1-2xy+y^2)\sum_{n=0}^{\infty} nP_n(x)y^{n-1} = 0.$$

This equation is a power series in y. We can expand it out to reveal

$$\sum_{m=0}^{\infty} mP_m(x)y^{m-1} - x\sum_{n=0}^{\infty}(2n+1)P_n(x)y^n + \sum_{k=0}^{\infty}(k+1)P_k(x)y^{k+1} = 0.$$

(We've switched the names of the dummy indices in several places to facilitate the next step.) Now shift indices: Let $m = n + 1$ and $k = n - 1$ so that

$$\sum_{n=-1}^{\infty}(n+1)P_{n+1}(x)y^n - x\sum_{n=0}^{\infty}(2n+1)P_n(x)y^n + \sum_{n=1}^{\infty} nP_{n-1}(x)y^n = 0.$$

Gather the terms into a single power series:

$$[P_1(x) - xP_0(x)]y^0 + \sum_{n=1}^{\infty}[(n+1)P_{n+1}(x) - (2n+1)xP_n(x) + nP_{n-1}(x)]y^n = 0.$$

Because all terms in y^n must independently vanish, we see that $P_1(x) - xP_0(x) = 0$ (obviously true) and

$$(2n+1)xP_n(x) = (n+1)P_{n+1}(x) + nP_{n-1}(x). \quad n = 1, 2, 3, \ldots \quad (6.25)$$

Equation (6.25) relates $P_{n+1}(x)$ to $P_n(x)$ and $P_{n-1}(x)$ (very useful in computer programs and often much easier to apply than Eqs. (6.7) or (6.10)). Numerous other recursion relations for Legendre polynomials and their derivatives can be derived, as shown in Exercise 6.18.

Example. Find $P_3(x)$ using the recursion formula Set $n = 2$ in Eq. (6.25): $5xP_2(x) = 3P_3(x) + 2P_1(x)$. Thus,

$$P_3(x) = \frac{1}{3}(5xP_2(x) - 2P_1(x)) = \frac{1}{2}(5x^3 - 3x).$$

Note that $P_3(1) = 1$.

Example. Find the Legendre series of the step function:

$$f(x) = \begin{cases} 1 & 0 < x < 1 \\ -1 & -1 < x < 0. \end{cases}$$

Because f is odd, all coefficients in Eq. (6.15) associated with even powers of x vanish, $a_{2n} = 0$. From Eq. (6.16),

$$a_{2n+1} = \frac{4n+3}{2}\int_{-1}^{1} P_{2n+1}(x)f(x)\,dx = (4n+3)\int_0^1 P_{2n+1}(x)\,dx. \quad (6.26)$$

The integral in Eq. (6.26) simplifies through the use of a recursion relation. From the recursion relation derived in Exercise 6.18b, $(4n+3)P_{2n+1}(x) = P'_{2n+2}(x) - P'_{2n}(x)$. Thus,

$$a_{2n+1} = \int_0^1 (P'_{2n+2}(x) - P'_{2n}(x))\,dx = P_{2n}(0) - P_{2n+2}(0).$$

From Eq. (6.7), we infer that $P_{2n}(0) = (-1)^n(2n)!/(2^{2n}(n!)^2)$. Thus,

$$a_{2n+1} = (-1)^n \frac{4n+3}{2^{n+1}(n+1)} \frac{(2n)!}{(n!)^2}.$$

The Legendre series is therefore

$$f(x) = \frac{1}{2}\sum_{n=0}^{\infty}(-1)^n \frac{4n+3}{2^n(n+1)}\frac{(2n)!}{(n!)^2}P_{2n+1}(x).$$

Note that at $x=0$, the series produces the value $f(0)=0$. From the theorem cited in Section 2.5, an eigenfunction expansion converges to the arithmetic mean of the values of the function on the two sides of a discontinuity, what we see at work here.

6.2 Associated Legendre functions, $P_l^m(x)$

We now have the tools to tackle the associated Legendre differential equation, Eq. (3.75). As noted previously, the roots of the indicial equation for expansions about $x=\pm 1$ have the values $\pm m/2$ (for convenience we now take m to be positive). Let's attempt to construct a solution finite at $x=\pm 1$ by building in that feature. Try

$$P(x) = (1-x)^{m/2}(1+x)^{m/2}F(x) = (1-x^2)^{m/2}F(x), \qquad (6.27)$$

where $F(x)$ is a function to be determined. Substitute Eq. (6.27) into Eq. (3.75). We find that F satisfies a differential equation analogous to the Legendre differential equation: (Exercise 6.19)

$$(1-x^2)F'' - 2(m+1)xF' + [\lambda - m(m+1)]F = 0. \qquad (6.28)$$

Differentiate Eq. (5.46) (the differential equaiton for $P_l(x)$) m times. We find (Exercise 6.20)

$$(1-x^2)P_l^{(m+2)} - 2(m+1)xP_l^{(m+1)} + [l(l+1) - m(m+1)]P_l^{(m)} = 0, \qquad (6.29)$$

where $f^{(k)}$ denotes the kth derivative of f. Comparing Eqs. (6.29) and (6.28), we can equate $F(x)$ with $P_l^{(m)}(x)$ if $\lambda = l(l+1)$. We thus have a solution of Eq. (3.75) finite at $x=\pm 1$ when $\lambda = l(l+1)$, denoted $P_l^m(x)$:

$$P_l^m(x) \equiv (1-x^2)^{m/2}\frac{\mathrm{d}^m}{\mathrm{d}x^m}P_l(x) = \frac{(1-x^2)^{m/2}}{2^l l!}\frac{\mathrm{d}^{m+l}}{\mathrm{d}x^{m+l}}(x^2-1)^l, \qquad (6.30)$$

where we've used the Rodrigues formula, Eq. (6.10). A study of the series solutions of Eq. (3.75) shows that only for $\lambda = l(l+1)$ do we obtain solutions finite at $x=\pm 1$. The functions $P_l^m(x)$ are known as the *associated Legendre functions.*[14] Examples are shown in Table 6.1. Note

associated Legendre functions

[14]Such functions are also (loosely) referred to as associated Legendre polynomials. The functions $P_l^m(x)$ are not in general polynomials; however, for odd values of m, there's a factor of $\sqrt{1-x^2}$.

Table 6.1 Associated Legendre functions $P_l^m(x)$ for $l = 1, 2, 3,\ 1 \le m \le l$.

$P_1^1(x) = \sqrt{1-x^2}$	$P_3^1(x) = \frac{3}{2}(5x^2-1)\sqrt{1-x^2}$
$P_2^1(x) = 3x\sqrt{1-x^2}$	$P_3^2(x) = 15x(1-x^2)$
$P_2^2(x) = 3(1-x^2)$	$P_3^3(x) = 15(1-x^2)^{3/2}$

that $P_l^0(x) \equiv P_l(x)$. Because x^l is the highest power in $P_l(x)$, if $m > l$ in Eq. (6.30), $P_l^m(x)$ is identically zero. Thus,[15] $m = 0, 1, \ldots, l$.

Orthogonality

The associated Legendre functions are thus solutions of a Sturm–Liouville system, Eq. (3.75)

$$\frac{\mathrm{d}}{\mathrm{d}x}\left((1-x^2)\frac{\mathrm{d}P_l^m(x)}{\mathrm{d}x}\right) - \frac{m^2}{1-x^2}P_l^m(x) = -l(l+1)P_l^m(x), \qquad (6.31)$$

together with the boundary condition that $P_l^m(x)$ is finite at $x = \pm 1$. As such, the eigenfunctions $P_l^m(x)$, corresponding to eigenvalues $l(l+1)$ (for $l = m, m+1, m+2, \ldots$) form a complete orthogonal set *for each* m, $\int_{-1}^{1} P_l^m(x)P_{l'}^m \,\mathrm{d}x = 0$, $l \neq l'$. For $l = l'$, the normalization integral is calculated in the same way as for the Legendre polynomials: repeated integration by parts. We omit the derivation and list the result:

orthogonality

$$\int_{-1}^{1} P_l^m(x)P_{l'}^m(x)\,\mathrm{d}x = \frac{2}{2l+1}\frac{(l+m)!}{(l-m)!}\delta_{ll'}. \qquad (6.32)$$

Note that Eq. (6.32) reduces to Eq. (6.14) for $m = 0$.

6.3 Spherical harmonic functions, $Y_l^m(\theta, \phi)$

We're finally ready to assemble the spherical harmonic function. From Eq. (6.1),

$$Y_l^m(\theta, \phi) = N_l^m P_l^{|m|}(\cos\theta)\mathrm{e}^{\mathrm{i}m\phi},$$

where N_l^m is a normalization factor and where we now allow negative values[16] of m. This factor is determined by requiring

[15] In quantum theory, the requirement that $m \le l$ has a physical consequence that the z-component of the angular momentum cannot exceed the magnitude of the total angular momentum.

[16] The quantity m in the associated Legendre function P_l^m is always positive. We must allow for *negative* values of m in $\mathrm{e}^{\mathrm{i}m\phi}$; however, because the completeness relation for complex exponentials, Eq. (4.11), requires negative as well as positive values of m.

Table 6.2 Spherical harmonic functions $Y_l^m(\theta,\phi)$ for $l = 0, 1, 2,\ 0 \le |m| \le l$.

$Y_0^0(\theta,\phi) = \frac{1}{\sqrt{4\pi}}$	$Y_2^2(\theta,\phi) = \frac{1}{4}\sqrt{\frac{15}{2\pi}}\sin^2\theta e^{2i\phi}$
$Y_1^1(\theta,\phi) = \sqrt{\frac{3}{8\pi}}\sin\theta e^{i\phi}$	$Y_2^1(\theta,\phi) = \sqrt{\frac{15}{8\pi}}\sin\theta\cos\theta e^{i\phi}$
$Y_1^0(\theta,\phi) = \sqrt{\frac{3}{4\pi}}\cos\theta$	$Y_2^0(\theta,\phi) = \frac{1}{2}\sqrt{\frac{5}{4\pi}}(3\cos^2\theta - 1)$
$Y_1^{-1}(\theta,\phi) = \sqrt{\frac{3}{8\pi}}\sin\theta e^{-i\phi}$	$Y_2^{-1}(\theta,\phi) = \sqrt{\frac{15}{8\pi}}\sin\theta\cos\theta e^{-i\phi}$
	$Y_2^{-2}(\theta,\phi) = \frac{1}{4}\sqrt{\frac{15}{2\pi}}\sin^2\theta e^{-i\phi}$

$$\begin{aligned}
&\int (Y_{l'}^{m'}(\theta,\phi))^* Y_l^m(\theta,\phi)\, d\Omega \\
&= N_{l'}^{m'} N_l^m \int_0^\pi P_{l'}^{|m'|}(\cos\theta) P_l^{|m|}(\cos\theta)\sin\theta\, d\theta \int_0^{2\pi} e^{i(m-m')\phi}\, d\phi \\
&= 2\pi\delta_{mm'} N_{l'}^m N_l^m \int_{-1}^{l} P_{l'}^{|m|}(x) P_l^{|m|}(x)\, dx = 2\pi\delta_{mm'}(N_l^m)^2 \frac{2}{2l+1}\frac{(l+|m|)!}{(l-|m|)!}\delta_{ll'} = \delta_{ll'}\delta_{mm'},
\end{aligned} \tag{6.33}$$

where $d\Omega \equiv \sin\theta\, d\theta\, d\phi$ is the differential element of solid angle, where we've let $x = \cos\theta$ in the second equality, and where we've used Eqs. (4.2) and (6.32).

The normalized spherical harmonic functions are therefore defined as[17]

$$Y_l^m(\theta,\phi) \equiv \sqrt{\frac{2l+1}{4\pi}\frac{(l-|m|)!}{(l+|m|)!}} P_l^{|m|}(\cos\theta) e^{im\phi}. \quad l = 0, 1, 2, \ldots, \quad |m| \le l \tag{6.34}$$

Examples are listed in Table 6.2. Note that $Y_l^0(\theta,\phi) = \sqrt{\frac{2l+1}{4\pi}} P_l(\cos\theta)$.

spherical harmonics

Completeness on the sphere

The functions $\{e^{im\phi}\}_{m=-\infty}^{\infty}$ are a complete set for $\phi \in [0, 2\pi]$, Eq. (4.11). The associated Legendre functions $\{P_l^m(\cos\theta)\}_{l=m}^{\infty}$ are a complete set for $\theta \in [0, \pi]$ for each m (Exercise 6.21). *Products* of two functions such as $P_l^m(\cos\theta)e^{im\phi}$, each of which is a member of a complete set, themselves form a complete set in the union of their domains, in this case the sphere $0 \le \phi \le 2\pi$, $0 \le \theta \le \pi$ [14, p. 56]. Thus, we can expand any function $f(\theta,\phi)$ in a series

$$f(\theta,\phi) = \sum_{l=0}^{\infty}\sum_{m=-l}^{l} A_{lm} Y_l^m(\theta,\phi). \tag{6.35}$$

[17] There are several other normalization factors for the spherical harmonics. Beware. In particular, a factor of $(-1)^m$ is often introduced for $m > 0$, a convention employed in the quantum theory of angular momentum.

The expansion coefficients are found using orthonormality, Eq. (6.33):

$$A_{lm} = \int (Y_l^m(\theta,\phi))^* f(\theta,\phi)\, \mathrm{d}\Omega. \tag{6.36}$$

Combining Eq. (6.36) with Eq. (6.35), we have the completeness relation:

completeness

$$\sum_{l=0}^{\infty}\sum_{m=-l}^{l} (Y_l^m(\theta',\phi'))^* Y_l^m(\theta,\phi) = \delta(\cos\theta - \cos\theta')\delta(\phi-\phi') = \frac{1}{\sin\theta}\delta(\theta-\theta')\delta(\phi-\phi'),$$

where we've used one of the properties of $\delta(x)$ listed in Section 4.5.

6.4 Addition theorem for $Y_l^m(\theta,\phi)$

We now derive an important property of the spherical harmonics known as the *addition theorem*. Referring to Figure 6.4, let $\hat{\boldsymbol{r}}$ and $\hat{\boldsymbol{r}}'$ be unit vectors in the directions of $\boldsymbol{r}$ and $\boldsymbol{r}'$. Expressed in terms of Cartesian unit vectors, $\hat{\boldsymbol{r}} = \sin\theta(\cos\phi\hat{\boldsymbol{x}} + \sin\phi\hat{\boldsymbol{y}}) + \cos\theta\hat{\boldsymbol{z}}$, and similarly for $\hat{\boldsymbol{r}}'$. The angle γ between $\boldsymbol{r}$ and $\boldsymbol{r}'$ is found from $\hat{\boldsymbol{r}}\cdot\hat{\boldsymbol{r}}' = \cos\gamma = \cos\theta\cos\theta' + \sin\theta\sin\theta'\cos(\phi-\phi')$. The addition theorem, what we're going to show, is[18]

addition theorem

$$P_n(\cos\gamma) = \frac{4\pi}{2n+1}\sum_{m=-n}^{n} Y_n^m(\theta,\phi)(Y_n^m(\theta',\phi'))^*. \tag{6.37}$$

An equivalent way of writing Eq. (6.37) is

$$P_n(\cos\gamma) = P_n(\cos\theta)P_n(\cos\theta') + 2\sum_{m=1}^{n}\frac{(n-m)!}{(n+m)!}P_n^m(\cos\theta)P_n^m(\cos\theta')\cos(m(\phi-\phi')). \tag{6.38}$$

These equations are highly useful in applications.

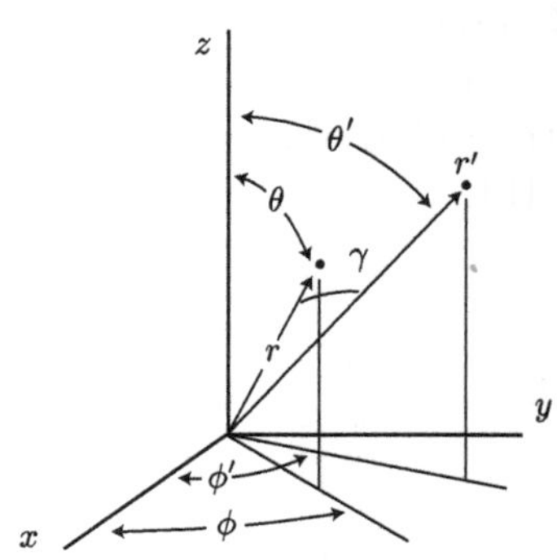

Figure 6.4 Coordinate system for the addition theorem.

[18]Equation (6.37) can be written with the complex conjugate on either factor.

Before getting to the addition theorem, we need to demonstrate that the Laplacian operator is invariant under rotations, and to do that requires a few steps. The reader uninterested in the details should skip to Eq. (6.41). We showed in Section 1.4.10 that a rigid rotation of coordinate axes is effected by an orthogonal matrix, $\boldsymbol{R}$ (a matrix such that $\boldsymbol{R}^T = \boldsymbol{R}^{-1}$). A version of $\boldsymbol{R}$ for a rotation of the (x, y) coordinate axes about the z-axis is given by the 2×2 matrix in Eq. (1.47). A rotation of all three coordinate axes about an *axis of rotation* (an arbitrary direction in three-dimensional space) is also effected by an orthogonal matrix. We won't write down the 3×3 matrix $[R_{ij}]$ for three-dimensional rotations as it's rather complicated[19]; what's important is that it's an orthogonal matrix – that's all we need to know. Write the components of the position vector $\boldsymbol{r}$ as $(x_1, x_2, x_3)^T$ (instead of $(x, y, z)^T$). The position vector $\boldsymbol{r}'$ in the rotated coordinate system has components related to those in the original coordinate system through the matrix relation, $\boldsymbol{r}' = \boldsymbol{R}\boldsymbol{r}$. The length of the position vector is invariant under such a transformation (referred to as *rotational symmetry*)[20,21]: $(\boldsymbol{r}')^T\boldsymbol{r}' = (\boldsymbol{R}\boldsymbol{r})^T\boldsymbol{R}\boldsymbol{r} = \boldsymbol{r}^T\boldsymbol{R}^T\boldsymbol{R}\boldsymbol{r} = \boldsymbol{r}^T\boldsymbol{R}^{-1}\boldsymbol{R}\boldsymbol{r} = \boldsymbol{r}^T\boldsymbol{r}$. In terms of components,

$$x_i' = \sum_{j=1}^{3} R_{ij}x_j. \qquad i = 1, 2, 3 \tag{6.39}$$

Let's apply this to the Laplacian. The inverse relation to Eq. (6.39) is

$$x_i = \sum_{j=1}^{3} (R^{-1})_{ij}x_j' = \sum_{j=1}^{3} R_{ji}x_j', \tag{6.40}$$

where we've used $\boldsymbol{R}^{-1} = \boldsymbol{R}^T$. From Eq. (6.40) and the chain rule, $\frac{\partial}{\partial x_j'} = \sum_{i=1}^{3} \frac{\partial x_i}{\partial x_j'}\frac{\partial}{\partial x_i} = \sum_{i=1}^{3} R_{ji}\frac{\partial}{\partial x_i}$. If we can do this once, we can do it twice:

$$\frac{\partial^2}{\partial x_j'^2} = \frac{\partial}{\partial x_j'}\left(\frac{\partial}{\partial x_j'}\right) = \sum_{i=1}^{3} R_{ji}\frac{\partial}{\partial x_i}\left(\sum_{k=1}^{3} R_{jk}\frac{\partial}{\partial x_k}\right) = \sum_{i=1}^{3}\sum_{k=1}^{3} R_{ji}R_{jk}\frac{\partial^2}{\partial x_i \partial x_k}.$$

By summing over j and using $\sum_k R_{ki}R_{kj} = \delta_{ij}$ (orthogonality of $\boldsymbol{R}$), we have that ∇^2 is invariant under rigid rotations of the coordinate axes:

$$(\nabla^2)' = \sum_{j=1}^{3} \frac{\partial^2}{\partial x_j'^2} = \sum_{j=1}^{3}\sum_{i=1}^{3}\sum_{k=1}^{3} R_{ji}R_{jk}\frac{\partial^2}{\partial x_i \partial x_k} = \sum_{i=1}^{3}\sum_{k=1}^{3} \delta_{ik}\frac{\partial^2}{\partial x_i \partial x_k} = \sum_{i=1}^{3}\frac{\partial^2}{\partial x_i^2} = \nabla^2. \tag{6.41}$$

[19] See for example Goldstein [21, p. 109].

[20] There are two aspects to *symmetry*: a transformation and something invariant under the transformation.

[21] The terms $\boldsymbol{r}^T\boldsymbol{r}$ generate the *quadratic form* $x^2 + y^2 + z^2$. Rotational symmetry can be stated alternatively as invariance of the quadratic form $\boldsymbol{r}^T\boldsymbol{r}$ under rotations.

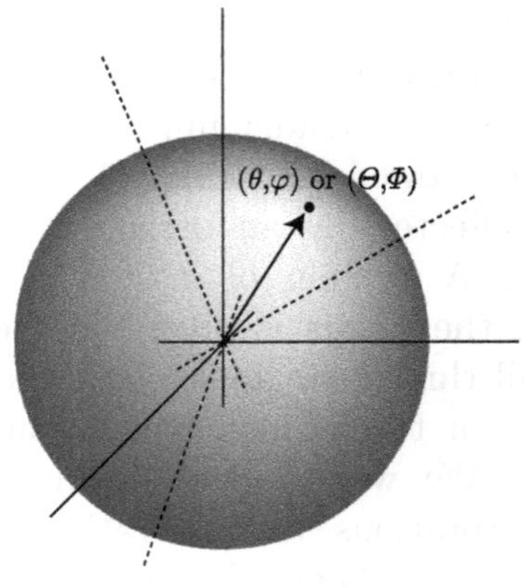

Figure 6.5 Point on sphere labeled by two sets of angles.

We've demonstrated Eq. (6.41) using rectangular coordinates, but the same conclusion holds in any orthogonal coordinate system.

The operator L^2 in Eq. (3.46) is therefore such that $(L^2)' = L^2$. Here $(L^2)'$ is the angular part of the Laplacian involving polar and azimuthal angles, call them Θ and Φ, referred to coordinate axes that have been rigidly rotated relative to a coordinate system for which the spherical angles are θ and ϕ (see Figure 6.5). The spherical harmonics are eigenfunctions of L^2: $L^2 Y_l^m(\theta,\phi) = -l(l+1)Y_l^m(\theta,\phi)$. Because $(L^2)' = L^2$, however, $(L^2)' Y_l^m(\theta,\phi) = -l(l+1)Y_l^m(\theta,\phi)$. Yet the eigenfunction of $(L^2)'$ corresponding to eigenvalue $l(l+1)$ is also $Y_l^{m'}(\Theta,\Phi)$, where we've deliberately written the magnetic quantum number[22] as m'. There's an ambiguity here[23]: There are $2l+1$ values of m associated with each value of l. Thus, there must be a linear connection between the eigenfunctions associated with the same eigenvalue at the *same* point on the sphere:

$$Y_l^m(\theta,\phi) = \sum_{m'=-l}^{l} A_{mm'} Y_l^{m'}(\Theta,\Phi), \tag{6.42}$$

where $A_{mm'}$ are expansion coefficients.[24]

To prove the addition theorem, consider that the coordinate system has been rotated so that the new polar axis (z-axis) lies along the vector $\boldsymbol{r}'$ in Figure 6.4. Relative to this coordinate system, the vector $\boldsymbol{r}$ has polar angle[25] γ and azimuthal angle Φ (which won't play a role here). For the same point on the sphere (which in the original coordinate system is labeled by angles (θ,ϕ)), we have, applying Eq. (6.42),

[22]In quantum mechanics, the integer m appearing in Y_l^m is referred to as the *magnetic quantum number* because of its importance in the theory of the Zeeman effect. The integer l is referred to as the *azimuthal* or *orbital-angular-momentum quantum number.*

[23]For a fixed value of l, the $2l+1$ functions Y_l^m are said to be $(2l+1)$-fold degenerate.

[24]Equation (6.42) could equally well be written "the other way," as an expansion of $Y_l^m(\Theta,\Phi)$ in terms of $Y_l^{m'}(\Theta,\Phi)$.

[25]That is, $\Theta = \gamma$.

$$Y_l^0(\gamma) = \sum_{m=-l}^{l} a_m Y_l^m(\theta, \phi). \tag{6.43}$$

If we could evaluate the expansion coefficient a_m, we'd be done (compare the form of Eq. (6.43) with Eq. (6.37)).

Utilizing orthonormality, Eq. (6.33), we have from Eq. (6.43):

$$a_m = \int Y_l^0(\gamma)(Y_l^m(\theta, \phi))^* \, \mathrm{d}\Omega. \tag{6.44}$$

Because $\cos\gamma = \cos\theta\cos\theta' + \sin\theta\sin\theta'\cos(\phi - \phi')$, we infer from Eq. (6.44) that[26] $a_m = a_m(\theta', \phi')$. Equivalently, we can use Eq. (6.42) to write

$$Y_l^m(\theta, \phi) = \sum_{m'=-l}^{l} b_{mm'} Y_l^{m'}(\gamma, \Phi). \tag{6.45}$$

Using orthonormality, we have from Eq. (6.45):

$$b_{mm'} = \int (Y_l^{m'}(\gamma, \Phi))^* Y_l^m(\theta, \phi) \, \mathrm{d}\Omega_\gamma, \tag{6.46}$$

where $\mathrm{d}\Omega_\gamma$ is the element of solid angle in the coordinate system with the polar axis along the vector $\boldsymbol{r}'$ in Figure 6.4. We now come to the key step that under a rigid rotation, the infinitesimal solid-angle element is invariant so that $\mathrm{d}\Omega_\gamma = \mathrm{d}\Omega$. Formally, the Jacobian is unity: The determinant of an orthogonal matrix is unity. Thus, we have from Eq. (6.46), specializing to $m' = 0$,

$$b_{m0} = \int Y_l^0(\gamma) Y_l^m(\theta, \phi) \, \mathrm{d}\Omega, \tag{6.47}$$

where we've used that Y_l^0 is real and depends only on the polar angle, $Y_l^0(\theta, \phi) = \sqrt{(2l + 1/4\pi)} P_l(\cos\theta)$. Comparing Eqs. (6.47) and (6.44), we infer that $a_m = b_{m0}^*$. We can get an independent expression for b_{m0} by considering the limit $\gamma \to 0$ in Eq. (6.45). First, for any spherical harmonic, $Y_l^m(\theta = 0, \phi) = \sqrt{(2l + 1/4\pi)}\delta_{m0}$. As $\gamma \to 0$, $(\theta, \phi) \to (\theta', \phi')$. Therefore, from Eq. (6.45), for $\gamma = 0$:

$$Y_l^m(\theta', \phi') = \sqrt{\frac{2l+1}{4\pi}} b_{m0}. \tag{6.48}$$

Utilizing $a_m = b_{m0}^*$ in Eq. (6.43), we have demonstrated the addition theorem, Eq. (6.37).

---

[26] That is, if we could actually do the integral in Eq. (6.44), which is why we're trying to prove the addition theorem, which would make integrals like that in Eq. (6.44) easy.

Example. Let's illustrate the addition theorem in a few cases. For $n = 0$, Eq. (6.37) has that $P_0(\cos\gamma) = 4\pi|Y_0^0|^2 = 1$, which is true. For $n = 1$, we have

$$\begin{aligned}P_1(\cos\gamma) &= \frac{4\pi}{3}[Y_1^1(\theta,\phi)Y_1^{1*}(\theta',\phi') + Y_1^0(\theta,\phi)Y_1^{0*}(\theta',\phi') + Y_1^{-1}(\theta,\phi)Y_1^{-1*}(\theta',\phi')]\\ &= \frac{4\pi}{3}\left[\frac{3}{8\pi}\sin\theta\sin\theta' \mathrm{e}^{\mathrm{i}(\phi-\phi')} + \frac{3}{4\pi}\cos\theta\cos\theta' + \frac{3}{8\pi}\sin\theta\sin\theta' \mathrm{e}^{-\mathrm{i}(\phi-\phi')}\right]\\ &= \cos\theta\cos\theta' + \sin\theta\sin\theta'\cos(\phi-\phi') \equiv \cos\gamma,\end{aligned}$$

which is correct. It's laborious, but it can be shown directly that Eq. (6.37) yields $P_2(\cos\gamma) = \frac{1}{2}(3\cos^2\gamma - 1)$ as required.

Example. Multipole expansion

The electrostatic potential φ at a distance r from a point charge varies as r^{-1} (Eq. (6.22)). What's the form of φ produced by more complicated arrangements of charge? The addition theorem can be used to develop a power-series representation, the *multipole expansion* for the potential due to an arbitrary distribution of charge characterized by a charge density function $\rho(\boldsymbol{r})$.

The potential at a point located by vector $\boldsymbol{r}$ (the *field point*) is obtained, by superposition, from a summation over the charges at the *source points* $\boldsymbol{r}'$:

$$\varphi(\boldsymbol{r}) = \frac{1}{4\pi\epsilon_0}\int_V \mathrm{d}\boldsymbol{r}' \frac{\rho(\boldsymbol{r}')}{|\boldsymbol{r}-\boldsymbol{r}'|}. \tag{6.49}$$

Equation (6.49) is what results by treating the infinitesimal charge surrounding $\boldsymbol{r}'$, $\rho(\boldsymbol{r}')\,\mathrm{d}\boldsymbol{r}'$, as a point charge. By the law of cosines, $|\boldsymbol{r}-\boldsymbol{r}'|^2 = r^2 + (r')^2 - 2rr'\cos\gamma$, where $\cos\gamma \equiv \hat{\boldsymbol{r}}\cdot\hat{\boldsymbol{r}}'$. For the field point *exterior* to V,

$$\frac{1}{|\boldsymbol{r}-\boldsymbol{r}'|} = \frac{1}{r}\frac{1}{\sqrt{1-2(r'/r)\cos\gamma + (r'/r)^2}} = \frac{1}{r}\sum_{n=0}^{\infty}\left(\frac{r'}{r}\right)^n P_n(\cos\gamma), \tag{6.50}$$

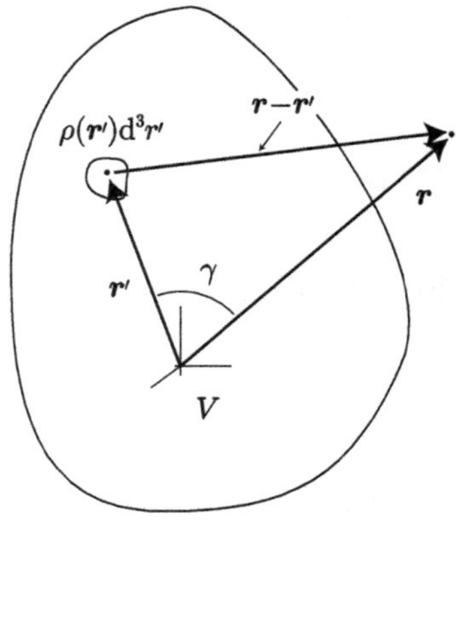

Figure 6.6 Potential exterior to charges in V.

where we've used the generating function, Eq. (6.19). The sum in Eq. (6.50) converges because $r > r'$ for every point in V. Combine Eqs. (6.50) and (6.49):

$$\varphi(\boldsymbol{r}) = \frac{1}{4\pi\epsilon_0 r}\sum_{n=0}^{\infty}\frac{1}{r^n}\int_V \mathrm{d}\boldsymbol{r}'\rho(\boldsymbol{r}')(r')^n P_n(\hat{\boldsymbol{r}}\cdot\hat{\boldsymbol{r}}'). \tag{6.51}$$

The integral for $n = 0$ in Eq. (6.51) (the *monopole*) represents the total charge, Q. The integrals for $n \geq 1$ characterize the distribution of charge; these are difficult integrals due to the variable angle between $\boldsymbol{r}'$ and $\boldsymbol{r}$. Fortunately, the addition theorem for Legendre polynomials allows us to separate the source and field-point coordinates in Eq. (6.51). Combining Eq. (6.38) with Eq. (6.51), we have, referring to the coordinates in Figure 6.4, the multipole expansion of the electrostatic potential:

$$\varphi(\boldsymbol{r}) = \frac{Q}{4\pi\epsilon_0 r}\left(1+\sum_{n=1}^{\infty}\left(\frac{a}{r}\right)^n\left[J_n P_n(\cos\theta)+\sum_{m=1}^{n}P_n^m(\cos\theta)(C_n^m\cos m\phi + S_n^m\sin m\phi)\right]\right), \tag{6.52}$$

multipole expansion

where the dimensionless moments of the charge distribution are

$$J_n \equiv \frac{1}{Qa^n}\int_V \mathrm{d}\boldsymbol{r}'(r')^n\rho(\boldsymbol{r}')P_n(\cos\theta')$$
$$\begin{Bmatrix} C_n^m \\ S_n^m \end{Bmatrix} \equiv \frac{2}{Qa^n}\frac{(n-m)!}{(n+m)!}\int_V \mathrm{d}\boldsymbol{r}'(r')^n\rho(\boldsymbol{r}')P_n^m(\cos\theta')\begin{Bmatrix} \cos m\phi' \\ \sin m\phi' \end{Bmatrix}, \tag{6.53}$$

with a a characteristic length associated with the charge distribution.

Having introduced a coordinate system to arrive at Eq. (6.52) (that characterized by unit vectors $\hat{\boldsymbol{x}}$, $\hat{\boldsymbol{y}}$, and $\hat{\boldsymbol{z}}$ associated with Figure 6.4), we must recognize that *the multipole moments depend on the coordinate system chosen.* The *simplest* coordinate systems are those in which *symmetries* are readily apparent. It's shown in Exercise 6.23 that for *axisymmetric* charges ($\rho(\boldsymbol{r})$ independent of ϕ), the moments C_n^m and S_n^m vanish; for spherically symmetric charges (ρ independent of θ and ϕ), the moments J_n vanish, and in a coordinate system attached to the center of charge ($\int_V \rho(\boldsymbol{r}')\boldsymbol{r}'\,\mathrm{d}\boldsymbol{r}' = 0$), the first-order moments vanish, $J_1 = C_1^1 = S_1^1 = 0$, for any charge distribution. Thus, for an axisymmetric charge with the coordinate system attached to the center of charge,

$$\varphi(\boldsymbol{r}) = \frac{Q}{4\pi\epsilon_0 r}\left(1 + J_2\left(\frac{a}{r}\right)^2 P_2(\cos\theta) + \cdots\right). \tag{6.54}$$

6.5 Laplace equation in spherical coordinates

We now consider solutions to the Laplace equation in spherical coordinates. We've already examined the Helmholtz equation in spherical coordinates in Section 3.2.3; the Laplace equation follows by setting the separation constant $k^2 = 0$ in Eq. (3.43). We now know that the single-valued, everywhere finite solutions of the angular equation are given by the spherical harmonic functions $Y_l^m(\theta, \phi)$. What we don't know are the solutions of the radial Eq. (3.47) with $\lambda = l(l+1)$:

$$R''(r) + \frac{2}{r}R'(r) - \frac{l(l+1)}{r^2}R(r) = 0. \tag{6.55}$$

The solution of Eq. (6.55) is, as can be verified,

$$R(r) = Ar^l + Br^{-(l+1)}, \tag{6.56}$$

where A and B are constants. The full solution to the Laplace equation is thus

$$\psi(r, \theta, \phi) = \sum_{l=0}^{\infty}\sum_{m=-l}^{l}[A_{lm}r^l + B_{lm}r^{-(l+1)}]Y_l^m(\theta, \phi), \tag{6.57}$$

where the constants $\{A_{lm}\}$ and $\{B_{lm}\}$ are found by fitting to the boundary conditions.

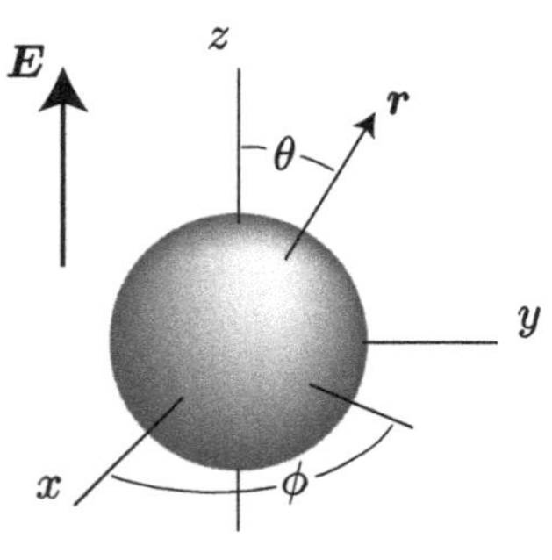

Figure 6.7 Conducting sphere in uniform external electric field.

Example. Conducting sphere in uniform electric field

Consider a perfectly conducting sphere of radius a placed in a static electric field $\boldsymbol{E}$ directed along the z-axis (see Figure 6.7). The nature of the interaction between electric fields and conductors is that certain boundary conditions must be satisfied by the field on the boundary of the conductor. Specifically, the conductor must be an equipotential surface. The presence of the sphere therefore locally affects the values of the electric field.

On the surface of the sphere, the potential is a constant, which we take to be zero. This boundary condition is written in spherical coordinates as

$$\varphi(a, \theta, \phi) = 0. \tag{6.58}$$

At sufficiently far distances, the perturbing effect of the sphere on the electric field will be very small. Because $\boldsymbol{E} = -\boldsymbol{\nabla}\varphi$, with φ satisfying the Laplace equation $\nabla^2\varphi = 0$, we require a boundary condition *at infinity*

$$\varphi(r, \theta, \phi) \overset{r\to\infty}{\to} -E_0 z = -E_0 r\cos\theta, \tag{6.59}$$

where E_0 denotes the unperturbed magnitude of $\boldsymbol{E}$.

Because this problem has azimuthal symmetry, the general solution to the Laplace equation is of the form of (6.57) with[27] $m = 0$:

$$\varphi(r,\theta) = \sum_{n=0}^{\infty} a_n r^n P_n(\cos\theta) + \sum_{n=0}^{\infty} b_n r^{-(n+1)} P_n(\cos\theta). \tag{6.60}$$

There are two boundary conditions, Eqs. (6.58) and (6.59). The boundary condition for $r \to \infty$ implies that all of the coefficients a_n must be zero except $a_1 = -E_0$. The other boundary condition at $r = a$ implies that

$$\begin{aligned} 0 &= -E_0 a\cos\theta + \sum_{n=0}^{\infty} b_n a^{-(n+1)} P_n(\cos\theta) \\ &= \left(\frac{b_1}{a^2} - E_0 a\right)\cos\theta + \sum_{\substack{n=0\\ n\neq 1}}^{\infty} b_n a^{-(n+1)} P_n(\cos\theta). \end{aligned}$$

Because the Legendre polynomials are linearly independent (they're an orthogonal set, Eq. (6.14)), we infer that $b_n = 0$, $n \neq 1$, and $b_1 = E_0 a^3$. The solution to Laplace's equation that satisfies the boundary conditions is therefore

$$\varphi(r,\theta,\phi) = -E_0\left(1 - \frac{a^3}{r^3}\right) r\cos\theta.$$

Summary

This chapter was devoted to the spherical harmonic functions $Y_l^m(\Theta, \Phi)$, which were introduced in Chapter 3 as part of the solution to the Helmholtz equation in spherical coordinates, but whose properties were not explored until now.

- The spherical harmonics are eigenfunctions of the angular part of the Laplacian operator $L^2 Y_l^m = -l(l+1)Y_l^m$ with the boundary condition that $Y_l^m(\theta,\phi)$ be single-valued and finite for $0 \leq \theta \leq \pi$ and $0 \leq \phi \leq 2\pi$, and where $l = 0, 1, 2, \ldots$, and $|m| \leq l$.
- $Y_l^m(\theta,\phi) = N_l^m P_l^{|m|}(\theta) e^{im\phi}$, where the normalization factor is specified in Eq. (6.34). They are normalized such that $\int (Y_{l'}^{m'})^* Y_l^m \, d\Omega = \delta_{ll'}\delta_{mm'}$, Eq. (6.33). The associated Legendre functions $P_l^m(\theta)$ are obtained from the Rodrigues formula Eq. (6.30).

[27] That is, the coefficients A_{lm} and B_{lm} in Eq. (6.57) vanish for $m \neq 0$ because the boundary conditions are independent of the azimuthal angle ϕ. We suppress the unnecessary index $m = 0$ in the coefficients a_n and b_n in Eq. (6.60).

- The functions $\{Y_l^m(\theta,\phi)\}$ are a complete orthonormal set with which any function on the surface of a sphere may be expanded: $f(\theta,\phi) = \sum_{l=0}^{\infty}\sum_{m=-l}^{l} A_{lm} Y_l^m(\theta,\phi)$.

Exercises

6.1. Show that $x = \pm 1$ are regular singular points of Eq. (3.75). Hint: Examine the limits $\lim_{x\to\pm 1}(x \mp 1)P(x)$ and $\lim_{x\to\pm 1}(x\mp 1)^2 Q(x)$.

6.2. Derive Eq. (6.2). This takes some work and goes beyond the analysis presented in Section 5.14. Write Eq. (3.75) in the form

$$P'' - \frac{2x}{1-x^2}P' + \frac{\lambda}{1-x^2}P - \frac{m^2}{(1-x^2)^2}P = 0. \tag{6.61}$$

Define a new variable $y = x - 1$, so that $x \to 1$ is equivalent to $y \to 0$. Under this substitution, Eq. (6.61) is equivalent to

$$P'' + f(y)\frac{1}{y}P' - \lambda g(y)\frac{1}{y}P - m^2 h(y)\frac{1}{y^2}P = 0, \tag{6.62}$$

where

$$f(y) \equiv \frac{2(1+y)}{2+y} \qquad g(y) \equiv \frac{1}{2+y} \qquad h(y) \equiv \frac{1}{(2+y)^2}.$$

Each of the functions f, g, and h have power-series expansions about $y = 0$: $f(y) = f_0 + f_1 y + f_2 y^2 + \cdots$, $g(y) = g_0 + g_1 y + g_2 y^2 + \cdots$, and $h(y) = h_0 + h_1 y + h_2 y^2 + \cdots$, respectively. Let $P(y) = \sum_{n=1}^{\infty} a_n y^{n+r}$. Substitute these expansions into Eq. (6.62). Gather all terms associated with the lowest power of y and equate to zero. You should find for the indicial equation:

$$y^{r-2}a_0[r(r-1) + f_0 r - m^2 h_0] = 0.$$

Note that the indicial equation is independent of the eigenvalue λ. Equation (6.2) then follows. Repeat this analysis for the other regular singular point, $x = -1$. You should find the same indicial equation. Note that we're not asking you to derive a recursion relation for the coefficients a_n.

6.3. Derive the recursion relation, Eq. (6.4). Substitute Eq. (6.3) into Eq. (5.46) and equate coefficients of powers of x to zero.

6.4. Verify Eq. (6.5) by showing that it satisfies Eq. (6.4).

6.5. Use Eq. (6.7) to derive expressions for P_0, P_1, P_2, and P_3. Next use the Rodrigues formula, Eq. (6.10), to derive the same quantities. Your results should agree with the expressions in Table 5.1.

6.6. The set of functions $\{x^n\}_{n=0,1,2,\ldots}$ forms a basis (it is, for example, the expansion basis for the Taylor series). The basis functions $\phi_n(x) = x^n$ are clearly linearly independent (uniqueness of power series), but they are also nonorthogonal and, consequently, they are generally difficult to apply in a basis expansion. Use the Gram–Schmidt orthogonalization process (see Chapter 1) to form a new basis from $\{x^n\}$ which is orthonormal on $[-1, 1]$. Do you recognize this basis?

6.7. Derive Eq. (6.9). Hint:

$$\frac{\mathrm{d}^l}{\mathrm{d}x^l}x^{2l-2k} = \frac{\mathrm{d}^{l-1}}{\mathrm{d}x^{l-1}}\left(\frac{\mathrm{d}}{\mathrm{d}x}x^{2l-k}\right) = (2l-2k)\frac{\mathrm{d}^{l-1}}{\mathrm{d}x^{l-1}}x^{2l-2k-1} = \cdots$$
$$= (2l-2k)(2l-2k-1)\cdots(l-2k+1)x^{l-2k}.$$

6.8. Show that $P_l(x)$ as given by the Rodrigues formula, Eq. (6.10), satisfies the Legendre differential Eq. (5.46). A direct attack on this problem is a challenging task. We offer a "guided derivation" that gets the job done in a more subtle way. Define the quantity $y \equiv (x^2-1)^l$. Denote the kth derivative of y as $y^{(k)}$. We're going to show, by induction, that $y^{(k)}$ satisfies the differential equation

$$(1-x^2)y^{(k+2)} + 2x(l-k-1)y^{(k+1)} + (2l-k)(k+1)y^{(k)} = 0. \qquad (6.63)$$

The first derivative $y' = 2xl(x^2-1)^{l-1}$, and thus $(1-x^2)y' + 2lxy = 0$ (show this). Differentiating again, we have (show this)

$$(1-x^2)y^{(2)} + 2(l-1)xy^{(1)} + 2ly^{(0)} = 0.$$

Thus, Eq. (6.63) holds for $k=0$ (no derivative, $y^{(0)} \equiv y$). Assume that Eq. (6.63) holds for $k-1$. Let $k \to k-1$ in Eq. (6.63),

$$(1-x^2)y^{(k+1)} + 2x(l-k)y^{(k)} + (2l-k+1)ky^{(k-1)} = 0. \qquad (6.64)$$

Show that Eq. (6.63) follows from differentiating Eq. (6.64). Now let $k=l$ in Eq. (6.63) and we're done: $P_l(x)$ as given by the Rodrigues formula satisfies the Legendre differential equation.

6.9. Show that $P_l(1) = 1$. Hint: From Eq. (6.10), evaluate the limit

$$P_l(1) = \frac{1}{2^l l!}\lim_{x\to 1}\frac{\mathrm{d}^l}{\mathrm{d}x^l}(x^2-1)^l.$$

As $x \to 1$, $(x^2-1)^l \approx 2^l(x-1)^l$. Use Eq. (6.9) (with $k=0$) to conclude that $P_l(1) = 1$.

6.10. Show that $\lim_{x\to 1}\frac{\mathrm{d}^{l-1}}{\mathrm{d}x^{l-1}}(x^2-1)^l = 2^l l! \lim_{x\to 1}(x-1) = 0$.

6.11. Show that $\frac{\mathrm{d}^{2l}}{\mathrm{d}x^{2l}}(x^2-1)^l = \frac{\mathrm{d}^{2l}}{\mathrm{d}x^{2l}}x^{2l} = (2l)!$.

6.12. Derive Eq. (6.16). Hint: Multiply Eq. (6.15) by $P_{l'}(x)$, integrate over $[-1,1]$, and make use of Eq. (6.14).

6.13. Derive Eq. (6.17), Parseval's theorem. Hint: Take "two copies" of Eq. (6.15), multiply them, integrate as in Eq. (6.17), and use Eq. (6.14).

6.14. Evaluate the integral in Eq. (6.21) and verify the Taylor series shown.

6.15. Expand the expression in Eq. (6.23) to second order in a/r to show that

$$\frac{1}{r_1} = \frac{1}{r}\left[1 + \frac{a}{r}\cos\theta + \frac{1}{2}(3\cos^2\theta - 1)\left(\frac{a}{r}\right)^2 + \cdots\right].$$

6.16. Derive the analog of Eq. (6.24) in the case where $r < a$. A:

$$\varphi = \frac{q}{4\pi\epsilon_0 a}\sum_{n=0}^{\infty}P_n(\cos\theta)\left(\frac{r}{a}\right)^n.$$

6.17. Show that $|P_n(\cos\theta)| \le 1$ (a guided exercise). Set $x = \cos\theta$ in the generating function Eq. (6.19). Derive the expansion

$$(1 - 2y\cos\theta + y^2)^{-1/2} = (1 - y\mathrm{e}^{\mathrm{i}\theta})^{-1/2}(1 - y\mathrm{e}^{-\mathrm{i}\theta})^{-1/2}$$
$$= \left(1 + \frac{1}{2}y\mathrm{e}^{\mathrm{i}\theta} + \frac{3}{8}y^2\mathrm{e}^{2\mathrm{i}\theta} + \frac{5}{16}y^3\mathrm{e}^{3\mathrm{i}\theta} + \cdots\right)$$
$$\times \left(1 + \frac{1}{2}y\mathrm{e}^{-\mathrm{i}\theta} + \frac{3}{8}y^2\mathrm{e}^{-2\mathrm{i}\theta} + \frac{5}{16}y^3\mathrm{e}^{-3\mathrm{i}\theta} + \cdots\right).$$

By multiplying these expressions and gathering all terms multiplying y^l, show that for the first few terms, $P_l(\cos\theta)$ (the coefficient of y^l) is a polynomial in $\mathrm{e}^{\mathrm{i}\theta}$ and $\mathrm{e}^{-\mathrm{i}\theta}$ with *positive coefficients.* Deduce that[28] $|P_l(\cos\theta)| \le \{$ the same polynomial with $\mathrm{e}^{\pm\mathrm{i}\theta}$ replaced by 1$\}$, i.e. the same polynomial evaluated at $\theta = 0$ or $x = \cos\theta = 1$. Thus, $|P_l(\cos\theta)| \le P_l(1) = 1$.

6.18. Fun with recursion relations:

(a) Differentiate the generating function $g(x, y)$, Eq. (6.19), with respect to x, to show that

$$P'_{n+1}(x) + P'_{n-1}(x) = P_n(x) + 2xP'_n(x). \tag{6.65}$$

(b) Show that

$$P'_{n+1}(x) - P'_{n-1} = (2n + 1)P_n(x). \tag{6.66}$$

Hint: Differentiate Eq. (6.25). Multiply that result by two and subtract from it $(2n + 1)$ times Eq. (6.65).

(c) Show that

$$xP'_n(x) - P'_{n-1} = nP_n(x). \tag{6.67}$$

Hint: Subtract Eq. (6.66) from Eq. (6.65) and divide by 2.

(d) Show that

$$P'_{n+1}(x) - xP'_n(x) = (n + 1)P_n(x).$$

Hint: Subtract Eq. (6.67) from Eq. (6.66).

6.19. Show that Eq. (6.28) follows from Eq. (6.27).

6.20. Derive Eq. (6.28) by differentiating Eq. (5.46) m times. Hint: The Leibniz rule for multiple derivatives of a product of functions is

$$\frac{\mathrm{d}^m}{\mathrm{d}x^m}(fg) = \sum_{k=0}^{m}\binom{m}{k} f^{(k)}g^{(m-k)}.$$

6.21. The associated Legendre functions $P_l^m(x)$ form a complete orthogonal set for each value of m. Any function $f(x)$ defined on $[-1, 1]$ has an expansion

$$f(x) = \sum_{l=m}^{\infty} a_l^m P_l^m(x). \qquad m = 0, 1, 2, \ldots$$

Derive an expression for the expansion coefficients. A: $a_l^m = \frac{2l+1}{2}\frac{(l-m)!}{(l+m)!}$

[28] The positivity of the coefficients is important. If $\sum_n c_n$ is a series with positive coefficients and is known to be convergent, then if a series $\sum_n a_n$ with positive coefficients such that $a_n \le c_n$, the series $\sum_n a_n$ is also convergent [10, p. 113].

6.22. Derive Eq. (6.38) starting from Eq. (6.37).

6.23. (a) Show that the moments C_n^m and S_n^m in Eq. (6.53) vanish for axisymmetric charges. Show that the moments J_n vanish for spherically symmetric charges.

(b) Show that in a coordinate system attached to the center of charge, the first-order moments vanish, $J_1 = C_1^1 = S_1^1 = 0$. For a vector to vanish, all its components must vanish. Use $\boldsymbol{r} = r\cos\theta\hat{\boldsymbol{z}} + r\sin\theta(\cos\phi\hat{\boldsymbol{x}} + \sin\phi\hat{\boldsymbol{y}})$.

(c) Work out the electric field $\boldsymbol{E} = -\boldsymbol{\nabla}\varphi$ associated with φ given by Eq. (6.54). Note that it's not purely radial.

6.24. In quantum mechanics, a *closed shell* has enough electrons to occupy all magnetic substates associated with a given angular momentum quantum number, l. The electron density function is related to the absolute value squared of its quantum-mechanical wavefunction.

(a) Show that the following holds:

$$\sum_{m=-l}^{m=l} |Y_l^m(\theta,\phi)|^2 = \frac{2l+1}{4\pi}.$$

Hint: Use Eq. (6.37).

(b) Verify explicitly that such a relation holds for $l = 1$.

7

Bessel functions

We first encountered Bessel functions in Chapter 3, where the radial equation in cylindrical and spherical coordinates was found to involve these functions (Eqs. (3.40) and (3.49)). Bessel functions occur in numerous applications. In many respects, they're similar to trigonometric functions, as we'll see. If you work in certain areas of applied science, you'll need to know their properties as well as you do trigonometric functions. In this chapter, we develop the properties of Bessel functions.

7.1 Small-argument and asymptotic forms

The Bessel differential Eq. (5.25) has two linearly independent solutions, the Bessel function $J_\nu(x)$, with power-series representation given by Eq. (5.28), and the Neumann function $N_\nu(x)$, defined by Eq. (5.32), with series representation Eq. (5.39). The Bessel function $J_\nu(x)$ is finite for $0 \le x < \infty$, while the Neumann function diverges as $x \to 0$, but is finite as $x \to \infty$.

7.1.1 Limiting forms for small argument

The behavior at small x for either kind of Bessel function is obtained from the series solutions. From Eq. (5.28), we have[1]

$$J_\nu(x) \overset{x\to 0}{\sim} \begin{cases} 1 - x^2/4 & \nu = 0 \\ \\ \left(\frac{x}{2}\right)^\nu \frac{1}{\Gamma(\nu+1)} & \nu \neq 0. \end{cases} \tag{7.1}$$

[1]We're using *asymptotic notation* where $f \sim g$ means that functions $f(x)$ and $g(x)$ are *equal asymptotically*, that the ratio $f(x)/g(x) \to 1$ as x approaches some value, usually zero or infinity. For example, $\sinh x \sim \frac{1}{2}e^x$.

Mathematical Methods in Physics, Engineering, and Chemistry, First Edition.
Brett Borden and James Luscombe.

For the Neumann function, we list only the small-x behavior for integer values of ν, which we find from Eq. (5.39): ($\gamma = 0.5772$; see appendix C)

$$N_m(x) \overset{x\to 0}{\sim} \begin{cases} \frac{2}{\pi}\left(\ln(\frac{x}{2}) + \gamma\right) & m = 0 \\ -\frac{\Gamma(m)}{\pi}\left(\frac{x}{2}\right)^{-m} & m \neq 0. \end{cases} \tag{7.2}$$

7.1.2 Asymptotic forms for large argument

The behavior of Bessel functions for *large* values of their arguments can't be found from their powers series, but instead must be inferred from an analysis of their integral representations, a process that we illustrate in Chapter 8. To derive the asymptotic forms for Bessel functions would take us far afield, and we simply state the results here [22, p. 364]:

$$\begin{aligned} J_\nu(x) &\overset{x\to\infty}{\sim} \sqrt{\frac{2}{(\pi x)}} \cos\left(x - \tfrac{1}{2}\nu\pi - \tfrac{1}{4}\pi\right) \\ N_\nu(x) &\overset{x\to\infty}{\sim} \sqrt{\frac{2}{(\pi x)}} \sin\left(x - \tfrac{1}{2}\nu\pi - \tfrac{1}{4}\pi\right). \end{aligned} \tag{7.3}$$

For large x, the two kinds of Bessel function are similar – they're both damped trigonometric functions.[2] There is a crossover from power-law behavior for small x (Eqs. (7.1) and (7.2)), to damped oscillatory behavior at large x, Eq. (7.3). As can be seen in Figures 5.1 and 5.2, the crossover occurs for $x \approx \nu$, when the argument is approximately equal to the order. Asymptotic formulae strictly valid for $x \to \infty$ often work when "they have no right to." Only experience can show for what values of the argument that asymptotic results become accurate.

7.1.3 Hankel functions

Hankel functions

Just as with trigonometric functions, where $e^{\pm ix} = \cos x \pm i \sin x$, for some problems it's useful to form complex linear combinations of the two kinds of Bessel functions. The *Hankel functions* of the first and second kind are defined as

$$\begin{aligned} H_\nu^{(1)}(x) &\equiv J_\nu(x) + iN_\nu(x) \\ H_\nu^{(2)}(x) &\equiv J_\nu(x) - iN_\nu(x). \end{aligned} \tag{7.4}$$

The Hankel functions are a linearly independent pair of solutions to the Bessel differential equation (Exercise 7.1). Using the results of Eq. (7.3), we have at large x,

[2]The Bessel and Neumann functions are cosines and sines "with a disease." Thanks to Professor Andrès Larraza for this joke. Physics jokes are hard to come by; cherish them well.

$$H_\nu^{(1)}(x) \overset{x\to\infty}{\sim} \sqrt{\frac{2}{(\pi x)}}\mathrm{e}^{-\mathrm{i}\pi/4}\mathrm{e}^{\mathrm{i}(x-\frac{1}{2}\nu\pi)}$$

$$H_\nu^{(2)}(x) \overset{x\to\infty}{\sim} \sqrt{\frac{2}{(\pi x)}}\mathrm{e}^{i\pi/4}\mathrm{e}^{-\mathrm{i}(x-\frac{1}{2}\nu\pi)}. \tag{7.5}$$

Hankel functions can be used to represent waves propagating in two dimensions.

7.2 Properties of the Bessel functions, $J_n(x)$

7.2.1 Series associated with the generating function

Bessel functions of *integer order* have a generating function $g(x, y)$ that allows us to infer many properties of these functions.[3] For all x and for $y \neq 0$, $g(x, y)$ is such that[4]

generating function

$$g(x,y) \equiv \exp\left(\frac{x}{2}(y - y^{-1})\right) = \sum_{n=-\infty}^{\infty} J_n(x)y^n. \tag{7.6}$$

To prove Eq. (7.6), multiply the series of the exponential functions:

$$\begin{aligned}
\exp\left(\frac{x}{2}y\right)\exp\left(-\frac{x}{2}y^{-1}\right) &= \sum_{m=0}^{\infty}\sum_{k=0}^{\infty}\frac{(-1)^k}{m!k!}\left(\frac{x}{2}\right)^{m+k} y^{m-k} \\
&= \sum_{k=0}^{\infty}\sum_{n=-k}^{\infty}\frac{(-1)^k}{k!(n+k)!}\left(\frac{x}{2}\right)^{2k+n} y^n \\
&= \sum_{n=-\infty}^{\infty} y^n \sum_{k=0}^{\infty}\frac{(-1)^k}{k!(n+k)!}\left(\frac{x}{2}\right)^{2k+n} = \sum_{n=-\infty}^{\infty} y^n J_n(x),
\end{aligned}$$

where in the second equality, we let $n = m - k$. Because $m, k = 0, 1, \ldots \infty$, the index n is such that $-\infty < n < \infty$. In the final equality, we used Eq. (5.28) for integer n.

Many properties of the functions $J_n(x)$ can be derived directly from $g(x, y)$, properties that might be difficult to establish from the series solution, Eq. (5.28). The generating function is such that $g^{-1}(x, y) = g(-x, y) = g(x, y^{-1})$.

---

[3]The generating function for Legendre polynomials, Eq. (6.19), involves quantities indexed by integers because the degree of a polynomial is always an integer. Bessel functions $J_\nu(x)$, however, are defined for arbitrary order. The generating function Eq. (7.6) pertains only to integer-order Bessel functions $J_n(x)$.

[4]Equation (7.6) is likely an unfamiliar type of expansion that has terms with negative powers, called a *Laurent expansion.* Laurent expansions are developed in Chapter 8.

1. Several useful results follow from substituting $y = e^{i\theta}$ in Eq. (7.6):

$$\exp\left(\frac{x}{2}(e^{i\theta} - e^{-i\theta})\right) = \exp(ix\sin\theta) = \cos(x\sin\theta) + i\sin(x\sin\theta)$$
$$= \sum_{n=-\infty}^{\infty} J_n(x)e^{in\theta} = \sum_{n=-\infty}^{\infty} J_n(x)[\cos(n\theta) + i\sin(n\theta)]. \tag{7.7}$$

Separating real and imaginary parts,

$$\cos(x\sin\theta) = \sum_{n=-\infty}^{\infty} J_n(x)\cos(n\theta) = J_0(x) + 2\sum_{n=1}^{\infty} J_{2n}(x)\cos(2n\theta)$$
$$\sin(x\sin\theta) = \sum_{n=-\infty}^{\infty} J_n(x)\sin(n\theta) = 2\sum_{n=0}^{\infty} J_{2n+1}(x)\sin((2n+1)\theta), \tag{7.8}$$

where we've invoked $J_{-n}(x) = (-1)^n J_n(x)$, Eq. (5.31). For $\theta = \pi/2$ in Eq. (7.8), we have

$$\cos x = J_0(x) + 2\sum_{n=1}^{\infty}(-1)^n J_{2n}(x) = J_0(x) - 2J_2(x) + 2J_4(x) - \cdots$$
$$\sin x = 2\sum_{n=0}^{\infty}(-1)^n J_{2n+1}(x) = 2J_1(x) - 2J_3(x) + 2J_5(x) - \cdots . \tag{7.9}$$

2. From $g(x,y) \times g(x,y^{-1}) = 1$, we have:

$$1 = \sum_{n=-\infty}^{\infty}\sum_{m=-\infty}^{\infty} y^{n-m} J_n(x)J_m(x).$$

Comparing like powers[5] of y, we conclude that

$$J_0^2(x) + 2\sum_{n=1}^{\infty} J_n^2(x) = 1 \qquad \sum_{n=-\infty}^{\infty} J_n(x)J_{n+k}(x) = 0. \qquad (k \neq 0) \tag{7.10}$$

3. Consider that $g(x,z) \times g(y,z) = g(x+y,z)$, implying:

$$\sum_{k=-\infty}^{\infty} J_k(x+y)z^k = \sum_{n=-\infty}^{\infty}\sum_{m=-\infty}^{\infty} J_n(x)J_m(y)z^{n+m}.$$

addition theorem

By changing dummy indices so that $m = k - n$ and equating coefficients of like powers of z, we conclude that

[5]Equating like powers of y is another way of invoking the uniqueness of power series, Section 5.1.

$$J_k(x+y) = \sum_{n=-\infty}^{\infty} J_n(x) J_{k-n}(y). \tag{7.11}$$

Equation (7.11) is the addition theorem for Bessel functions of integer order.

4. Because $g(-x, y^{-1}) = g(x, y)$, we have

$$\sum_{n=-\infty}^{\infty} J_n(-x) y^{-n} = \sum_{n=-\infty}^{\infty} J_n(x) y^n = \sum_{n=-\infty}^{\infty} J_{-n}(x) y^{-n},$$

where we let $n \to -n$ in the second equality. Comparing like powers of y,

$$J_n(-x) = J_{-n}(x) = (-1)^n J_n(x), \tag{7.12}$$

where we've used Eq. (5.31). Equation (7.12) is technically an extension, or *analytic continuation*, of $J_n(x)$ (which is defined only for $x \geq 0$) to $x < 0$. We see from Eq. (7.9) that Bessel functions of even (odd) order are associated with an even (odd) function.

7.2.2 Recursion relations

Recursion relations can be developed for Bessel functions by taking derivatives of the generating function (just as with the Legendre polynomials). Differentiate Eq. (7.6) with respect to x:

$$\begin{aligned}\sum_{n=-\infty}^{\infty} J_n'(x) y^n &= \frac{1}{2}(y - y^{-1}) \mathrm{e}^{x(y-y^{-1})/2} = \frac{1}{2}(y - y^{-1}) \sum_{n=-\infty}^{\infty} y^n J_n(x) \\ &= \frac{1}{2} \sum_{n=-\infty}^{\infty} y^n [J_{n-1}(x) - J_{n+1}(x)],\end{aligned}$$

where we've played the usual game with indices. Comparing like powers of y,

$$2J_n'(x) = J_{n-1}(x) - J_{n+1}(x). \tag{7.13}$$

By differentiating Eq. (7.6) with respect to y, we find that[6]

$$\frac{2}{x} n J_n(x) = J_{n+1}(x) + J_{n-1}(x). \tag{7.14}$$

[6]We note that the generating function, Eq. (7.6), can be *derived* (instead of just stating it as we have done) by first deriving the recursion relation Eq. (7.14) directly from the power series for the Bessel function, Eq. (5.28). Multiplying Eq. (7.14) by y^n and summing over all n, one eventually arrives at a PDE for $g(x, y)$, which can be solved to yield Eq. (7.6).

There are many recursion relations that can be derived for the Bessel functions and their derivatives. For example, it can be shown that (Exercise 7.6)

$$\frac{\mathrm{d}}{\mathrm{d}x}(x^n J_n(x)) = x^n J_{n-1}(x). \tag{7.15}$$

Note: While *derived* from the generating function, which involves Bessel functions of integer order, the recursion relations are actually more general and hold for Bessel functions of arbitrary order. For example, Eq. (7.14) generalizes:

$$J_{\nu+1}(x) = \frac{2\nu}{x} J_\nu(x) - J_{\nu-1}(x). \tag{7.16}$$

To show that the recursion relations are satisfied by Bessel functions of arbitrary order, it's necessary to use the series representation, Eq. (5.28). Moreover, the recursion relations for Bessel functions hold for any of the function J_ν, Y_ν, $H_\nu^{(1)}$, $H_2^{(2)}$, or any linear combination of these functions, the coefficients in which are independent of x and ν.

Example. Wronskian for Bessel functions

The Wronskian $W(J_\nu, J_{-\nu})$ for Bessel functions,

$$W(J_\nu, J_{-\nu}) = \begin{vmatrix} J_\nu & J_{-\nu} \\ J'_\nu & J'_{-\nu} \end{vmatrix} = J_\nu J'_{-\nu} - J_{-\nu} J'_\nu,$$

can be simplified using the recursion relations. By using Eq. (7.13) (for arbitrary order ν) and Eq. (7.16), it can be shown that

$$W(J_\nu, J_{-\nu}) = J_\nu(x) J_{-(\nu+1)}(x) + J_{-\nu}(x) J_{\nu+1}(x). \tag{7.17}$$

Is there a magic bullet that can simplify Eq. (7.17)? Unfortunately not. We know, however, using Eq. (2.8) that the Wronskian for the solutions to the Bessel differential equation is of the form $W = C/x$, where C is a constant. We can evaluate C using the small-argument form of the Bessel functions. We find, using Eq. (7.1),

$$J_\nu(x) J_{-(\nu+1)}(x) + J_{-\nu}(x) J_{\nu+1}(x) \overset{x\to 0}{\sim} \frac{2}{x} \frac{1}{\Gamma(-\nu)\Gamma(\nu+1)}. \tag{7.18}$$

By the Euler reflection property, Eq. (C.8), $\Gamma(-\nu)\Gamma(1+\nu) = -\pi/(\sin \pi\nu)$. Therefore, $C = -2\sin \pi\nu/\pi$, the same as given in Eq. (5.30).

7.2.3 Integral representation

Equation (7.7) is in the form of a complex Fourier series (Eq. (4.9)):

$$\exp(\mathrm{i}x \sin\theta) = \sum_{n=-\infty}^{\infty} J_n(x) \mathrm{e}^{\mathrm{i}n\theta}. \tag{7.19}$$

As such, we can isolate the Fourier coefficients (in this case $J_n(x)$) using the orthogonality of the complex exponentials, as in Eq. (4.10):

$$J_n(x) = \frac{1}{2\pi}\int_{-\pi}^{\pi} \mathrm{e}^{-\mathrm{i}n\theta}\mathrm{e}^{\mathrm{i}x\sin\theta}\,\mathrm{d}\theta. \tag{7.20}$$

Equation (7.20) is an example of an *integral representation*: The value of $J_n(x)$ is obtained from an integral, as opposed to a power series. It can be simplified:

$$\begin{aligned} J_n(x) &= \frac{1}{2\pi}\left[\int_{-\pi}^{\pi}\cos(x\sin\theta - n\theta)\,\mathrm{d}\theta + \mathrm{i}\int_{-\pi}^{\pi}\sin(x\sin\theta - n\theta)\,\mathrm{d}\theta\right] \\ &= \frac{1}{\pi}\int_0^{\pi}\cos(x\sin\theta - n\theta)\,\mathrm{d}\theta, \end{aligned} \tag{7.21}$$

where we've written $\int_{-\pi}^{\pi} = \int_0^{\pi} + \int_{-\pi}^{0}$ and have let $\theta \to -\theta$ in the second integral.[7] The case of $J_0(x)$ in particular can be written in various ways:

$$J_0(x) = \frac{1}{2\pi}\int_0^{2\pi}\mathrm{e}^{\mathrm{i}x\sin\theta}\,\mathrm{d}\theta = \frac{1}{2\pi}\int_0^{2\pi}\mathrm{e}^{\mathrm{i}x\cos\theta}\,\mathrm{d}\theta. \tag{7.22}$$

Example. Circular, coherently illuminated aperture

In the theory of diffraction of waves by an aperture, it's shown that the diffracted field $U(P)$ at a point P is proportional to a surface integral over the aperture:

$$U(P) \sim \int_{\text{aperture}} \frac{\mathrm{e}^{\mathrm{i}kr}}{r}\,\mathrm{d}s,$$

where k is a constant and r is the distance from a point on the aperture. If the aperture lies in the y-z plane and the point P is located at (X, Y, Z), then the distance from P to a source element $\mathrm{d}s$ at (x, y, z) is (referring to Figure 7.1)

$$r = \sqrt{X^2 + (Y - y)^2 + (Z - z)^2} \approx R - \frac{Yy + Zz}{R}$$

where $R^2 = X^2 + Y^2 + Z^2$ and we have performed a truncated binomial series expansion appropriate for large R, $(R \gg y, z)$. Consequently, we can write

$$U(P) \approx \frac{\mathrm{e}^{\mathrm{i}kR}}{R}\int_{\text{aperture}} \mathrm{e}^{-\mathrm{i}k(Yy+Zz)/R}\,\mathrm{d}s.$$

Referring to Figure 7.1, introduce polar coordinates (ρ, φ) in the aperture $z = \rho\cos\varphi$, $y = \rho\sin\varphi$, $\mathrm{d}s = \rho\,\mathrm{d}\rho\,\mathrm{d}\varphi$. We also set $Z = q\cos\Phi$ and $Y = q\sin\Phi$. Then,

---

[7] It can be inferred from Eq. (7.21), using the comparison theorem for integrals, that $|J_n(x)| \leq 1$.

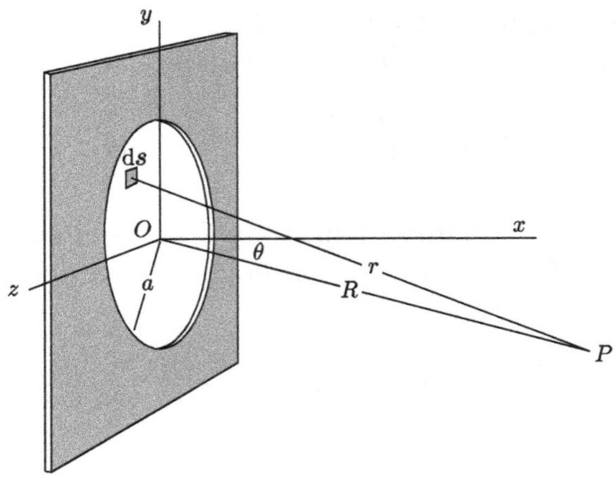

Figure 7.1 Diffraction from a circular aperture.

$$\int_{\text{aperture}} e^{-ik(Yy+Zz)/R}\, ds = \int_0^{2\pi}\int_0^a e^{-i(k\rho q/R)(\sin\varphi\sin\Phi+\cos\varphi\cos\Phi)}\rho\, d\rho\, d\varphi.$$

Note that $\sin\varphi\sin\Phi + \cos\varphi\cos\Phi = \cos(\varphi - \Phi)$. By geometric symmetry, however, this diffraction pattern is independent of Φ (which acts only as an initial angle and can be set to zero). If we first do the integral over φ, it's in the form of Eq. (7.22):

$$\int_0^{2\pi} e^{-i(k\rho q/R)\cos\varphi}\, d\varphi = 2\pi J_0\left(\frac{k\rho q}{R}\right).$$

The remaining integral over ρ is then $\int_0^a J_0\left(\frac{k\rho q}{R}\right)\rho\, d\rho$. This integral can be addressed by recognizing that the integrand is a special case of Eq. (7.15) with $n = 1$:

$$xJ_0(x) = \frac{d}{dx}[xJ_1(x)].$$

Thus,

$$\int_0^a J_0\left(\frac{k\rho q}{R}\right)\rho\, d\rho = \frac{R\rho}{kq}J_1\left(\frac{k\rho q}{R}\right)\Bigg|_0^a = \frac{Ra}{kq}J_1\left(\frac{kaq}{R}\right),$$

and so

$$|U(P)| \sim \frac{a}{kq}J_1\left(\frac{kaq}{R}\right).$$

7.3 Orthogonality

The Bessel differential equation, when placed in self-adjoint form, Eq. (2.16), can be written in the form of a Sturm–Liouville equation:

$$\frac{d}{d\rho}(\rho R') - \frac{\nu^2}{\rho}R = -\alpha^2\rho R. \tag{7.23}$$

For fixed ν, there is an infinite sequence of eigenvalue–eigenfunction pairs (Section 2.3). How is the dependence on eigenvalue indicated notationally in Bessel functions? Recall that the Bessel equation as it was first derived, Eq. (3.38), was cast into standard form, Eq. (5.25), by changing to a dimensionless variable $x = \alpha\rho$. Let's "reinstate" the factor of α and write the solutions of the Bessel equation as $J_\nu(\alpha\rho)$. *The eigenvalue resides in the argument of the Bessel function.*[8] As solutions of a Sturm–Liouville equation, we expect there to be, from Eq. (2.28), an orthogonality relation between eigenfunctions (for fixed ν) corresponding to different eigenvalues, α_1^2 and α_2^2:

$$(\alpha_2^2 - \alpha_1^2)\int_a^b \rho J_\nu(\alpha_1\rho)J_\nu(\alpha_2\rho)\,\mathrm{d}\rho = \rho[(J_\nu(\alpha_1\rho))'J_\nu(\alpha_2\rho) - J_\nu(\alpha_1\rho)(J_\nu(\alpha_2\rho))']\bigg|_a^b . \tag{7.24}$$

It's time to talk again about boundary conditions. We "want" the terms on the right side of Eq. (7.24) to vanish. Let's take $a = 0$, so that these terms automatically vanish at that point. In Chapter 2, we showed that terms of the form as on the right of Eq. (7.24) vanish for separated boundary conditions, Eq. (2.19). Thus, we want the Bessel function, or its derivative, or some combination of the two, to vanish at $\rho = b$.

For simplicity, assume Dirichlet conditions at $\rho = b$, so that $J_\nu(\alpha_2 b) = J_\nu(\alpha_1 b) = 0$. *The quantity αb must coincide with a zero of $J_\nu(x)$.* Where *are* the zeros of Bessel functions?[9] Denote the nth zero of J_ν as $\alpha_{\nu,n}$. (There are an infinite number of zeros for each Bessel function, as we see from Eq. (7.3)). Thus, $J_\nu(\alpha_{\nu,n}) = 0$, $n = 1, 2, 3, \ldots$. The zeros of $J_n(x)$ are widely tabulated [22, p. 409]; a few are shown in Table 7.1.

The eigenvalues are therefore[10] $\alpha_{\nu,n}/b$, $n = 1, 2, 3, \ldots$. With the eigenvalues determined, the eigenfunctions $J_\nu(\alpha_{\nu,n}\rho/b)$ are simply "snips" of $J_\nu(x)$, with the nth zero occurring at $\rho = b$. The orthogonality condition Eq. (7.24) then becomes (with $\alpha_1 = \alpha_{\nu,n}/b$ and $\alpha_2 = \alpha_{\nu,m}/b$)

$$\int_0^b \rho J_\nu\left(\alpha_{\nu,n}\frac{\rho}{b}\right) J_\nu\left(\alpha_{\nu,m}\frac{\rho}{b}\right)\,\mathrm{d}\rho = 0. \qquad (n \neq m) \tag{7.25}$$

Table 7.1 Zeros $\alpha_{\nu,m}$ for which $J_\nu(\alpha_{\nu,m}) = 0$, $\nu = 0, 1, 2$, $m = 1, 2, 3$.

ν	$m = 1$	$m = 2$	$m = 3$
$\nu = 0$	2.405	5.520	8.654
$\nu = 1$	3.832	7.016	10.173
$\nu = 2$	5.136	8.417	11.620

[8]That makes it different from, say, $P_l^m(x)$ (which satisfies the Sturm–Liouville equation, Eq. (6.31)), where for fixed m, there are an infinite number of eigenfunctions for $l \geq m$.
[9]We know, for example, precisely where the zeros of the sine function are: $\sin(n\pi) = 0$.
[10]For the first example in Section 2.3, the eigenvalues are determined by the condition that $\sqrt{\lambda}b$ coincide with a zero of the sine function, $\sqrt{\lambda}b = n\pi$.

Equation (7.25) can be rewritten through the change of variables $x = \rho/b$:

$$\int_0^1 xJ_\nu(\alpha_{\nu,n}x)J_\nu(\alpha_{\nu,m}x)\,\mathrm{d}x = 0. \qquad (n \neq m) \tag{7.26}$$

The normalization integral follows from Eq. (7.24) after substituting $x = \rho/b$:

$$\int_0^1 x[J_\nu(\alpha_{\nu,n}x)]^2\,\mathrm{d}x$$
$$= \lim_{\alpha_{\nu,m}\to\alpha_{\nu,n}} \frac{[\alpha_{\nu,n}J'_\nu(\alpha_{\nu,n})J_\nu(\alpha_{\nu,m}x) - \alpha_{\nu,m}J_\nu(\alpha_{\nu,n}x)J'_\nu(\alpha_{\nu,m})]|_{x=1}}{\alpha^2_{\nu,m} - \alpha^2_{\nu,n}}.$$

Such an expression leads to the indeterminate form 0/0, so we invoke l'Hopital's rule. Define $y \equiv \alpha_{\nu,m} - \alpha_{\nu,n}$, differentiate with respect to y, and let $y \to 0$. We find:

$$\int_0^1 x[J_\nu(\alpha_{\nu,n}x)]^2\,\mathrm{d}x = \frac{1}{2}[J'_\nu(\alpha_{\nu,n})]^2 = \frac{1}{2}[J_{\nu+1}(\alpha_{\nu,n})]^2, \tag{7.27}$$

i.e. the normalization integral requires the *slope* of J_ν at its zero, information that's tabulated for integer ν. The second equality in Eq. (7.27) follows from the recursion relations (Exercise 7.10). Restoring the dimensional factors,

orthogonality

$$\int_0^b \rho J_\nu\left(\frac{\alpha_{\nu,n}\rho}{b}\right) J_\nu\left(\frac{\alpha_{\nu,m}\rho}{b}\right)\,\mathrm{d}\rho = \delta_{m,n}\frac{b^2}{2}[J_{\nu+1}(\alpha_{\nu,n})]^2. \tag{7.28}$$

7.4 Bessel series

Arbitrary functions can be represented as infinite linear combinations of the eigenfunctions of Sturm–Liouville differential operators (expansion theorem), and Bessel functions are no exception. Thus, for a function $f(\rho)$ defined on $[0, b]$, we can write, for fixed ν,

Bessel series

$$f(\rho) = \sum_{n=1}^{\infty} a_{\nu,n}J_\nu\left(\frac{\alpha_{\nu,n}\rho}{b}\right), \qquad (0 \le \rho \le b) \tag{7.29}$$

where the coefficients are obtained using orthogonality, Eq. (7.28):

$$a_{\nu,n} = \frac{2}{b^2[J_{\nu+1}(\alpha_{\nu,n})]^2}\int_0^b \rho J_\nu\left(\frac{\alpha_{\nu,n}\rho}{b}\right) f(\rho)\,\mathrm{d}\rho. \tag{7.30}$$

Equation (7.29) is known as a *Bessel series*. In most applications, ν is an integer m, which can be traced back to the azimuthal variation $\mathrm{e}^{im\phi}$ in the solution of the Helmholtz equation, Eq. (3.30). Let's see how this works.

Example. Laplace equation in cylindrical coordinates

For a cylinder of height h and radius r (see Figure 7.2), find the solution of the Laplace equation when $V(\rho, \phi, z)$ satisfies the boundary conditions:

$$V(\rho, \phi, 0) = 0 \quad V(r, \phi, z) = 0 \quad V(\rho, \phi, h) = f(\rho, \phi),$$

where $f(\rho, \phi)$ is a prescribed function. From Section 3.2.2, $V(\rho, \phi, z) = R(\rho)\mathrm{e}^{\mathrm{i}m\phi}Z(z)$, where m is an integer. The function $Z(z)$ is, from Eq. (3.37), $Z(z) = A\sinh(\alpha z) + B\cosh(\alpha z)$ (because we seek the solution of the Laplace equation, not the Helmholtz equation, the separation constant $k^2 = 0$). The quantity α is an eigenvalue of the differential equation for $J_m(x)$. The first boundary condition is satisfied with $B = 0$, and the second with $\alpha = \alpha_{m,n}/r$. Thus, for given $m \geq 0$,

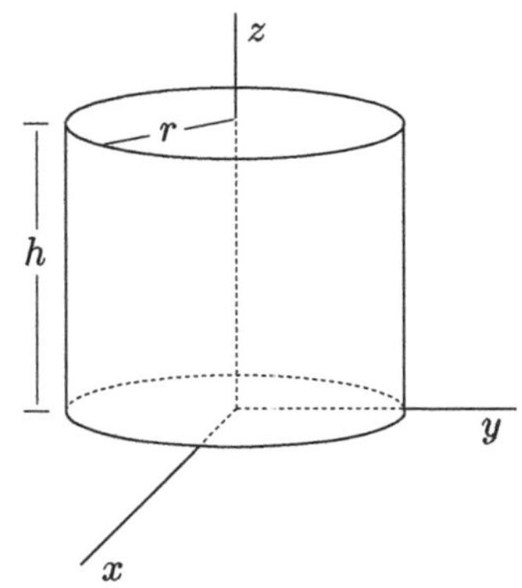

Figure 7.2 Cylindrical geometry for the Laplace equation.

$$V(\rho, \phi, z) = J_m\left(\frac{\alpha_{m,n}\rho}{r}\right)\sinh\left(\frac{\alpha_{m,n}z}{r}\right)\mathrm{e}^{\mathrm{i}m\phi}, \qquad n = 1, 2, 3, \ldots$$

where the Neumann function is excluded because the solution domain includes $\rho = 0$. The most general solution thus has the form:

$$V(\rho, \phi, z) = \sum_{m=-\infty}^{\infty}\sum_{n=1}^{\infty} a_{mn} J_{|m|}\left(\frac{\alpha_{|m|,n}\rho}{r}\right)\sinh\left(\frac{\alpha_{|m|,n}z}{r}\right)\mathrm{e}^{\mathrm{i}m\phi}. \tag{7.31}$$

Note that we must sum over positive and negative values of m because of the completeness properties of the functions $\{\mathrm{e}^{\mathrm{i}m\phi}\}$, Eq. (4.11).

To satisfy the boundary condition at $z = h$, we expand $f(\rho, \phi)$ in the complete set of functions on the domain $0 \leq \rho \leq r$ and $0 \leq \phi \leq 2\pi$:

$$f(\rho, \phi) = \sum_{m=-\infty}^{\infty}\sum_{n=1}^{\infty} a_{mn} J_{|m|}\left(\frac{\alpha_{|m|,n}\rho}{r}\right)\sinh\left(\frac{\alpha_{|m|,n}h}{r}\right)\mathrm{e}^{\mathrm{i}m\phi}.$$

The coefficients are found using orthogonality, Eqs. (4.2) and (7.28):

$$a_{mn}\sinh\left(\frac{\alpha_{|m|,n}h}{r}\right) = \frac{1}{\pi r^2 J^2_{|m|+1}(\alpha_{|m|,n})}\int_0^{2\pi}\int_0^r \rho\mathrm{e}^{-\mathrm{i}m\phi} J_{|m|}\left(\frac{\alpha_{|m|,n}\rho}{r}\right) f(\rho, \phi)\,\mathrm{d}\rho\,\mathrm{d}\phi$$

By combining the coefficients a_{mn} with Eq. (7.31), we obtain the solution of the Laplace equation that meets all boundary conditions. We won't write down the complicated expression that ensues.

Example. Vibrating circular membrane

Consider a vibrating circular membrane of radius a ("drumhead"), where, working in circular polar coordinates (r, θ), the amplitude of vibration $\psi(r, \theta, t)$ is described by the two-dimensional wave equation, Eq. (3.13). Assume a Dirichlet condition at $r = a$, with $\psi(a, \theta, t) = 0$. Separating variables with $\psi = R(r, \theta)T(t)$; the time-varying solution is given by Eq. (3.22). The spatial part satisfies the Helmholtz equation, (3.20), which in circular polar coordinates is given by

$$\left[\frac{1}{r}\frac{\partial}{\partial r}\left(r\frac{\partial}{\partial r}\right) + \frac{1}{r^2}\frac{\partial^2}{\partial \theta^2} + k^2\right] R(r, \theta) = 0,$$

where k^2 is a separation constant. Writing $R(r, \theta) = f(r)\Theta(\theta)$, we have in the usual way $\Theta(\theta) = \mathrm{e}^{im\theta}$, where m is an integer. The radial equation is then

$$f'' + \frac{1}{r}f'(r) + \left(k^2 - \frac{m^2}{r^2}\right) f(r) = 0,$$

which we recognize as the Bessel differential equation, Eq. (3.38). The solutions of the Helmholtz equation for this two-dimensional example thus have the form

$$[A_m J_m(kr) + B_m N_m(kr)][C_m \cos(m\theta) + D_m \sin(m\theta)].$$

At some point, it's easier to restrict $m \geq 0$ and use sines and cosines than to allow negative values of m and use $\mathrm{e}^{im\theta}$. We can immediately eliminate the Neumann function because the drumhead is centered about $r = 0$. Thus, $B_m = 0$. The boundary condition determines the values of k through the requirement that $J_m(ka) = 0$; thus, $k_n = \alpha_{m,n}/a$ for $n = 1, 2, 3, \ldots$. We can now assemble the space and time-dependent solution:

$$\psi(r, \theta, t) = \sum_{m=0}^{\infty}\sum_{n=1}^{\infty} J_m\left(\frac{\alpha_{m,n}r}{a}\right)\Big[[A_{mn}\cos(m\theta) + B_{mn}\sin(m\theta)]\cos(\omega_{mn}t)$$
$$+ [C_{mn}\cos(m\theta) + D_{mn}\sin(m\theta)]\sin(\omega_{mn}t)\Big],$$

where $\omega_{mn} \equiv \alpha_{m,n}c/a$. The coefficients A_{mn}, B_{mn}, C_{mn}, and D_{mn} are determined by the initial conditions, as in the example of the vibrating string (last example of Section 4.3). At $t = 0$,

$$\psi(r, \theta, 0) = \sum_{m=0}^{\infty}\sum_{n=1}^{\infty} J_m\left(\frac{\alpha_{m,n}r}{a}\right)[A_{mn}\cos(m\theta) + B_{mn}\sin(m\theta)]$$

$$\left.\frac{\partial\psi}{\partial t}\right|_{t=0} = \sum_{m=0}^{\infty}\sum_{n=1}^{\infty} \omega_{mn} J_m\left(\frac{\alpha_{m,n}r}{a}\right)[C_{mn}\cos(m\theta) + D_{mn}\sin(m\theta)].$$

The coefficients are isolated by using the orthogonality properties of the trigonometric and Bessel functions, Eqs. (3.38) and (7.28).

7.5 The Fourier-Bessel transform

Functions defined on finite intervals $0 \leq \rho \leq b$ can thus be represented with Bessel series, Eq. (7.29). What can we say if the interval is infinite, $b \to \infty$? We essentially asked (and answered) the same question in Section 4.6, in passing from a Fourier series to a Fourier integral. As b gets progressively larger, the difference between successive eigenvalues of the Bessel equation gets progressively smaller, until we can speak of a continuous eigenvalue parameter. We can see this using Eq. (7.3), from which the zeros of Bessel functions are found to asymptotically approach the values

$$\alpha_{\nu,n} \overset{n\to\infty}{\sim} \pi\left(n + \frac{\nu}{2} + \frac{3}{4}\right) \overset{n\gg\nu}{\approx} n\pi. \tag{7.32}$$

The difference between successive zeros $\Delta\alpha_{\nu,n} \equiv \alpha_{\nu,n+1} - \alpha_{\nu,n}$ therefore approaches π for large[11] n, implying that the difference between successive eigenvalues $\Delta\alpha \equiv (\Delta\alpha_{\nu,n})/b = \pi/b$. Let α denote the continuous variable that the eigenvalues "pack down" to as $b \to \infty$, $(\alpha_{\nu,n}/b) \to \alpha$.

Before we can replace the Bessel series in Eq. (7.29) with an integral, we must analyze the behavior of the expansion coefficients $a_{\nu,n}$, given by Eq. (7.30). In particular, we must analyze the prefactor in Eq. (7.30),[12] $2/[bJ'_\nu(\alpha_{\nu,n})]^2$. Using Eq. (7.3), we have

$$J'_\nu(\alpha_{\nu,n}) \overset{n\to\infty}{\sim} -\sqrt{\frac{2}{\pi\alpha_{\nu,n}}}\sin\left(\alpha_{\nu,n} - \frac{1}{2}\nu\pi - \frac{1}{4}\pi\right) = \pm\sqrt{\frac{2}{\pi\alpha_{\nu,n}}}, \tag{7.33}$$

because $\sin x = \pm 1$ at the zeros of $\cos x$. Thus,

$$b^2[J'_\nu(\alpha_{\nu,n})]^2 \overset{n\to\infty}{\sim} \frac{2b^2}{\pi\alpha_{\nu,n}} = \frac{2}{\alpha\Delta\alpha}. \tag{7.34}$$

We therefore have for the expansion coefficients:

$$a_{\nu,n} \overset{n\to\infty}{\sim} \alpha\Delta\alpha \int_0^b \rho J_\nu(\alpha\rho) f(\rho)\,\mathrm{d}\rho. \tag{7.35}$$

By combining Eq. (7.35) with Eq. (7.29),

$$f(\rho) = \sum_{n=1} a_{\nu,n} J_\nu\left(\frac{\alpha_{\nu,n}\rho}{b}\right) \xrightarrow{b\to\infty} \int_0^\infty \alpha g(\alpha) J_\nu(\rho\alpha)\,\mathrm{d}\alpha, \tag{7.36}$$

where

$$g(\alpha) \equiv \int_0^\infty \rho J_\nu(\alpha\rho) f(\rho)\,\mathrm{d}\rho. \tag{7.37}$$

The function $g(\alpha)$ is known as the *Fourier-Bessel transform* of $f(\rho)$, and Eqs. (7.36) and (7.37) are Fourier-Bessel transform pairs.

Fourier-Bessel transform

[11] From Table 7.1, the difference between zeros is *roughly* π even for the first few zeros.

[12] Its more convenient here to use $[J_{\nu+1}(\alpha_{\nu,n})]^2 = [J'_\nu(\alpha_{\nu,n})]^2$.

7.6 Spherical Bessel functions

Solutions to the homogeneous Helmholtz equation in spherical coordinates have the form (see Section 3.2.3),

$$\psi(r,\theta,\phi)=\sum_{l=0}^{\infty}\sum_{m=-l}^{l}\left(A_{l,m}\frac{J_{l+\frac{1}{2}}(kr)}{\sqrt{kr}}+B_{l,m}\frac{N_{l+\frac{1}{2}}(kr)}{\sqrt{kr}}\right)Y_l^m(\theta,\phi), \quad (7.38)$$

where k is the separation constant. Bessel functions of half-integer order, together with the square root factor, occur so often in certain problems that it's convenient to consider them as distinct functions. The *spherical* Bessel and Neumann functions are defined as

spherical Bessel and Neumann functions

$$j_l(x)\equiv\sqrt{\frac{\pi}{2x}}J_{l+\frac{1}{2}}(x)$$

$$n_l(x)\equiv\sqrt{\frac{\pi}{2x}}N_{l+\frac{1}{2}}(x). \quad (7.39)$$

These functions are also referred to as the spherical Bessel functions of the first and second kinds. The spherical Hankel functions, (also known as the spherical Bessel functions of the third kind), are defined as in Eq. (7.4):

spherical Hankel functions

$$h_l^{(1)}(x)\equiv j_l(x)+\mathrm{i}n_l(x)=\sqrt{\frac{\pi}{2x}}H^{(1)}_{l+\frac{1}{2}}(x)$$

$$h_l^{(2)}(x)\equiv j_l(x)-\mathrm{i}n_l(x)=\sqrt{\frac{\pi}{2x}}H^{(2)}_{l+\frac{1}{2}}(x). \quad (7.40)$$

The pairs of functions $(j_l(x), n_l(x))$ and $(h_l^{(1)}(x), h_l^{(2)}(x))$ are linearly independent (Exercise 7.18).

7.6.1 Reduction to elementary functions

The spherical Bessel functions are combinations of trigonometric functions and inverse powers of x (see Table 7.2). Consider $j_0(x)$. Using Eq. (5.28),

$$j_0(x)=\sqrt{\frac{\pi}{2x}}J_{\frac{1}{2}}(x)=\frac{\sqrt{\pi}}{2}\sum_{k=0}^{\infty}\frac{(-1)^k}{k!\Gamma(k+\frac{3}{2})}\left(\frac{x}{2}\right)^{2k}. \quad (7.41)$$

Use the recursion relation for the Γ function, Eq. (C.2):

$$\Gamma\left(k+\frac{3}{2}\right)=\left(k+\frac{1}{2}\right)\left(k-\frac{1}{2}\right)\cdots\frac{1}{2}\Gamma\left(\frac{1}{2}\right)$$

$$=\frac{(2k+1)(2k-1)\cdots 1}{2^{k+1}}\sqrt{\pi}=\frac{(2k+1)!}{2^{2k+1}k!}\sqrt{\pi}.$$

Table 7.2 Spherical Bessel and Neumann functions, $j_l(x)$ and $n_l(x)$ for $l = 0, 1, 2$.

$j_0(x) = \frac{\sin x}{x}$	$n_0(x) = -\frac{\cos x}{x}$
$j_1(x) = \frac{\sin x}{x^2} - \frac{\cos x}{x}$	$n_1(x) = -\frac{\cos x}{x^2} - \frac{\sin x}{x}$
$j_2(x) = \left(\frac{3}{x^3} - \frac{1}{x}\right)\sin x - \frac{3}{x^2}\cos x$	$n_2(x) = -\left(\frac{3}{x^3} - \frac{1}{x}\right)\cos x - \frac{3}{x^2}\sin x$

Combine this result with Eq. (7.41),

$$j_0(x) = \frac{1}{x}\sum_{k=0}^{\infty}\frac{(-1)^k x^{2k+1}}{(2k+1)!} = \frac{\sin x}{x}. \tag{7.42}$$

In a similar way it can be shown that (Exercise 7.1)

$$n_0(x) = -\frac{\cos x}{x}. \tag{7.43}$$

Example. Using Eqs. (7.42) and (7.43), $h_0^{(1)}(x) = -\mathrm{i}\frac{\mathrm{e}^{\mathrm{i}x}}{x}$ and $h_0^{(2)}(x) = \mathrm{i}\frac{\mathrm{e}^{-\mathrm{i}x}}{x}$.

To calculate $j_l(x)$ for $l \geq 1$, we rely on an identity:

$$\left(\frac{1}{x}\frac{\mathrm{d}}{\mathrm{d}x}\right)^l [x^{-\nu}J_\nu(x)] = (-1)^l x^{-(\nu+l)} J_{\nu+l}(x). \tag{7.44}$$

The derivation of Eq. (7.44) is based on a trick. Let $y \equiv x^2/2$. From the chain rule,

$$\frac{\mathrm{d}}{\mathrm{d}y} = \frac{\mathrm{d}x}{\mathrm{d}y}\frac{\mathrm{d}}{\mathrm{d}x} = \frac{1}{\mathrm{d}y/\,\mathrm{d}x}\frac{\mathrm{d}}{\mathrm{d}x} = \frac{1}{x}\frac{\mathrm{d}}{\mathrm{d}x}.$$

Then, using the series representation for $J_\nu(x)$, Eq. (5.28):

$$\left(\frac{1}{x}\frac{\mathrm{d}}{\mathrm{d}x}\right)^l [x^{-\nu}J_\nu(x)] = \frac{1}{2^\nu}\frac{\mathrm{d}^l}{\mathrm{d}y^l}\sum_{k=0}^{\infty}\frac{(-1)^k}{k!2^k\Gamma(k+\nu+1)}y^k.$$

It's straightforward to show that, for integer k:

$$\frac{\mathrm{d}^l}{\mathrm{d}y^l}y^k = \begin{cases}\frac{k!}{(k-l)!}y^{k-l} & k \geq l \\ 0 & k < l.\end{cases}$$

Thus,

$$\left(\frac{1}{x}\frac{\mathrm{d}}{\mathrm{d}x}\right)^l [x^{-\nu}J_\nu(x)] = \frac{1}{2^\nu}\sum_{k=l}^{\infty}\frac{(-1)^k}{2^k\Gamma(k+\nu+1)(k-l)!}y^{k-l}$$

$$= \frac{(-1)^l}{2^{\nu+l}}\sum_{m=0}^{\infty}\frac{(-1)^m}{m!\Gamma(m+l+\nu+1)}\left(\frac{y}{2}\right)^m = (-1)^l x^{-(\nu+l)}J_{\nu+l}(x),$$

which is Eq. (7.44). With $\nu = \frac{1}{2}$ in Eq. (7.44), we find

$$j_l(x) = (-x)^l \left(\frac{1}{x}\frac{\mathrm{d}}{\mathrm{d}x}\right)^l j_0(x). \qquad l = 0, 1, 2, \ldots \tag{7.45}$$

By a similar calculation, it can be shown that (Exercise 7.14)

$$n_l(x) = (-x)^l \left(\frac{1}{x}\frac{\mathrm{d}}{\mathrm{d}x}\right)^l n_0(x). \tag{7.46}$$

A few expressions for $j_l(x)$ and $n_l(x)$ are shown in Table 7.1. The spherical Hankel functions can be expressed, using Eqs. (7.46) and (7.40):

$$\begin{aligned} h_l^{(1)}(x) &= -\mathrm{i}(-x)^l \left(\frac{1}{x}\frac{\mathrm{d}}{\mathrm{d}x}\right)^l \left(\frac{\mathrm{e}^{\mathrm{i}x}}{x}\right) \\ h_l^{(2)}(x) &= \mathrm{i}(-x)^l \left(\frac{1}{x}\frac{\mathrm{d}}{\mathrm{d}x}\right)^l \left(\frac{\mathrm{e}^{-\mathrm{i}x}}{x}\right). \end{aligned} \tag{7.47}$$

7.6.2 Small-argument forms

The small-argument forms of $j_l(x)$ and $n_l(x)$ are obtained from the series representations of $J_{\pm(l+\frac{1}{2})}(x)$ (Exercise 7.15)

$$\begin{aligned} j_l(x) &\xrightarrow{x\to 0} \begin{cases} 1 - x^2/6 & l = 0 \\ \frac{2^l l!}{(2l+1)!} x^l & l \neq 0 \end{cases} \\ n_l(x) &\xrightarrow{x\to 0} -\frac{(2l)!}{2^l l!}\frac{1}{x^{l+1}}. \end{aligned} \tag{7.48}$$

7.6.3 Asymptotic forms

For the asymptotic forms of these functions we could use the results given in Eqs. (7.3) and (7.5). It's easier, however, to use Eq. (7.47) directly. As $x \to \infty$, the largest contribution is obtained by applying the derivatives to *just the complex exponentials*, ignoring the inverse powers of x. Thus,

$$\begin{aligned} h_l^{(1)}(x) &\overset{x\to\infty}{\sim} \frac{1}{x}\exp\left(\mathrm{i}\left[x - \frac{\pi}{2}(l+1)\right]\right) \\ h_l^{(2)}(x) &\overset{x\to\infty}{\sim} \frac{1}{x}\exp\left(-\mathrm{i}\left[x - \frac{\pi}{2}(l+1)\right]\right). \end{aligned} \tag{7.49}$$

The spherical Hankel functions of the first kind, $h_l^{(1)}(x)$, are in the form of outgoing spherical waves, while those of the second kind, $h_l^{(2)}(x)$, are in the form of incoming spherical waves. We then have the asymptotic forms of $j_l(x)$ and $n_l(x)$:

$$j_l(x) = \frac{1}{2}(h_l^{(1)}(x) + h_l^{(2)}(x)) \overset{x\to\infty}{\sim} \frac{1}{x}\cos\left(x - \frac{\pi}{2}(l+1)\right)$$
$$n_l(x) = \frac{1}{2\mathrm{i}}(h_l^{(1)}(x) - h_l^{(2)}(x)) \overset{x\to\infty}{\sim} \frac{1}{x}\sin\left(x - \frac{\pi}{2}(l+1)\right). \qquad (7.50)$$

7.6.4 Orthogonality and completeness

The spherical Bessel functions satisfy the differential equation, Eq. (3.47) (in the radial coordinate r). Equation (3.47), however, is not in self-adjoint form. It can be put in that form by multiplying through by r^2, which we discover using Eq. (2.15). Equation (3.47) expressed as a Sturm–Liouville equation, with $\lambda = l(l+1)$ and $x = kr$, is:

$$x^2 j_l''(x) + 2x j_l'(x) - l(l+1)j_l(x) = -x^2 j_l(x). \qquad l = 0, 1, 2, \ldots$$

The functions $j_l(kr)$ (restoring the factor of k), being solutions of a Sturm–Liouville equation, therefore satisfy an orthogonality condition[13] with respect to the weight function r^2:

$$\int_0^b r^2 j_l(k_1 r) j_l(k_2 r)\,\mathrm{d}r = 0, \qquad k_1 \neq k_2$$

when kb is a zero of $J_{l+\frac{1}{2}}$. Let's denote the zeros of $j_l(x)$ as $\beta_{l,n} \equiv \alpha_{l+\frac{1}{2},n}$; $j_l(\beta_{l,n}) = 0$. The normalization integral can be evaluated using Eq. (7.28):

$$\int_0^b r^2\left[j_l\left(\frac{\beta_{l,n}r}{b}\right)\right]^2 \mathrm{d}r = \frac{\pi b}{2\beta_{l,n}}\int_0^b r\left[J_{l+\frac{1}{2}}\left(\frac{\beta_{l,n}r}{b}\right)\right]^2 \mathrm{d}r = \frac{\pi b^3}{4\beta_{l,n}}\left[J'_{l+\frac{1}{2}}(\beta_{l,n})\right]^2$$
$$= \frac{1}{2}b^3[j_l'(\beta_{l,n})]^2.$$

The orthogonality relation is therefore

orthogonality

$$\int_0^b r^2 j_l\left(\frac{\beta_{l,n}r}{b}\right) j_l\left(\frac{\beta_{l,m}r}{b}\right)\mathrm{d}r = \delta_{n,m}\frac{1}{2}b^3[j_l'(\beta_{l,n})]^2 \quad . \qquad (7.51)$$

Spherical Bessel functions are thus an orthogonal set, and the expansion theorem can be used to represent functions $f(r)$ of the radial coordinate:

spherical Bessel series

$$f(r) = \sum_{n=1}^{\infty} c_n j_l\left(\frac{\beta_{l,n}r}{b}\right), \qquad 0 \le r \le b \qquad (7.52)$$

where the expansion coefficients are given by $c_n = 2\int_0^b r^2 j_l(\beta_{l,n}r/b) f(r)\,\mathrm{d}r/(b^3[j_l'(\beta_{l,n})]^2)$.

By letting $b \to \infty$, one obtains a Fourier-Bessel transform in terms of spherical Bessel functions. We can use what we derived in Section 7.5 (instead of repeating the analysis). By letting $\nu = l + \frac{1}{2}$ in Eqs. (7.36) and (7.37), it can be shown that:

spherical Bessel-Fourier transform

[13] See discussion in Section 7.3.

$$f(r) = \int_0^\infty \beta^2 h(\beta) j_l(\beta r) \, \mathrm{d}\beta$$

$$h(\beta) = \frac{2}{\pi} \int_0^\infty r^2 j_l(\beta r) f(r) \, \mathrm{d}r. \tag{7.53}$$

completeness relation

We then obtain in the usual way the completeness relation:

$$\int_0^\infty \beta^2 j_l(\beta r) j_l(\beta r') \, \mathrm{d}\beta = \frac{\pi}{2r^2} \delta(r - r'). \tag{7.54}$$

7.7 Expansion of plane waves in spherical harmonics

plane waves

At a fixed instant of time, *plane waves* of wavelength λ are described by $\psi = \mathrm{e}^{\mathrm{i}\boldsymbol{k}\cdot\boldsymbol{r}}$, where the wavevector $\boldsymbol{k} = k\hat{\boldsymbol{n}}$, with $\hat{\boldsymbol{n}}$ the direction of propagation and $k = 2\pi/\lambda$. We can see why plane waves are so named (see Figure 7.3). For a vector $\boldsymbol{r}$ locating positions on a plane perpendicular to $\boldsymbol{k}$, the construct $\boldsymbol{k}\cdot\boldsymbol{r} = kr\cos\theta$ has the same value at all points of the plane. The locus of points having the same phase of a wave is called a *wavefront*. If we add in time dependence, $\mathrm{e}^{\mathrm{i}(\boldsymbol{k}\cdot\boldsymbol{r}-\omega t)}$ describes the motion of a plane wavefront in the direction $\hat{\boldsymbol{n}}$ traveling at speed c when $\omega = ck$. The function $\psi = \mathrm{e}^{\mathrm{i}\boldsymbol{k}\cdot\boldsymbol{r}}$ is a solution of the Helmholtz equation, $(\nabla^2 + k^2)\psi = 0$.

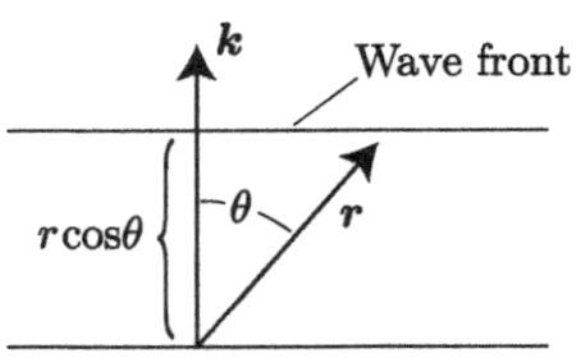

Figure 7.3 Geometry of a plane wave.

In many circumstances, it's useful to have a representation of the plane-wave factor $\mathrm{e}^{\mathrm{i}\boldsymbol{k}\cdot\boldsymbol{r}}$ in spherical coordinates. As a solution of the Helmholtz equation, $\mathrm{e}^{\mathrm{i}\boldsymbol{k}\cdot\boldsymbol{r}}$ can be expanded in terms of the functions in Eq. (7.38) (solutions of the Helmholtz equation in spherical coordinates). We must exclude the spherical Neumann functions, however – plane waves are finite at $r = 0$. We're free to align the polar axis of our coordinate system in the direction of $\boldsymbol{k}$, and thus in that coordinate system $\mathrm{e}^{\mathrm{i}\boldsymbol{k}\cdot\boldsymbol{r}} = \mathrm{e}^{\mathrm{i}kr\cos\theta}$. Because there's no dependence on the azimuthal angle, only terms with $m = 0$ in Eq. (7.38) contribute. As $Y_l^0(\cos\theta) \propto P_l(\cos\theta)$, we try the expansion:

$$\mathrm{e}^{\mathrm{i}kr\cos\theta} = \sum_{l=0}^\infty c_l j_l(kr) P_l(\cos\theta). \tag{7.55}$$

As usual, we can isolate the expansion coefficient c_l exploiting the orthogonality of Legendre polynomials. From Eq. (6.16), with $x = \cos\theta$, we find:

$$c_l j_l(kr) = \frac{2l+1}{2} \int_{-1}^1 \mathrm{e}^{\mathrm{i}krx} P_l(x) \, \mathrm{d}x = \frac{2l+1}{2} \frac{1}{2^l l!} \int_{-1}^1 \mathrm{e}^{\mathrm{i}krx} \frac{\mathrm{d}^l}{\mathrm{d}x^l} (x^2-1)^l, \tag{7.56}$$

where we've used the Rodrigues formula, Eq. (6.10). Integrating by parts l times the integral on the right of Eq. (7.56):

$$\int_{-1}^1 \mathrm{e}^{\mathrm{i}krx} \frac{\mathrm{d}^l}{\mathrm{d}x^l} (x^2-1)^l \, \mathrm{d}x = (-\mathrm{i}kr)^l \int_{-1}^1 \mathrm{e}^{\mathrm{i}krx} (x^2-1)^l \, \mathrm{d}x. \tag{7.57}$$

Because evidently we really like integrating by parts, keep going. For the integral on the right of Eq. (7.57), integrating by parts:

$$\int_{-1}^{1} \mathrm{e}^{ikrx}(x^2-1)^l \,\mathrm{d}x = \frac{-2l}{\mathrm{i}kr}\int_{-1}^{1} \mathrm{e}^{ikrx}x(x^2-1)^{l-1}\,\mathrm{d}x$$
$$= \frac{2l}{kr}\frac{\mathrm{d}}{\mathrm{d}(kr)}\int_{-1}^{1} \mathrm{e}^{ikrx}(x^2-1)^{l-1}\,\mathrm{d}x.$$

Repeating this process,

$$\int_{-1}^{1} \mathrm{e}^{ikrx}(x^2-1)^l \,\mathrm{d}x = 2^l l!\left(\frac{1}{kr}\frac{\mathrm{d}}{\mathrm{d}(kr)}\right)^l \int_{-1}^{1} \mathrm{e}^{ikrx}\,\mathrm{d}x$$
$$= 2^{l+1} l!\left(\frac{1}{kr}\frac{\mathrm{d}}{\mathrm{d}(kr)}\right)^l \left(\frac{\sin(kr)}{kr}\right)$$

Combining the aforementioned results,

$$c_l j_l(kr) = \mathrm{i}^l(2l+1)(-kr)^l\left(\frac{1}{kr}\frac{\mathrm{d}}{\mathrm{d}(kr)}\right)^l \left(\frac{\sin(kr)}{kr}\right) = \mathrm{i}^l(2l+1)j_l(kr)$$

where we've used Eq. (7.46). Thus, $c_l = \mathrm{i}^l(2l+1)$. By combining with Eq. (7.55), we have the desired result:

plane-wave expansion

$$\mathrm{e}^{ikr\cos\theta} = \sum_{l=0}^{\infty} \mathrm{i}^l(2l+1)j_l(kr)P_l(\cos\theta). \tag{7.58}$$

Equation (7.58) should be contrasted with the result of Exercise 7.4: One is the expansion of a plane wave in spherical coordinates, the other is an expansion in cylindrical coordinates.

Equation (7.58) applies for plane waves propagating along the direction of the polar axis, but it can readily be generalized to arbitrary directions. Referring to Figure 6.4, $\mathrm{e}^{\mathrm{i}\boldsymbol{k}\cdot\boldsymbol{r}} = \mathrm{e}^{ikr\cos\gamma}$ when $\boldsymbol{k}$ is aligned with the direction of $\boldsymbol{r}'$. Utilizing Eq. (7.58),

$$\mathrm{e}^{ikr\cos\gamma} = \sum_{l=0}^{\infty} \mathrm{i}^l(2l+1)j_l(kr)P_l(\cos\gamma)$$
$$= 4\pi\sum_{l=0}^{\infty}\sum_{m=-l}^{l} \mathrm{i}^l j_l(kr)Y_l^m(\theta,\phi)(Y_l^m(\theta',\phi'))^*, \tag{7.59}$$

where we've used the addition theorem for Legendre polynomials, Eq. (6.37).

Summary

This chapter was devoted to the properties of Bessel functions (beyond the power series representation given in Chapter 5) and those of spherical Bessel functions.

- Bessel functions, being the solutions of a Sturm–Liouville equation, are a set of orthogonal functions. Bessel functions thus form (by the expansion theorem, Section 2.5) a complete set, just as with the other orthogonal functions developed in Chapter 4 (complex exponentials) and in Chapter 6 (Legendre polynomials and spherical harmonics). The orthogonality relation, Eq. (7.28), involves the zeros of Bessel functions, $\alpha_{\nu,m}$. One can develop Bessel series, Eq. (7.29), to represent functions defined on finite intervals $[0, b]$ and Fourier-Bessel transforms for functions defined on $[0, \infty)$, Eq. (7.36).
- Bessel functions of integer order can be obtained from a generating function, Eq. (7.6). Just as with Legendre polynomials, which can also be obtained from a generating function, Bessel functions satisfy various recursion relations, results that are straightforward to prove using the generating function.
- Bessel functions possess integral representations, as in Eq. (7.20). The values of Bessel functions could be computed from an integral, rather than from a power series, but the main use of integral representations is in proving the asymptotic results given in this chapter. Given the parallels between Bessel functions and Legendre polynomials as orthogonal functions and in possessing generating functions, it's natural to ask if Legendre polynomials have integral representations. They do, as shown in Exercises 8.23 and 8.24.
- Spherical Bessel functions are related to Bessel functions of half-integer order, Eq. (7.39). They can be related to combinations of trigonometric functions and inverse powers of x. Spherical Bessel functions comprise a complete orthogonal set, with the orthogonality property Eq. (7.51), and with functions defined on finite intervals represented by spherical Bessel series, Eq. (7.52), and spherical Bessel-Fourier transforms, Eq. (7.53), for functions defined on the semi-infinite interval $[0, \infty)$.

Exercises

7.1. Show that the Hankel functions are linearly independent. Hint: The Wronskian for the Bessel and Neumann functions is given by Eq. (5.33).

7.2. Justify the steps leading to Eq. (7.8).

7.3. Show that $1 = J_0(x) + 2\sum_{n=1}^{\infty} J_{2n}(x) = J_0(x) + 2J_2(x) + 2J_4(x) + \cdots$. Hint: Set $\theta = 0$ in Eq. (7.8).

7.4. Show that $\exp(\mathrm{i}x\cos\theta) = \sum_{n=-\infty}^{\infty} \mathrm{i}^n \mathrm{e}^{\mathrm{i}n\theta} J_n(x)$. Hint: Let $\theta \to \theta + \pi/2$ in Eq. (7.7). This relation expresses a plane wave in terms of cylindrical coordinates.

7.5. Derive the recursion relation Eq. (7.14), by differentiating the generating function with respect to y.

7.6. Derive the recursion relation Eq. (7.15). Hint: Add Eqs. (7.13) and (7.14) and multiply by x^n.

7.7. Derive the results in Eq. (7.10) using the addition theorem, Eq. (7.11). Hint: Set $y = -x$ in Eq. (7.11) and make use of Eq. (7.12).

7.8. Derive Eq. (7.17) for the Wronskian.

7.9. Show that $\int_0^\pi \cos(x\sin\theta)\,\mathrm{d}\theta = \int_0^\pi \cos(x\cos\theta)\,\mathrm{d}\theta$. Hint: Break up the integral $\int_0^\pi = \int_0^{\pi/2} + \int_{\pi/2}^{\pi}$ and use $\sin(\theta \pm \pi/2) = \pm\cos\theta$.

7.10. Show that $[J_\nu'(\alpha_{\nu,n})]^2 = [J_{\nu+1}(\alpha_{\nu,n})]^2$. Hint: Use the recursion relations. Subtract Eq. (7.14) from Eq. (7.13) and evaluate the result at $x = \alpha_{\nu,n}$.

7.11. Derive Eq. (7.32). Use Eq. (7.3).

7.12. Combine Eq. (7.37) with Eq. (7.36) to derive the completeness relation for Bessel functions:

$$\int_0^\infty \alpha J_\nu(\alpha\rho)J_\nu(\alpha\rho')\,\mathrm{d}\alpha = \frac{1}{\rho'}\delta(\rho-\rho').$$

7.13. Show that $n_0(x) = -\cos x/x$.

(a) First show that

$$n_0(x) = \sqrt{\frac{\pi}{2x}}N_{\frac{1}{2}}(x) = -\sqrt{\frac{\pi}{2x}}J_{-\frac{1}{2}}(x) = -\frac{\sqrt{\pi}}{x}\sum_{k=0}^{\infty}\frac{(-1)^k}{k!\Gamma(k+\frac{1}{2})}\left(\frac{x}{2}\right)^{2k}.$$

(b) Then show that $n_0(x) = -\frac{1}{x}\sum_{k=0}^{\infty}\frac{(-1)^k x^{2k}}{(2k)!} = -\frac{\cos x}{x}$. Hint: Show that for integer k, $\Gamma(k+\frac{1}{2}) = \frac{(2k)!}{2^{2k}k!}\Gamma(\frac{1}{2})$.

7.14. Derive Eq. (7.46).

(a) Derive the identity similar to Eq. (7.5):

$$\left(\frac{1}{x}\frac{\mathrm{d}}{\mathrm{d}x}\right)^l [x^{-\nu}J_{-\nu}(x)] = x^{-(\nu+l)}J_{-(\nu+l)}(x). \tag{7.60}$$

Hint: Show that for ν not equal to an integer, and for m an integer,

$$\frac{\mathrm{d}^l}{\mathrm{d}y^l}y^{m-\nu} = \frac{\Gamma(m-\nu+1)}{\Gamma(m-\nu-l+1)}y^{m-\nu-l}.$$

(b) Show that $n_l(x) = (-1)^{l+1}\sqrt{\frac{\pi}{2x}}J_{-(l+\frac{1}{2})}(x)$.

(c) Show that Eq. (7.46) follows by setting $\nu = \frac{1}{2}$ in Eq. (7.60).

7.15. Derive the results shown in Eq. (7.48). Hint: Use the small-argument form of $J_{\pm(l+\frac{1}{2})}(x)$ at lowest order in x. Use the result for $\Gamma(l+\frac{3}{2})$ derived. Use the Euler reflection property, Eq. (C.8), to relate $\Gamma(1-(l+\frac{1}{2}))$ to $\Gamma(l+\frac{1}{2})$, and use the expression for $\Gamma(l+\frac{1}{2})$ given in Exercise 7.13.

7.16. Derive the integral representation for the spherical Bessel function $j_l(x)$:

$$j_l(x) = \frac{1}{2}(-\mathrm{i})^l\int_{-1}^{1} P_l(z)\mathrm{e}^{\mathrm{i}xz}\,\mathrm{d}z.$$

Hint: Use the expansion for plane waves, Eq. (7.58), and the orthogonality properties of the Legendre polynomials, Eq. (6.14).

7.17. An important tool in going between momentum space (in the context of quantum mechanics) and position space is the expansion of a plane wave in spherical harmonics, Eq. (7.59). In quantum mechanics, momentum $\boldsymbol{p}$ is related to the wavenumber through the de Broglie relationship, $\boldsymbol{p} = \hbar \boldsymbol{k}$, and a plane wave is represented by $\exp(\mathrm{i}\boldsymbol{p} \cdot \boldsymbol{r})$ where $\boldsymbol{p} \cdot \boldsymbol{r}$ corresponds to $kr \cos\gamma$ in Eq. (7.59). The real part of the plane wave is $\cos(\boldsymbol{p} \cdot \boldsymbol{r})$. Identify the real part of the expansion on the right side of Eq. (7.59) as a way of obtaining an expansion of $\cos(\boldsymbol{k} \cdot \boldsymbol{r})$.

7.18. Derive the Wronskian relations:

$$W(j_l(x), n_l(x)) = \frac{1}{x^2} \qquad W(h_l^{(1)}(x), h_l^{(2)}(x)) = -2\mathrm{i}\frac{1}{x^2}.$$

7.19. The Helmholtz equation $\nabla^2 R + k^2 R = 0$, when separated in the cylindrical coordinate system, has solutions involving the Bessel and Neumann functions, $J_\nu(k\rho)$ and $N_\nu(k\rho)$; see Eq. (3.38). Under the (seemingly silly) substitution $k \to \mathrm{i}k$, the Helmholtz equation takes the form of the diffusion equation, $\nabla^2 R = k^2 R$. There are, accordingly, applications involving Bessel functions of purely imaginary argument, which have special names and symbols. Solutions to the Bessel differential equation for $x \to \mathrm{i}x$ are known as *modified Bessel functions*, defined as

$$I_\nu(x) \equiv \mathrm{e}^{-\mathrm{i}\nu\pi/2} J_\nu(\mathrm{i}x)$$

$$K_\nu(x) \equiv \frac{\pi}{2}\frac{I_{-\nu}(x) - I_\nu(x)}{\sin \pi\nu} = \frac{\pi}{2}\frac{\mathrm{e}^{\mathrm{i}\pi\nu/2} J_{-\nu}(\mathrm{i}x) - \mathrm{e}^{-\mathrm{i}\nu\pi/2} J_\nu(\mathrm{i}x)}{\sin \pi\nu}.$$

(a) Show, starting from Eq. (5.25), that under the substitution $x \to \mathrm{i}x$, modified Bessel functions satisfy the differential equation

$$x^2 y''(x) + xy' - (x^2 + \nu^2) y(x) = 0.$$

(b) Show that $K_\nu(x) = \mathrm{i}(\pi/2)\mathrm{e}^{\mathrm{i}\nu\pi/2} H_\nu^{(1)}(\mathrm{i}x)$.

(c) Show from Eq. (5.28) that

$$I_\nu(x) = \left(\frac{x}{2}\right)^\nu \sum_{k=0}^{\infty} \frac{(x^2/4)^k}{k!\Gamma(\nu + k + 1)}.$$

Show, using the power series and the properties of the gamma function, that for $\nu = n =$ integer, $I_{-n}(x) = I_n(x)$

(d) Show that the generating function for the functions $I_n(x)$ is

$$\exp\left(\frac{x}{2}(y + y^{-1})\right) = \sum_{n=-\infty}^{\infty} I_n(x) y^n. \qquad (y \neq 0)$$

Start with Eq. (7.6), let $x \to \mathrm{i}x$ and $y \to -\mathrm{i}y$.

(e) Show that

$$\mathrm{e}^{x \cos\theta} = I_0(x) + 2\sum_{k=1}^{\infty} I_k(x) \cos k\theta.$$

8

Complex analysis

Complex analysis is the study of *analytic functions of a complex argument.*[1] Analytic functions, as we'll see, are functions having a well-defined derivative throughout some portion, possibly the entirety, of the *complex plane.* This seemingly simple attribute has far-reaching consequences in determining a class of functions with remarkable properties. Much advanced work in physics makes use of analytic functions.

8.1 Complex functions

Complex numbers can be placed in one-to-one correspondence with the points of a plane, the *complex plane* or the *z-plane*, with every complex number $z = x + iy$ corresponding to a point whose abscissa is the real part $x = \text{Re}\, z$ and whose ordinate is the imaginary part $y = \text{Im}\, z$. Functions of complex variables $f(z)$ (henceforth referred to as complex functions) associate a new complex number w with every z in a suitably defined domain of the z-plane: $w = f(z)$. The values of $f(z)$ are said to be in the *w-plane* (see Figure 8.1). By separating z and w into real and imaginary parts, $z = x + iy$, $w = u + iv$, the relation $w = f(z)$ can be interpreted to mean that the pair of real numbers (x, y) is associated with two new real numbers (u, v), where $u = u(x, y)$ and $v = v(x, y)$, and thus we can write

$$f(z) = u(x, y) + iv(x, y), \tag{8.1}$$

where u is called the real part of f and v the imaginary part. Example: For $f(z) = z^2$, $f(z) = x^2 - y^2 + 2ixy$. The ability to resolve complex functions

[1]Analytic functions were defined in Chapter 5 as possessing convergent power-series representations, a statement that holds for real or complex analytic functions. *Complex* analytic functions exhibit additional properties not shared by real analytic functions.

Mathematical Methods in Physics, Engineering, and Chemistry, First Edition.
Brett Borden and James Luscombe.

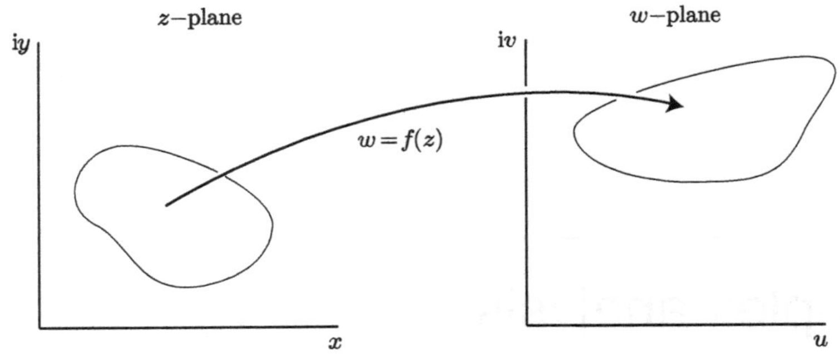

Figure 8.1 Complex-valued functions map complex numbers into complex numbers.

into real and imaginary parts can prove quite useful in calculations. Yet, this property can also obscure the fact that f is a function of a *single* complex variable z, and in many cases, it's advantageous *not* to make use of Eq. (8.1).

Complex analysis is a subject where it pays to be more attentive to points of rigor than one might otherwise be accustomed. We require (at a minimum) that complex functions be *single valued.* We'll see in later sections how complex functions can be *multivalued,* implying that $f(z)$ must be restricted to a region of the z-plane for which the mapping makes sense. The *domain* $\mathcal{D}$ of a complex function is defined as the set of points $z \in \mathcal{D} \subset \mathbb{C}$ for which it's single valued.[2]

One way to define complex functions is as *extensions* of real-valued functions. Starting from real-valued functions such as e^x, $\ln x$, and $\sin x$, we can obtain complex functions by letting $x \to z$ to obtain[3] e^z, $\ln z$, etc. Such a step represents the extension of a function $f(x)$, defined over an interval $\mathcal{R}$ of the real line $x \in \mathcal{R} \subset \mathbb{R}$, to a function $f(z)$ defined over a domain in the complex plane $z \in \mathcal{D} \subset \mathbb{C}$, where $\mathcal{R} \subset \mathcal{D}$, and where we follow the usual practice in physics of using the same symbol to denote functions defined on different domains. For example, when $\exp(x) = \sum_{n=0}^{\infty} x^n/n!$ is considered as a function with imaginary argument, we obtain the useful result, the Euler formula:

$$\exp(\mathrm{i}y) = \sum_{n=0}^{\infty} \frac{(\mathrm{i}y)^n}{n!} = \sum_{n=0}^{\infty} \frac{(-1)^n y^{2n}}{(2n)!} + \mathrm{i} \sum_{n=0}^{\infty} \frac{(-1)^n y^{2n+1}}{(2n+1)!} = \cos y + \mathrm{i} \sin y.$$

Alternatively, real-valued functions $f(x)$ can be considered as complex functions $f(z)$ *restricted to the real line.* We'll see that operations

[2]A domain is an open, connected point set. *Open* sets do not include their boundaries; each point of an open set is an *interior point,* where for each point there is an $\epsilon > 0$ such that every point within a distance ϵ belongs to the set. Open sets are said to be *connected* if any two points can be joined by an arc segment of points belonging entirely to the set. In many cases, it suffices to think of the domain as the interior of a circle.

[3]The substitution $x \to z$ in real-valued functions is not the *only* way to obtain complex functions, but, as discussed in Section 8.6, it's a good one.

on functions with real arguments – in particular, integration of functions – can often yield surprising and useful results when functions are considered as defined on the z-plane.

8.2 Analytic functions: differentiable in a region

Single-valuedness alone prescribes a class of functions too broad to be governed by the theorems developed in this chapter. We require (in addition to single-valuedness) that complex functions be *differentiable*. Differentiability of complex functions is defined the same as it is for real functions and presupposes the concept of *continuity*.[4]

8.2.1 Continuity, differentiability, and analyticity

Definition. *A function $w = f(z)$, single-valued in $\mathcal{D}$, is* continuous *at point $\zeta \in \mathcal{D}$ if $\lim_{z\to\zeta} f(z)$ exists and is equal to $f(\zeta)$. For any $\epsilon > 0$, there exists $\delta(\epsilon) > 0$ such that $|f(z) - f(\zeta)| < \epsilon$ for all z for which $|z - \zeta| < \delta$.*

This definition is formally the same as the definition of continuity for real functions. The only difference is the use of the modulus or absolute value of complex numbers, $|z|$. Thus, all z for which $|z - \zeta| < \delta$ implies that z lies within a *circle* of radius δ centered at ζ in the z-plane. Likewise, the values of $f(z)$ lie within a circle of radius ϵ centered at $f(\zeta)$ in the w-plane. Continuity of f implies that neighboring points in the z-plane correspond to neighboring points in the w-plane for which $w = f(z)$.

Definition. *A function $w = f(z)$, single-valued in $\mathcal{D}$, is* differentiable *at point $\zeta \in \mathcal{D}$ if $\lim_{z\to\zeta}[(f(z) - f(\zeta))/(z - \zeta)]$ exists. The limit is denoted $f'(\zeta)$ and is called the derivative of $f(z)$ at the point ζ.*

For f to be differentiable at a point ζ, we must be able to associate a new number $f'(\zeta)$ with ζ so that for any $\epsilon > 0$, we can find $\delta(\epsilon) > 0$ such that

$$\left| \frac{f(z) - f(\zeta)}{z - \zeta} - f'(\zeta) \right| < \epsilon \tag{8.2}$$

for all z for which $|z - \zeta| < \delta$.

All the usual rules of differentiation (i.e. those for real functions) apply to complex functions, including the product rule and the chain rule (see Exercise 8.2).

---

[4]Differentiable functions are continuous, but continuous functions are not necessarily differentiable. Example: $f(x) \equiv |x|$ is continuous but not differentiable at $x = 0$.

Example. Let $f(z) = z^2$; by the rules of calculus, $f'(z) = 2z$. Consider the quotient

$$\frac{f(\zeta) - f(z)}{\zeta - z} = \frac{\zeta^2 - z^2}{\zeta - z} = \zeta + z.$$

Thus, $|\zeta + z - 2z| = |\zeta - z| < \epsilon$ provided $|\zeta - z| < \delta = \epsilon$. The derivative $f'(z) = 2z$.

---

complex analytic functions

Functions that are differentiable *throughout a region* are sufficiently important to get their own name.

Definition. *A complex function $f(z)$ is* analytic[5] *throughout a region $\mathcal{D}$ if it's defined and differentiable at each point of $\mathcal{D}$.*

8.2.2 Cauchy–Riemann conditions

An issue arises with the concept of differentiability of complex functions and that is the question of uniqueness of the derivative. For a function to be differentiable, the limit specified by Eq. (8.2) must *exist* in the sense that it's *well defined.* For functions of a real variable x, one speaks of *one-sided* limits as x approaches a specified point either from below or from above. In the complex plane, z can approach ζ in an *unlimited* number of ways. For the derivative of a complex function to have an unambiguous, i.e. unique value, the limit in Eq. (8.2) must exist independently of *how* the limit is taken. This requirement (uniqueness of derivative) adds a new twist to the traditional definition appropriate to real-valued functions and is an important new restriction on what are – and are not – differentiable functions.

Example. Show that $f(z) = x^2 - y^2 + 2\mathrm{i}xy$ is differentiable, but $f(z) = 2y + \mathrm{i}x$ is not.

Solution: For $f(z) = z^2 = x^2 - y^2 + 2\mathrm{i}xy$ with $\Delta z = \Delta x + \mathrm{i}\Delta y$, we have

$$\begin{aligned} f'(z) &= \lim_{\Delta z \to 0} \frac{(x+\Delta x)^2 - (y+\Delta y)^2 + 2\mathrm{i}(x+\Delta x)(y+\Delta y) - (x^2 - y^2 + 2\mathrm{i}xy)}{\Delta x + \mathrm{i}\Delta y} \\ &= \lim_{\Delta z \to 0} \frac{2x\Delta x + (\Delta x)^2 - 2y\Delta y - (\Delta y)^2 + 2\mathrm{i}(x\Delta y + y\Delta x + \Delta x\Delta y)}{\Delta x + \mathrm{i}\Delta y}. \end{aligned}$$

To check for "isotropy" in the complex plane, let Δy be related to Δx so that they specify the line $\Delta y = m\Delta x$. Then

[5] Analytic functions are also called *regular* or *holomorphic.*

$$
\begin{aligned}
f'(z) &= \lim_{\Delta x \to 0} \frac{2x\Delta x + (\Delta x)^2 - 2ym\Delta x - (m\Delta x)^2 + 2\mathrm{i}(mx\Delta x + y\Delta x + m(\Delta x)^2)}{\Delta x(1 + \mathrm{i}m)} \\
&= \lim_{\Delta x \to 0} \frac{2x + \Delta x - 2ym - m^2\Delta x + 2\mathrm{i}(mx + y + m\Delta x)}{1 + \mathrm{i}m} \\
&= \frac{2x(1 + \mathrm{i}m) + 2\mathrm{i}y(1 + \mathrm{i}m)}{1 + \mathrm{i}m} = 2(x + \mathrm{i}y) = 2z.
\end{aligned}
$$

The limit is independent of m; $f(z) = z^2$ is differentiable (and hence analytic) for all z.

For $f(z) = 2y + \mathrm{i}x$, however, we have

$$
\begin{aligned}
f'(z) &= \lim_{\Delta z \to 0} \frac{2(y + \Delta y) + \mathrm{i}(x + \Delta x) - (2y + \mathrm{i}x)}{\Delta x + \mathrm{i}\Delta y} \\
&= \lim_{\Delta z \to 0} \frac{2\Delta y + \mathrm{i}\Delta x}{\Delta x + \mathrm{i}\Delta y} = \lim_{\Delta x \to 0} \frac{(2m + \mathrm{i})\Delta x}{(1 + \mathrm{i}m)\Delta x} = \frac{2m + \mathrm{i}}{1 + \mathrm{i}m}.
\end{aligned}
$$

The value of the limit depends on m and is therefore not unique. *The function is not differentiable anywhere in the complex plane*; it is nowhere analytic.

Proving differentiability for a given complex function can thus be a tedious chore. Fortunately, there's a simple test for differentiability that makes use of Eq. (8.1). Consider the difference quotient

$$
\begin{aligned}
\frac{\Delta f}{\Delta z} &= \frac{f(z + \Delta z) - f(z)}{\Delta z} \\
&= \frac{u(x + \Delta x, y + \Delta y) + \mathrm{i}v(x + \Delta x, y + \Delta y) - (u(x, y) + \mathrm{i}v(x, y))}{\Delta x + \mathrm{i}\Delta y} \\
&\approx \frac{(\partial u/\partial x)\Delta x + (\partial u/\partial y)\Delta y + \mathrm{i}[(\partial v/\partial x)\Delta x + (\partial v/\partial y)\Delta y]}{\Delta x + \mathrm{i}\Delta y} \\
&= \frac{[(\partial u/\partial x) + \mathrm{i}(\partial v/\partial x)]\Delta x + \mathrm{i}[(\partial v/\partial y) - \mathrm{i}(\partial u/\partial y)]\Delta y}{\Delta x + \mathrm{i}\Delta y} \\
&= \frac{(\partial u/\partial x) + \mathrm{i}(\partial v/\partial x)}{1 + \mathrm{i}(\Delta y/\Delta x)} \left[1 + \mathrm{i}(\Delta y/\Delta x)\left(\frac{(\partial v/\partial y) - \mathrm{i}(\partial u/\partial y)}{(\partial u/\partial x) + \mathrm{i}(\partial v/\partial x)}\right)\right],
\end{aligned}
\tag{8.3}
$$

where we have replaced $u(x + \Delta x, y + \Delta y)$ and $v(x + \Delta x, y + \Delta y)$ with their first-order Taylor expansions. We're presuming therefore that the first-order partial derivatives of u and v *exist*, which they must *if* $f(z)$ is differentiable (see Eq. (8.5)).

Equation (8.3) shows that $\lim_{\Delta z \to 0} \Delta f/\Delta z$ generally depends on $\Delta y/\Delta x$, i.e. on how the limit is taken. There's an exception, however: The limit will be independent of the manner by which $\Delta z \to 0$ if it's the case that

$$
\frac{\partial v}{\partial y} - \mathrm{i}\frac{\partial u}{\partial y} = \frac{\partial u}{\partial x} + \mathrm{i}\frac{\partial v}{\partial x},
$$

or, equating real and imaginary parts, if

Cauchy–Riemann conditions

$$\frac{\partial u}{\partial x} = \frac{\partial v}{\partial y} \qquad \frac{\partial u}{\partial y} = -\frac{\partial v}{\partial x}. \tag{8.4}$$

These differential equations are known as the *Cauchy–Riemann conditions*. When the Cauchy–Riemann conditions are met,[6] $f'(z)$ has the value (from Eq. (8.3))

$$f'(z) = \frac{\partial}{\partial x}(u + \mathrm{i}v) = \frac{\partial}{\partial(\mathrm{i}y)}(u + \mathrm{i}v). \tag{8.5}$$

Equation (8.5) is to be expected if $f'(z)$ is independent of how $\Delta z \to 0$: Take the derivative with respect to x, holding y fixed, or with respect to $\mathrm{i}y$, holding x fixed.

Example. Consider $f(z) = z^2 = x^2 - y^2 + 2\mathrm{i}xy$. The real part $u(x,y) = x^2 - y^2$ and the imaginary part $v(x,y) = 2xy$. Check the Cauchy–Riemann conditions. Using Eq. (8.4),

$$\frac{\partial u}{\partial x} = 2x = \frac{\partial v}{\partial y} \qquad \frac{\partial u}{\partial y} = -2y = -\frac{\partial v}{\partial x}.$$

Thus, the Cauchy–Riemann conditions are met, which must be the case if $f(z) = z^2$ is differentiable. Using Eq. (8.5), $f'(z) = 2(x + \mathrm{i}y) = 2z$, in agreement with what we obtained in the previous example.

Example. Is $f(z) = z^{-1}$ an analytic function? Let's check the Cauchy–Riemann conditions. What are the real and imaginary parts?

$$\frac{1}{z} = \frac{1}{x + \mathrm{i}y} = \frac{x}{x^2 + y^2} - \mathrm{i}\frac{y}{x^2 + y^2}.$$

Evaluating the partial derivatives,

$$\frac{\partial u}{\partial x} = \frac{y^2 - x^2}{(x^2 + y^2)^2} = \frac{\partial v}{\partial y} \qquad \frac{\partial u}{\partial y} = -\frac{2xy}{(x^2 + y^2)^2} = -\frac{\partial v}{\partial x}.$$

Thus, the real and imaginary parts of $f(z) = z^{-1}$ satisfy the Cauchy–Riemann conditions wherever the partial derivatives exist and

[6]The Cauchy–Riemann conditions are necessary for the existence of the derivative of a complex function, but they're also sufficient. If $f(z) = u + \mathrm{i}v$ is differentiable at a point, the Cauchy–Riemann conditions must hold at that point; conversely, if the Cauchy–Riemann conditions hold at a point, the function is differentiable at that point. See for example [23, p. 66].

are continuous. What about at $z = 0$? The function is not defined at the origin nor are the partial derivatives. The function $f(z) = z^{-1}$ is analytic everywhere in the complex plane *except* at the origin.

To summarize, differentiability at a point requires that the Cauchy–Riemann conditions be met at that point. Analytic functions are (by definition) differentiable throughout a region $\mathcal{D}$. Thus, *the real and imaginary parts of analytic functions satisfy the Cauchy–Riemann conditions throughout* $\mathcal{D}$. Conversely, if the four partial derivatives of u and v with respect to x and y satisfy the Cauchy–Riemann conditions throughout $\mathcal{D}$, then $f(z) = u(x, y) + \mathrm{i}v(x, y)$ is analytic in $\mathcal{D}$.

8.2.3 Analytic functions are functions only of $z = x + \mathrm{i}y$

Consider a complex function $f(z) = u(x, y) + \mathrm{i}v(x, y)$. We now show that $\partial f/\partial z^* = 0$ ($z^* \equiv x - \mathrm{i}y$ is the complex conjugate of z), if and only if the Cauchy–Riemann conditions hold. The import of this result is that *analytic functions are functions of* z *only.*

To show this, note that

$$x = \frac{1}{2}(z^* + z) \qquad y = \frac{\mathrm{i}}{2}(z^* - z). \tag{8.6}$$

Then, using the fact that z and z^* are linearly independent (show this)

$$\begin{aligned}\frac{\partial f}{\partial z^*} &= \frac{\partial f}{\partial x}\frac{\partial x}{\partial z^*} + \frac{\partial f}{\partial y}\frac{\partial y}{\partial z^*} = \frac{1}{2}\left(\frac{\partial u}{\partial x} + \mathrm{i}\frac{\partial v}{\partial x}\right) + \frac{\mathrm{i}}{2}\left(\frac{\partial u}{\partial y} + \mathrm{i}\frac{\partial v}{\partial y}\right)\\ &= \frac{1}{2}\left(\frac{\partial u}{\partial x} - \frac{\partial v}{\partial y}\right) + \frac{\mathrm{i}}{2}\left(\frac{\partial v}{\partial x} + \frac{\partial u}{\partial y}\right).\end{aligned} \tag{8.7}$$

If the Cauchy-Riemann conditions are met, then $\partial f/\partial z^* = 0$. Conversely, if $\partial f/\partial z^* = 0$, then the vanishing of the real and imaginary parts in Eq. (8.7) imply the Cauchy–Riemann conditions. Consequently, *analytic functions are independent of* z^*; they are functions of z *only*. Said differently, analytic functions are functions solely of the single complex variable z, not complex functions of two real variables. We saw in the example in Section 8.2.1 that $f(x, y) = 2y + \mathrm{i}x$ is nowhere differentiable, and hence not analytic. Using Eq. (8.6), $f(x, y) = 2y + \mathrm{i}x = \frac{\mathrm{i}}{2}(3z^* - z)$; which is not a function solely of z.

8.2.4 Useful definitions

Singularities

It may happen that a function, analytic throughout a region, fails to be analytic at isolated points of the region. Such points are called *singular points* of the function. For example, $f(z) = z^{-1}$ is analytic everywhere in the complex plane *except* at $z = 0$. In any neighborhood of $z = 0$,

$f(z) = z^{-1}$ is analytic, but not at $z = 0$. In contrast, $f(z) = |z|^2 \equiv zz^*$ is *nowhere* analytic because it's differentiable *only* at $z = 0$ and not in any neighborhood of $z = 0$. As we'll see, understanding the behavior of analytic functions near their singularities is important (actually, it's essential) in evaluating complex-valued integrals (see Section 8.8).

Point at infinity, finite complex plane

Mathematicians refer to $z \to \infty$ as the *point at infinity* (even though it's not a point). The idea makes sense in a certain way. The transformation $w = z^{-1}$ establishes a one-to-one correspondence between points in the z-plane and points in the w-plane. Where is the image of the *unique* point $z = 0$ under this transformation? The point at infinity. Where is that? Answer: The image of $w = 0$ under $z = w^{-1}$. The *finite complex plane* is defined as the complex plane "minus" the point at infinity.

Entire functions

An *entire function* is a function analytic everywhere throughout the *finite* complex plane. For example, $f(z) = \sin z$ is an entire function. How is $\sin z$ defined? We show in Section 8.6 that the substitution $x \to z$ in a real-valued analytic function produces an analytic function in the complex plane. Thus, $\sin z$ can be defined analogously to its real-valued counterpart, $\sin z = (\mathrm{e}^{\mathrm{i}z} - \mathrm{e}^{-\mathrm{i}z})/(2\mathrm{i})$. Substituting $z = x + \mathrm{i}y$, we find that $\sin z = \sin x \cosh y + \mathrm{i} \cos x \sinh y$ (see Exercise 8.3). The Cauchy–Riemann conditions are met by the real and imaginary parts of $\sin z$ everywhere in the finite plane (check it!). In the finite complex plane, we don't have to grapple with constructs such as $\lim_{x\to\infty} \cos x$. Using Eq. (8.5), it's straightforward to show that $\mathrm{d} \sin z / \mathrm{d}z = \cos z$. Thus, the derivative exists everywhere throughout the finite complex plane, implying that $f(z) = \sin z$ is an entire function. Examples of entire functions are polynomials $P_n(z) = \sum_{m=0}^{n} a_m z^m$, and certain transcendental functions such as $\sin z$, $\cos z$, e^z, and Bessel functions $J_n(z)$ (there are many others). Other functions are not entire, e.g. $f(z) = \sqrt{z}$ and $f(z) = \ln z$.

8.3 Contour integrals

Now that we understand differentiating complex functions, what about integrating them? The integral of a complex function $f(z)$ between points z_0 and Z (such as shown in Figure 8.2) is defined in an analogous manner as the way in which integrals of real-valued functions are defined: as the limit of a sequence of Riemann sums. First, however, we must say something about the path of integration. For the integral of a function of a real variable x, there is but one integration path, the interval of the x-axis between the limits of integration. In the complex plane, the

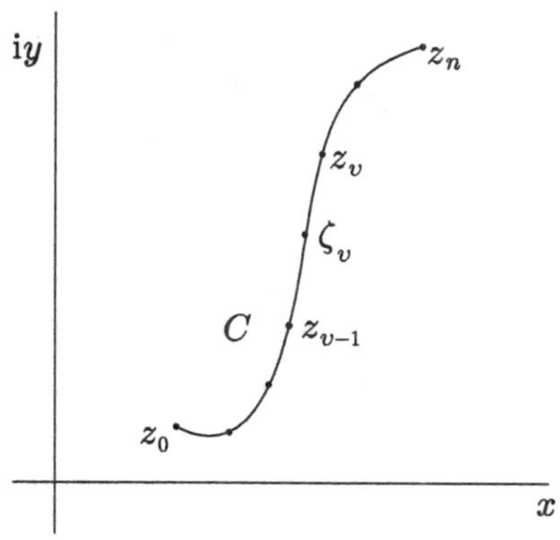

Figure 8.2 Integration path C in the complex plane.

number of possible paths C connecting fixed endpoints is unlimited. *The path C must be specified or the integral is not defined.* For that reason, integrals of complex functions are called *contour integrals*. A continuous curve in the complex plane can be parameterized by two continuous real functions $x(t)$ and $y(t)$, $t_0 \leq t \leq T$, so that $z = z(t) = x(t) + \text{i}y(t)$, with $z(t_0) = z_0$ and $z(T) = Z$. We restrict the class of curves for contour integrals to be *Jordan curves*: those for which distinct points (x, y) on the curve correspond to distinct values of t, i.e. to *curves that do not intersect themselves*. Jordan curves are *oriented*: It's always clear given two points on the curve which precedes the other, and which parts of the path are between them. This property is possible because of the one-to-one correspondence between points of the curve and the parameter t. A Jordan curve "imports" the *ordering* of the real line to produce an ordering of the curve in the complex plane. Contour integrals are thus formally similar to integrals of real-valued functions in the plane, with $f(x, y) \to f(z)$ under the subsitution $z = x + \text{i}y$. A *closed* Jordan curve is a continuous curve for which $x(t_0) = x(T)$ and $y(t_0) = y(T)$, but otherwise does not intersect itself. The *Jordan curve theorem* states that a closed Jordan curve divides the complex plane into two disjoint regions, one inside and the other outside the curve.[7]

Jordan curves

Let $f(z)$ be continuous for $z \in \mathcal{D}$. Let z_0 and Z be arbitrary points of $\mathcal{D}$ connected by a path C lying entirely within $\mathcal{D}$. Divide the path C into n segments (such as shown in Figure 8.2) delimited by points (as they occur along C), $z_0, z_1, \ldots, z_{n-1}, z_n \equiv Z$. On each segment, choose an arbitrary point ζ_ν. Form the Riemann sum

$$I_n \equiv \sum_{\nu=1}^{n} (z_\nu - z_{\nu-1}) f(\zeta_\nu). \tag{8.8}$$

[7]While intuitively obvious, the Jordan curve theorem is difficult to prove and has a rich history.

For $f(z)$ continuous on C, $\lim_{n\to\infty} I_n = I$ exists[8] and is denoted $I \equiv \int_C f(z)\ \mathrm{d}z$. The limit exists independently of how the segments and the intermediate points ζ_ν are chosen. Generally, the limit is not independent of the path C, however.

Example. Let $f(z) = 1$, $\mathcal{D} = \mathbb{C}$, where C is any path from z_0 to Z. In this case, the Riemann sum is trivial, $I_n = Z - z_0$, as can be seen from Eq. (8.8). Thus, $\int_{z_0,C}^{Z} \mathrm{d}z = Z - z_0$. If $Z = z_0$, we have the *closed* contour integral, $\oint_C \mathrm{d}z = 0$.

Example. Let $f(z) = z$, $\mathcal{D} = \mathbb{C}$, where C is any path from z_0 to Z. Evaluate the Riemann sum in two ways. First, let $\zeta_\nu = z_{\nu-1}$; in that case, $I_n^{(1)} \equiv z_0(z_1 - z_0) + \cdots + z_{n-1}(Z - z_{n-1})$. Next, let $\zeta_\nu = z_\nu$; in that case, $I_n^{(2)} \equiv z_1(z_1 - z_0) + \cdots + Z(Z - z_{n-1})$. Add the two: $I_n^{(1)} + I_n^{(2)} = Z^2 - z_0^2$. By the fundamental theorem of calculus, the limit $n \to \infty$ is independent of how the points ζ_ν are chosen: $\lim_{n\to\infty} I_n^{(1)} = \lim_{n\to\infty} I_n^{(2)}$. Thus $\int_{z_0,C}^{Z} z\ \mathrm{d}z = \frac{1}{2}(Z^2 - z_0^2)$. If $Z = z_0$, $\oint_C z\ \mathrm{d}z = 0$.

In these two examples, the value of the contour integral is *independent* of C. This result is not an accident. We show in Section 8.4 that contour integrals of *analytic functions* (such as $f(z) = 1$ and $f(z) = z$) are independent of the path of integration (Cauchy's theorem). Contour integrals of non analytic functions depend on the path C.

By separating f and z into real and imaginary parts, the Riemann sum can be written

$$
\begin{aligned}
I_n &= \sum_{\nu=1}^{n}(z_\nu - z_{\nu-1})f(\zeta_\nu) = \sum_{\nu=1}^{n}[(x_\nu - x_{\nu-1}) + \mathrm{i}(y_\nu - y_{\nu-1})][u(\zeta_\nu) + \mathrm{i}v(\zeta_\nu)] \\
&= \sum_{\nu=1}^{n}\{[u(\zeta_\nu)(x_\nu - x_{\nu-1}) - v(\zeta_\nu)(y_\nu - y_{\nu-1})] \\
&\quad + \mathrm{i}[v(\zeta_\nu)(x_\nu - x_{\nu-1}) + u(\zeta_\nu)(y_\nu - y_{\nu-1})]\},
\end{aligned}
$$

where by $u(\zeta_\nu)$ and $v(\zeta_\nu)$ is meant $u(x, y)$ and $v(x, y)$ evaluated at the (x, y) values associated with ζ_ν. Thus, the Riemann sum associated with a contour integral implies four *real* Riemann sums. As $n \to \infty$, $|x_\nu - x_{\nu-1}|$ and $|y_\nu - y_{\nu-1}|$ tend to zero, and we have that the contour integral can be expressed in terms of real curvilinear integrals

$$
I = \int_C f(z)\ \mathrm{d}z = \int_C [u(x,y)\ \mathrm{d}x - v(x,y)\ \mathrm{d}y] + \mathrm{i}\int_C [v(x,y)\ \mathrm{d}x + u(x,y)\ \mathrm{d}y]. \tag{8.9}
$$

---

[8]The fundamental theorem of integral calculus applies in the complex plane [24, p. 78].

For the curve C parameterized by t for $t_0 \le t \le T$, Eq. (8.9) can be reduced to integrals with respect to t:

$$I = \int_{t_0}^{T} \left[u\frac{\mathrm{d}x}{\mathrm{d}t} - v\frac{\mathrm{d}y}{\mathrm{d}t} \right] \mathrm{d}t + \mathrm{i} \int_{t_0}^{T} \left[v\frac{\mathrm{d}x}{\mathrm{d}t} + u\frac{\mathrm{d}y}{\mathrm{d}t} \right] \mathrm{d}t. \tag{8.10}$$

Because $\mathrm{d}z/\mathrm{d}t = \mathrm{d}x/\mathrm{d}t + \mathrm{i}\,\mathrm{d}y/\mathrm{d}t$, we have from Eq. (8.10),

$$I = \int_{t_0}^{T} (u + \mathrm{i}v)\frac{\mathrm{d}z}{\mathrm{d}t}\,\mathrm{d}t = \int_{t_0}^{T} f(z(t))\frac{\mathrm{d}z}{\mathrm{d}t}\,\mathrm{d}t. \tag{8.11}$$

Equation (8.11) is a useful formula for evaluating definite integrals of complex functions.

Example. Let $f(z) = z^{-1}$ and let the integration path C be given in the parameterized form, $z(t) = \cos t + \mathrm{i}\sin t$ for $0 \le t < 2\pi$, i.e. C is the unit circle in the complex plane traversed in the counterclockwise direction. Using Eq. (8.11),

$$\int_C \frac{\mathrm{d}z}{z} = \int_0^{2\pi} \frac{1}{\cos t + \mathrm{i}\sin t}(-\sin t + \mathrm{i}\cos t)\,\mathrm{d}t = \mathrm{i}\int_0^{2\pi} \mathrm{d}t = 2\pi\mathrm{i}. \tag{8.12}$$

Note that $f(z)$ need be continuous only *on* the path of integration for the integral to exist. Another way of evaluating Eq. (8.12) is to use the *polar form* of a complex variable. Specify C by $z = \mathrm{e}^{\mathrm{i}\theta}$ for $0 \le \theta < 2\pi$. On C, $\mathrm{d}z = \mathrm{i}\mathrm{e}^{\mathrm{i}\theta}\,\mathrm{d}\theta = \mathrm{i}z\,\mathrm{d}\theta \Rightarrow \mathrm{d}z/z = \mathrm{i}\,\mathrm{d}\theta$. The integral $\int_C \mathrm{d}z/z$ is therefore equivalent to $\int_0^{2\pi} \mathrm{i}\,\mathrm{d}\theta$, the same as Eq. (8.12).

Example. Let $f(z) = \mathrm{Re}\, z$. (Is this function analytic? No, check it!) Evaluate the contour integral $\int_C f(z)\,\mathrm{d}z$ between $z_0 = 0$ and $Z = 1 + \mathrm{i}$ for two paths of integration.

1. C_1 is the line segment $z = (1 + \mathrm{i})t$ for $0 \le t \le 1$. On this path, $f(z) = \mathrm{Re}\,(1 + \mathrm{i})t = t$ and $\mathrm{d}z = (1 + \mathrm{i})\,\mathrm{d}t$. From Eq. (8.11),

$$\int_{C_1} f(z)\,\mathrm{d}z = \int_0^1 t(1 + \mathrm{i})\,\mathrm{d}t = \frac{1}{2}(1 + \mathrm{i}).$$

2. C_2 is comprised of a segment of the real axis to $x = 1$ ($z = t, 0 \le t \le 1$, $\mathrm{d}z = \mathrm{d}t$) and then the straight line from $z = 1$ to $z = 1 + \mathrm{i}$ ($z = 1 + \mathrm{i}t, 0 \le t \le 1$, $\mathrm{d}z = \mathrm{i}\,\mathrm{d}t$). Thus,

$$\int_{C_2} f(z)\,\mathrm{d}z = \int_0^1 t\,\mathrm{d}t + \int_0^1 \mathrm{i}\,\mathrm{d}t = \frac{1}{2} + \mathrm{i}.$$

Contour integrals of non analytic functions depend upon the path of integration.

Example. Evaluate $\oint_C z^n \, \mathrm{d}z$ where n is an integer and C is a circle of radius R centered on $z = 0$ taken in the counterclockwise direction.

Solution: Parameterize C with $z = R\mathrm{e}^{\mathrm{i}\theta}$ for $0 \le \theta \le 2\pi$. Then $\mathrm{d}z = \mathrm{i}R\mathrm{e}^{\mathrm{i}\theta}\,\mathrm{d}\theta$ and

$$\oint_C z^n \, \mathrm{d}z = \mathrm{i}R^{n+1} \int_0^{2\pi} \mathrm{e}^{\mathrm{i}(n+1)\theta} \, \mathrm{d}\theta = \begin{cases} 0 & n \neq -1 \\ 2\pi\mathrm{i} & n = -1. \end{cases} \tag{8.13}$$

The integral *is independent of the radius R of the circle and is nonzero only for* $n = -1$. Equation (8.13) is one of the fundamental results of the theory of complex functions. We'll use it often.

It will be useful in what follows to place an upper bound on the modulus of a contour integral. Assume that $|f(z)|$ is bounded on C, i.e. $|f(z)| \le M$ on C. An integral I is the limit of the Riemann sums, $I_n = \sum_{\nu=1}^{n}(z_\nu - z_{\nu-1})f(\zeta_\nu)$. Using the triangle inequality,

$$|I_n| \le \sum_{\nu=1}^{n} |z_\nu - z_{\nu-1}||f(\zeta_\nu)| \le M \sum_{\nu=1}^{n} |z_\nu - z_{\nu-1}|.$$

For finite n, the sum $\sum_{\nu=1}^{n} |z_\nu - z_{\nu-1}|$ is the length of a polygon inscribed by the curve C and is therefore smaller than the arc length L of C. Thus, for all n, $|I_n| \le ML$. Taking the limit $n \to \infty$, we have the *Darboux inequality* (which we'll use frequently)

Darboux inequality

$$\left| \int_C f(z) \, \mathrm{d}z \right| \le ML. \tag{8.14}$$

As an example, for the integral around the unit circle it follows from the Darboux inequality that $|\oint(\,\mathrm{d}z/z)| \le 2\pi$. The upper bound in this case is the absolute value of the integral (see Eq. (8.13)).

8.4 Integrating analytic functions

Contour integrals generally depend on the integration path (as we've seen). We now show that integrals of *analytic* functions are *independent* of the path connecting the endpoints *when the domain of analyticity is "simply connected."* This statement is the content of *Cauchy's integral theorem*, a result of *fundamental importance* to the theory of complex functions.

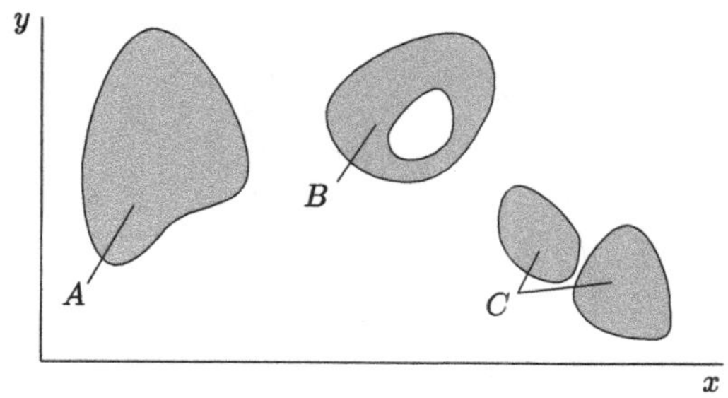

Figure 8.3 Region A is simply connected; regions B and C are not.

simply connected

Definition. *A region $\mathcal{D}$ of the complex plane is* simply connected *if every closed path lying within it can be shrunk continuously to a point without passing outside of $\mathcal{D}$.*

Region A in Figure 8.3 is simply connected, while B and C are not. A closed path in region B surrounding the "hole" cannot be shrunk to a point and still be in the region.

Cauchy integral theorem

Theorem 8.4.1. (Cauchy integral theorem) *If $f(z)$ is analytic in a simply-connected region $\mathcal{D}$, then*

$$\oint_C f(z)\,\mathrm{d}z = 0, \tag{8.15}$$

where C is an arbitrary closed curve lying within $\mathcal{D}$.

Proof. Cauchy's theorem follows from Green's theorem in a plane, applied to analytic functions. To recount Green's theorem (see Appendix A): If functions $u(x, y)$ and $v(x, y)$ have continuous first-order partial derivatives within and on a closed contour C bounding a simply connected region $\mathcal{D}$ in the xy-plane, then

$$\oint_C (u\,\mathrm{d}y - v\,\mathrm{d}x) = \iint_{\mathcal{D}} \left(\frac{\partial u}{\partial x} + \frac{\partial v}{\partial y}\right) \mathrm{d}x\,\mathrm{d}y. \tag{8.16}$$

Apply Eq. (8.16) to Eq. (8.9):

$$\begin{aligned}\oint_C f(z)\,\mathrm{d}z &= \oint_C (u\,\mathrm{d}x - v\,\mathrm{d}y) + \mathrm{i}\oint_C (v\,\mathrm{d}x + u\,\mathrm{d}y)\\ &= \iint_{\mathcal{D}} \left(-\frac{\partial u}{\partial y} - \frac{\partial v}{\partial x}\right) \mathrm{d}x\,\mathrm{d}y + \mathrm{i}\iint_{\mathcal{D}} \left(-\frac{\partial v}{\partial y} + \frac{\partial u}{\partial x}\right) \mathrm{d}x\,\mathrm{d}y.\end{aligned} \tag{8.17}$$

For $f(z)$ analytic in $\mathcal{D}$, u and v satisfy the Cauchy–Riemann conditions, which when combined with Eq. (8.17) implies $\oint_C f(z)\,\mathrm{d}z = 0$. □

Conversely, if $\oint_C f(z)\,\mathrm{d}z = 0$ for every closed curve C in $\mathcal{D}$, the real and imaginary parts of Eq. (8.17) separately vanish, implying the Cauchy–Riemann conditions are met, and thus $f(z)$ is analytic

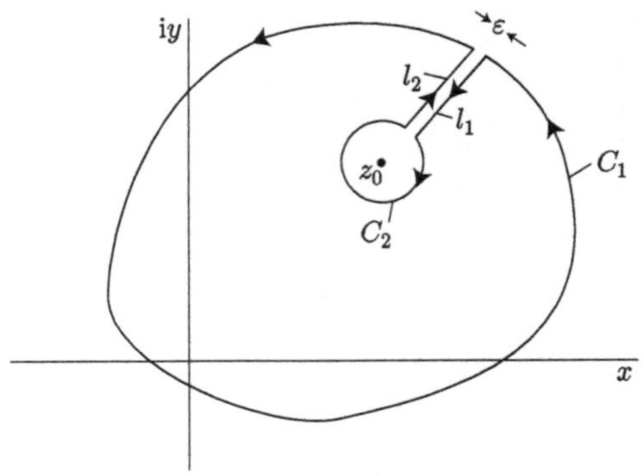

Figure 8.4 Paths of integration.

throughout $\mathcal{D}$. The converse of Cauchy's theorem is known as *Morera's theorem.* We discuss Morera's theorem further in Section 8.5.2.

Morera's theorem

Theorem 8.4.2. (Morera's theorem) *If $f(z)$ is continuous in the simply connected region $\mathcal{D}$, and if $\oint_C f(z)\,\mathrm{d}z = 0$ for any closed contour C lying within $\mathcal{D}$, then f is analytic in $\mathcal{D}$.*

Thus, the integral $\oint_C f(z)\,\mathrm{d}z$ over a closed path surrounding a simply-connected region throughout which the integrand is analytic, vanishes identically. We found in the last example of Section 8.3 a particular instance of this result: z^n is analytic everywhere for $n \geq 0$, but not in a region containing $z = 0$ when n is negative, for which the integral is nonzero. This last point raises the question: Can any use be made of Cauchy's theorem when the integrand is *not* analytic throughout the region bounded by C?

To address this question, consider the contour in Figure 8.4. Suppose $f(z)$ is analytic everywhere *except at $z = z_0$*. Then $f(z)$ is analytic within the region bounded by a contour C composed of the paths C_1, l_1, C_2, and l_2. By Cauchy's theorem,

$$\oint_C f(z)\,\mathrm{d}z = 0 = \int_{C_1} f(z)\,\mathrm{d}z + \int_{l_1} f(z)\,\mathrm{d}z + \oint_{C_2} f(z)\,\mathrm{d}z + \int_{l_2} f(z)\,\mathrm{d}z, \tag{8.18}$$

where $\oint_{C_2}$ indicates that C_2 in Figure 8.4 is taken in the clockwise direction. It's a convention that closed contour integrals are taken in the counterclockwise direction.[9] If the direction of the path of integration is reversed, the sign of the integral reverses (Exercise 8.7). Thus, $\oint_{C_2} = -\oint_{C_2}$, where we *agree that C_2 is now to be taken in the counterclockwise direction* (the opposite of what's shown in Figure 8.4). If $f(z)$ is single-valued

[9]We noted in Section 8.3 that Jordan curves are orientable and that closed Jordan curves divide the complex plane into two disjoint regions, one inside and the other outside the curve. The orientation of a closed Jordan curve C is by convention considered *positive* if the interior region lies to the *left* of C, i.e. the integration proceeds along C in a counterclockwise fashion; otherwise the orientation is considered negative.

(except possibly at z_0), then the contributions from the l_1 and l_2 legs will be nearly equal in magnitude and opposite in sign (owing to the different directions of these paths). In the limit $\varepsilon \to 0$, $\int_{l_1} = -\int_{l_2}$, and we conclude from Eq. (8.18)

$$\oint_{C_1} f(z)\,\mathrm{d}z = \oint_{C_2} f(z)\,\mathrm{d}z, \tag{8.19}$$

where C_1 and C_2 are taken in the counterclockwise sense. *If the region between C_1 and C_2 lies entirely within the domain of analyticity $\mathcal{D}$ of the function $f(z)$, and both paths are oriented in the same sense, the closed-path integrals are equal regardless of whether the interior of C_2 belongs entirely to $\mathcal{D}$.* The practice of replacing contour C_1 with C_2 is referred to as *deforming* the contour.

deforming the contour

Example. Consider the closed contour integral

$$I \equiv \oint_C \frac{1}{z - z_0}\,\mathrm{d}z,$$

where C is an arbitrary contour surrounding the point z_0. Clearly the integrand is not analytic at $z = z_0$. By what we've just shown, I is equivalent to a contour integral with C replaced by a circle of radius R centered on z_0. The last example of Section 8.3 then allows us to conclude

$$\oint_C \frac{1}{z - z_0}\,\mathrm{d}z = 2\pi\mathrm{i} \tag{8.20}$$

for *any* contour C surrounding z_0.

For $f(z)$ analytic throughout a simply connected region $\mathcal{D}$, Cauchy's theorem states that $\oint_C f(z)\,\mathrm{d}z = 0$ for every closed contour C in $\mathcal{D}$. Conversely, by Morera's theorem, for $f(z)$ continuous in $\mathcal{D}$ such that $\oint_C f(z)\,\mathrm{d}z = 0$ for every closed curve C in $\mathcal{D}$, then $f(z)$ is analytic in $\mathcal{D}$. *Another way of characterizing analytic functions is as those for which* $\oint_C f(z)\,\mathrm{d}z = 0$. These considerations show *there is a connection between the differential and integral properties of analytic functions.*[10] For example, we'll show in Section 8.5 that the derivative of an analytic function is analytic, but to do so *requires the use of complex integration.* Many important properties of analytic functions can be shown only using integration – such as the existence of higher order *derivatives.* That we must use integration to develop properties which are seemingly the province of differential calculus underscores the deep-rooted differences between real and complex functions.

[10] Analytic functions were defined in Section 8.2 as possessing a derivative throughout a region $\mathcal{D}$.

8.5 Cauchy integral formulas

In this section, we derive what is arguably the most important consequence of Cauchy's theorem, the *Cauchy integral representation* of analytic functions.[11] We state the fundamental formula in the form of a theorem, and then present its derivation.

Cauchy integral formula

Theorem 8.5.1. (Cauchy integral formula) *Let $f(z)$ be analytic throughout a simply-connected region $\mathcal{D}$. If C is a closed curve within $\mathcal{D}$ and z_0 is not a point on C, then*

$$\frac{1}{2\pi \mathrm{i}} \oint_C \frac{f(z)}{z - z_0}\, \mathrm{d}z = \begin{cases} f(z_0) & \text{if } z_0 \text{ is interior to } C \\ 0 & \text{if } z_0 \text{ is exterior to } C, \end{cases} \tag{8.21}$$

where the integration along C is taken in the counterclockwise sense.

Equation (8.21) is a remarkable result. If a function f is known to be analytic within $\mathcal{D}$, and if its values are known along C, then its values interior to C are completely determined.[12] Said differently, there is no freedom in choosing the value of an analytic function in the interior of C if its values have been specified on C. This feature underscores the *nonlocal*, connected character of analytic functions mentioned in Section 8.2. The function cannot be altered in the interior of C without altering its values on C.

To derive Eq. (8.21), consider that

$$\oint_C \frac{f(z)}{z - z_0}\, \mathrm{d}z = \oint_C \frac{f(z_0)}{z - z_0}\, \mathrm{d}z + \oint_C \frac{f(z) - f(z_0)}{z - z_0}\, \mathrm{d}z \equiv I_1 + I_2.$$

By Eq. (8.20), $I_1 = 2\pi \mathrm{i} f(z_0)$. Equation (8.21) thus implies $I_2 = 0$. By Eq. (8.19), the contour C may be replaced by any other closed contour surrounding z_0 which is interior to C. Let that contour be specified by $z = z_0 + \rho \mathrm{e}^{\mathrm{i}\theta}$, with ρ a real constant. In that case, $I_2 = \mathrm{i} \int_0^{2\pi} [f(z_0 + \rho \mathrm{e}^{\mathrm{i}\theta}) - f(z_0)]\, \mathrm{d}\theta$. But f is continuous, implying that for any $\epsilon > 0$, there exists a $\rho > 0$ such that $|f(z_0 + \rho \mathrm{e}^{\mathrm{i}\theta}) - f(z_0)| < \epsilon$. By the Darboux inequality, $|I_2| < 2\pi\epsilon$. Thus, $|I_2|$, a nonnegative quantity, is less than an arbitrarily small positive number; I_2 must equal zero. By Eq. (8.19), the conclusion $I_2 = 0$ holds for any other contour surrounding z_0. If z_0 is *exterior* to C, the integrand of Eq. (8.21) is analytic for all z on and within C, and thus the right side of Eq. (8.21) vanishes by Cauchy's theorem.

[11] Equation (8.21) *represents* the value of f at point z_0 interior to C in terms of its values on the contour C.

[12] Imagine walking completely around a mountain at a constant elevation. By so doing, could you infer the height of the mountain? Yet that's what you can do with an analytic function.

8.5.1 Derivatives of analytic functions

With Eq. (8.21) established, we can use it to provide an integral formula for the derivative of f at z_0, $f'(z_0)$. Because f is analytic at z_0, it is differentiable there. By formally differentiating Eq. (8.21) with respect to z_0 under the integral sign (for z_0 interior to C), we have

$$f'(z_0) = \frac{1}{2\pi i}\oint_C \frac{f(z)}{(z-z_0)^2}\,dz. \tag{8.22}$$

To verify Eq. (8.22), form the difference quotient using Eq. (8.21):

$$\begin{aligned}\frac{f(z_0+\Delta z)-f(z_0)}{\Delta z} &= \frac{1}{2\pi i\Delta z}\oint_C f(z)\left[\frac{1}{z-z_0-\Delta z}-\frac{1}{z-z_0}\right]dz\\ &= \frac{1}{2\pi i}\oint_C \frac{f(z)}{(z-z_0)(z-z_0-\Delta z)}\,dz,\end{aligned}$$

where Δz is sufficiently small that $z_0+\Delta z$ is interior to C. Let d be the smallest distance from z_0 to points z on C, so that $0<|\Delta z|<d$. Then, using Eq. (8.22),

$$\begin{aligned}\left|\frac{f(z_0+\Delta z)-f(z_0)}{\Delta z}-f'(z_0)\right| &= \frac{1}{2\pi}\left|\oint_C f(z)\left[\frac{1}{(z-z_0)(z-z_0-\Delta z)}-\frac{1}{(z-z_0)^2}\right]dz\right|\\ &= \frac{|\Delta z|}{2\pi}\left|\oint_C \frac{f(z)}{(z-z_0)^2(z-z_0-\Delta z)}\,dz\right|\\ &\le \frac{|\Delta z|}{2\pi}\frac{M\cdot L}{d^2|d-\Delta z|},\end{aligned}$$

where we've used $|z-z_0-\Delta z|\ge d-|\Delta z|>0$ (because $|z-z_0|\ge d$) and where M denotes the maximum value of $|f(z)|$ on C, and L is the length of C. We can thus let $\Delta z\to 0$, establishing Eq. (8.22) as an integral representation of $f'(z_0)$.

Equation (8.21) can therefore be differentiated (under the integral sign) with respect to z_0 to generate an integral representation for the derivative, $f'(z_0)$. Equation (8.22), however, does not establish that $f'(z_0)$ is analytic. To do that, we must show that the derivative $f''(z_0)$ of $f'(z_0)$ exists over some region of the complex plane.[13] *The process of differentiating under the integral sign can be repeated any number of times to obtain an integral formula for every higher order derivative.* It can be shown that[25, p. 299]

$$f^{(n)}(z_0) \equiv \left.\frac{d^n}{dz^n}f(z)\right|_{z=z_0} = \frac{n!}{2\pi i}\oint_C \frac{f(z)}{(z-z_0)^{n+1}}\,dz. \quad n=0,1,2,\ldots \tag{8.23}$$

Thus, *if $f(z)$ is analytic in a domain $\mathcal{D}$, then $f(z)$ is infinitely differentiable on $\mathcal{D}$.*

[13] Because z_0 is an arbitrary point interior to C, we can take it as a variable.

8.5.2 Consequences of the Cauchy formulas

Morera's theorem

We can now sketch the proof Morera's theorem. A function $f(z)$ for which $\oint_C f(z)\, dz = 0$ for every closed contour C lying in $\mathcal{D}$ implies that the value of the integral $\int_a^z f(z')\, dz'$ is independent of the path connecting points a and z in $\mathcal{D}$. For fixed a, $F(z) \equiv \int_a^z f(z')\, dz'$ is an analytic function, the *antiderivative*[14] of f, such that $f(z) = dF(z)/\,dz$ in $\mathcal{D}$. By Eq. (8.23), the second derivative $F''(z) = f'(z)$ exists throughout $\mathcal{D}$, implying that $f(z)$ is analytic in $\mathcal{D}$.

Liouville's theorem

Theorem 8.5.2. *(Liouville) A bounded, entire function is a constant.*

Proof. Let the contour in Eq. (8.22) be a circle of radius R centered at z_0, i.e. $z = z_0 + Re^{i\theta}$ for z on the contour. Using Eq. (8.22), $f'(z_0) = (1/(2\pi R)) \int_0^{2\pi} e^{-i\theta} f(z_0 + Re^{i\theta})\, d\theta$, and thus, $|f'(z_0)| = (1/(2\pi R)) \left|\int_0^{2\pi} f(z_0 + Re^{i\theta}) e^{-i\theta}\, d\theta\right| \le M/R$ (Darboux inequality), with M the bound on f; by assumption, $|f(z)| \le M$ for all z in the finite complex plane. For R sufficiently large, therefore, $|f'(z_0)| < \epsilon$ for any preassigned ϵ. Thus, $|f'(z_0)| = 0$, implying that $f'(z_0) = 0$ for all z_0, and hence, $f(z) = $ constant. □

Note that $f(z) = \sin z$ is a *unbounded* entire function, and it's not a constant.[15] The converse of Liouville's theorem holds: A *non constant* entire function $f(z)$ is unbounded: $|f(z)|$ assumes arbitrarily large values for $|z|$ sufficiently large.[16]

The fundamental theorem of algebra

The *fundamental theorem of algebra* states that a polynomial of degree n, $P_n(z) = \sum_{k=0}^{n} a_k z^k$ has at least one, and at most n roots. A polynomial is a non constant entire function. If $P_n(z) \neq 0$ for all z, then $g(z) \equiv 1/P_n(z)$ is a nonconstant entire function, which by Liouville's theorem is

[14] It can be shown that analytic functions possess antiderivatives, analytic functions $F(z)$ such that $f(z) = dF(z)/\,dz$.

[15] Beware of the distinction between the finite and infinite complex plane. We have defined entire functions as analytic throughout the *finite* complex plane (Section 8.2). Some books state Liouville's theorem as applying to functions analytic and bounded throughout the *infinite* complex plane, implying that unless a function $f(z)$ is a constant, it must have singularities. This form of Liouville's theorem (not what we have presented) rules out functions such as $f(z) = \sin z$ that are *analytic everywhere* but not bounded as $z \to \infty$. Liouville's theorem indicates that we should *generally* expect singularities in complex functions; singularities are a way of life.

[16] To show the converse of Liouville's theorem, we need to know that complex analytic functions possess Taylor series expansions, which is established in Section 8.6.

unbounded, i.e. for $|z|$ sufficiently large, $|g(z)|$ is arbitrarily large. Hence, $P_n(z) = 1/g(z)$ becomes arbitrarily small for some z, contradicting the assumption that $P_n(z) \neq 0$ for all z. Thus, $P_n(z) = 0$ has at least one solution. Let $P_n(z_0) = 0$. Write $P_n(z)$ in the form $P_n(z) = (z - z_0)G_{n-1}(z)$. Repeat the argument for $G_{n-1}(z)$, a polynomial of degree $n - 1$. Note that *transcendental* entire functions need not have zeros; $f(z) = \mathrm{e}^z$ has no zeros, but $f(z) = \sin z$ does.

8.6 Taylor and Laurent series

Complex functions can be defined as extensions of real-valued functions $f(x)$ into the complex plane (Section 8.1). Applied blindly, the method is not particularly useful for our purposes. There are an infinite number of ways by which a given real function could be extended, but which of these extensions lead to analytic functions? From $f(x) = x$, for example, we could define $\tilde{f}(x, y) = x - iy$, which is defined in the complex plane and is equivalent to $f(x)$ when restricted to the real line ($y = 0$). It is, however, not analytic[17] and so Cauchy's theorem does not apply – and we'll see that Cauchy's theorem is pretty useful when it applies.

We want a way for functions $f(x)$ to be *analytically* extended into the complex plane with $\tilde{f}(x, 0) = f(x)$ such that $\tilde{f}(x, y)$ is analytic. There are several schemes for performing this kind of special extension, but the easiest makes use of the power-series representation of $f(x)$. This is the approach we now examine.

8.6.1 Taylor series

Start with a function $f(z)$, analytic throughout a region $\mathcal{D}$. By Cauchy's integral formula, for C a closed circular contour, centered on $z_0 \in \mathcal{D}$ and lying entirely within $\mathcal{D}$, we have

$$f(z) = \frac{1}{2\pi \mathrm{i}} \oint_C \frac{f(\xi)}{\xi - z} \, \mathrm{d}\xi \tag{8.24}$$

for any z inside C, where ξ lies on C. Because $|z - z_0| < |\xi - z_0|$ for all ξ on C, we can expand $(\xi - z)^{-1}$ in a geometric series:

$$\frac{1}{\xi - z} = \frac{1}{(\xi - z_0) - (z - z_0)} = \frac{1}{\xi - z_0} \frac{1}{1 - (z - z_0)/(\xi - z_0)} = \frac{1}{\xi - z_0} \sum_{n=0}^{\infty} \left(\frac{z - z_0}{\xi - z_0} \right)^n .$$

Substitute this result into Eq. (8.24) and make use of Eq. (8.23) to obtain

$$f(z) = \frac{1}{2\pi \mathrm{i}} \sum_{n=0}^{\infty} (z - z_0)^n \oint_C \frac{f(\xi)}{(\xi - z_0)^{n+1}} \, \mathrm{d}\xi = \sum_{n=0}^{\infty} \frac{1}{n!} f^{(n)}(z_0)(z - z_0)^n . \tag{8.25}$$

[17] It is not a function solely of z; Section 8.2.

Equation (8.25) is the Taylor series of an analytic function.[18] *Any analytic function can be expanded in a power series about any point z_0 interior to the domain of analyticity,* a representation that is *unique*. As with any infinite series, the Taylor series is convergent only in a neighborhood of z_0 specified by its *radius of convergence*, R, the distance to the nearest point where the function is no longer analytic, i.e. a singularity of f. We see explicitly from Eq. (8.25) that analytic functions are *infinitely differentiable* (within the radius of convergence), a property shared by real and complex analytic functions.[19]

When the Taylor series for $f(z)$ in Eq. (8.25) is restricted to the real line, it becomes the Taylor series of $f(x)$ in powers of $(x - x_0)$ (because differentiation is independent of direction in the complex plane). Consequently, the complex power series provides a scheme for *analytically* extending $f(x) \to f(z)$. Any real analytic function can be extended to a complex analytic function[20] through the substitution $x \to z$.

Example. Develop the Taylor series for $f(z) = \sin z$. Using Eq. (8.25) with $z_0 = 0$, and using $f'(z) = \cos z$, $f''(z) = -\sin z$, etc.,

$$\sin z = z - \frac{1}{3!}z^3 + \frac{1}{5!}z^5 + \cdots \qquad |z| < \infty$$

This expansion is valid throughout the finite complex plane; it's an entire function. It is the same as the Taylor series for $\sin x$, with $x \to z$.

Example. Consider the function

$$f(z) \equiv \int_{1,C}^{z} \frac{1}{\xi}\, \mathrm{d}\xi, \tag{8.26}$$

for z in the interior of the right-half of the complex plane and where C is any path extending from $1 + 0\mathrm{i}$ to z. The integral $F(z) \equiv \int_{z_0}^{z} g(z')\, \mathrm{d}z'$ of an analytic function $g(z)$ is an analytic function of the upper limit of integration, with $F'(z) = g(z)$. Moreover, $g(z) = z^{-1}$ is analytic everywhere except for $z = 0$. By what we've established in this section, $f(z)$ (defined by Eq. (8.26)) admits a power-series expansion. Let's develop an expansion around $z_0 = 1$. We need:

$$f(1) = 0, \quad f'(z) = \frac{1}{z}, \quad f''(z) = -\frac{1}{z^2}, \cdots, \quad f^{(n)}(z) = (-1)^{n-1}\frac{(n-1)!}{z^n}, \cdots$$

---

[18] Note that the Taylor series of an analytic function $f(z)$ has been established using its *integral properties*.

[19] Real analytic functions are *defined* as possessing a convergent power-series representation.

[20] The extension is "local," that is within the domain of convergence of the expansion, which, for entire functions, could be the entire complex plane.

Thus, using Eq. (8.25) with $z_0 = 1$,

$$f(z) = (z-1) - \frac{1}{2}(z-1)^2 + \cdots + \frac{(-1)^{n-1}}{n}(z-1)^n + \cdots$$

The expansion is unique; it's the only possible expansion about $z_0 = 1$. We see that the radius of convergence $R = 1$.

8.6.2 The zeros of analytic functions are isolated

Let $f(z)$ be analytic in a region $\mathcal{D}$, and suppose that $f(z_0) = 0$ for $z_0 \in \mathcal{D}$. We now show that *the zeros of analytic functions are isolated*, meaning there is a neighborhood of z_0 throughout which[21] $f(z) \neq 0$.

Because f is analytic throughout $\mathcal{D}$, it has a Taylor series expansion about $z_0 \in \mathcal{D}$: $f(z) = \sum_{n=k}^{\infty} a_n (z-z_0)^n$, where $a_k \neq 0$, for $|z - z_0| < R$, the radius of convergence. If $k = 1$, z_0 is known as a *simple zero* of f. We can rewrite the Taylor series,

$$f(z) = (z-z_0)^k \sum_{n=0}^{\infty} a_{n+k}(z-z_0)^n \equiv (z-z_0)^k g(z).$$

The function $g(z)$ is analytic for $|z - z_0| < R$, and hence, it is continuous at $z = z_0$ with $\lim_{z\to z_0} g(z) = g(z_0) = a_k \neq 0$. For any $\epsilon > 0$, there exists a $0 < \delta < R$ so that $|g(z) - a_k| < \epsilon$ for $|z - z_0| < \delta$. Take $\epsilon = |a_k|/2$, $\Rightarrow$ $|g(z) - a_k| < |a_k|/2$. The function $g(z)$ is thus nonzero for $|z - z_0| < \delta$; otherwise, we would have $|a_k| < \frac{1}{2}|a_k|$. Because $g(z) \neq 0$ in a neighborhood of z_0, z_0 is the only zero of $f(z)$ in that neighborhood.

8.6.3 Laurent series

Taylor series apply to functions analytic throughout a simply-connected region $\mathcal{D}$, with $z_0 \in \mathcal{D}$. We now develop a power-series expansion for functions having singular points in the interior of $\mathcal{D}$. In order to keep the presentation definite, consider a function $f(z)$ analytic throughout a closed annular region bounded by closed circular contours C_1 and C_2 both centered on point z_0. We can use the argument that led to Eq. (8.19) to develop a series representation for $f(z)$. Referring to Figure 8.4, consider a point z that lies outside C_2 and inside C_1, where both circles are centered on $z = z_0$. If $f(z)$ is analytic in the region between C_1 and C_2, then through the application of Cauchy's integral formula,

$$f(z) = \frac{1}{2\pi \mathrm{i}} \oint_{C_1} \frac{f(\xi)}{\xi - z}\, \mathrm{d}\xi - \frac{1}{2\pi \mathrm{i}} \oint_{C_2} \frac{f(\xi)}{\xi - z}\, \mathrm{d}\xi, \tag{8.27}$$

where C_1 and C_2 are taken in the counterclockwise sense, and where we have suppressed the two integrals over the paths l_1 and l_2 will cancel in

[21] Unless, of course, $f(z) = 0$ everywhere

the limit as $\varepsilon \to 0$. In the first integral, we note that z lies within C_1 and so $|z - z_0| < |\xi - z_0|$ (for ξ restricted to C_1). This is the case we examined in developing the Taylor series expansion. In the second integral, z lies outside $C_2 \Rightarrow |\xi - z_0| < |z - z_0|$, and so we must expand $(\xi - z)^{-1}$ as

$$\frac{1}{\xi - z} = \frac{1}{(\xi - z_0) - (z - z_0)} = \frac{1}{z - z_0}\frac{1}{(\xi - z_0)/(z - z_0) - 1} = \frac{-1}{z - z_0}\sum_{n=0}^{\infty}\left(\frac{\xi - z_0}{z - z_0}\right)^n.$$

Substituting the series expansions into (8.27):

$$2\pi i f(z) = \sum_{n=0}^{\infty}(z - z_0)^n \oint_{C_1}\frac{f(\xi)}{(\xi - z_0)^{n+1}}\,\mathrm{d}\xi + \sum_{n=0}^{\infty}\frac{1}{(z - z_0)^{n+1}}\oint_{C_2} f(\xi)(\xi - z_0)^n\,\mathrm{d}\xi.$$

The first series results in the same (Taylor) series as in Eq. (8.25). In the second series change the dummy index $n + 1 \to -n$:

$$\sum_{n=0}^{\infty}\frac{1}{(z - z_0)^{n+1}}\oint_{C_2} f(\xi)(\xi - z_0)^n\,\mathrm{d}\xi = \sum_{n=-\infty}^{-1}(z - z_0)^n \oint_{C_2}\frac{f(\xi)}{(\xi - z_0)^{n+1}}\,\mathrm{d}\xi.$$

Combining these results, we obtain the *Laurent series expansion* for $f(z)$:

$$f(z) = \sum_{n=-\infty}^{\infty} a_n (z - z_0)^n \quad \text{where} \quad a_n = \frac{1}{2\pi i}\oint_C \frac{f(\xi)}{(\xi - z_0)^{n+1}}\,\mathrm{d}\xi, \qquad (8.28)$$

Laurent series

where C is any contour within the region between C_1 and C_2. The Laurent series generalizes a Taylor series to include *negative* powers of $(z - z_0)$; the Laurent series is the extension of a Taylor series to multiply connected regions.

The terms of this series with $n \geq 0$ are collectively called *the analytic part*, whereas the remaining terms (negative n) comprise *the principal part*. By construction, we expect the principal part to converge only for z lying *outside* some circle centered on z_0, whereas the analytic part will converge for z lying *inside* some different circle centered on z_0.

Example. Consider the function

$$f(z) = \frac{1}{(z - 2)(z - 3)}.$$

The function is analytic for $|z| < 2$, $2 < |z| < 3$, and $|z| > 3$. Suppose $2 < |z| < 3$. By the method of partial-fraction decomposition,

$$\frac{1}{(z - 2)(z - 3)} = -\frac{1}{z - 2} - \frac{1}{3 - z} = -\frac{1}{z}\left(\frac{1}{1 - 2/z}\right) - \frac{1}{3}\left(\frac{1}{1 - z/3}\right).$$

Applying the geometric series to both terms, we have a Laurent series centered at $z_0 = 0$, with its principal part (converges for $|z| > 2$) and analytic part (converges for $|z| < 3$):

$$\frac{1}{(z-2)(z-3)} = -\frac{1}{z}\sum_{n=0}^{\infty}\left(\frac{2}{z}\right)^n - \frac{1}{3}\sum_{n=0}^{\infty}\left(\frac{z}{3}\right)^n. \qquad 2 < |z| < 3$$

For $|z| > 3$,

$$\frac{1}{(z-2)(z-3)} = -\frac{1}{z-2} + \frac{1}{z-3} = -\frac{1}{z}\left(\frac{1}{1-2/z}\right) + \frac{1}{z}\left(\frac{1}{1-3/z}\right)$$
$$= -\frac{1}{z}\sum_{n=0}^{\infty}\left(\frac{2}{z}\right)^n + \frac{1}{z}\sum_{n=0}^{\infty}\left(\frac{3}{z}\right)^n = \sum_{n=1}^{\infty}\frac{3^n - 2^n}{z^{n+1}}, \qquad |z| > 3$$

a Laurent series convergent for $|z| > 3$.

8.7 Singularities and residues

8.7.1 Isolated singularities, residue theorem

We noted in Section 8.2 that complex functions, otherwise analytic throughout a region, can fail to be analytic at isolated points of that region, *isolated singularities* of the function. As we now show, integrals of complex functions are often *determined* by their behavior in the vicinity of such singularities.

Referring to Eq. (8.28), we see that singularities in $f(z)$ are associated with the principal part of the Laurent expansion – terms of the form $1/(z - z_0)^n$.

Definition. *When a function $f(z)$ has the form*

$$f(z) = \frac{g(z)}{(z - z_0)^n}$$

where n is a positive integer and $g(z)$ is analytic at all points in a neighborhood containing z_0 with $g(z_0) \neq 0$, then we say that $f(z)$ has a pole *of order n at $z = z_0$. First-order poles are termed* simple poles. *If n is not finite, then all the terms in the principal part of the Laurent series must be included, and we say that $f(z)$ has an essential singularity at $z = z_0$.* poles

We're now going to formally repeat the last example of Section 8.3 and obtain one of our main results. Suppose a function $f(z)$ has a pole of order no higher than m at $z = z_0$. In that case, the Laurent series expansion for f has the form

$$f(z) = \sum_{n=-m}^{\infty} a_n(z - z_0)^n.$$

Let C be a closed contour surrounding $z = z_0$ but no other singular points. By Cauchy's theorem, the integral $I = \oint_C f(z)\, \mathrm{d}z$ will have the same value when C is replaced with a circular path of radius R centered on z_0, with $z = z_0 + R\mathrm{e}^{\mathrm{i}\theta}, \quad 0 \leq \theta < 2\pi$. Thus,

$$I = \sum_{n=-m}^{\infty} a_n \int_0^{2\pi} (R\mathrm{e}^{\mathrm{i}\theta})^n \mathrm{i}R\mathrm{e}^{\mathrm{i}\theta}\, \mathrm{d}\theta = \mathrm{i} \sum_{n=-m}^{\infty} a_n R^{n+1} \int_0^{2\pi} \mathrm{e}^{\mathrm{i}(n+1)\theta}\, \mathrm{d}\theta = 2\pi\mathrm{i}\, a_{-1}. \tag{8.29}$$

Evaluating the integral has thus been reduced to finding the value of just one of the expansion coefficients, a_{-1}. Equation (8.29) generalizes Eq. (8.20).

residue

Definition. *The coefficient a_{-1} is called the* residue *of the pole at z_0 and is denoted* $\mathrm{Res}(z_0)$.

It is straightforward to extend Eq. (8.29) to the case where there are many isolated singularities inside the contour:

Theorem 8.7.1. (Residue theorem) *For $f(z)$ continuous and single-valued within and on a closed contour C and analytic except for a finite number of poles within C at points $z = z_j$, then*

$$\oint_C f(z)\, \mathrm{d}z = 2\pi\mathrm{i} \sum_j \mathrm{Res}(z_j) \tag{8.30}$$

where $\sum_j \mathrm{Res}(z_j)$ is the sum of the residues of the poles of $f(z)$ lying within C.

The residue theorem is one of the most important results of the theory of analytic functions. It shows that the calculation of path integrals can, in many cases, be reduced to the calculation of residues. There are several ways to calculate the residue of a pole. Two of the most common methods are:

Laurent series method

If the Laurent series expansion about the pole z_0 is easy to calculate, then the residue is just the coefficient of the $(z - z_0)^{-1}$ term.

Example. Let

$$f(z) = \frac{1}{z(z-2)^3}.$$

The poles are (by inspection) $z = 0$ (order 1) and $z = 2$ (order 3). Expanding about $z = 0$, we rewrite $f(z)$ in the form

$$f(z) = \frac{-1}{z(2-z)^3} = \frac{-1}{8z(1-z/2)^3}$$

and expand in a geometric series to obtain

$$f(z) = \frac{-1}{8z}\left(1 + (-3)\left(\frac{-z}{2}\right) + \frac{(-3)(-4)}{2!}\left(\frac{-z}{2}\right)^2 + \cdots\right).$$

The coefficient of z^{-1} is $\text{Res}(0) = -1/8$.

Expanding about $z = 2$, we change variables $\xi = z - 2$ and write

$$f(z) = \frac{1}{\xi^3(2+\xi)} = \frac{1}{2\xi^3(1+\xi/2)} = \frac{1}{2\xi^3}\left(1 - \left(\frac{\xi}{2}\right) + \left(\frac{\xi}{2}\right)^2 - \left(\frac{\xi}{2}\right)^3 + \cdots\right).$$

The coefficient of the ξ^{-1} term is $\text{Res}(2) = (1/2)(1/2)^2 = 1/8$.

Differentiation method

The Laurent series method can become algebraically complicated if the integrand contains multiple functional factors (such as $\exp(z)/(1+z^2)$) because each factor must be expanded in its series representation and multiplied. Fortunately, there is a more direct approach.

If $f(z)$ has a pole at z_0 of order no higher than m, then multiply $f(z)$ by $(z-z_0)^m$ to obtain

$$(z-z_0)^m f(z) = a_{-m} + a_{-m+1}(z-z_0) + \cdots + a_{-1}(z-z_0)^{m-1} + a_0(z-z_0)^m + \cdots$$

Differentiate both sides of this result $m-1$ times to obtain

$$\frac{\mathrm{d}^{m-1}}{\mathrm{d}z^{m-1}}((z-z_0)^m f(z)) = (m-1)!\ a_{-1} + m!\ a_0(z-z_0) + \cdots$$

Evaluate this derivative at $z = z_0$, which isolates the term a_{-1}. The residue associated with an mth-order pole is therefore

$$a_{-1} = \frac{1}{(m-1)!}\lim_{z\to z_0}\left[\frac{\mathrm{d}^{m-1}}{\mathrm{d}z^{m-1}}((z-z_0)^m f(z))\right]. \tag{8.31}$$

Equation (8.31) is quite useful in applications.

Example. Let

$$f(z) = \frac{\exp(\mathrm{i}z)}{(z^2+1)^2} = \frac{\exp(\mathrm{i}z)}{(z+\mathrm{i})^2(z-\mathrm{i})^2}.$$

This function has the two poles $z_0 = \pm\mathrm{i}$ (each of order $m = 2$).

Applying Eq. (8.31),

$$\frac{1}{1!}\frac{\mathrm{d}}{\mathrm{d}z}((z\pm \mathrm{i})^2 f(z)) = \begin{cases} \dfrac{\mathrm{d}}{\mathrm{d}z}\dfrac{\exp(\mathrm{i}z)}{(z-\mathrm{i})^2} = \dfrac{\exp(\mathrm{i}z)}{(z-\mathrm{i})^3}(\mathrm{i}(z-\mathrm{i})-2) \text{ if } z_0 = -\mathrm{i} \\ \dfrac{\mathrm{d}}{\mathrm{d}z}\dfrac{\exp(\mathrm{i}z)}{(z+\mathrm{i})^2} = \dfrac{\exp(\mathrm{i}z)}{(z+\mathrm{i})^3}(\mathrm{i}(z+\mathrm{i})-2) \text{ if } z_0 = +\mathrm{i} \end{cases}$$

Thus,

$$\mathrm{Res}(-\mathrm{i}) = 0 \quad \text{and} \quad \mathrm{Res}(+\mathrm{i}) = \frac{-\mathrm{i}}{2\mathrm{e}}.$$

8.7.2 Multivalued functions, branch points, and branch cuts

We need to clean up a detail that we have so far ignored. In the derivation of the Laurent series (and, for that matter, in the last example of section 8.3), we argued that the contributions of the contour integrals from legs l_1 and l_2 cancel and can be ignored (refer to Figure 8.4). This assertion follows if the integrand is *single-valued*: only then are the contributions of each integral guaranteed to be equal in magnitude and opposite in sign (direction) in the limit $\varepsilon \to 0$. This argument is no longer valid when the integrand is not single-valued and, in this case, the contributions from the two legs l_1 and l_2 must be explicitly calculated.

To see how non-single-valued functions arise, consider $f(z) = z^\alpha$, where $\alpha \in \mathbb{R}$ *is a non integer*. For z constrained to move around a unit circle centered on the origin, we find that $z = \mathrm{e}^{\mathrm{i}\theta}$ starts at $z = \mathrm{e}^{\mathrm{i}0}$ and ends at $z = \mathrm{e}^{\mathrm{i}2\pi} = \mathrm{e}^{\mathrm{i}0}$ (same place). The function $f(z)$, however, has initial value $f(\theta = 0) = \mathrm{e}^{\mathrm{i}\alpha 0}$ but ends at $f(\theta = 2\pi) = \mathrm{e}^{\mathrm{i}\alpha 2\pi}$, and this is not the same value as $f(\theta = 0)$ because α is not an integer. We say that $f(z)$ is a *multivalued function*. Another example of a multivalued function is the natural logarithm, $f(z) = \ln z$. If $z = R\exp(\mathrm{i}(\theta + 2n\pi))$ (where n is an integer so that z repeats itself around a circle centered on the origin), then the natural logarithm of z is $\ln(z) = \ln(R) + \mathrm{i}(\theta + 2n\pi)$, which is clearly multivalued. In the case of the logarithm, we say that its principal value is determined by $n = 0$ and $-\pi < \theta \le \pi$ and we write $\ln(z) = \ln(R) + \mathrm{i}\theta$.

multivalued functions

There is a standard technique for dealing with multivalued functions. When $f(z)$ is not single-valued in a closed path around a point $z = z_0$, we say that z_0 is a *branch point* of $f(z)$. Contour integrals of functions around branch points will not necessarily obey Eq. (8.19), and $f(z)$ may not have a Laurent series representation. To address this situation we introduce a "barrier" known as a *branch cut*, connected to the branch point, and prohibit integration paths from crossing this barrier. (There is some freedom in the choice of location of this barrier as is illustrated in the next example.)

branch point and branch cut

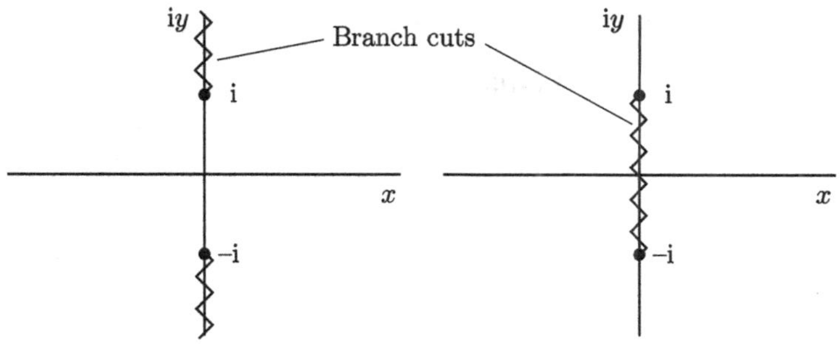

Figure 8.5 Possible branch cuts for $f(z) = \sqrt{z^2+1}$.

Example. Find the branch points for $f(z) = \sqrt{z^2+1}$ and suggest several branch cuts.

Solution: Because $f(z) = \sqrt{z^2+1} = \sqrt{z+\mathrm{i}}\sqrt{z-\mathrm{i}} = (z+\mathrm{i})^{1/2}(z-\mathrm{i})^{1/2}$, we see that $z = \pm\mathrm{i}$ are candidates for branch points. To check these points, let $z - \mathrm{i} = R_1 \mathrm{e}^{\mathrm{i}\theta_1}$ and $z + \mathrm{i} = R_2 \mathrm{e}^{\mathrm{i}\theta_2}$.

Then

$$f(z) = \sqrt{R_1 R_2} \exp\left(\frac{\mathrm{i}(\theta_1 + \theta_2)}{2}\right)$$

There are several cases for enclosing contours C:

1. If C encloses neither branch point, then $\theta_1 \to \theta_1$, $\theta_2 \to \theta_2$ and so $f(z) \to f(z)$;
2. If C encloses $z = \mathrm{i}$ but not $z = -\mathrm{i}$, then $\theta_1 \to \theta_1 + 2\pi$, $\theta_2 \to \theta_2$ and so $f(z) \to -f(z)$;
3. If C encloses $z = -\mathrm{i}$ but not $z = \mathrm{i}$, then $\theta_1 \to \theta_1$, $\theta_2 \to \theta_2 + 2\pi$ and so $f(z) \to -f(z)$;
4. If C encloses both branch points, then $\theta_1 \to \theta_1 + 2\pi$, $\theta_2 \to \theta_2 + 2\pi$ and so $f(z) \to f(z)$.

Thus, $f(z)$ changes value around loops containing either $z = \mathrm{i}$ or $z = -\mathrm{i}$ (but not both). Therefore, we must choose branch cuts that prevent us from making a complete loop around either branch point. Two possible branch cuts are illustrated in Figure 8.5

8.8 Definite integrals

There are a variety of definite integrals that can be performed by considering them to be contour integrals of analytic functions in the complex plane.

8.8.1 Integrands containing $\cos\theta$ and $\sin\theta$

This is perhaps the easiest application of contour integration because, in some sense, the integral "already is" a contour integral. Suppose the integral is of the form

$$I = \int_0^{2\pi} F(\cos\theta, \sin\theta)\, \mathrm{d}\theta.$$

Applying the substitution $z = \mathrm{e}^{\mathrm{i}\theta}$ so that

$$\cos\theta = \frac{z + z^{-1}}{2}, \quad \sin\theta = \frac{z - z^{-1}}{2\mathrm{i}}, \quad \mathrm{d}\theta = -\mathrm{i}z^{-1}\,\mathrm{d}z$$

may allow us to interpret the integral as a contour integral over the unit circle which will be determined by the poles lying inside.

Example. Evaluate

$$I = \int_0^{2\pi} \frac{\cos 2\theta}{a^2 + b^2 - 2ab\cos\theta}\, \mathrm{d}\theta \quad b > a > 0.$$

Solution: We first note that $\cos 2\theta = (z^2 + z^{-2})/2$. With C the unit circle centered on the origin of the complex plane, we have

$$I = -\mathrm{i}\oint_C \frac{z^2 + z^{-2}}{2z(a^2 + b^2 - ab(z + z^{-1}))}\, \mathrm{d}z = \frac{\mathrm{i}}{2ab}\oint_C \frac{z^4 + 1}{z^2(z - a/b)(z - b/a)}\, \mathrm{d}z.$$

The integrand has three poles, *but only two of them lie within* C (because $b/a > 1$). The residue from the double pole at $z = 0$ is

$$\mathrm{Res}(0) = \lim_{z\to 0} \frac{\mathrm{d}}{\mathrm{d}z}(z^2 f(z)) = \frac{\mathrm{i}}{2ab}\frac{a^2 + b^2}{ab}.$$

The residue from the pole at $z = a/b$ is

$$\mathrm{Res}\frac{a}{b} = \lim_{z\to a/b}(z - \frac{a}{b})f(z) = \frac{\mathrm{i}}{2ab}\frac{a^4 + b^4}{ab(a^2 - b^2)}.$$

By the residue theorem, we have

$$I = 2\pi\mathrm{i}\left(\mathrm{Res}(0) + \mathrm{Res}\frac{a}{b}\right) = \frac{2\pi a^2}{b^2(b^2 - a^2)}.$$

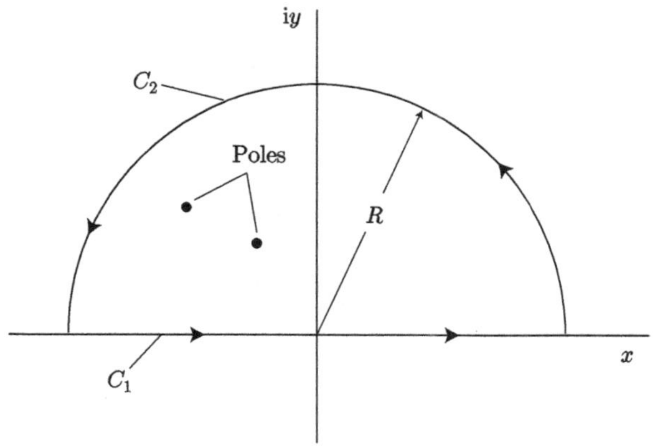

Figure 8.6 "Closing the contour" in the upper half-plane.

8.8.2 Infinite integrals

It is sometimes possible to consider integrals of the form $I = \int_{-\infty}^{\infty} f(x)\, \mathrm{d}x$ to be one leg of a closed contour on which $f(z)$ is restricted to the real axis. Consider, in Figire 8.6, the closed contour composed of the paths C_1 and C_2. We say that the contour C_1 has been "closed" by the contour C_2. The closing contour C_2 need not be a semicircle in the upper half-plane – it could be any closing path. The point is that we should choose a closing contour along which the integral $\int_{C_2} f(z)\, \mathrm{d}z$ is easy to evaluate. The entire closed contour can be evaluated by the residue theorem, and we can write $I = 2\pi\mathrm{i} \sum_j \mathrm{Res}(z_j) - \int_{C_2} f(z)\, \mathrm{d}z$.

An especially convenient choice for the closing contour C_2 is a path along which $f(z) = 0$. This will occur when $f(z)$ decays[22] to zero as $R \to \infty$ everywhere on C_2. A sufficient condition for this kind of decay is that

1. $zf(z) \to 0$ as $|z| \to \infty$
2. $\int_{-\infty}^{0} f(x)\, \mathrm{d}x$ and $\int_{0}^{\infty} f(x)\, \mathrm{d}x$ both exist.

For example

$$f(z) = \frac{\exp(\mathrm{i}z)}{1+z^2} = \frac{\mathrm{e}^{\mathrm{i}x}\mathrm{e}^{-y}}{1+x^2-y^2+2\mathrm{i}xy}$$

decays properly for C_2 in the upper half-plane because for $y = 0$, the integrand falls off as x^{-2}, whereas for small $|x|$, the integrand has exponential decay in y. Note that this behavior is not exhibited by choosing a closing contour in the lower half-plane. If, however, we had $f(z) = \exp(-\mathrm{i}z)/(1+z^2)$, then we would need to choose C_2 to be a semicircle in the lower half-plane to ensure proper decay.

[22]Because, in this case, arc length scales as $|z|$, we actually require $f(z)$ to decay faster than $1/|z|$ as $|z| \to \infty$.

Example. Evaluate

$$I = \int_0^\infty \frac{dx}{1+x^2}.$$

Solution: The integrand is an even function and so we note that

$$I = \frac{1}{2}\int_{-\infty}^\infty \frac{dx}{1+x^2}$$

therefore, we examine

$$\frac{1}{1+z^2} = \frac{1}{(z+i)(z-i)}.$$

This integrand has two simple poles $z = \pm i$. If we complete the contour with an infinite semicircle in the upper half-plane, then we need only consider the residue from $z = +i$. If, however, we complete the contour in the lower half-plane, then we need the residue from $z = -i$ only. Both completing contours work, and we choose to complete in the lower half-plane. The residue is

$$\text{Res}(-i) = \lim_{z\to -i}(z+i)f(z) = \frac{-1}{2i}.$$

From the residue theorem, we have

$$2I = -2\pi i\left(\frac{-1}{2i}\right) = \pi \quad \Rightarrow \quad I = \frac{\pi}{2}$$

where the minus sign comes from the fact that the contour in the lower half plane is *traversed in the clockwise direction.*

Example. Evaluate

$$I = \int_0^\infty \frac{dx}{(x+a)^3 x^{1/2}}. \qquad (a > 0)$$

Solution: We consider the integrand

$$f(z) = \frac{1}{(z+a)^3 z^{1/2}}$$

and note that $z = 0$ is a branch point. We introduce a branch cut along the positive real axis and consider the closed contour of Figure 8.7.

We will eventually let the radius of C_2 become infinite and the radius of C_1 shrink to zero. Because $|zf(z)| \to 0$ on C_2 as $R \to \infty$ *and* $|zf(z)| \to 0$ on C_1 as $\rho \to 0$, it's clear that neither C_2 nor C_1 make a contribution to the contour integral.

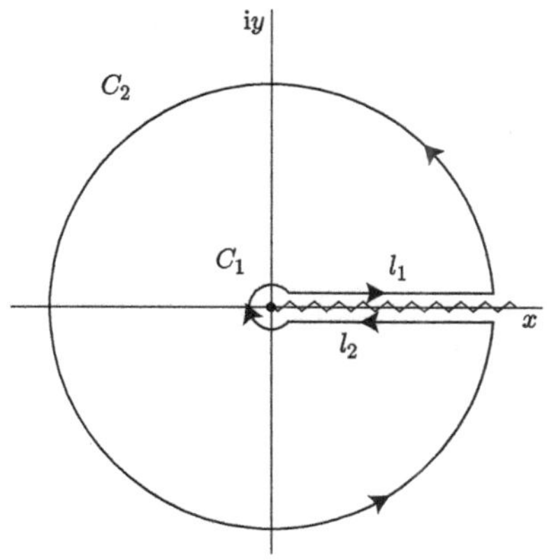

Figure 8.7 Contour geometry.

The only pole inside the contour occurs at $z = -a$. This is a pole of order $m = 3$ and the residue is

$$\operatorname{Res}(-a) = \lim_{z\to -a} \frac{1}{2}\frac{\mathrm{d}^2}{\mathrm{d}z^2}((z+a)^3 f(z)) = \lim_{z\to -a} \frac{3}{8} z^{-5/2} = \frac{-3\mathrm{i}}{8a^{5/2}}.$$

The residue theorem yields

$$\int_{C_1} f(z)\,\mathrm{d}z + \int_{C_2} f(z)\,\mathrm{d}z + \int_{l_1} f(z)\,\mathrm{d}z + \int_{l_2} f(z)\,\mathrm{d}z = 2\pi\mathrm{i}\left(\frac{-3\mathrm{i}}{8a^{5/2}}\right) = \frac{3\pi}{4a^{5/2}}.$$

In the limit as C_1 shrinks to zero and C_2 expands to infinity, the first two of these integrals vanishes, and we are left with two line integrals: On l_1, we set $z = x$, while on l_2, we set $z = x\mathrm{e}^{\mathrm{i}2\pi}$ (to account for the multivalued integrand). Then,

$$\int_0^\infty \frac{\mathrm{d}x}{(x+a)^3 x^{1/2}} + \int_\infty^0 \frac{\mathrm{d}x}{(x\mathrm{e}^{\mathrm{i}2\pi}+a)^3(x\mathrm{e}^{\mathrm{i}2\pi})^{1/2}} = \frac{3\pi}{4a^{5/2}}$$

$$\Rightarrow \quad \left(1 - \frac{1}{\mathrm{e}^{\mathrm{i}\pi}}\right)\int_0^\infty \frac{\mathrm{d}x}{(x+a)^3 x^{1/2}} = \frac{3\pi}{4a^{5/2}}$$

$$\Rightarrow \quad I = \int_0^\infty \frac{\mathrm{d}x}{(x+a)^3 x^{1/2}} = \frac{3\pi}{4a^{5/2}} \times \left(1 - \frac{1}{\mathrm{e}^{\mathrm{i}\pi}}\right)^{-1} = \frac{3\pi}{8a^{5/2}}.$$

We note that in the special case, the integral is of the form

$$I = \int_{-\infty}^{\infty} f(z)\mathrm{e}^{\mathrm{i}ax}\,\mathrm{d}x \quad \text{where } a \text{ is real-valued}$$

(i.e. this is a Fourier transform) we can relax the "strong decay" requirement on the integrand a little.

Assuming $a > 0$, we choose a closing contour in the upper half-plane. Then, the contribution from C_2 is

$$I_2 = \int_0^{\pi} f(Re^{i\theta})e^{iaR\cos\theta}e^{-aR\sin\theta}iRe^{i\theta}\,d\theta.$$

If $f(z)$ decays so that $\lim_{|z|\to 0}f(z) = 0$ (when $0 \le \arg z \le \pi$), then

$$|I_2| \le \varepsilon R\int_0^{\pi} e^{-aR\sin\theta}\,d\theta = 2\varepsilon R\int_0^{\pi/2} e^{-aR\sin\theta}\,d\theta.$$

for R large enough that $|f(Re^{i\theta})| < \varepsilon$. But

$$\frac{2\theta}{\pi} \le \sin\theta \text{ when } 0 \le \theta \le \pi/2$$

$$\Rightarrow |I_2| \le 2\varepsilon R\int_0^{\pi/2} e^{-aR2\theta/\pi}\,d\theta = 2\varepsilon R\frac{1-e^{-aR}}{aR2/\pi}.$$

In the limit as $R \to \infty$ we see that

$$\lim_{R\to\infty}|I_2| \le \frac{\pi\varepsilon}{a}.$$

Consequently, if we simply require that $\lim_{|z|\to 0}f(z) = 0$ (restricted to C_2), we are guaranteed that the contribution from the closing contour at $R \to \infty$ will vanish. This is known as *Jordan's lemma.*[23]

Jordan's lemma

8.8.3 Poles on the contour of integration

It sometimes happens that a pole will sit on the contour of integration itself. When we have the freedom to choose the contour (as in closing contours), we can usually avoid this situation – but what if the pole is located on the real axis, and we need to evaluate $I = \int_{-\infty}^{\infty} f(x)\,dx$? This situation occurs frequently and, just as in the case of improper integrals in "ordinary" calculus, we must approach these poles in a limiting sense.

Figure 8.8 illustrates this situation and displays two possible means of deforming the contour around a pole. Note that in the half-plane ($x < 0$), we deformed the contour to exclude the pole from the region enclosed by the contour of integration. This means that the residue from this pole will not contribute through residue sum, but the contribution from the small semicircle must be calculated separately. (Also note that this small semicircle is traversed in a clockwise sense.) In the right half-plane, however, we deformed the contour to include the pole, and its residue will contribute to the residue sum.

[23] Jordan's lemma is similar to the Darboux inequality in that they are both *boundedness formulas.*

Let each deformation be a semicircle of radius ρ centered on the poles (located at $z = x_-$ and $z = x_+$, respectively). Then, the contributions along the real axis are of the form

$$\int_{-R}^{x_- - \rho} f(x)\,\mathrm{d}x - \int_0^{\pi} f(\rho e^{i\theta}) i\rho e^{i\theta}\,\mathrm{d}\theta + \int_{x_- + \rho}^{x_+ - \rho} f(x)\,\mathrm{d}x$$
$$+ \int_0^{\pi} f(\rho e^{i\theta}) i\rho e^{i\theta}\,\mathrm{d}\theta + \int_{x_+ + \rho}^{R} f(x)\,\mathrm{d}x$$

where the second integral is the contribution from the path C_3, and the minus sign accounts for the clockwise traverse direction. The fourth integral is the contribution from the integral over C_4. We've actually done these integrals several times before:

$$-\int_0^{\pi} f(\rho e^{i\theta}) i\rho e^{i\theta}\,\mathrm{d}\theta = -i\pi \mathrm{Res}(-)$$

where Res(−) is the residue due to the pole at $z = x_-$. Similarly, the contribution from the integral over C_4 (in the right half-plane) is $+i\pi\mathrm{Res}(+)$. In the case illustrated in Figure 8.8 with no other poles in the upper half plane and assuming that $f(z)$ decays strongly (so that the contribution from the closing contour vanishes as $R \to \infty$), we obtain

$$\oint_{C_{\mathrm{tot}}} f(z)\,\mathrm{d}z = \int_{-\infty}^{x_- - \rho} f(x)\,\mathrm{d}x - i\pi\mathrm{Res}(-) + \int_{x_- + \rho}^{x_+ - \rho} f(x)\,\mathrm{d}x$$
$$+ i\pi\mathrm{Res}(+) + \int_{x_+ + \rho}^{\infty} f(x)\,\mathrm{d}x = 2\pi i\mathrm{Res}(+).$$

This result follows from the residue theorem and only the residues within C_{tot} contribute to the residue sum. It follows that

$$\oint_{C_{\mathrm{tot}}} f(z)\,\mathrm{d}z = \int_{-\infty}^{x_- - \rho} f(x)\,\mathrm{d}x + \int_{x_- + \rho}^{x_+ - \rho} f(x)\,\mathrm{d}x + \int_{x_+ + \rho}^{\infty} f(x)\,\mathrm{d}x$$
$$= \pi i(\mathrm{Res}(-) + \mathrm{Res}(+))$$

and so each pole x_j on the contour can be considered to contribute $\pi i \times \mathrm{Res}(z_j)$ regardless of whether it lies inside or outside the deformation (i.e. exactly half of what it would have contributed were it to lie within C_{tot}).

We denote the *principal value of the integral* by[24]

$$P\int_{-R}^{R} f(x)\,\mathrm{d}x = \lim_{\rho \to 0}\left\{\int_{-R}^{x_j - \rho} f(x)\,\mathrm{d}x + \int_{x_j + \rho}^{R} f(x)\,\mathrm{d}x\right\}.$$

[24] $P\int \cdots$ is also sometimes denoted by $⨍ \cdots$.

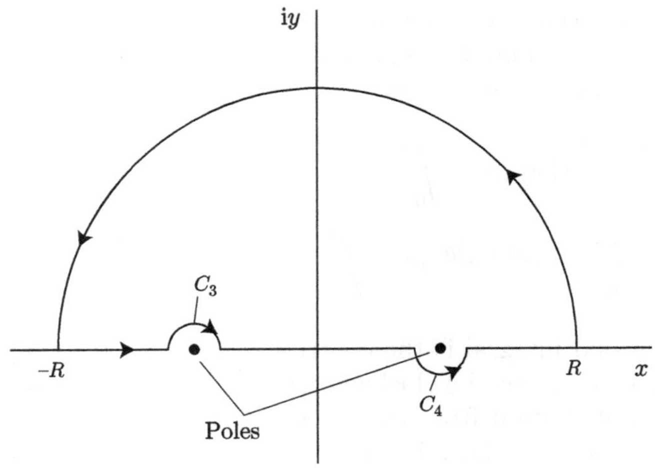

Figure 8.8 Deformation of a contour to avoid poles.

Example. Find the principal value of

$$\int_{-\infty}^{\infty} \frac{\cos mx}{x-a}\,\mathrm{d}x \quad a \mathrm{is real\text{-}valued and} m > 0$$

Solution: This integral is the real part of

$$\int_{-\infty}^{\infty} \frac{\exp(\mathrm{i}mx)}{x-a}\,\mathrm{d}x$$

and so (by Jordan's Lemma) we can easily see that the contribution from a closing contour consisting of a semicircle of infinite radius in the upper half-plane will make no contribution to the closed contour integral. Therefore, we can write

$$P\int_{-\infty}^{\infty} \frac{\exp(\mathrm{i}mz)}{z-a}\,\mathrm{d}z = \pi\mathrm{i}\mathrm{Res}(a)$$

where $\mathrm{Res}(a)$ is the residue from the single pole at $x = a$. This residue is $\mathrm{Res}(a) = \exp(\mathrm{i}ma)$ and so

$$P\int_{-\infty}^{\infty} \frac{\cos mx}{x-a}\,\mathrm{d}x = \mathrm{Re}\{\pi\mathrm{i}\exp(\mathrm{i}ma)\} = -\pi\sin ma.$$

8.9 Meromorphic functions

A class of functions with wide application in science and engineering are *rational functions*, functions $f(z)$ that occur in the form of a ratio of functions $f(z) = P(z)/Q(z)$, where P and Q are polynomials.[25] Thus, rational functions can be expressed

[25] To cite just one example, rational functions are used in numerical analysis for the interpolation of functions, such as with the Padé approximation technique.

$$f(z) = \frac{a_0 + a_1 z + \cdots + a_m z^m}{b_0 + b_1 z + \cdots + b_k z^k}.$$

A rational function can be singular only at those points where the denominator vanishes, and these occur at isolated points (Section 8.6). These considerations lead us to define a class of functions that share with rational functions the property of having isolated poles.

Definition. *A function is said to be* meromorphic *in a domain $\mathcal{D}$ if at every point of $\mathcal{D}$ it is either analytic or has isolated poles.*

Example. The function $f(z) = 1/(\sin z)$ is a meromorphic function. It has isolated poles at $z = n\pi$, for all integer n.

Mittag-Leffler theorem

One of the principal theorems on meromorphic functions is the *Mittag-Leffler theorem*, which states that any meromorphic function can be expressed as a sum (in general, infinite) of partial fractions. The Mittag-Leffler theorem is quite general, and presenting it in full generality would take us beyond the intended purpose of this book.[26] We present a version that will allow us to quickly derive useful results.

Suppose a function $f(z)$ has no singularities except for a finite number of poles at $z = z_j$, $j = 1, \ldots, n$ not including $z = 0$ and is bounded at infinity. Applying the Cauchy integral formula to a contour C large enough to include all the poles:

$$f(z) = \frac{1}{2\pi \mathrm{i}} \oint_C \frac{f(\xi)}{\xi - z}\,\mathrm{d}\xi - \frac{1}{2\pi \mathrm{i}} \sum_{k=1}^{n} \oint_{C_k} \frac{f(\xi)}{\xi - z}\,\mathrm{d}\xi, \tag{8.32}$$

where z is a point within C and the contours $C_1, \ldots, C_n$ are small circles about the poles taken in the counterclockwise direction. The first integral in Eq. (8.32) is independent of C as long as C surrounds all the poles of $f(\xi)/(\xi - z)$. Take C to be arbitrarily large. Then, the first integral is then a *constant* because f is presumed bounded; call it K. Replace $f(\xi)$ in the vicinity of each pole with its Laurent expansion about that pole. Apply the residue theorem to each of the integrals about the poles. We find

$$f(z) = K + \sum_{k=1}^{n} \frac{\mathrm{Res}(z_k)}{z - z_k}. \tag{8.33}$$

Equation (8.33) is the basic idea behind the Mittag-Leffler theorem: *A meromorphic function can be replaced by the sum of the principal parts of the Laurent expansions at the poles, plus a constant.* The constant

---

[26]The "full strength" Mittag-Leffler theorem can be found in[26, p. 299].

K can be eliminated by evaluating the function at $z = 0$: $f(0) = K - \sum_{k=1}^{n}(\text{Res}(z_k)/z_k)$. Thus, from Eq. (8.33),

$$f(z) = f(0) + \sum_k \text{Res}(z_k)\left(\frac{1}{z - z_k} + \frac{1}{z_k}\right). \tag{8.34}$$

Equation (8.34) is the Mittag-Leffler expansion in its most elemental form. It can be extended to functions having an infinite number of poles.

Example. Consider $f(z) = \csc z - z^{-1}$. The function vanishes at the origin, $f(0) = 0$. It has simple poles at $z = n\pi$ for all integer $n \neq 0$, for which the residues are

$$\text{Res}(n\pi) = \lim_{z \to n\pi}(z - n\pi)\left(\frac{1}{\sin z} - \frac{1}{z}\right) = (-1)^n.$$

Thus, from Eq. (8.34),

$$\frac{1}{\sin z} - \frac{1}{z} = \left(\sum_{n=1}^{\infty} + \sum_{n=-1}^{-\infty}\right)(-1)^n\left(\frac{1}{z - n\pi} + \frac{1}{n\pi}\right) = 2z\sum_{n=1}^{\infty}\frac{(-1)^n}{z^2 - (n\pi)^2}. \tag{8.35}$$

8.10 Approximation of integrals

An important consequence of Cauchy's theorem is that when $f(z)$ is analytic throughout a region $\mathcal{D}$, we can *deform* the contour of integration to any other contour with the same endpoints (within $\mathcal{D}$) without changing the value of $\int_C f(z)\,\mathrm{d}z$ (Exercise 8.11). This is illustrated in Figure 8.9, which shows several deformations C_i of the contour C connecting points z_1 and z_2. Let the contour $\tilde{C}$ bound the region between C and C_1. From Cauchy's theorem,

$$\oint_{\tilde{C}} f(z)\,\mathrm{d}z = \int_{C_1} f(z)\,\mathrm{d}z - \int_C f(z)\,\mathrm{d}z = 0 \Rightarrow \int_C f(z)\,\mathrm{d}z = \int_{C_1} f(z)\,\mathrm{d}z.$$

The minus sign accounts for the fact that the path C is traversed backwards.

This observation holds true for any deformation that does not enclose a singularity of $f(z)$ within the region bounded by the original path and the deformed path. If $z = a$ is the only singular point of $f(z)$, for example, then (refer to Figure 8.9)

$$\int_C f(z)\,\mathrm{d}z = \int_{C_2} f(z)\,\mathrm{d}z \quad \text{but} \quad \int_C f(z)\,\mathrm{d}z \neq \int_{C_3} f(z)\,\mathrm{d}z.$$

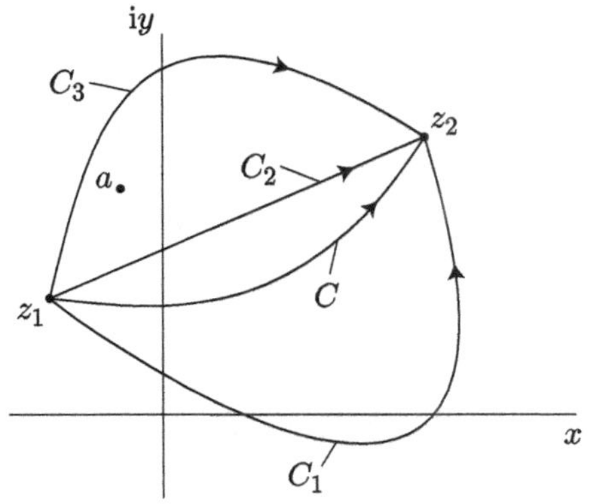

Figure 8.9 Deformed paths of integration.

The ability to deform contours sometimes allows us to choose an alternate path of integration (with the same starting and ending points as the original contour) that may make the evaluation of the integral easier. Of particular interest in many practical situations are integrals of the form

$$\int_C g(z)\exp(f(z))\,\mathrm{d}z \tag{8.36}$$

where both $g(z)$ and $f(z)$ are analytic and $g(z)$ varies slowly in comparison with $f(z)$. (Such integrals appear often in physical applications.) The general method we shall employ is known as the "saddle point method" and exploits properties of analytic functions $f(z)$ in the neighborhood of points z_0 for which $f'(z_0) = 0$.

For functions u and v satisfying the Cauchy–Riemann conditions (f is assumed analytic), the second-order partial derivatives have the property

$$\frac{\partial^2 u}{\partial x^2} = \frac{\partial}{\partial x}\left(\frac{\partial u}{\partial x}\right) = \frac{\partial^2 v}{\partial x \partial y} \qquad \frac{\partial^2 u}{\partial y^2} = \frac{\partial}{\partial y}\left(\frac{\partial u}{\partial y}\right) = -\frac{\partial^2 v}{\partial y \partial x}. \tag{8.37}$$

Because f' is analytic, the partial derivatives of *its* real and imaginary parts are continuous,[27] i.e. the *second* order partial derivatives of u and v are continuous. By adding the equations in Eq. (8.37), we then have

$$\frac{\partial^2 u}{\partial x^2} + \frac{\partial^2 u}{\partial y^2} = 0.$$

In the same way, it can be shown that

$$\frac{\partial^2 v}{\partial x^2} + \frac{\partial^2 v}{\partial y^2} = 0.$$

Thus, u and v (in addition to satisfying the Cauchy–Riemann conditions) satisfy the two-dimensional Laplace equation.

[27]From Eq. (8.5), $f'(z) = (\partial u/\partial x) + \mathrm{i}(\partial v/\partial x)$.

Consider a function $f(z) = u(x,y) + \mathrm{i}v(x,y)$ analytic in a region $\mathcal{D}$. The points in the complex plane where $f(z)$ is *stationary* are those for which $f'(z) = 0$. From Eq. (8.5), these occur when

$$\frac{\partial u}{\partial x} = \frac{\partial u}{\partial y} = \frac{\partial v}{\partial x} = \frac{\partial v}{\partial y} = 0.$$

This result is the same as one would obtain from "ordinary" calculus. As with ordinary calculus, we can classify stationary points as local maxima/minima or saddle points by examining the second partial derivatives. Because u and v satisfy the two-dimensional Laplace equation, it must be that the second partial derivatives of $u(x,y)$ and $v(x,y)$ *are of opposite sign*. Consequently, *the stationary points of an analytic function cannot be local maxima or local minima of the function and, instead, must always be saddle points.*

saddle point method

Armed with this knowledge, we can expand $f(z)$ in a Taylor series about a stationary point z_0 (a saddle point for which $f'(z_0) = 0$) to obtain

$$f(z) = f(z_0) + 0 + \frac{1}{2}f''(z_0)(z - z_0)^2 + O((z - z_0)^3)$$

for all points $z \in \mathcal{D}$. If we restrict z to a straight path through the saddle point in the angular direction θ, so that $z - z_0 = s\mathrm{e}^{\mathrm{i}\theta}$, and set $f''(z_0) = A\mathrm{e}^{\mathrm{i}\alpha}$ (a general complex number), then

$$\begin{aligned} f(z) &= f(z_0) + \frac{1}{2}As^2\mathrm{e}^{\mathrm{i}(\alpha+2\theta)} + O(s^3) \\ &= f(z_0) + \frac{1}{2}As^2(\cos(\alpha + 2\theta) + \mathrm{i}\sin(\alpha + 2\theta)) + O(s^3). \end{aligned}$$

The integrand of Eq. (8.36) – when restricted to this straight path – has the form

$$\begin{aligned} g(z)\exp(f(z)) &= g(z_0 + s\mathrm{e}^{\mathrm{i}\theta})\exp(f(z_0)) \\ &\exp\left(\frac{1}{2}As^2(\cos(\alpha + 2\theta) + \mathrm{i}\sin(\alpha + 2\theta)) + O(s^3)\right). \end{aligned} \tag{8.38}$$

We aren't free to deform our original path C into an arbitrary straight path oriented at angle θ, of course, because our deformed path must still start and end at the same points as C. We are, however, free to deform the path so that it is *locally* straight and at angle θ in a region around z_0. It turns out that if we're careful how we choose θ, then it's possible to force the main contribution to the integral to come only from this locally straight path segment.

8.10.1 The method of steepest descent

In Eq. (8.38), choose θ so that $\sin(\alpha + 2\theta) = 0$, implying $\alpha + 2\theta = \pm n\pi$, $n = 0, 1, 2, \ldots$, and thus, $\cos(\alpha + 2\theta) = \cos(\pm n\pi) = (-1)^n$. Then, with $n = 1$,

$$\theta = \frac{1}{2}(\pm\pi - \alpha)$$

and Eq. (8.38) reduces to

$$g(z)\exp(f(z)) = g(z_0 + s\mathrm{e}^{\mathrm{i}\theta})\exp(f(z_0))\exp\left(-\frac{1}{2}As^2 + O(s^3)\right).$$

This direction for θ is known as the direction of "steepest descent" and when A is sufficiently large, the factor $\exp(-As^2/2)$ decays rapidly while $g(z)$ is only slowly varying. In this case, we can approximate

path of steepest descent

$$g(z)\exp(f(z)) \approx g(z_0)\exp(f(z_0))\exp\left(-\frac{1}{2}As^2\right)$$

on the deformed path in the vicinity of the saddle point. It's also clear that the phase of $\exp(f(z))$ is constant along this path of steepest descent and only its magnitude changes – that is, $|\exp(f(z))|$ decreases most rapidly along this path (to second order in path length s). Moreover, owing to the exponential decay of the integrand, by the time the deformed contour starts to deviate from the straight path, the additional contribution to the integral will be negligible. (Provided, of course, that the contributions from regions near the endpoints can be neglected.)

Using $\mathrm{d}z = \mathrm{e}^{\mathrm{i}\theta}\,\mathrm{d}s$, the integral Eq. (8.36) can then be approximated as

$$\int_C g(z)\exp(f(z))\,\mathrm{d}z \approx g(z_0)\exp(f(z_0))\,\mathrm{e}^{\mathrm{i}\theta}\int_{C'}\exp\left(-\frac{1}{2}As^2\right)\,\mathrm{d}s$$

where C' is a deformation of the original contour C which passes through the saddle point z_0 in the direction $\theta = (\pm\pi - \alpha)/2$. The integral over C' can often be performed directly, for example:

$$\int_{-\infty}^{\infty}\exp\left(-\frac{1}{2}As^2\right)\,\mathrm{d}s = \sqrt{\frac{\pi}{A}} \Rightarrow \int_0^{\infty}\exp\left(-\frac{1}{2}As^2\right)\,\mathrm{d}s = \frac{1}{2}\sqrt{\frac{\pi}{A}}.$$

Example. The *Airy function*,[28] denoted $\mathrm{Ai}(x)$, has the integral representation

$$\mathrm{Ai}(x) = \frac{1}{\pi}\int_0^{\infty}\cos\left(\frac{t^3}{3} + xt\right)\,\mathrm{d}t.$$

Develop an approximate expression for large positive values of $x \in \mathbb{R}$.

Solution: Because the integrand is even, we can write the integral as the real part of

[28]The Airy function of the first kind, $\mathrm{Ai}(x)$, is a linearly independent solution of the differential equation $\psi'' = x\psi$ that decays to zero as $x \to +\infty$. As $x \to -\infty$, $\mathrm{Ai}(x)$ is oscillatory.

$$\frac{1}{2\pi}\int_{-\infty}^{\infty}\exp\left(\mathrm{i}\left(\frac{t^3}{3}+xt\right)\right)\,\mathrm{d}t.$$

Consequently, our "original" contour C is just the real axis from $z=-\infty$ to $z=\infty$ and $f(z)=\mathrm{i}(z^3/3+xz)$.

The saddle points occur at

$$f'(z)=\mathrm{i}(z^2+x)=0 \Rightarrow z_0=\pm\mathrm{i}\sqrt{x}=\sqrt{x}\,\mathrm{e}^{\pm\mathrm{i}\pi/2}\ \text{for which}\ f(z_0)=\mp 2x^{3/2}/3.$$

We assume that x is sufficiently large that the saddle points (both on the imaginary axis) are separated enough so that one does not influence the geometry of $f(z)$ in the region around the other. We are therefore free to choose either saddle point. If we choose $z_0=\sqrt{x}\,\mathrm{e}^{+\mathrm{i}\pi/2}$, then

$$f''(z_0)=\mathrm{i}z_0=\sqrt{x}\,\mathrm{e}^{\mathrm{i}\pi}\ \Rightarrow\ A=\sqrt{x}\,,\quad \alpha=\pi\ \Rightarrow\ \theta=\frac{\pm\pi-\alpha}{2}=0\,,-\pi.$$

If, however, we choose $z_0=\sqrt{x}\,\mathrm{e}^{-\mathrm{i}\pi/2}$, then

$$f''(z_0)=\mathrm{i}z_0=\sqrt{x}\quad\Rightarrow\quad A=\sqrt{x}\,,\quad \alpha=0\quad\Rightarrow\quad\theta=\pm\pi/2.$$

We choose to deform the contour so that it passes through the saddle point $z_0=\sqrt{x}\,\mathrm{e}^{+\mathrm{i}\pi/2}$ because the line of steepest descent through this saddle point is co linear with the original contour (which is directed in the $\theta=0$ direction).

It follows that

$$\frac{1}{2\pi}\int_{-\infty}^{\infty}\exp\left(\mathrm{i}\left(\frac{t^3}{3}+xt\right)\right)\mathrm{d}t\approx\frac{\exp(f(z_0))}{2\pi}\,\mathrm{e}^{\mathrm{i}\theta}\int_{-\infty}^{\infty}\exp\left(-\frac{1}{2}As^2\right)\,\mathrm{d}s$$
$$=\frac{\exp(-2x^{3/2}/3)}{2\pi}\sqrt{\frac{\pi}{\sqrt{x}}}=\frac{\exp(-2x^{3/2}/3)}{2\sqrt{\pi\sqrt{x}}}.$$

Example. The Gamma function $\Gamma(x)$ (see Appendix C) is defined for $x>-1$ through the integral

$$\Gamma(x+1)=\int_0^{\infty}t^x\mathrm{e}^{-t}\,\mathrm{d}t\equiv\int_0^{\infty}\mathrm{e}^{f(t)}\,\mathrm{d}t,$$

where $f(t)=x\ln t-t$. The function $f(t)$ has a maximum (for fixed x) at $t=x$. In the neighborhood of $t=x$,

$$f(t)\approx x\ln x-x-\frac{1}{2x}(t-x)^2+\cdots$$

Thus, for large x, we can approximate the integral,

$$\Gamma(x+1)\approx\mathrm{e}^{x\ln -x}\int_0^{\infty}\mathrm{e}^{-(t-x)^2/(2x)}\,\mathrm{d}t\approx\mathrm{e}^{x\ln -x}\int_{-\infty}^{\infty}\mathrm{e}^{-(t-x)^2/(2x)}\,\mathrm{d}t$$
$$=\sqrt{2\pi x}x^x\mathrm{e}^{-x},\qquad(8.39)$$

a result known as *Stirling's approximation.* One usually sees the Stirling approximation in the form

$$\ln \Gamma(x+1) \approx x(\ln x - 1) + \frac{1}{2}\ln(2\pi x) + \cdots$$

For $x \gg 1$, the second logarithmic term can be ignored.

8.10.2 The method of stationary phase

If, instead of choosing θ so that the phase of $\exp(f(z))$ remains constant along the deformed contour, we were to select θ in Eq. (8.38) so that $\cos(\alpha + 2\theta) = 0$, then $\alpha + 2\theta = \pm\pi/2\,, \pm 3\pi/2$. In either case, we have $\sin(\alpha + 2\theta) = \pm 1$ and so – along these lines – Eq. (8.38) reduces to

$$g(z)\exp(f(z)) = g(z_0 + s\mathrm{e}^{\mathrm{i}\theta})\exp(f(z_0))\exp\left(\pm \mathrm{i}\frac{1}{2}As^2 + O(s^3)\right)$$

$$\text{where}\quad \theta = \frac{\pm\pi/2 - \alpha}{2} \quad\text{or}\quad \theta = \frac{\pm 3\pi/2 - \alpha}{2}.$$

When z is restricted to this deformed contour, it's clear that $|\exp(f(z))|$ is held constant. (These lines are known as "level lines" and are obviously orthogonal to the lines of steepest descent/ascent.) It's also clear that the phase of $\exp(f(z))$ varies most rapidly along level lines, and when A is sufficiently large, the rapid oscillations of the $\exp(\pm \mathrm{i}As^2/2)$ phase factor will alternately *cancel* the contributions of the slowly varying $g(z)$ factor to the integral (almost). Consequently, the only significant contribution to an integral restricted to a level line through a saddle point z_0 will come from the immediate neighborhood of z_0 (for which $s \sim 0 \Rightarrow As^2$ is very small). level lines

In this case, the integral (8.36) can be approximated

$$\int_C g(z)\exp(f(z))\,\mathrm{d}z \approx g(z_0)\exp(f(z_0))\,\mathrm{e}^{\mathrm{i}\theta}\int_{C'}\exp\left(\pm \mathrm{i}\frac{1}{2}As^2\right)\,\mathrm{d}s$$

where C' is a deformation of the original contour which passes through the saddle point z_0 in the direction $\theta = (\pm\pi/2 - \alpha)/2$ or $\theta = (\pm 3\pi/2 - \alpha)/2$. The integral over C' can often be expressed in terms of the Fresnel integrals[29] $\mathcal{C}(x)$ and $\mathcal{S}(x)$, for example:

$$\int_{-\infty}^{\infty}\exp\left(\mathrm{i}\frac{1}{2}As^2\right)\,\mathrm{d}s = 2\sqrt{\frac{\pi}{A}}[\mathcal{C}(\infty) + \mathrm{i}\mathcal{S}(\infty)] = \sqrt{\frac{2\pi}{A}}\,\mathrm{e}^{\mathrm{i}\pi/4}.$$

[29]The Fresnel integrals $\mathcal{C}(x) \equiv \int_0^x \cos(\pi t^2/2)\,\mathrm{d}t$ and $\mathcal{S}(x) \equiv \int_0^x \sin(\pi t^2/2)\,\mathrm{d}t$. As $x \to \infty$, $\mathcal{C}(x) \to \frac{1}{2}$ and $\mathcal{S}(x) \to \frac{1}{2}$.

8.11 The analytic signal

Solutions to the partial differential equations of physics which display harmonic time dependence are of particular practical interest. These are of the form

$$\psi(\boldsymbol{r},t) = R(\boldsymbol{r})\mathrm{e}^{-\mathrm{i}\omega t}, \tag{8.40}$$

which, as we saw in Section 2.1.2, occur through the method of separation of variables in the reduction of the wave equation to the Helmholtz equation. Time-harmonic solutions are sufficiently common in signal processing that they have fostered their own terminology.

When the angular frequency ω is a fixed constant, the signal $\psi(\boldsymbol{r},t) = R(\boldsymbol{r})\mathrm{e}^{-\mathrm{i}\omega t}$ is said to be "monochromatic." This especially simple case – it focusses on a single component of the Fourier time transform representation of $\psi(\boldsymbol{r},t)$ – allows us to define the *phase* of the signal as

$$\phi = \arctan\frac{\mathrm{Im}\ R}{\mathrm{Re}\ R}.$$

phasors

The quantity ϕ is time-independent because $R(\boldsymbol{r})$ is. The factor $R(\boldsymbol{r})$ is called a "phasor," and (in the monochromatic case) the phasor is the central focus of analysis (because *all* system time dependence is of the form $\mathrm{e}^{-\mathrm{i}\omega t}$). The practical consequence of this approach is enormous: sines and cosines are much harder to work with than complex exponentials, and the (eventual) solution of physical interest (which should be real-valued) can be determined simply by discarding the imaginary part of the phasor.

In actuality, however, no physical signal is monochromatic. Natural and man-made signals must start at a certain time and, consequently, must be composed of more than one frequency (see Section 4.6).But the notions of phase and frequency are sufficiently powerful that we want to preserve them as primitive concepts, yet be appropriately generalized for real-world signals. One way is to extend the phasor definition. Let

$$\psi(\boldsymbol{r},t) = R(\boldsymbol{r},t)\mathrm{e}^{-\mathrm{i}\omega t}$$

(compare with Eq. (8.40)), where now $R(\boldsymbol{r},t)$ is assumed to be *slowly varying in time* (in comparison with the $\exp(-\mathrm{i}\omega t)$, dependence). The frequency spectrum of $R(\boldsymbol{r},t)$ will therefore be narrow (in comparison with the frequency ω), and such functions are referred to as "narrow band." The approximation has, effectively, split the field description into a product of a slowly-varying *envelope* $R(\boldsymbol{r},t)$ and a carrier $\exp(-\mathrm{i}\omega t)$.

A limitation of this approach is that we cannot simply apply phasor calculus to the analysis of the problem because the appropriate phasors are time-varying. In this section, we show how analytic function theory allows us to analyze signals with time-variable parameters in a way much more general than the narrow-band approximation.

8.11.1 The Hilbert transform

We start with the Fourier analysis of *real-valued* functions[30] $f(t)$. For such functions, using Eq. (4.38),

$$f(t) = \frac{1}{2\pi}\int_{-\infty}^{\infty} F(\omega)\mathrm{e}^{\mathrm{i}\omega t}\,\mathrm{d}\omega = f^*(t) = \frac{1}{2\pi}\int_{-\infty}^{\infty} F^*(\omega)\mathrm{e}^{-\mathrm{i}\omega t}\,\mathrm{d}\omega$$
$$= \frac{1}{2\pi}\int_{-\infty}^{\infty} F^*(-\omega)\mathrm{e}^{\mathrm{i}\omega t}\,\mathrm{d}\omega.$$

Thus, the Fourier transform of a real-valued function has the property[31] $F(\omega) = F^*(-\omega)$. This means that for $f(t)$ real valued, we can write its Fourier representation in the form

$$f(t) = \frac{1}{2\pi}\left(\int_{-\infty}^{0} F(\omega)\mathrm{e}^{\mathrm{i}\omega t}\,\mathrm{d}\omega + \int_{0}^{\infty} F(\omega)\mathrm{e}^{\mathrm{i}\omega t}\,\mathrm{d}\omega\right)$$
$$= \frac{1}{2\pi}\left(\int_{0}^{\infty} F(-\omega)\mathrm{e}^{-\mathrm{i}\omega t}\,\mathrm{d}\omega + \int_{0}^{\infty} F(\omega)\mathrm{e}^{\mathrm{i}\omega t}\,\mathrm{d}\omega\right)$$
$$= \frac{1}{2\pi}\int_{0}^{\infty} \left(F^*(\omega)\mathrm{e}^{-\mathrm{i}\omega t} + F(\omega)\mathrm{e}^{\mathrm{i}\omega t}\right)\mathrm{d}\omega = \frac{1}{2\pi}\int_{0}^{\infty} \left((F(\omega)\mathrm{e}^{\mathrm{i}\omega t})^* + F(\omega)\mathrm{e}^{\mathrm{i}\omega t}\right)\mathrm{d}\omega.$$

The final result here is an integral over positive frequencies only, and, consequently, *real-valued functions are completely determined by their positive frequency content* (the negative frequencies yield redundant information about $F(\omega)$).

To exploit this fact, we define a special Fourier-domain function:

$$Z_f(\omega) \equiv F(\omega) + \mathrm{sgn}(\omega)F(\omega) = \begin{cases} 2F(\omega) & \omega \geq 0 \\ 0 & \omega < 0, \end{cases}$$

where $F(\omega)$ is the Fourier transform of a real-valued function $f(t)$ and

$$\mathrm{sgn}(\omega) \equiv \begin{cases} -1 & \text{if } \omega < 0 \\ 1 & \text{if } \omega \geq 0 \end{cases}$$

is the "signum" function. Thus, $Z_f(\omega)$ has enough information to determine real-valued functions $f(t)$. The inverse Fourier transform of $Z_f(\omega)$ is (notation: $Z_f(\omega)$, $z_f(t)$)

$$z_f(t) = \frac{1}{2\pi}\int_{-\infty}^{\infty} Z_f(\omega)\mathrm{e}^{\mathrm{i}\omega t}\,\mathrm{d}\omega = \frac{1}{2\pi}\int_{-\infty}^{\infty} (F(\omega) + \mathrm{sgn}(\omega)F(\omega))\mathrm{e}^{\mathrm{i}\omega t}\,\mathrm{d}\omega$$
$$= f(t) + \frac{1}{2\pi}\int_{-\infty}^{\infty} \mathrm{sgn}(\omega)F(\omega)\mathrm{e}^{\mathrm{i}\omega t}\,\mathrm{d}\omega. \tag{8.41}$$

[30] We suppress the spatial dependence of f on $\boldsymbol{r}$ for the present discussion. Actual solutions will, of course, really be functions of the form $f(\boldsymbol{r}, t)$.

[31] Functions with the property $F(\omega) = F^*(-\omega)$ are said to be *Hermitian symmetric* or *conjugate symmetric*.

The remaining integral term in Eq. (8.41) is the inverse Fourier transform of the product of functions $\operatorname{sgn}(\omega)$ and $F(\omega)$, perfect for an application of the convolution theorem (see Section 4.7).The inverse Fourier transformation of $F(\omega)$ is clearly $f(t)$; the inverse Fourier transform of $\operatorname{sgn}(\omega)$ is found by applying a limit "trick." Write

$$\operatorname{sgn}(\omega) = \lim_{\alpha\to 0}\begin{cases} -\mathrm{e}^{\alpha\omega} & \text{if } \omega < 0 \\ \mathrm{e}^{-\alpha\omega} & \text{if } \omega \geq 0 \end{cases} \qquad (\alpha > 0)$$

so that its inverse Fourier transform, which we denote $\Sigma(t)$, is

$$\begin{aligned}
\Sigma(t) &\equiv \frac{1}{2\pi}\int_{-\infty}^{\infty} \operatorname{sgn}(\omega)\mathrm{e}^{\mathrm{i}\omega t}\,\mathrm{d}\omega \\
&= \frac{1}{2\pi}\lim_{\alpha\to 0}\left(-\int_{-\infty}^{0} \mathrm{e}^{(\alpha+\mathrm{i}t)\omega}\,\mathrm{d}\omega + \int_{0}^{\infty} \mathrm{e}^{-(\alpha-\mathrm{i}t)\omega}\,\mathrm{d}\omega\right) \\
&= \frac{1}{2\pi}\lim_{\alpha\to 0}\left(-\left.\frac{\mathrm{e}^{(\alpha+\mathrm{i}t)\omega}}{\alpha+\mathrm{i}t}\right|_{-\infty}^{0} - \left.\frac{\mathrm{e}^{-(\alpha-\mathrm{i}t)\omega}}{\alpha-\mathrm{i}t}\right|_{0}^{\infty}\right) \\
&= \frac{1}{2\pi}\lim_{\alpha\to 0}\left(-\frac{1}{\alpha+\mathrm{i}t} + \frac{1}{\alpha-\mathrm{i}t}\right) \\
&= \frac{1}{2\pi}\lim_{\alpha\to 0}\frac{2\mathrm{i}t}{\alpha^2+t^2} = \frac{\mathrm{i}}{\pi}\frac{1}{t}.
\end{aligned}$$

By the Fourier convolution theorem, $\Sigma(t) * f(t) = \mathcal{F}^{-1}(\operatorname{sgn}(\omega)F(\omega))$, and thus

$$\frac{1}{2\pi}\int_{-\infty}^{\infty} \operatorname{sgn}(\omega)F(\omega)\mathrm{e}^{\mathrm{i}\omega t}\,\mathrm{d}\omega = \int_{-\infty}^{\infty} \Sigma(t-t')f(t')\,\mathrm{d}t'.$$

Combining these results with Eq. (8.41), we have

$$z_f(t) = \frac{1}{2\pi}\int_{-\infty}^{\infty} Z_f(\omega)\mathrm{e}^{\mathrm{i}\omega t}\,\mathrm{d}\omega = f(t) + \frac{\mathrm{i}}{\pi}P\int_{-\infty}^{\infty}\frac{f(t')}{t-t'}\,\mathrm{d}t',$$

where the singularity in the integral at $t' = t$ is handled through the Cauchy principle value. The *Hilbert transform* of a real-valued function $f(t)$ is the real-valued function $\hat{f}(t)$ defined by

Hilbert transform

$$\hat{f}(t) = \{\mathcal{H}f\}(t) \equiv \frac{1}{\pi}P\int_{-\infty}^{\infty}\frac{f(t')}{t-t'}\,\mathrm{d}t'. \tag{8.42}$$

The quantity

$$z_f(t) = f(t) + \mathrm{i}\hat{f}(t)$$

analytic signal

is known as the *analytic signal* associated with the real function $f(t)$. The analytic signal is a complex-valued quantity having no negative frequency components; it provides a way to speak of the phase of a signal that is not monochromatic (see in the following).

Example. Hilbert transform of circular functions
The Hilbert transform of a function is conveniently calculated in the Fourier domain: owing to the derivation leading to Eq. (8.42), we can write

$$\mathrm{i}\{\mathcal{H}f\}(t) = \frac{1}{2\pi}\int_{-\infty}^{\infty} \mathrm{sgn}(\omega)F(\omega)\mathrm{e}^{\mathrm{i}\omega t}\,\mathrm{d}\omega.$$

Then, using our previous result for the Fourier transform of a cosine (Eq. (4.40)), we obtain

$$\begin{aligned}\{\mathcal{H}\cos(\omega_0 t)\} &= \frac{-\mathrm{i}}{2\pi}\int_{-\infty}^{\infty} \mathrm{sgn}(\omega)\pi(\delta(\omega-\omega_0)+\delta(\omega+\omega_0))\mathrm{e}^{\mathrm{i}\omega t}\,\mathrm{d}\omega \\ &= \frac{-\mathrm{i}}{2}[\mathrm{sgn}(\omega_0)\mathrm{e}^{\mathrm{i}\omega_0 t}+\mathrm{sgn}(-\omega_0)\mathrm{e}^{-\mathrm{i}\omega_0 t}] = \frac{-\mathrm{i}}{2}\mathrm{sgn}(\omega_0)[\mathrm{e}^{\mathrm{i}\omega_0 t}-\mathrm{e}^{-\mathrm{i}\omega_0 t}] \\ &= \frac{-\mathrm{i}}{2}\mathrm{sgn}(\omega_0)[2\mathrm{i}\sin(\omega_0 t)] = \mathrm{sgn}(\omega_0)\sin(\omega_0 t) = \sin(\omega_0 t).\end{aligned}$$

The Hilbert transform of a real-valued monochromatic solution to the wave equation of the form $f(t) = A\cos(\omega_0 t)$ will result in the imaginary part of $A\mathrm{e}^{-\mathrm{i}\omega_0 t}$. Consequently, the Hilbert transform is consistent with our phasor representation of wave fields.

8.11.2 Paley–Wiener and Titchmarsh theorems

Is the Fourier transform, $F(\omega)$, an analytic function when we let frequencies have complex values, $\omega = \omega_r + \mathrm{i}\omega_i$? The answer to this question is the province of the *Paley–Weiner theorem.*[32] The gist of the idea is as follows.

Consider the Fourier transform (Eq. (4.39)) defined as a function of complex frequency:

$$F(\omega) = \int_{-\infty}^{\infty} f(t)\mathrm{e}^{-\mathrm{i}(\omega_r+\mathrm{i}\omega_i)t}\,\mathrm{d}t = \int_{-\infty}^{\infty} f(t)\mathrm{e}^{-\mathrm{i}\omega_r t}\mathrm{e}^{\omega_i t}\,\mathrm{d}t. \tag{8.43}$$

The $\mathrm{e}^{\omega_i t}$ term is badly behaved when $\omega_i t \to \infty$ (the integral blows up). The only way Eq. (8.43) can exist mathematically is if arbitrarily large values of t can be *prevented* from contributing to the integral because the function $f(t)$ is such that $f(t) \to 0$ sufficiently rapidly as $t \to \infty$. A class of functions for which this condition is met are those with *finite support*, where $f(t)$ is bounded and *nonzero* only over a finite portion of the t-axis. Suppose $f(t)$ is zero[33] outside of the region $(-L, L)$. Equation (8.43) becomes

[32] A "Paley–Weiner theorem" is one of a *host* of theorems developed by R. Paley and N. Weiner relating the properties of a function $f(t)$ as $t \to \infty$ to the analyticity of its Fourier transform [27].

[33] A function f with finite support cannot be analytic because the zeros of analytic functions occur only at isolated points (Section 8.6): If f is analytic and zero over a finite line, it must be zero everywhere.

$$F(\omega) = F_r + \mathrm{i}F_i = \int_{-L}^{L} f(t)\cos(\omega_r t)\mathrm{e}^{\omega_i t}\,\mathrm{d}t - \mathrm{i}\int_{-L}^{L} f(t)\sin(\omega_r t)\mathrm{e}^{\omega_i t}\,\mathrm{d}t. \tag{8.44}$$

Differentiating F_r in Eq. (8.44) with respect to ω_r, and F_i with respect to ω_i:

$$\frac{\partial F_r}{\partial \omega_r} = \int_{-L}^{L} f(t)(-t)\sin(\omega_r t)\mathrm{e}^{\omega_i t}\,\mathrm{d}t = \frac{\partial F_i}{\partial \omega_i}. \tag{8.45}$$

In a similar way, by differentiating Eq. (8.44),

$$\frac{\partial F_r}{\partial \omega_i} = -\frac{\partial F_i}{\partial \omega_r}. \tag{8.46}$$

Equations (8.45) and (8.46) are the Cauchy–Riemann conditions. Thus, $F(\omega)$ is analytic for all finite complex ω: *the Fourier transform of a function with finite support is an entire function.* This result is generally known as the Paley–Wiener theorem.

Of course, the argument works both ways, and if $F(\omega)$ is zero except on a finite set $\omega \in (\omega_1, \omega_2)$ (i.e. $F(\omega)$ is "band-limited"), then the $f(t)$ formed by the inverse Fourier transform is analytic everywhere.[34]

It's worth mentioning that because an object function $f(t)$ with finite support in the time (or space) domain is analytic everywhere in the Fourier domain, then its Fourier transform cannot be zero on a finite region of the Fourier domain (or else it would be zero everywhere). Consequently, any linear and shift-invariant measurement kernel that attempts to faithfully collect all the information about $f(t)$ *cannot be zero on any finite region in the Fourier domain* because the convolution theorem requires

$$M(\omega) = K(\omega)F(\omega)\ .$$

(These terms are introduced in Section 4.7.)But nature precludes this behavior in a measurement kernel because such data collection systems would, for example, be required to respond instantaneously to an arbitrarily fast change in a (time-varying) object function (i.e. one with arbitrarily large frequency). Realizable systems are said to be "band-limited" and will have finite support *in the Fourier domain.*

A deeper connection between physical signals and analyticity – and one which helps explain why analyticity is so important – is provided by the Titchmarsh theorem[35]:

Theorem 8.11.1. (Titchmarsh theorem) *If $f(t)$ is square-integrable over the real t-axis (i.e. if $f(t)$ has finite energy), then the following statements are equivalent.*

[34] Most functions we encounter in physics can be treated as analytic. Even for functions that start and stop in the time domain, the Fourier transforms of such functions (in the frequency domain) are analytic.

[35] What we have referred to as "Titchmarsh's theorem" is contained in theorem 95 of [28, p. 128].

1. *The Fourier transform of $f(t)$ is zero for $\omega < 0$.*
2. *Replacing t by $z = t + iy$, the function $f(z)$ is analytic in the complex plane for $y > 0$ and approaches $f(t)$ almost everywhere as $y \to 0$.*
3. *The real and imaginary parts of $f(z)$ are Hilbert transforms of each other.*

8.11.3 Is the analytic signal, analytic?

We can examine the analyticity of $z_f(t) = f(t) + i\hat{f}(t)$ by replacing t in Eq. (8.41) with $t + iy$ and using the fact that $Z_f(\omega)$ is zero for $\omega < 0$. Then

$$\begin{aligned} z_f(t, y) &= \frac{1}{2\pi}\int_0^\infty Z_f(\omega)e^{i\omega t}e^{-\omega y}\,d\omega \\ &= \frac{1}{2\pi}\int_0^\infty Z_f(\omega)\cos(\omega t)e^{-\omega y}\,d\omega + \frac{i}{2\pi}\int_0^\infty Z_f(\omega)\sin(\omega t)e^{-\omega y}\,d\omega. \end{aligned}$$

These integrals exist when $y > 0$. Moreover, a straightforward calculation shows that the Cauchy–Riemann conditions are satisfied by z_f (when $y > 0$). Consequently, the function $z_f(t, y)$ *is analytic on the upper half of the complex plane* (for which y is positive).

Example. Envelope and phase

Using Euler's formula, it is always possible to rewrite the analytic signal in polar form: $z_f(t) = f(t) + i\hat{f}(t) = |z_f(t)|e^{i\Phi(t)}$. The amplitude of the analytic signal is called the *complex envelope:*

$$|z_f(t)| = \sqrt{f^2(t) + \hat{f}^2(t)}. \tag{8.47}$$

The argument of the complex exponential defines the *phase:*

$$\Phi(t) = \arctan\frac{\hat{f}(t)}{f(t)} \tag{8.48}$$

Because the Hilbert transform of $\cos\omega_0 t$ yields $\sin\omega_0 t$ (and vise-versa), it can be implemented as a simple phase shift (i.e. by multiplying by $e^{\pm i\pi/2}$) *when the function $f(t)$ is of the form of $f(t) = a\cos\omega_0 t + b\sin\omega_0 t$.* Such functions will be described by only one Fourier component and will have zero bandwidth (they are said to be "pure tones").

Example. Dispersion

The Titchmarsh theorem works for frequency domain functions as well. For a homogeneous, isotropic, and linear dielectric, we know from electrodynamics that the electric and displacement fields are proportionally related as

$$\boldsymbol{D}(\omega) = \epsilon(\omega)\boldsymbol{E}(\omega)$$

where ϵ characterizes the material and is known as the permittivity. (A product in the *frequency domain.*)

But in the time domain, the permittivity is zero at times for which the electric field is turned off. (The system is said to be *causal.*) This means that the inverse Fourier transform of $\epsilon(\omega)$ is zero for $t < 0$ and, consequently, the real and imaginary parts of $\epsilon(\omega)$ are Hilbert transforms of each other. (A result known as the *Kramers-Kronig relations.*)

8.12 The Laplace transform

The Laplace transform of a function $f(t)$ is denoted $F(s) = \mathcal{L}\{f\}(s)$ and is a special case of Eq. (8.43) with $\omega_r = 0$ and $\omega_i = -s$:

$$F(s) = \int_0^\infty \mathrm{e}^{-st} f(t)\, \mathrm{d}t\, .$$

This integral transform is defined provided $f(t)$ does not grow faster than $\exp(s_0 t)$. Several examples of Laplace transforms are shown in Table 8.1.

In addition, the Laplace transform of the n-th derivative of $f(t)$ obeys

$$\mathcal{L}\{f^{(n)}(t)\} = s^n\, \mathcal{L}\{f(t)\} - s^{n-1} f(0) - s^{n-2} f'(0) - \cdots - f^{(n-1)}(0)\, , \tag{8.49}$$

(where $f^{(n)}(t)$ means "differentiate $f(t)$ n-times").[36] And, like the Fourier transform, the Laplace transform obeys a convolution theorem:

$$\mathcal{L}\{f(t) * g(t)\} = F(s)G(s) \tag{8.50}$$

where

$$f(t) * g(t) \equiv \int_0^t f(t')g(t-t')\, \mathrm{d}t'\, . \tag{8.51}$$

Because of the form of the transform of a derivative, Eq. (8.49), the Laplace transform can be used to express an initial value problem (consisting of an ordinary differential equation and the initial conditions of the system it describes) in the form of an algebraic equation. Consequently, Laplace transforms are well-suited for analyzing the time-evolution of dynamical systems.

The utility of the Laplace transform is evident from the following example.

[36]The Laplace transform of the integral of f obeys

$$\mathcal{L}\left\{\int_0^t f(t')\, \mathrm{d}t'\right\} = \frac{1}{s}\mathcal{L}\{f(t)\}\, .$$

Table 8.1 Some important Laplace transforms.

$f(t)$	$F(s)$		$f(t)$	$F(s)$
1	$\frac{1}{s}$		$\cos(kt)$	$\frac{s}{s^2+k^2}$
$t^n, \quad n=0,1,2,\ldots$	$\frac{n!}{s^{n+1}}$,	$\text{Re}\, s > 0$	$\sin(kt)$	$\frac{k}{s^2+k^2}$
e^{kt}	$\frac{1}{s-k}$		$f(t)\, e^{at}$	$F(s-a)$

Example. The damped oscillator.

$$m\ddot{x}(t) + b\dot{x}(t) + kx(t) = 0, \quad x(0) = x_0, \dot{x}(0) = 0.$$

The Laplace transform of the differential equation yields

$$ms^2X(s) - msx_0 + bsX(s) - bx_0 + kX(s) = 0$$

$$\Rightarrow X(s) = x_0 \frac{ms+b}{ms^2+bs+k} = x_0 \frac{s+b/m}{(s+b/2m)^2+\omega_1^2}$$
$$= x_0 \left\{ \frac{s+b/2m}{(s+b/2m)^2+\omega_1^2} + \frac{b/2m}{(s+b/2m)^2+\omega_1^2} \right\}$$

where $\omega_1^2 = k/m - (b/2m)^2$ is assumed to be > 0.

Then, the inverse transform yields (from Table 8.1)

$$x(t) = x_0\, e^{-\frac{b}{2m}t} \left[\cos(\omega_1 t) + \frac{b}{2m\omega_1} \sin(\omega_1 t) \right].$$

This oscillator will be damped provided $b > 0$.

Inverse Laplace transform

In the last example, we saw that the original initial value problem was transformed, in the Laplace domain, to an algebraic equation which is easily solved. The more algebraically challenging step in the solution required us to use Table 8.1 to inverse transform our solution back to the time domain. This step usually requires heroic application of partial fractions. But there is another way.

We begin by observing that the Laplace transform is related to the Fourier transform when $f(t) = 0$ for $t < 0$ and s is pure imaginary. (The idea here is that we know what the inverse Fourier transform looks like.) Laplace transforms, however, allow $f(t)$ to grow exponentially, and this is not allowed in Fourier transforms (which, for the most part, requires functions to be integrable: i.e. $\int |f(t)|\, dt < \infty$). We can deal with this limitation by examining, instead, the function

$$g(t) = e^{-\gamma t} f(t) \quad \text{where } \gamma > s_0$$

where $g(t) = 0$ for $t < 0$. Because $f(t)$ does not grow faster than $\exp(s_0 t)$, the function $g(t)$ is guaranteed to be integrable and its Fourier transform

$$G(\omega) = \int_{-\infty}^{\infty} g(t) e^{-i\omega t}\, dt$$

exists. The inverse Fourier transform of $G(\omega)$ is

$$g(t) = \frac{1}{2\pi} \int_{-\infty}^{\infty} G(\omega) e^{i\omega t}\, d\omega$$

from which we conclude

$$\begin{aligned} e^{-\gamma t} f(t) &= \frac{1}{2\pi} \int_{-\infty}^{\infty} G(\omega) e^{i\omega t}\, d\omega = \frac{1}{2\pi} \int_{-\infty}^{\infty} \left(\int_{-\infty}^{\infty} g(t') e^{-i\omega t'}\, dt' \right) e^{i\omega t}\, d\omega \\ &= \frac{1}{2\pi} \int_{-\infty}^{\infty} e^{i\omega t} \int_{0}^{\infty} g(t') e^{-i\omega t'}\, dt'\, d\omega \quad (\text{because } g(t) = 0 \text{ for } t < 0) \\ &= \frac{1}{2\pi} \int_{-\infty}^{\infty} e^{i\omega t} \int_{0}^{\infty} f(t') e^{-\gamma t'} e^{-i\omega t'}\, dt'\, d\omega \\ &= \frac{1}{2\pi} \int_{-\infty}^{\infty} e^{i\omega t} \int_{0}^{\infty} f(t') e^{-(\gamma + i\omega) t'}\, dt'\, d\omega. \end{aligned}$$

Now, let $s = \gamma + i\omega \Rightarrow d\omega = ds/i$ (γ is a constant), to obtain

$$\begin{aligned} e^{-\gamma t} f(t) &= \frac{1}{2\pi i} \int_{\gamma - i\infty}^{\gamma + i\infty} e^{(s-\gamma)t} \int_{0}^{\infty} f(t') e^{-st'}\, dt'\, ds \\ &= \frac{1}{2\pi i} \int_{\gamma - i\infty}^{\gamma + i\infty} e^{(s-\gamma)t} F(s)\, ds = \frac{e^{-\gamma t}}{2\pi i} \int_{\gamma - i\infty}^{\gamma + i\infty} F(s) e^{st}\, ds. \end{aligned}$$

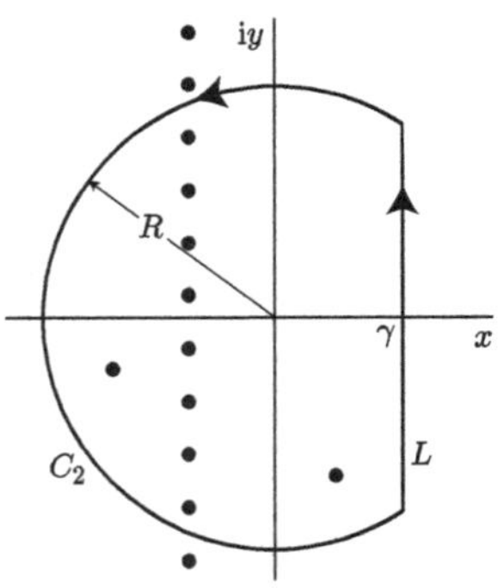

Figure 8.10 Contour for calculating the inverse Laplace transform.

This result yields the inverse Laplace transform:

$$f(t) = \frac{1}{2\pi i} \int_{\gamma - i\infty}^{\gamma + i\infty} F(s) e^{st}\, ds = \sum (\text{residues for } \text{Re}(s) < \gamma), \tag{8.52}$$

a contour integral along the vertical path L with $\text{Re}(s) = \gamma$ in the complex plane (see Figure 8.10). By construction γ is chosen so that $e^{-\gamma t} f(t)$ is guaranteed to decay. This choice is equivalent to requiring that all singularities of $F(s)$ appear to the left of the contour path. Equation (8.52) is known as the Bromwich integral.

Example. Find the function $f(t)$ which has the Laplace transform $F(s) = a/(s^2 - a^2)$, valid for $s > a$. It's straightforward to show that $f(t) = \sinh at$ has the Laplace transform in question. $F(s)$ has simple poles at $s = \pm a$. The residues of $e^{st} F(s)$ at $s = \pm a$ are $\text{Res}(\pm a) = \pm \frac{1}{2} e^{\pm at}$. Equation (8.52) yields $f(t) = \sinh at$.

Summary

- In this chapter, we examined the curious and useful behavior of analytic functions of a complex variable. These functions are infinitely smooth (in the sense that they are infinitely differentiable throughout their domain of analyticity) and, as a consequence, can be represented as a Taylor series. If the function fails to be analytic at isolated points, then it may be represented as a Laurent series, Eq. (8.28). The analyticity of a function can be determined from the Cauchy–Riemann conditions, Eq. (8.4).
- *Analyticity* imposes some pretty severe restrictions on functions that fit the classification:
  - They can vanish only at isolated points;
  - Their values within a region bounded by a contour C are determined by their values on C.
  - Line integrals of complex functions which are analytic in a region except at isolated points (singularities) are often determined by their local behavior at these singular points through the residue theorem, Eq. (8.7.1).
- The residue theorem can be used to represent many integrals in terms of the integrand's behavior within a closed contour in the complex plane (see Section 8.8). Moreover, the value of integrals can sometimes be approximated by carefully deforming the integration path within an analytic region (see Section 8.10).
- The phasor representation of (monochromatic) time-harmonic signals can be generalized to arbitrary time varying signals by introducing the "analytic signal." Real-valued, finite energy signals $f(t)$ have analytic extension $z_f(t)$ whose real and imaginary parts are Hilbert transform pairs, which can be used to define a signal envelope and phase. The Fourier transform $Z_f(\omega)$ of the analytic signal obeys: $Z_f(\omega) = 0$ for $\omega < 0$.
- The Bromwich integral, Eq. (8.52), yields a representation of the inverse Laplace transform (which is useful in the solution of initial-value ordinary differential equations [ODEs]).

Exercises

8.1. What curves in the complex plane are implied by the relations

(a) $\left|\frac{z-1}{z+1}\right| = 1$

(b) $|z-1| = 2$.

8.2. Show that if complex functions $f_1(z)$ and $f_2(z)$ are differentiable at point z, the product rule of calculus applies at that point. That is, show that

$$\frac{\mathrm{d}}{\mathrm{d}z}[f_1(z)f_2(z)] = f_1(z)f_2'(z) + f_1'(z)f_2(z).$$

Hint: Systematically apply Eq. (8.2), the definition of differentiability.

8.3. Show that the real and imaginary parts of $f(z) = \sin z$ satisfy the Cauchy–Riemann conditions and the Laplace equation. Hint: Show that $\sin(x + \mathrm{i}y) = \sin x \cosh y + \mathrm{i} \cos x \sinh y$.

8.4. Is the function $f(x, y) = x^2 + y + \mathrm{i}(y^2 - x)$ analytic?

8.5. Is $f(x, y) = x^3 - 3xy^2 + \mathrm{i}(3x^2y - y^3)$ analytic? If yes, express $f(x, y)$ as a function of z only.

8.6. For $f(z)$ analytic throughout a domain $\mathcal{D}$, show that

(a) If f is real valued, then f must be a constant.

(b) If $|f|$ is a constant, then f is a constant.

8.7. Using the definition of Riemann sum, Eq. (8.8), show that if one integrates over a given path, once in one direction and then in the opposite direction, the two integrals are the same except for a sign, i.e.

$$\int_{z_0,C}^{Z} = -\int_{Z,C}^{z_0}.$$

8.8. Evaluate the integral

$$\int_{-2}^{2} \frac{2z - 3}{z} \, \mathrm{d}z$$

when the path of integration is the upper half of the circle $|z| = 2$. A: $8 + 3\mathrm{i}\pi$

8.9. Show explicitly that $\oint_C z \, \mathrm{d}z = 0$ where C is around the unit square having one vertex at the origin and whose sides are parallel to the coordinate axes.

8.10. Evaluate $\int_0^{2+\mathrm{i}} z^2 \, \mathrm{d}z$ where the integration path is along the straight line from 0 to $2 + \mathrm{i}$. A: $\frac{1}{3}(2 + \mathrm{i})^3 = \frac{1}{3}(2 + 11\mathrm{i})$. Try some other integration paths between the same endpoints. What value of the integral do you find?

8.11. We know from Cauchy's theorem (Eq. (8.15)), that for a function $f(z)$ analytic within a simply connected region $\mathcal{D}$, $\oint_C f(z) \, \mathrm{d}z = 0$, where C is a closed path lying within $\mathcal{D}$. Show that an implication of Cauchy's theorem is that the integral $\int_{z_0}^{Z} f(z) \, \mathrm{d}z$ has the same value along every integration path joining z_0 and Z that lies entirely within $\mathcal{D}$. Hint: Let C_1 and C_2 be arbitrary paths extending from z_0 to Z and lying within $\mathcal{D}$. Let $-C_2$ be joined to C_1 to form a closed path C.

8.12. Show, for α an arbitrary real constant, that

$$\int_{-\infty}^{\infty} \mathrm{e}^{-(x+\mathrm{i}\alpha)^2} \, \mathrm{d}x = \sqrt{\pi}.$$

We'll guide you through this problem in steps.

(a) First show that $\int_{-\infty}^{\infty} \mathrm{e}^{-x^2} \, \mathrm{d}x = \sqrt{\pi}$. Hint: There is a standard trick for evaluating this integral, one that you may already know. Define $I \equiv \int_{-\infty}^{\infty} \mathrm{e}^{-x^2} \, \mathrm{d}x$. Consider $I^2 = \int_{-\infty}^{\infty} \int_{-\infty}^{\infty} \mathrm{e}^{-(x^2+y^2)} \, \mathrm{d}x \, \mathrm{d}y$. Convert from Cartesian coordinates to plane polar coordinates with $r^2 \equiv x^2 + y^2$ and $\mathrm{d}x \, \mathrm{d}y = r \, \mathrm{d}r \, \mathrm{d}\theta$.

(b) The function e^{-z^2} is an entire function, and thus from Cauchy's theorem, $\oint \mathrm{e}^{-z^2} \, \mathrm{d}z = 0$ for any closed contour in the complex plane. Let the contour consist a rectangle with a segment along the x-axis, from $-R$ to R, a segment from $z = R$ to $z = R + \mathrm{i}\alpha$, from $z = R + \mathrm{i}\alpha$

to $z=-R+\mathrm{i}\alpha$, and finally from $z=-R+\mathrm{i}\alpha$ to $z=-R$. In the limit $R\to\infty$, show that

$$\int_{-\infty}^{\infty} \mathrm{e}^{-x^2}\,\mathrm{d}x = \int_{-\infty}^{\infty} \mathrm{e}^{-(x+\mathrm{i}\alpha)^2}\,\mathrm{d}x.$$

8.13. Show that the Fourier transform of a Gaussian function is itself a Gaussian, i.e. show that the Fourier transform of $f(x)=\mathrm{e}^{-x^2}$ is $F(\omega)=\sqrt{\pi}\exp(-\omega^2/4)$, using the definition of Fourier transform, Eq. (4.39). Hint: The most direct way to show this result is to complete the square in the argument of the exponential and make use of the result of Exercise 8.12. One can also prove this result through integration by parts. Yet another way is to separate the complex exponential into real and imaginary parts and integrate term by term the Taylor series expansion of the cosine function.

8.14. Evaluate the integral

$$I=\int_0^{2\pi}\frac{1}{2+\cos t}\,\mathrm{d}t. \qquad \mathrm{A:}=\frac{2}{3}\pi\sqrt{3}.$$

8.15. Show that

$$\int_0^{\infty}\frac{\sqrt{x}}{x^2+1}=\frac{\pi}{2}\sqrt{2}.$$

8.16. Show that the step function $\theta(t)$ can be represented by a complex integral (for $\epsilon>0$),

$$\frac{1}{2\pi\mathrm{i}}\int_{-\infty}^{\infty}\frac{\mathrm{e}^{\mathrm{i}\omega t}}{\omega-\mathrm{i}\epsilon}\,\mathrm{d}\omega=\theta(t)=\begin{cases}1 & t>0\\ 0 & t<0.\end{cases}$$

The function is discontinuous at $t=0$. Hint: Close the integral in the upper half plane, then in the lower half plane.

8.17. Show that

$$\oint_{|z|=1}\frac{\cosh z}{z^4}\,\mathrm{d}z=0.$$

8.18. Find the location and order of the branch points for $f(z)=(z-1)^{1/3}$. Suggest a branch cut.

8.19. Show that the Mittag-Leffler expansion of $f(z)=\cot z-z^{-1}$ is

$$\cot z-\frac{1}{z}=2z\sum_{n=1}^{\infty}\frac{1}{z^2-(n\pi)^2}.$$

Hint: You should find that $f(0)=0$ and that the residue at all poles, $\mathrm{Res}(n\pi)=1$.

8.20. We can use the result of the previous problem to derive the *infinite product representation* of $\mathrm{sinc}(x)$.

(a) First, by a cosmetic change on the result of the previous problem, show that

$$\pi\cot\pi x=\frac{1}{x}+2x\sum_{n=1}^{\infty}\frac{1}{x^2-n^2}.$$

(b) Then show that

$$\frac{\mathrm{d}}{\mathrm{d}x}\ln(\sin\pi x) = \frac{1}{x} + \sum_{n=1}^{\infty}\frac{\mathrm{d}}{\mathrm{d}x}\ln\left(1-\frac{x^2}{n^2}\right).$$

(c) Now integrate the formula you just derived to show that

$$\sin\pi x = Cx\prod_{n=1}^{\infty}\left(1-\frac{x^2}{n^2}\right).$$

where C is a constant.

(d) Use the small-x approximation for the sine function to argue that $C = \pi$, and thus

$$\sin\pi x = \pi x\prod_{n=1}^{\infty}\left(1-\frac{x^2}{n^2}\right). \tag{8.53}$$

8.21. Evaluate the Laplace transforms shown in Table 8.1.

8.22. Derive Eq. (8.50), the Laplace transform of the convolution of two functions.

8.23. Show, using the Rodrigues formula for the Legendre polynomial,

$$P_n(x) = \frac{1}{2^n n!}\frac{\mathrm{d}^n}{\mathrm{d}x^n}(x^2-1)^n,$$

and the Cauchy integral formula

$$\left.\frac{\mathrm{d}^n}{\mathrm{d}z^n}f(z)\right|_{z=z_0} = \frac{n!}{2\pi\mathrm{i}}\oint_C \frac{f(z)}{(z-z_0)^{n+1}}\,\mathrm{d}z,$$

that we obtain an integral representation for Legendre polynomials

$$P_n(z) = \frac{1}{2^n 2\pi\mathrm{i}}\oint_C \frac{(t^2-1)^n}{(t-z)^{n+1}}\,\mathrm{d}t,$$

where the contour C makes one counter-clockwise turn around z. This result is known as the *Schläfli integral representation* for $P_n(z)$.

8.24. Let the contour in the Schläfli integral be a circle centered on z with radius $|\sqrt{z^2-1}|$. Let $t = z + \sqrt{z^2-1}\mathrm{e}^{\mathrm{i}\phi}$, where ϕ runs from 0 to 2π. Show that an equivalent integral representation for the Legendre polynomials is

$$P_n(z) = \frac{1}{\pi}\int_0^{\pi}(z+\sqrt{z^2-1}\cos\phi)^n\,\mathrm{d}\phi.$$

This result is known as the *Laplace integral representation* of $P_n(z)$.

8.25. Consider the real-valued signal $f(t) = a(t)\cos(\omega_0 t)$ with $\omega_0 > 0$. Show that if the Fourier transform $A(\omega)$ of $a(t)$ is band limited (so that $A(\omega) = 0$ for $|\omega| > \omega_0$) then the complex signal

$$z_f(t) = a(t)\mathrm{e}^{\mathrm{i}\omega_0 t}$$

is the analytic signal.

8.26. Three species of radioactive nuclei have a parent-daughter relationship described by the coupled differential equations:

$$\frac{dN_1}{dt} = -\lambda_1 N_1$$
$$\frac{dN_2}{dt} = \lambda_1 N_1 - \lambda_2 N_2$$
$$\frac{dN_3}{dt} = \lambda_2 N_2 - \lambda_3 N_3,$$

where $\lambda_1, \lambda_2, \lambda_3$ are known constants. If at $t = 0$, $N_1(0) = N$, $N_2(0) = 0$, and $N_3(0) = 0$, find $N_1(t)$, $N_2(t)$, and $N_3(t)$ using Laplace transforms.

8.27. For the Laplace transform $F(s)$ of the function $f(t)$, show that

$$\lim_{s\to\infty} sF(s) = \lim_{t\to 0^+} f(t).$$

Hint: Assume that $f(t)$ has a power-series representation, $f(t) = \sum_{n=0}^{\infty} a_n t^n$.

9

Inhomogeneous differential equations

In this chapter, we take up *inhomogeneous* problems, partial differential equations (PDEs) with nonzero source terms (see Table 3.1). We develop the method of *Green functions* for handling such problems. Green functions feature prominently in advanced physics, such as quantum field theory.[1] One can safely say that the analysis of many problems in science and engineering would not be possible without Green functions. We begin with inhomogeneous ordinary differential equations (ODEs) and then consider inhomogeneous PDEs.

9.1 The method of Green functions

Consider an inhomogeneous differential equation defined on $a \leq x \leq b$:

$$\mathcal{L}f \equiv \frac{\mathrm{d}}{\mathrm{d}x}\left(p(x)\frac{\mathrm{d}f(x)}{\mathrm{d}x}\right) + r(x)f(x) = \psi(x), \tag{9.1}$$

where ψ is a known function and we've used the notation of linear operators introduced in Section 1.4. There's no loss of generality in writing $\mathcal{L}$ in self-adjoint form, as in Eq. (9.1), because any second-order linear differential operator can be put in that form (Section 2.1.2). For $\mathcal{L}$ self-adjoint, we have from Eq. (2.21) that for any functions[2] f, g:

$$\int_a^b [g\mathcal{L}f - (\mathcal{L}g)f]\,\mathrm{d}x = p(gf' - g'f)\Big|_a^b. \tag{2.21}$$

[1]The list of topics making use of Green functions could be expanded greatly. Hopefully you get the point.

[2]Equation (2.21) is a property of self-adjoint operators $\mathcal{L}$; it holds regardless of the type of differential equation that $\mathcal{L}f$ might be part of, homogeneous or inhomogeneous.

Mathematical Methods in Physics, Engineering, and Chemistry, First Edition.
Brett Borden and James Luscombe.

When f and g are restricted to satisfy the boundary conditions discussed in Section 2.2, the right side of Eq. (2.21) vanishes.

To solve Eq. (9.1), we first define a new function *associated with the operator* $\mathcal{L}$, the Green function $G(x, x')$ – a function of *two* variables – such that

$$\mathcal{L}_x G(x, x') = \delta(x - x'), \tag{9.2}$$

where we've written $\mathcal{L}_x$ to indicate that $\mathcal{L}$ acts only on the variable x in $G(x, x')$. The significance of the Green function is that *if one has found* $G(x, x')$ *satisfying Eq.* (9.2), *then one has found a solution of Eq.* (9.1) *in the form of an integral:*

$$f(x) = \int_a^b G(x, x')\psi(x')\, \mathrm{d}x'. \tag{9.3}$$

To show that Eq. (9.3) solves Eq. (9.1), use Eq. (9.2):

$$\mathcal{L}f = \int_a^b \mathcal{L}_x G(x, x')\psi(x')\, \mathrm{d}x' = \int_a^b \delta(x - x')\psi(x')\, \mathrm{d}x' = \psi(x).$$

two-step process

The Green function is in essence the *inverse*[3] of $\mathcal{L}$ (a differential operator) in the form of an integral. The Green function method is a two-step process: Find the Green function, then do an integral. *Finding* the Green function can be a laborious task, but (fortunately) Green functions for many differential operators are known.

It's useful to think about the function f in Eq. (9.1) as the *response* of a system to the source function ψ. If, for example, Eq. (9.1) describes the flow of heat in a system, $f(x)$ might represent the temperature distribution arising from a heat source $\psi(x)$. Given this interpretation, $G(x, x')$ can be considered the response at x to a *point source* at x' with *unit strength.* Equation (9.3) then provides, by superposition, the net response at x to a distribution of point sources each having strength $\psi(x')$.

9.1.1 Boundary conditions

The differential equation satisfied by $G(x, x')$, Eq. (9.2), must be supplemented with boundary conditions. For ODEs, *Green functions can always be taken to satisfy homogeneous boundary conditions.*[4] To show this, assume that the solution f to Eq. (9.1) satisfies the Dirichlet condition $f(a) = 0$ at $x = a$. We would then have from Eq. (9.3),

$$0 = \int_a^b G(a, x')\psi(x')\, \mathrm{d}x'.$$

[3]Equation (9.1) has the form $\mathcal{L}f = \psi$; its solution can be written $f = \mathcal{L}^{-1}\psi \equiv \int G(x, x')\psi(x')\, \mathrm{d}x'$. The inverse is unique: G is specific to the operator $\mathcal{L}$.

[4]Homogeneous boundary conditions were introduced in Section 2.2. The boundary conditions for Green functions associated with PDEs are more involved.

The only way such a requirement can be met for all functions ψ is if $G(a,x') = 0$. The same holds for homogeneous mixed boundary conditions, Eq. (2.19):

boundary conditions on $G(x,x')$

$$\alpha_1 f(a) + \alpha_2 f'(a) = 0 \quad \Longrightarrow \quad \alpha_1 G(a,x') + \alpha_2 \frac{\mathrm{d}}{\mathrm{d}x} G(x,x')\bigg|_{x=a} = 0. \tag{9.4}$$

For solutions of Eq. (9.1) satisfying homogeneous boundary conditions, *the boundary conditions are built into the associated Green function.*

Remarkably, for solutions of Eq. (9.1) satisfying *nonhomogeneous* boundary conditions, we can still use Green functions satisfying homogeneous boundary conditions.[5] To show this, express the solution of $\mathcal{L}f = \psi$ as[6] $f = u + v$, where u is the solution of $\mathcal{L}u = \psi$ *subject to homogeneous boundary conditions* and v is such that $\mathcal{L}v = 0$ *and* satisfies the same nonhomogeneous boundary conditions as does f. In that case, Eq. (9.3) must be replaced with

$$f(x) = \int_a^b G(x,x')\psi(x')\,\mathrm{d}x' + v(x), \tag{9.5}$$

where $G(x,x')$ in Eq. (9.5) satisfies homogeneous boundary conditions. The solutions of second-order inhomogeneous differential equations are *unique*: There is at most one solution of Eq. (9.1) satisfying given boundary conditions [19, p. 34]. If you've found a solution to an inhomogeneous ODE, you've found *the* solution.

uniqueness

9.1.2 Reciprocity relation: $G(x,x') = G(x',x)$

In Eq. (2.21), let $f(x) = G(x,x')$ and $g(x) = G(x,x'')$. We then have, by using Eq. (9.2):

$$G(x',x'') - G(x'',x') = p(x)\left[G(x,x'')\frac{\mathrm{d}}{\mathrm{d}x}G(x,x') - G(x,x')\frac{\mathrm{d}}{\mathrm{d}x}G(x,x'')\right]\Bigg|_a^b. \tag{9.6}$$

For $G(x,x')$ satisfying homogeneous boundary conditions, Eq. (9.4), the right side of Eq. (9.6) vanishes, implying the *reciprocity relation*,

reciprocity relation

$$G(x,x') = G(x',x). \tag{9.7}$$

Conversely, if the reciprocity relation is assumed on the left of Eq. (9.6), the right side must vanish, implying that $G(x,x')$ satisfies homogeneous boundary conditions.

---

[5] $G(x,x')$ is often *defined* as satisfying Eq. (9.2) subject to homogeneous boundary conditions.

[6] We're free do this because $\mathcal{L}$ is linear.

The reciprocity relation can be used to derive the form of Eq. (9.5). In Eq. (2.21), let $g = G(x, x')$ and let f be such that $\mathcal{L}f = \psi$. We find the intermediate result

$$f(x') = \int_a^b G(x,x')\psi(x)\,\mathrm{d}x - p(x)\left[G(x,x')f'(x) - f(x)\frac{\mathrm{d}}{\mathrm{d}x}G(x,x')\right]\Bigg|_{x=a}^{x=b}.$$

Interchange $x \leftrightarrow x'$, use Eq. (9.7), and we have a solution in the form of Eq. (9.5),

$$f(x) = \int_a^b G(x,x')\psi(x')\,\mathrm{d}x' - p(x')\left[G(x,x')f'(x') - f(x')\frac{\mathrm{d}}{\mathrm{d}x'}G(x,x')\right]\Bigg|_{x'=a}^{x'=b}. \tag{9.8}$$

9.1.3 Matching conditions

From Eqs. (9.1) and (9.2), we have an explicit expression for the differential equation satisfied by $G(x, x')$:

$$\frac{\mathrm{d}}{\mathrm{d}x}\left(p(x)\frac{\mathrm{d}}{\mathrm{d}x}G(x,x')\right) + r(x)G(x,x') = \delta(x-x'). \tag{9.9}$$

Equation (9.9) is a relation between $G(x, x')$, its first two derivatives, and $\delta(x - x')$. When $x \neq x'$, Eq. (9.9) is homogeneous and there's not much to learn. We need to understand how $G(x, x')$ behaves as $x \to x'$. Moreover, since we know (from Eq. (4.35)) that the derivative of a discontinuity at $x = x'$ is proportional to $\delta(x - x')$, it's reasonable to expect $\mathrm{d}G/\,\mathrm{d}x$ to be discontinuous at $x = x'$ (so the second derivative of G would result in the delta function in Eq. (9.9)). Note that if G was discontinuous at $x = x'$ then, while the first derivative of G would yield a delta function, the second derivative would result in the derivative of $\delta(x - x')$, and we see no evidence of that behavior in Eq. (9.9). Consequently, we require $G(x, x')$ to be continuous at $x = x'$ and its first derivative to be discontinuous at the same point.

Let's integrate[7] Eq. (9.9) over the interval $x = x' - \epsilon$ to $x = x' + \epsilon$, where ϵ is a small positive quantity:

$$\int_{x'-\epsilon}^{x'+\epsilon} \frac{\mathrm{d}}{\mathrm{d}x}\left(p(x)\frac{\mathrm{d}}{\mathrm{d}x}G(x,x')\right)\,\mathrm{d}x + \int_{x'-\epsilon}^{x'+\epsilon} r(x)G(x,x')\,\mathrm{d}x$$
$$= \int_{x'-\epsilon}^{x'+\epsilon} \delta(x-x')\,\mathrm{d}x.$$

[7]The differential equation for any Green function contains a singularity (associated with the Dirac delta function), which is of such a type that it can be integrated over (not all singularities can be integrated over). By integrating differential equations containing Dirac delta functions (because delta "functions" are only defined under integral signs), we discover properties to be satisfied by their solutions.

We therefore have, for $\epsilon > 0$:

$$p(x)\frac{\mathrm{d}}{\mathrm{d}x}G(x,x')\bigg|_{x=x'-\epsilon}^{x=x'+\epsilon} + \int_{x'-\epsilon}^{x'+\epsilon} r(x)G(x,x')\,\mathrm{d}x = 1. \tag{9.10}$$

Because $G(x,x')$ is continuous at $x = x'$ (and, presumably, so is $r(x)$ which is independent of x') it's clear that

$$\lim_{\epsilon\to 0}\int_{x'-\epsilon}^{x'+\epsilon} r(x)G(x,x')\,\mathrm{d}x = 0.$$

Consequently, Eq. (9.10) implies that

jump discontinuity

$$\lim_{\epsilon\to 0}\left(\frac{\mathrm{d}}{\mathrm{d}x}G(x,x')\bigg|_{x=x'-\epsilon}^{x=x'+\epsilon}\right) = \frac{1}{p(x')}, \tag{9.11}$$

assuming $p(x)$ is continuous at x'. The *discontinuity* in the derivative of G at $x = x'$ described by Eq. (9.11) is the *jump discontinuity* condition.[8] The two conditions:

$$G(x'+\epsilon,x') = G(x'-\epsilon,x') \qquad \epsilon\to 0 \tag{9.12}$$

and the jump discontinuity described by Eq. (9.11) are termed the *matching conditions.*

matching conditions

9.1.4 Direct construction of $G(x,x')$

The differential equation satisfied by the Green function, Eq. (9.9), is *homogeneous* for $x \neq x'$, which suggests a strategy for obtaining $G(x,x')$: Directly solve the homogeneous differential equation $\mathcal{L}_x G(x,x') = 0$ in the regions $a \le x < x'$ and $x' < x \le b$. For each region there are two integration constants associated with $\mathcal{L}_x G(x,x') = 0$. The two matching conditions, Eqs. (9.11) and (9.12) at $x = x'$, together with the two boundary conditions at $x = a$ and $x = b$, suffice to determine the four unknown constants, as we illustrate with the following example.

Example. Consider a string under tension between the points $x = 0$ and $x = l$, subject to a forcing function $S(x,t)$ that acts transverse to the string. The transverse displacement of the string, $\psi(x,t)$, obeys the inhomogeneous wave equation

$$\frac{\partial^2\psi(x,t)}{\partial x^2} - \frac{1}{c^2}\frac{\partial^2\psi(x,t)}{\partial t^2} = S(x,t),$$

where c is the speed of propagation. Assume that the forcing function acts with a single frequency ω, with $S(x,t) = \psi(x)\mathrm{e}^{-\mathrm{i}\omega t}$. In the steady-state,

[8]At a jump discontinuity, the one-sided limits exist and are finite but are not equal.

then, the forced response of the string has the same time dependence so that $\psi(x,t) = f(x)e^{-i\omega t}$. Under these conditions, the wave equation reduces to the inhomogeneous Helmholtz equation in one dimension:

$$\frac{d^2 f(x)}{dx^2} + k^2 f(x) = \psi(x), \tag{9.13}$$

where $k \equiv \omega/c$. (In general, one should add to the forced response the *free* response of the string; we concentrate here on the solution of the inhomogeneous wave equation, assuming that all transients associated with the free response have "died out.")

The Green function associated with Eq. (9.13) satisfies the differential equation

$$\left(\frac{d^2}{dx^2} + k^2\right) G(x,x') = \delta(x-x').$$

For $x \neq x'$, we can write

$$G(x,x') = \begin{cases} A\sin kx + B\cos kx & x < x' \\ C\sin kx + D\cos kx & x > x'. \end{cases}$$

The ends of the string are fixed ($f(0) = f(l) = 0$), and thus $G(x,x')$ mirrors those boundary conditions: $G(0,x') = G(l,x') = 0$. We have (Exercise 9.2)

$$G(x,x') = \begin{cases} A\sin kx & x < x' \\ E\sin k(x-l) & x > x', \end{cases}$$

where E is a constant.[9] The matching conditions, Eqs. (9.11) and (9.12) imply that

$$A\sin kx' - E\sin k(x'-l) = 0 \qquad kE\cos k(x'-l) - kA\cos kx' = 1.$$

Solving these equations, we find

$$A = \frac{\sin k(x'-l)}{k\sin kl} \qquad E = \frac{\sin kx'}{k\sin kl}.$$

The Green function is thus

$$G(x,x') = \frac{1}{k\sin kl} \times \begin{cases} \sin kx \, \sin k(x'-l) & x < x' \\ \sin kx' \, \sin k(x-l) & x > x'. \end{cases} \tag{9.14}$$

[9]Note that the boundary conditions do not determine the parameter k in this case, as with the *free* vibration of the string (Section 2.3). Here, k is related to the frequency of the driving force, $k = \omega/c$.

The solution to Eq. (9.13) is then obtained using Eq. (9.3):

$$f(x) = \int_0^l G(x,x')\psi(x')\,\mathrm{d}x'$$
$$= \frac{1}{k\sin kl}\left[\sin k(x-l)\int_0^x \sin kx'\psi(x')\,\mathrm{d}x' + \sin kx\int_x^l \sin k(x'-l)\psi(x')\,\mathrm{d}x'\right]. \tag{9.15}$$

The reader is urged to verify that Eq. (9.15) is correct. Note that if the driving frequency ω is such that $k=\omega/c=n\pi/l$, for $n=1,2,3,\ldots,$ the response $f(x)\to\infty$. If we drive the system at one of its resonant frequencies (determined from the homogeneous equation $\mathcal{L}f=0$), we get a big response.

9.1.5 Eigenfunction expansions

Another method for obtaining $G(x,x')$ makes use of the fact that eigenfunctions of self-adjoint operators are complete orthonormal sets of functions. From Chapter 2, the solutions $\{\phi_n\}$ of the Sturm–Liouville eigenvalue problem (Eq. (2.18)),

$$\mathcal{L}\phi_n + \lambda_n w(x)\phi_n(x) = 0, \qquad n=1,2,3,\ldots \tag{2.18}$$

subject to homogeneous boundary conditions are a complete set of orthonormal functions as specified by Eqs. (2.38) and (2.41). Any piecewise continuous function f can be represented in terms of the eigenfunctions $\{\phi_n\}$ (expansion theorem, Section 2.5)[10]:

$$f(x) = \sum_{n=1}^{\infty} c_n\phi_n(x). \tag{2.37}$$

Consider the inhomogeneous version of the Sturm–Liouville equation,

$$\mathcal{L}f + \lambda w(x)f(x) = \psi(x), \tag{9.16}$$

where λ is a parameter (that could for example come from the Helmholtz equation). To solve Eq. (9.16), find $G(x,x')$ satisfying

$$\mathcal{L}_x G(x,x') + \lambda w(x)G(x,x') = \delta(x-x'). \tag{9.17}$$

Expand the Green function in the eigenfunctions of $\mathcal{L}$ (the solutions of Eq. (2.18)):

$$G(x,x') = \sum_{n=1}^{\infty} c_n\phi_n(x), \tag{9.18}$$

[10]Equation (2.37) indicates a summation for $n\geq 1$. In any given case, Eq. (2.37) should involve whatever labels are required to effect the completeness relation for the solutions of the Sturm–Liouville eigenproblem.

where the coefficients c_n are functions of x'. Combining Eqs. (9.18) and (9.17) yields

$$(\mathcal{L}_x + \lambda w(x))G(x,x') = \sum_{n=1}^{\infty} c_n w(x)(\lambda - \lambda_n)\phi_n(x) = \delta(x - x').$$

Comparing with the completeness relation, Eq. (2.41), we obtain $c_n = \phi_n(x')/(\lambda - \lambda_n)$. The Green function associated with Eq. (9.16) therefore has the representation

$$G(x,x') = \sum_{n=1}^{\infty} \frac{\phi_n(x)\phi_n(x')}{\lambda - \lambda_n}. \tag{9.19}$$

The reciprocity relation is manifestly obeyed by $G(x,x')$ in this form.[11] The Green function in the form of Eq. (9.19) satisfies homogeneous boundary conditions because the eigenfunctions $\phi_n(x)$ themselves satisfy homogeneous boundary conditions.

Example. Vibrating string (Helmholtz equation in a bounded one-dimensional domain)

The normalized eigenfunctions of the operator $\mathcal{L} = \mathrm{d}^2/\,\mathrm{d}x^2$ for a string of length l with clamped boundary conditions (Dirichlet boundary conditions), are

$$\phi_n(x) = \sqrt{\frac{2}{l}} \sin\left(\frac{n\pi x}{l}\right) \qquad n = 1, 2, 3, \ldots$$

with eigenvalues $\lambda_n = (n\pi/l)^2$. Making use of Eq. (9.19), the Green function associated with the Helmholtz equation for the vibrating string is, with $\lambda = k^2$,

$$G(x,x') = \frac{2}{l} \sum_{n=1}^{\infty} \frac{\sin(n\pi x/l)\sin(n\pi x'/l)}{k^2 - (n\pi/l)^2}. \tag{9.20}$$

Equation (9.20), an infinite series, must be equivalent to Eq. (9.14), a closed-form expression since they both satisfy the same differential equation satisfying the same boundary conditions. To *show* the equivalence explicitly, we can use the Mittag-Leffler theorem (Section 8.9). Equation (9.14), considered as function of k as a *complex variable*, is a meromorphic function. It has simple, isolated poles at $k = n\pi/l$ for all integer $n \neq 0$ and is finite for $k = 0$. Using Eq. (9.14), the limit $k \to 0$ is

[11] If the eigenfunctions $\{\phi_n\}$ are complex valued, one of the terms $\phi_n(x)$ or $\phi_n(x')$ would be complex conjugated in Eq. (9.19), and the reciprocity condition must be modified to $G(x,x') = G^*(x',x)$.

$$G_{k=0}(x,x') \equiv \lim_{k\to 0} G(x,x') = \begin{cases} \frac{x}{l}(x'-l) & x<x' \\ \frac{x'}{l}(x-l) & x>x'. \end{cases} \quad \text{(from Eq. (9.14))} \tag{9.21}$$

The residue at $k = n\pi/l$ (required for the Mittag-Leffler expansion) is readily found:

$$\begin{aligned} \text{Res}\left(\frac{n\pi}{l}\right) &= \lim_{k\to n\pi/l}\left(\frac{k-n\pi/l}{k\sin kl}\sin kx \sin k(x'-l)\right) \\ &= \frac{1}{n\pi}\sin\left(\frac{n\pi x}{l}\right)\sin\left(\frac{n\pi x'}{l}\right). \end{aligned} \tag{9.22}$$

Note that the residue is symmetric in x and x'. Combining Eq. (9.22) with Eq. (8.34),

$$\begin{aligned} G(x,x') &= G_{k=0}(x,x') + \frac{2l}{\pi^2}\sum_{n=1}^{\infty}\frac{\sin(n\pi x/l)\sin(n\pi x'/l)}{n^2} \\ &+ \frac{2}{l}\sum_{n=1}^{\infty}\frac{\sin(n\pi x/l)\sin(n\pi x'/l)}{k^2-(n\pi/l)^2}. \end{aligned}$$

It can be shown (but not here) that the first and second terms in this equation cancel. The third term in this equation is the same as Eq. (9.20). Thus, Eq. (9.14) implies Eq. (9.20) through the Mittag-Leffler expansion.

The case of zero eigenvalue

A problem arises if $\mathcal{L}$ has an eigenfunction ϕ_0 corresponding to eigenvalue[12] $\lambda_0 = 0$. With an eigenvalue $\lambda_0 = 0$ in Eq. (9.19),[13] the Green function for $\lambda = 0$ doesn't exist.[14,15] Fortunately there's a work-around.

Define the function $H(x,x') \equiv -\sum_{n\neq 0}^{\infty}\phi_n(x)\phi_n(x')/\lambda_n$, the *generalized Green function.* This function satisfies the differential equation

generalized Green function

$$\begin{aligned} \mathcal{L}_x H(x,x') &= -\mathcal{L}_x\sum_{n\neq 0}^{\infty}\frac{1}{\lambda_n}\phi_n(x)\phi_n(x') = \sum_{n\neq 0}^{\infty} w(x)\phi_n(x)\phi_n(x') \\ &= \delta(x-x') - w(x)\phi_0(x)\phi_0(x'), \end{aligned} \tag{9.23}$$

[12]Exercise 9.5 provides an example of a system having an eigenfunction corresponding to zero eigenvalue.

[13]Equation (9.19) is nominally written as a sum for $n = 1, 2, \ldots$; for a given problem it would include a term for every member of the complete set of solutions to the Sturm–Liouville equation.

[14]The parameter λ in the Sturm–Liouville problem is often a separation constant introduced in the separation of variables method. The point is, λ is known – we're not solving for λ.

[15]We're referring to the Green function associated with $\mathcal{L}$, not that associated with Eq. (9.16).

where we've used Eq. (2.18) and the completeness relation, Eq. (2.41). $H(x, x')$ is not a Green function because the right side of Eq. (9.23) is not solely $\delta(x - x')$. How does it help us solve inhomogeneous differential equations?

Theorem 9.1.1. *If $\mathcal{L}$ has an eigenfunction ϕ_0 corresponding to zero eigenvalue, $\mathcal{L}\phi_0 = 0$, the problem $\mathcal{L}f = \psi$ (defined on the interval I) has a solution if and only if*

$$\int_I w(x)\phi_0(x)\psi(x)\,\mathrm{d}x = 0. \tag{9.24}$$

When a solution exists, it's of the form

$$f(x) = C\phi_0(x) + \int_I H(x, x')\psi(x')\,\mathrm{d}x', \tag{9.25}$$

where C is a constant.

Thus, a solution exists when the source function ψ has no projection along the function ϕ_0 that makes the Green function for $\mathcal{L}$ diverge.

Proof. Sufficiency (*if*) is easy to how; Eq. (9.24) is necessary (*only if*) as well [14, p. 356]. When $\mathcal{L}$ acts on f in the form of Eq. (9.25), we have ($\mathcal{L}$ "wipes out" ϕ_0)

$$\begin{aligned}\mathcal{L}f &= \int_I \mathcal{L}_x H(x, x')\psi(x')\,\mathrm{d}x' = \int_I [\delta(x - x') - w(x')\phi_0(x)\phi_0(x')]\psi(x')\,\mathrm{d}x' \\ &= \psi(x) - \phi_0(x)\int_I w(x')\phi_0(x')\psi(x')\,\mathrm{d}x',\end{aligned}$$

where we've used Eq. (9.23). If Eq. (9.24) is satisfied, then f in the form of Eq. (9.25) solves the inhomogeneous differential equation $\mathcal{L}f = \psi$. □

9.2 Poisson equation

Consider the Poisson equation, Eq. (3.10),[16]

$$\nabla^2\psi(\boldsymbol{r}) = -\alpha\rho(\boldsymbol{r}), \tag{9.26}$$

where the sources are confined to a region V bounded by the surface S. The Poisson equation is an elliptic PDE, the solutions of which satisfy Dirichlet or Neumann boundary conditions, *but not both*, on closed boundaries (see Appendix D). When dealing with second-order PDEs in

[16]The source function $\rho(\boldsymbol{r})$ most commonly encountered is the charge density function. The constant α in Eq. (9.26) depends on the units of charge employed – $\alpha = 1/\epsilon_0$ for SI units and $\alpha = 4\pi$ for cgs units.

two or more independent variables, Neumann boundary conditions consist of a specification of the *normal derivative* at S, $\hat{\boldsymbol{n}} \cdot \boldsymbol{\nabla}\psi(\boldsymbol{r})$, where $\hat{\boldsymbol{n}}$ is a unit vector locally normal to the boundary.

Let's find the Green function for the Poisson equation (actually, the Green function associated with the Laplacian operator). We seek the solution to

$$\nabla^2 G(\boldsymbol{r}, \boldsymbol{r}') = \delta(\boldsymbol{r} - \boldsymbol{r}'), \tag{9.27}$$

where $\delta(\boldsymbol{r} - \boldsymbol{r}')$ is the *three-dimensional Dirac delta function*,[17] subject to the appropriate boundary conditions.

9.2.1 Boundary conditions and reciprocity relations

We start with *Green's identity*[18] for functions $f(\boldsymbol{r})$ and $g(\boldsymbol{r})$ defined in a volume V that's bounded by surface S:

Green's identity

$$\int_V \left(f\nabla^2 g - g\nabla^2 f\right) \, \mathrm{d}^3 r = \int_S [f\boldsymbol{\nabla} g - g\boldsymbol{\nabla} f] \cdot \hat{\boldsymbol{n}} \, \mathrm{d}s. \tag{9.28}$$

Let $f = \psi(\boldsymbol{r})$ and $g = G(\boldsymbol{r}, \boldsymbol{r}')$ in Eq. (9.28) and make use of Eqs. (9.26) and (9.27):

$$\begin{aligned}&\int_V \left(\psi(\boldsymbol{r})\delta(\boldsymbol{r} - \boldsymbol{r}') + \alpha G(\boldsymbol{r}, \boldsymbol{r}')\rho(\boldsymbol{r})\right) \, \mathrm{d}^3 r \\ &= \int_S [\psi(\boldsymbol{r})\boldsymbol{\nabla}_r G(\boldsymbol{r}, \boldsymbol{r}') - G(\boldsymbol{r}, \boldsymbol{r}')\boldsymbol{\nabla}_r \psi(\boldsymbol{r})] \cdot \hat{\boldsymbol{n}} \, \mathrm{d}s.\end{aligned}$$

Interchange $\boldsymbol{r} \leftrightarrow \boldsymbol{r}'$ in this equation, and integrate over the delta function:

$$\psi(\boldsymbol{r}) = -\alpha \int_V G(\boldsymbol{r}', \boldsymbol{r})\rho(\boldsymbol{r}') \, \mathrm{d}^3 r' + \int_S [\psi(\boldsymbol{r}')\boldsymbol{\nabla}_{r'} G(\boldsymbol{r}', \boldsymbol{r}) - G(\boldsymbol{r}', \boldsymbol{r})\boldsymbol{\nabla}_{r'} \psi(\boldsymbol{r}')] \cdot \hat{\boldsymbol{n}} \, \mathrm{d}s'. \tag{9.29}$$

Equation (9.29) is the generalization of Eq. (9.8) to three dimensions; we can use it to determine the boundary conditions for G. Note that we have not invoked the reciprocity relation in Eq. (9.29); that will be developed momentarily.

Dirichlet boundary conditions

If $\psi(\boldsymbol{r}')$ is given on S (not necessarily a homogeneous Dirichlet condition), choose

$$G(\boldsymbol{r}', \boldsymbol{r}) = 0 \qquad (\boldsymbol{r}' \text{ on } S). \tag{9.30}$$

[17]The three-dimensional delta function is defined analogously to the one-dimensional delta function $\delta(x)$ (Section 2.4): $\int_V \delta(\boldsymbol{r}) \, \mathrm{d}^3 r = 1$ for some volume V. In Cartesian coordinates, $\delta(\boldsymbol{r}) = \delta(x)\delta(y)\delta(z)$, sometimes written $\delta^3(\boldsymbol{r})$; a notation we don't use.

[18]Apply the divergence theorem to $f\boldsymbol{\nabla} g - g\boldsymbol{\nabla} f$; note that $\boldsymbol{\nabla} \cdot (f\boldsymbol{\nabla} g - g\boldsymbol{\nabla} f) = f\nabla^2 g - g\nabla^2 f$.

In this way, the normal derivative, *which cannot be independently specified*[19] on S, is eliminated from Eq. (9.29). The solution then has the form

$$\psi(\boldsymbol{r}) = -\alpha \int_V G(\boldsymbol{r}',\boldsymbol{r})\rho(\boldsymbol{r}')\,\mathrm{d}^3r' + \int_S \psi(\boldsymbol{r}')\boldsymbol{\nabla}_{r'}G(\boldsymbol{r}',\boldsymbol{r})\cdot\hat{\boldsymbol{n}}\,\mathrm{d}s'. \tag{9.31}$$

If $\psi(\boldsymbol{r}') = 0$ (the homogeneous Dirichlet condition), the surface integral vanishes.

Neumann boundary conditions

For $\hat{\boldsymbol{n}}\cdot\boldsymbol{\nabla}\psi$ specified on S (see Exercise 9.6 for a constraint on the normal derivative), we *cannot* choose $\hat{\boldsymbol{n}}\cdot\boldsymbol{\nabla}_{r'}G(\boldsymbol{r}',\boldsymbol{r}) = 0$ because from the divergence theorem,

$$\int_S \boldsymbol{\nabla}_{r'}G(\boldsymbol{r}',\boldsymbol{r})\cdot\hat{\boldsymbol{n}}\,\mathrm{d}s' = \int_V \nabla^2_{r'}G(\boldsymbol{r}',\boldsymbol{r})\,\mathrm{d}^3r' = \int_V \delta(\boldsymbol{r}-\boldsymbol{r}')\,\mathrm{d}^3r' = 1. \tag{9.32}$$

The following boundary condition is *consistent* with Eq. (9.32):

$$\hat{\boldsymbol{n}}\cdot\boldsymbol{\nabla}_{r'}G(\boldsymbol{r}',\boldsymbol{r}) = \frac{1}{A}, \tag{9.33}$$

where A is the surface area of S. The solution for Neumann boundary conditions is then of the form, from Eq. (9.29):

$$\psi(\boldsymbol{r}) = -\alpha \int_V G(\boldsymbol{r}',\boldsymbol{r})\rho(\boldsymbol{r}')\,\mathrm{d}^3r' - \int_S G(\boldsymbol{r}',\boldsymbol{r})\boldsymbol{\nabla}_{r'}\psi(\boldsymbol{r}')\cdot\hat{\boldsymbol{n}}\,\mathrm{d}s' + \frac{1}{A}\int_S \psi(\boldsymbol{r}')\,\mathrm{d}s'.$$

The last term is a constant which can be dropped.[20]

Reciprocity relations

The reciprocity relation for the Green function associated with elliptic PDEs takes on a special form for Neumann boundary conditions. To see this, let $f = G(\boldsymbol{r},\boldsymbol{r}')$ and $g = G(\boldsymbol{r},\boldsymbol{r}'')$ in Eq. (9.28). We find

$$G(\boldsymbol{r}'',\boldsymbol{r}') - G(\boldsymbol{r}',\boldsymbol{r}'') = \int_S [G(\boldsymbol{r},\boldsymbol{r}')\boldsymbol{\nabla}_r G(\boldsymbol{r},\boldsymbol{r}'') - G(\boldsymbol{r},\boldsymbol{r}'')\boldsymbol{\nabla}_r G(\boldsymbol{r},\boldsymbol{r}')]\cdot\hat{\boldsymbol{n}}\,\mathrm{d}s. \tag{9.34}$$

Equation (9.34) is the generalization of Eq. (9.6) to three dimensions. For Dirichlet conditions, the right side of Eq. (9.34) vanishes because of Eq. (9.30), implying the reciprocity relation

$$G(\boldsymbol{r},\boldsymbol{r}') = G(\boldsymbol{r}',\boldsymbol{r}). \qquad \text{(Dirichlet boundary conditions on } \psi\text{)}$$

---

[19]To have unique solutions to the Poisson equation, one can specify Dirichlet or Neumann conditions on the boundary, but not both. See Appendix D.

[20]The solution of elliptic PDEs with Neumann boundary conditions is unique only up to a constant (see Appendix D).

For Neumann conditions, however, using Eq. (9.33) in Eq. (9.34) we have

$$G(\boldsymbol{r}'', \boldsymbol{r}') - G(\boldsymbol{r}', \boldsymbol{r}'') = \frac{1}{A}\int_S [G(\boldsymbol{r}, \boldsymbol{r}') - G(\boldsymbol{r}, \boldsymbol{r}'')] \, \mathrm{d}s \equiv \gamma(\boldsymbol{r}') - \gamma(\boldsymbol{r}'').$$

If $\int_S G(\boldsymbol{r}, \boldsymbol{r}') \, \mathrm{d}s$ is a constant, then we have the usual reciprocity relation. If not, we have a generalized reciprocity relation

$$G(\boldsymbol{r}', \boldsymbol{r}) + \gamma(\boldsymbol{r}') = G(\boldsymbol{r}, \boldsymbol{r}') + \gamma(\boldsymbol{r}). \text{ (Neumann boundary conditions on } \psi)$$

Thus, *Neumann boundary conditions on do not force the Green function to be a symmetric function of its arguments* (unless $\int_S G(\boldsymbol{r}, \boldsymbol{r}') \, \mathrm{d}s =$ constant).

9.2.2 So, what's the Green function?

Anyone who has studied electrostatics already knows the Green function in this case! The scalar potential ψ at location $\boldsymbol{r}$, due to a point charge q at $\boldsymbol{r}'$, is (from the Poisson equation[21] $\nabla^2\psi(\boldsymbol{r}) = -\alpha q \delta(\boldsymbol{r} - \boldsymbol{r}')$) the Coulomb potential $\psi(\boldsymbol{r}) = \alpha q/(4\pi|\boldsymbol{r} - \boldsymbol{r}'|)$. The Green function for the Laplacian is therefore[22]:

$$G_0(\boldsymbol{r}, \boldsymbol{r}') = -\frac{1}{4\pi}\frac{1}{|\boldsymbol{r} - \boldsymbol{r}'|}. \tag{9.35}$$

Note the subscript in Eq. (9.35): G_0 is termed the *free-space* Green function; it contains the basic singularity of the Green function. The function G_0 solves Eq. (9.27), but it does not satisfy the boundary conditions. We can add to Eq. (9.35) an additional term $F(\boldsymbol{r}, \boldsymbol{r}')$, symmetric in its arguments, that satisfies the Laplace equation $\nabla^2 F(\boldsymbol{r}, \boldsymbol{r}') = 0$,

free-space Green function

$$G(\boldsymbol{r}, \boldsymbol{r}') = -\frac{1}{4\pi}\frac{1}{|\boldsymbol{r} - \boldsymbol{r}'|} + F(\boldsymbol{r}, \boldsymbol{r}'), \tag{9.36}$$

such that $G(\boldsymbol{r}, \boldsymbol{r}')$ in Eq. (9.36) satisfies the appropriate boundary conditions.[23] The difficulty in constructing $G(\boldsymbol{r}, \boldsymbol{r}')$ is usually that of determining[24] $F(\boldsymbol{r}, \boldsymbol{r}')$. Note that G_0 depends only on the distance $|\boldsymbol{r} - \boldsymbol{r}'|$.

[21]The charge "density" function for a *point charge* is, through the miracle of the delta function, $q\delta(\boldsymbol{r} - \boldsymbol{r}')$. The three-dimensional delta function has the dimension of $(\text{volume})^{-1}$.

[22]Some find the minus sign in Eq. (9.35) to be an esthetic irritant. For that reason, the Green function is sometimes defined with a minus sign, $\nabla^2 G(\boldsymbol{r}, \boldsymbol{r}') = -\delta(\boldsymbol{r} - \boldsymbol{r}')$.

[23]Thus, G denotes the solution of $\mathcal{L}G(\boldsymbol{r}, \boldsymbol{r}') = \delta(\boldsymbol{r} - \boldsymbol{r}')$ that satisfies the boundary conditions.

[24]G_0 is variously referred to as the *principal solution*, the *fundamental solution*, or the *elementary solution*. The term $F(\boldsymbol{r}, \boldsymbol{r}')$ used to meet boundary conditions is called the *regular* part of the Green function.

Example. Green function interior to a sphere

Let's find the solution to Eq. (9.27) for the interior of a sphere of radius a, on which the potential $\psi(r,\theta,\phi)$ has the prescribed form $\psi(a,\theta,\phi)$ (nonhomogeneous Dirichlet condition), and show that it occurs in the form of Eq. (9.36). From Eq. (9.30), we impose the boundary condition $G(a\hat{\boldsymbol{r}},\boldsymbol{r}')=0$ (this is the boundary condition on the *Green function*, not the potential ψ – make sure you understand this point).

Start by writing the three-dimensional delta function in spherical coordinates:

$$\begin{aligned}\delta(\boldsymbol{r}-\boldsymbol{r}') &= \frac{1}{r^2\sin\theta}\delta(r-r')\delta(\theta-\theta')\delta(\phi-\phi')\\ &= \frac{1}{r^2}\delta(r-r')\sum_{l=0}^{\infty}\sum_{m=-l}^{l}(Y_l^m(\theta',\phi'))^*Y_l^m(\theta,\phi).\end{aligned} \tag{9.37}$$

The first equality in Eq. (9.37) follows by definition, $\int_V\delta(\boldsymbol{r})\,\mathrm{d}^3r=1$ (using the volume element for spherical coordinates $\mathrm{d}^3\boldsymbol{r}=r^2\,\mathrm{d}r\sin\theta\,\mathrm{d}\theta\,\mathrm{d}\phi$), while the second follows from the completeness property of the spherical harmonics (Section 6.3). The form of Eq. (9.37) suggests we try an expansion of the Green function in the same form,

$$G(\boldsymbol{r},\boldsymbol{r}')=\sum_{l=0}^{\infty}\sum_{m=-l}^{l}G_{lm}(r,\boldsymbol{r}')Y_l^m(\theta,\phi), \tag{9.38}$$

because the Y_l^m are a complete set for representing functions of (θ,ϕ), where the coefficients $G_{lm}(r,\boldsymbol{r}')$ are functions of the other four variables: the radial coordinate r and the three spherical coordinates (r',θ',ϕ'). The Laplacian operator in spherical coordinates can be written $\nabla^2=\nabla_r^2+(1/r^2)\nabla^2_{\theta,\phi}$, where $\nabla_r^2\equiv(1/r^2)(\partial/\partial r)(r^2\partial/\partial r)$, and the angular part $\nabla^2_{\theta,\phi}$, shown in Eq. (3.46), is such that $\nabla^2_{\theta,\phi}Y_l^m=-l(l+1)Y_l^m$. Thus, from Eq. (9.38),

$$\nabla^2G(\boldsymbol{r},\boldsymbol{r}')=\sum_{l=0}^{\infty}\sum_{m=-l}^{l}\left(\nabla_r^2-\frac{l(l+1)}{r^2}\right)G_{lm}(r,\boldsymbol{r}')Y_l^m(\theta,\phi). \tag{9.39}$$

Comparing Eqs. (9.39) and (9.37) (which represent the two sides of Eq. (9.27)), we have, for each l and m,

$$\left(\nabla_r^2-\frac{l(l+1)}{r^2}\right)G_{lm}(r,\boldsymbol{r}')=\frac{1}{r^2}(Y_l^m(\theta',\phi'))^*\delta(r-r'). \tag{9.40}$$

Because the differential operator acts only on the coordinate r, let

$$G_{lm}(r,\boldsymbol{r}')=g_l(r,r')(Y_l^m(\theta',\phi'))^*, \tag{9.41}$$

and thus, $g_l(r, r')$ satisfies the ODE,

$$\left(\frac{\mathrm{d}^2}{\mathrm{d}r^2} + \frac{2}{r}\frac{\mathrm{d}}{\mathrm{d}r} - \frac{l(l+1)}{r^2}\right) g_l(r, r') = \frac{1}{r^2}\delta(r - r'). \tag{9.42}$$

When $r \neq r'$, Eq. (9.42) reduces to the Laplace equation in spherical coordinates, Eq. (6.55). Using Eq. (6.56),

$$g_l(r, r') = \begin{cases} Ar^l + Br^{-(l+1)} & r < r' \\ Cr^l + Dr^{-(l+1)} & r > r'. \end{cases} \tag{9.43}$$

Because the domain includes the origin, we must set $B = 0$ (to guarantee that g_l remains finite). The boundary condition $g_l(a, r') = 0$ implies that $D = -Ca^{2l+1}$. Equation (9.43) therefore reduces to

$$g_l(r, r') = \begin{cases} Ar^l & r < r' \\ Cr^l \left[1 - \left(\frac{a}{r}\right)^{2l+1}\right] & r > r'. \end{cases} \tag{9.44}$$

Now apply the matching conditions (Section 9.1). We find: (Exercise 9.7)

$$A = \frac{(r')^l}{(2l+1)a^{2l+1}} \left[1 - \left(\frac{a}{r'}\right)^{2l+1}\right] \qquad C = \frac{(r')^l}{(2l+1)a^{2l+1}}. \tag{9.45}$$

We therefore have an expression for g_l:

$$g_l(r, r') = \frac{(rr')^l}{(2l+1)a^{2l+1}} \times \begin{cases} 1 - (a/r')^{2l+1} & r < r' \\ 1 - (a/r)^{2l+1}. & r > r' \end{cases} \tag{9.46}$$

We can now assemble the pieces. Combining Eq. (9.41) with Eq. (9.38),

$$\begin{aligned} G(\boldsymbol{r}, \boldsymbol{r}') &= \sum_{l=0}^{\infty} \sum_{m=-l}^{l} g_l(r, r')(Y_l^m(\theta', \phi'))^* Y_l^m(\theta, \phi) \\ &= \frac{1}{4\pi} \sum_{l=0}^{\infty} (2l+1) g_l(r, r') P_l(\hat{\boldsymbol{r}} \cdot \hat{\boldsymbol{r}}'), \end{aligned} \tag{9.47}$$

where we've used the addition theorem for spherical harmonics, Eq. (6.37). Having assembled the pieces, we now separate them. Using Eq. (9.46), it can be shown that

$$G(\boldsymbol{r}, \boldsymbol{r}') = \frac{1}{4\pi} \sum_{l=0}^{\infty} \frac{(rr')^l}{a^{2l+1}} P_l(\hat{\boldsymbol{r}} \cdot \hat{\boldsymbol{r}}') - \frac{1}{4\pi} \frac{1}{|\boldsymbol{r} - \boldsymbol{r}'|}, \tag{9.48}$$

(see Exercise 9.8) which is precisely in the form of Eq. (9.36). The regular part in Eq. (9.48) satisfies the Laplace equation, as advertised (Exercise 9.8).

9.2.2.1 *Free-space Green function in two dimensions*

Equation (9.35) was "derived" by appealing to the Coulomb potential of a point charge, which of course generates a three-dimensional electric field vector. There are cases, however, where the field lines are confined to a plane.[25] Let's derive G_0 in two dimensions.

In circular polar coordinates, $\nabla^2 = (1/r)(\partial/\partial r)(r\partial/\partial r) + (1/r^2)\partial^2/\partial\theta^2$ (Section 7.4). Because G_0 depends only on distance, we can discard the angular derivative. Thus, we seek $G_0(\boldsymbol{r}, \boldsymbol{r}')$ such that ($\delta(\boldsymbol{r} - \boldsymbol{r}')$ is a two-dimensional delta function)

$$\frac{1}{r}\frac{\mathrm{d}}{\mathrm{d}r}\left(r\frac{\mathrm{d}G_0(\boldsymbol{r}, \boldsymbol{r}')}{\mathrm{d}r}\right) = \delta(\boldsymbol{r} - \boldsymbol{r}') = \frac{1}{r}\delta(r - r')\delta(\theta - \theta'). \tag{9.49}$$

For $r \neq r'$, the solution of Eq. (9.49) is $A\ln r + B$, where A and B are constants. We can set $B = 0$ – it doesn't contribute to a singularity as $r \to 0$. Thus, we try $G_0(\boldsymbol{r}, \boldsymbol{r}') = A\ln|\boldsymbol{r} - \boldsymbol{r}'|$. Because we'll need it momentarily, the gradient $\boldsymbol{\nabla} G_0 = A\hat{\boldsymbol{r}}/|\boldsymbol{r} - \boldsymbol{r}'|$. The constant A can be determined by integrating Eq. (9.49) over a cylinder of height[26] h and radius ϵ centered on $\boldsymbol{r}'$, and applying the divergence theorem:

$$\int_V \nabla^2 G_0 \,\mathrm{d}^3r = \int_S \boldsymbol{\nabla} G_0 \cdot \hat{\boldsymbol{n}}\,\mathrm{d}s \overset{\epsilon\to 0}{\to} 2\pi\epsilon h \cdot \frac{A}{\epsilon} = h \quad \Longrightarrow \quad A = \frac{1}{2\pi}.$$

Thus,[27]

$$G_0(\boldsymbol{r}, \boldsymbol{r}') = \frac{1}{2\pi}\ln|\boldsymbol{r} - \boldsymbol{r}'| \qquad \text{(two dimensions).} \tag{9.50}$$

9.3 Helmholtz equation

Helmholtz operator

The inhomogeneous Helmholtz equation can be written in terms of a *Helmholtz operator*

$$\mathcal{L}\psi = (\nabla^2 + k^2)\psi(\boldsymbol{r}) = \rho(\boldsymbol{r}).$$

---

[25] The electric field associated with a line distribution of charge exhibits cylindrical symmetry.

[26] We've "gone" into the third dimension here, momentarily. On the top and bottom faces, $\boldsymbol{\nabla} G_0 \cdot \hat{\boldsymbol{n}} = 0$.

[27] One should review the logic used to arrive at Eq. (9.50). Guessing $G_0(\boldsymbol{r}, \boldsymbol{r}') = A\ln|\boldsymbol{r} - \boldsymbol{r}'|$ gets right what we want for $r \neq r'$, "half" the requirement on a delta function. By appropriately integrating the differential equation for G_0, we get right the other half, that $\int_V \delta(\boldsymbol{r} - \boldsymbol{r}')\,\mathrm{d}^3r = 1$. Truth in advertising: You should be wary of equations such as Eq. (9.50). Logarithms (and other transcendental functions) must have *dimensionless* arguments. To have a valid formula, there must be some other term in the equation that when combined with the logarithm leads to a dimensionless argument. We could for example take the constant B (which we discarded) to have the value $B = -(1/(2\pi))\ln a$, where a is some natural length scale in the problem.

To find the associated Green function, we seek the solution of

$$(\nabla^2 + k^2)G(\boldsymbol{r}, \boldsymbol{r}') = \delta(\boldsymbol{r} - \boldsymbol{r}') \tag{9.51}$$

subject to appropriate boundary conditions.

9.3.1 Green function for two-dimensional problems

We start with the free-space Green function for systems involving circular symmetry. The free-space Green function depends only on distance, and thus we seek the solution of the generalization of Eq. (9.49):

$$G_0'' + \frac{1}{r}G_0' + k^2 G_0 = \delta(\boldsymbol{r} - \boldsymbol{r}'), \tag{9.52}$$

where $\delta(\boldsymbol{r} - \boldsymbol{r}')$ is the two-dimensional delta function. For $r \neq r'$, Eq. (9.52) is the Bessel differential equation of order zero, Eq. (3.38), having linearly independent solutions $J_0(kr)$ and $N_0(kr)$. Of these, J_0 is regular at the origin while N_0 is singular, with $N_0(kr) \overset{r\to 0}{\sim} (2/\pi)\ln(kr)$ (see Eq. (7.2); we've kept only the most singular term). Let's try $G_0(\boldsymbol{r}, \boldsymbol{r}') = AN_0(k|\boldsymbol{r} - \boldsymbol{r}'|)$. Because we'll need it momentarily, $\boldsymbol{\nabla} G_0(kr) = -AkN_1(kr)\hat{\boldsymbol{r}}$ (from Eqs. (7.15) and (7.12)). Using Eq. (7.2), $\boldsymbol{\nabla} G_0(kr) \overset{r\to 0}{\sim} (2A/\pi r)\hat{\boldsymbol{r}}$. The constant A can be determined, as in the previous section, by integrating Eq. (9.52) over the volume of a cylinder of unit height and radius ϵ centered at $\boldsymbol{r}'$:

$$\begin{aligned}\int_V (\nabla^2 + k^2)G_0 \,\mathrm{d}^3r &= \int_S \boldsymbol{\nabla} G_0 \cdot \hat{\boldsymbol{n}} \,\mathrm{d}s + k^2 \int_V G_0 \,\mathrm{d}^3r \\ &\overset{\epsilon\to 0}{\to} 2\pi\epsilon \cdot \frac{2A}{\pi\epsilon} + 2A(k\epsilon)^2 \ln(k\epsilon) + O(k\epsilon)^2 = 1,\end{aligned}$$

where we've used the integral $\int x \ln x \,\mathrm{d}x = \frac{1}{2}x^2 \ln x - \frac{1}{4}x^2$. The second term can be ignored because[28] $\lim_{x\to 0} x^2 \ln x = 0$. Thus,[29] $A = 1/4$ and

$$G_0(\boldsymbol{r}, \boldsymbol{r}') = \frac{1}{4}N_0(k|\boldsymbol{r} - \boldsymbol{r}'|) \quad \text{(free-space Green function in two dimensions).} \tag{9.53}$$

Equation (9.53) can be expressed in terms of the Hankel function $H_0^{(1)}(x) = J_0(x) + \mathrm{i}N_0(x)$ (Eq. (7.4)), with $G_0(\boldsymbol{r}, \boldsymbol{r}') = (-\mathrm{i}/4)H_0^{(1)}(k|\boldsymbol{r} - \boldsymbol{r}'|)$. Including the Bessel function J_0 (regular at the origin) does not affect the basic singularity in $G_0(\boldsymbol{r}, \boldsymbol{r}')$ as given by N_0.

---

[28] We can apply l'Hopital's rule. Write $x^2 \ln x$ as $\ln x / x^{-2}$; this has the indeterminate form $-\infty/\infty$ as $x \to 0$. Applying l'Hopital's rule, however, the limit exists as $x \to 0$, and has the value zero.

[29] Note that the argument of N_0 in Eq. (9.53) properly involves a dimensionless argument: k from the Helmholtz equation has the dimension of inverse length.

Example. Green function for the vibration of a circular drumhead.

Let's find the Green function for the interior of a circular geometry (see Figure 9.1), such as would be associated with a vibrating circular membrane of radius a, where the amplitude of vibration $\psi(\boldsymbol{r})$ vanishes at $r=a$. With the membrane "clamped" at its outer boundary, we seek the solution to Eq. (9.51) in plane polar coordinates subject to the boundary condition $G(a,\theta,\boldsymbol{r}')=0$,

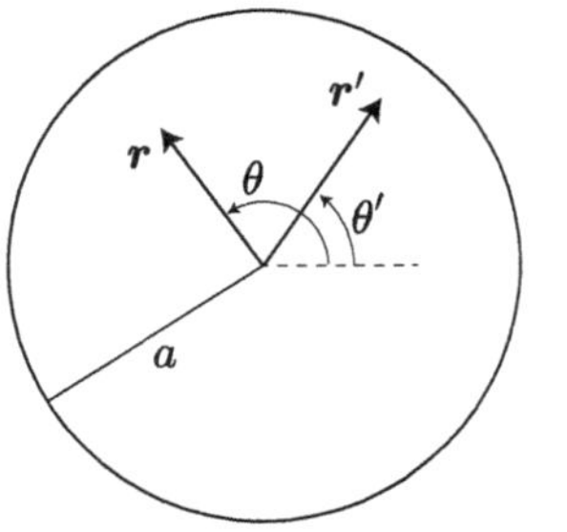

Figure 9.1 Circular drumhead geometry.

$$(\nabla^2+k^2)G(\boldsymbol{r},\boldsymbol{r}')=\left(\frac{1}{r}\frac{\partial}{\partial r}\left(r\frac{\partial}{\partial r}\right)+\frac{1}{r^2}\frac{\partial^2}{\partial\theta^2}+k^2\right)G(\boldsymbol{r},\boldsymbol{r}')=\delta(\boldsymbol{r}-\boldsymbol{r}'). \tag{9.54}$$

Start by expressing the two-dimensional delta function in the form

$$\delta(\boldsymbol{r}-\boldsymbol{r}')=\frac{1}{r}\delta(r-r')\delta(\theta-\theta')=\frac{1}{2\pi r}\delta(r-r')\sum_{n=-\infty}^{\infty}\mathrm{e}^{\mathrm{i}n(\theta-\theta')}, \tag{9.55}$$

where we've used the completeness relation, Eq. (4.11). Equation (9.55) suggests we try a similar expansion of the Green function; let[30]

$$G(\boldsymbol{r},\boldsymbol{r}')=\frac{1}{2\pi}\sum_{n=-\infty}^{\infty}g_n(r,r')\mathrm{e}^{\mathrm{i}n(\theta-\theta')}=\frac{1}{2\pi}\sum_{n=-\infty}^{\infty}g_n(r,r')\cos(n(\theta-\theta')), \tag{9.56}$$

where the second equality follows because $g_n=g_{|n|}$ (as we'll see). Substituting Eq. (9.56) into Eq. (9.54), $g_n(r,r')$ satisfies the ODE

$$\left(\frac{\mathrm{d}^2}{\mathrm{d}r^2}+\frac{1}{r}\frac{\mathrm{d}}{\mathrm{d}r}-\frac{n^2}{r^2}+k^2\right)g_n(r,r')=\frac{1}{r}\delta(r-r'). \tag{9.57}$$

We recognize Eq. (9.57) (for $r\neq r'$) as the Bessel differential equation, Eq. (3.38). We can therefore "build" the Green function in the usual way:

$$g_n(r,r')=\begin{cases}AJ_{|n|}(kr)+BN_{|n|}(kr) & r<r'\\ CJ_{|n|}(kr)+DN_{|n|}(kr) & r>r'.\end{cases} \tag{9.58}$$

Because the circular membrane is centered on the origin, we set $B=0$. The boundary condition at $r=a$ requires that $g_n(a,r')=0$. Equation (9.58) thus reduces to

$$g_n(r,r')=\begin{cases}AJ_{|n|}(kr) & r<r'\\ E[N_{|n|}(ka)J_{|n|}(kr)-J_{|n|}(ka)N_{|n|}(kr)] & r>r'.\end{cases} \tag{9.59}$$

[30]Note that Eq. (9.56) is a Fourier series; we have periodicity in θ of period 2π.

The constants A and E are determined from the matching conditions. We find:

$$A = -\frac{\pi}{2J_{|n|}(ka)}[N_{|n|}(ka)J_{|n|}(kr') - J_{|n|}(ka)N_{|n|}(kr')]$$
$$E = -\frac{\pi}{2}\frac{J_{|n|}(kr')}{J_{|n|}(ka)} \tag{9.60}$$

(see Exercise 9.10). The Green function $g_n(r, r')$ is then

$$g_n(r,r') = \frac{\pi}{2J_{|n|}(ka)} \times \begin{cases} [J_{|n|}(ka)N_{|n|}(kr') - N_{|n|}(ka)J_{|n|}(kr')]J_{|n|}(kr) & r < r' \\ [J_{|n|}(ka)N_{|n|}(kr) - N_{|n|}(ka)J_{|n|}(kr)]J_{|n|}(kr') & r > r'. \end{cases} \tag{9.61}$$

The complete Green function, the solution of Eq. (9.54), is obtained by combining Eqs. (9.61) and (9.56). Can we "dig out" the free-space Green function (Eq. (9.53)) from the rather complicated expression that ensues? We can, by invoking a property of Bessel functions. The *Graf addition theorem* [29, p. 361] states that, referring to Figure 9.2,

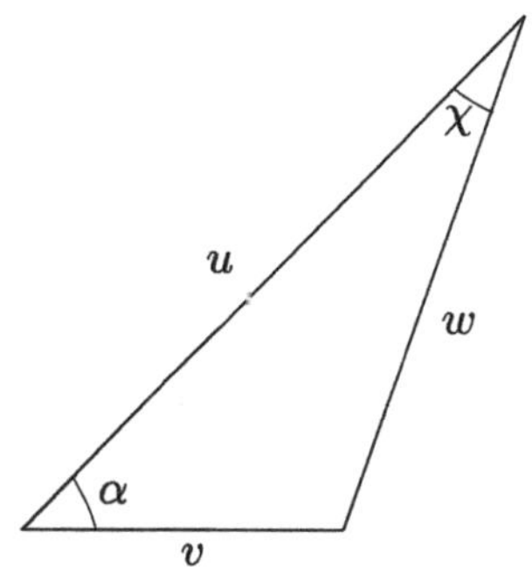

Figure 9.2 Geometry for the Graf addition theorem.

$$\mathcal{Q}_\nu(w)\left\{\begin{array}{c}\cos\nu\chi \\ \sin\nu\chi\end{array}\right\} = \sum_{n=-\infty}^{\infty} \mathcal{Q}_{\nu+n}(u)J_n(v)\left\{\begin{array}{c}\cos n\alpha \\ \sin n\alpha\end{array}\right\}, \tag{9.62}$$

where $\mathcal{Q}$ denotes any of the solutions of the Bessel differential equation: J, N, $H^{(1)}$, $H^{(2)}$, or any linear combination of these functions. Set $\nu = 0$ in Eq. (9.62) and take $\mathcal{Q} = N$. We thus have the identity[31]

$$N_0(k|\boldsymbol{r} - \boldsymbol{r}'|) = \sum_{n=-\infty}^{\infty} N_n(kr_>)J_n(kr_<)\cos(n(\theta - \theta')), \tag{9.63}$$

where $r_>$ ($r_<$) denotes the larger (lesser) of r, r'. Combining Eqs. (9.61) and (9.56), and making use of Eq. (9.63), we find

$$G(\boldsymbol{r}, \boldsymbol{r}') = \frac{1}{4}N_0(k|\boldsymbol{r} - \boldsymbol{r}'|) - \frac{1}{4}\sum_{n=-\infty}^{\infty}\frac{N_{|n|}(ka)}{J_{|n|}(ka)}J_{|n|}(kr)J_{|n|}(kr')\cos(n(\theta - \theta')). \tag{9.64}$$

The regular part satisfies the homogeneous Helmholtz equation (Exercise 9.11).

[31] In arriving at Eq. (9.48) (Green function for a spherical geometry), we made use of the addition theorem for spherical harmonics, Eq. (6.37), proven in Chapter 6. The Graf addition theorem, Eq. (9.62), used in finding the Green function for a circular geometry, can be derived using contour integration methods [29, p. 361].

9.3.2 Free-space Green function for three dimensions

It can be shown that (Exercise 9.9)

$$(\nabla^2 + k^2)\left(\frac{e^{ik|\boldsymbol{r}-\boldsymbol{r}'|}}{|\boldsymbol{r}-\boldsymbol{r}'|}\right) = -4\pi e^{ik|\boldsymbol{r}-\boldsymbol{r}'|}\delta(\boldsymbol{r}-\boldsymbol{r}'), \tag{9.65}$$

where we're now working in spherical coordinates. While not precisely in the form of a Green function because of the exponential factor on the right side of Eq. (9.65), the exponential has the value unity where the delta function has a singularity. Let's try $G_0(\boldsymbol{r}, \boldsymbol{r}') = A\exp(ik|\boldsymbol{r}-\boldsymbol{r}'|)/|\boldsymbol{r}-\boldsymbol{r}'|$. The gradient $\boldsymbol{\nabla} G_0 = Ae^{ikr}(ik/r - 1/r^2)\hat{\boldsymbol{r}}$. Determine A by integrating Eq. (9.51) over a small sphere of radius ϵ centered on $\boldsymbol{r}'$:

$$\begin{aligned}\int_V (\nabla^2 + k^2)G_0 \,\mathrm{d}^3r &= \int_S \boldsymbol{\nabla} G_0 \cdot \hat{\boldsymbol{n}}\,\mathrm{d}s + k^2\int_V G_0\,\mathrm{d}^3r \\ &\overset{\epsilon\to 0}{\to} Ae^{ik\epsilon}\left(\frac{ik}{\epsilon} - \frac{1}{\epsilon^2}\right)\cdot 4\pi\epsilon^2 - 4\pi A((ik\epsilon - 1)e^{ik\epsilon} + 1) = 1.\end{aligned}$$

We identify $A = -1/(4\pi)$. Thus,

$$G_0(\boldsymbol{r}, \boldsymbol{r}') = -\frac{1}{4\pi}\frac{e^{ik|\boldsymbol{r}-\boldsymbol{r}'|}}{|\boldsymbol{r}-\boldsymbol{r}'|} \quad \text{(free-space Green function in three dimensions).} \tag{9.66}$$

Note that Eq. (9.66) agrees with Eq. (9.35) for $k = 0$.

9.3.3 Expansion in spherical harmonics

It's often the case that we require an expression for $G_0(\boldsymbol{r}, \boldsymbol{r}')$ in a form where the role of the coordinates of $\boldsymbol{r}$ and $\boldsymbol{r}'$ occur in a separated form, such as we have in the addition theorem for spherical harmonics, Eq. (6.37). In this section, we show that G_0 in the form of Eq. (9.66) can be written

$$\frac{\exp(ik|\boldsymbol{r}-\boldsymbol{r}'|)}{ik|\boldsymbol{r}-\boldsymbol{r}'|} = 4\pi\sum_{l=0}^{\infty}\sum_{m=-l}^{l} j_l(kr_<)h_l^{(1)}(kr_>)Y_l^m(\theta,\phi)(Y_l^m(\theta',\phi'))^*, \tag{9.67}$$

where $j_l, h_l^{(1)}$ are spherical Bessel and Hankel functions, and (θ, ϕ), (θ', ϕ') are the spherical angles associated with $\boldsymbol{r}$ and $\boldsymbol{r}'$.

We start by noting, from Eqs. (7.42) and (7.43), that

$$h_0^{(1)}(kr) \equiv j_0(kr) + in_0(kr) = \frac{\sin kr}{kr} - i\frac{\cos kr}{kr} = \frac{e^{ikr}}{ikr},$$

is a solution of the homogeneous Helmholtz equation in spherical coordinates (see Exercise 10.1). Thus, the function

$$F(\boldsymbol{r}, \boldsymbol{r}') \equiv \frac{\exp(ik|\boldsymbol{r}-\boldsymbol{r}'|)}{ik|\boldsymbol{r}-\boldsymbol{r}'|} = \frac{\exp(ik\sqrt{r^2 + r'^2 - 2rr'\cos\gamma})}{ik\sqrt{r^2 + r'^2 - 2rr'\cos\gamma}} \tag{9.68}$$

is a solution of the homogeneous Helmholtz equation when $\boldsymbol{r}' \neq \boldsymbol{r}$, where γ is the angle between $\boldsymbol{r}$ and $\boldsymbol{r}'$; see Figure 6.4. As such, referring to Eq. (7.38), we can try an expansion of $F(\boldsymbol{r}, \boldsymbol{r}')$ in the form

$$F(\boldsymbol{r}, \boldsymbol{r}') = \sum_{l=0}^{\infty} f_l(r, r') P_l(\cos\gamma). \tag{9.69}$$

Equation (9.69) involves only the $m = 0$ terms that one would have from the general solution of the Helmholtz equation, Eq. (7.38); $F(\boldsymbol{r}, \boldsymbol{r}')$ involves only the distance $|\boldsymbol{r} - \boldsymbol{r}'|$; there is rotational symmetry about the line $\boldsymbol{r} - \boldsymbol{r}'$, and hence, we need only the $m = 0$ terms, $Y_l^0 \sim P_l(\cos\gamma)$. Comparing Eq. (9.69) with Eq. (7.38), the expansion coefficients $f_l(r, r')$ must be a combination of spherical Bessel and Neumann functions. The particular combination can be inferred by noting that $F(\boldsymbol{r}, \boldsymbol{r}')$ is symmetric in $\boldsymbol{r}$ and $\boldsymbol{r}'$. Because we want $F(\boldsymbol{r}, \boldsymbol{r}')$ to be finite as $r \to 0$, $f_l(r, r')$ can only be a function of $j_l(kr)$ for $r < r'$. For $r \to \infty$, however, we want outgoing wave solutions, and thus, only $h_l^{(1)}(kr)$ is allowed, and not $h_l^{(2)}(kr)$ (the Hankel functions are linearly independent, Exercise 7.18). Thus, $f_l(r, r')$ must be such that

$$f_l(r, r') = \begin{cases} a_l j_l(kr') h_l^{(1)}(kr) & r > r' \\ a_l h_l^{(1)}(kr') j_l(kr) & r < r' \end{cases} \equiv a_l j_l(kr_<) h_l^{(1)}(kr_>). \tag{9.70}$$

The coefficients a_l can be found by matching the asymptotic form of $F(\boldsymbol{r}, \boldsymbol{r}')$ with that of the Hankel functions as $r \to \infty$. Use the fact that for $r \to \infty$, $|\boldsymbol{r} - \boldsymbol{r}'| \sim r - \hat{\boldsymbol{n}} \cdot \boldsymbol{r}'$, where $\hat{\boldsymbol{n}} \equiv \boldsymbol{r}/r$. Thus,

$$\frac{\mathrm{e}^{\mathrm{i}k|\boldsymbol{r}-\boldsymbol{r}'|}}{|\boldsymbol{r} - \boldsymbol{r}'|} \sim \frac{\mathrm{e}^{\mathrm{i}kr}}{r} \mathrm{e}^{-\mathrm{i}\boldsymbol{k}' \cdot \boldsymbol{r}'}, \tag{9.71}$$

where $\boldsymbol{k}' \equiv k\hat{\boldsymbol{n}}$. Combining Eq. (9.71) with Eq. (9.68) and the expansion of plane waves in spherical harmonics, Eq. (7.58), we have that as $r \to \infty$,

$$F(\boldsymbol{r}, \boldsymbol{r}') \underset{r\to\infty}{\to} \frac{\mathrm{e}^{ikr}}{ikr} \mathrm{e}^{-ikr'\cos\gamma} = \frac{\mathrm{e}^{ikr}}{ikr} \sum_{l=0}^{\infty} (-\mathrm{i})^l (2l+1) j_l(kr') P_l(\cos\gamma). \tag{9.72}$$

By combining Eqs. (9.69) and (9.70), and comparing with Eq. (9.72), we want a_l such that, as $r \to \infty$,

$$a_l h_l^{(1)}(kr) \underset{r\to\infty}{\sim} \frac{\mathrm{e}^{ikr}}{ikr} (-\mathrm{i})^l (2l+1) = (2l+1) \frac{\mathrm{e}^{ikr}}{kr} \mathrm{e}^{-\mathrm{i}\pi(l+1)/2}. \tag{9.73}$$

Comparing with the asymptotic form of $h_l^{(1)}$, Eq. (7.49), this requirement is met with $a_l = 2l + 1$. Combining with Eqs. (9.70) and (9.69), we have

$$F(\boldsymbol{r}, \boldsymbol{r}') = \sum_{l=0}^{\infty} (2l+1) j_l(kr_<) h_l^{(1)}(kr_>) P_l(\cos\gamma). \tag{9.74}$$

Finally, the addition theorem for spherical harmonics, Eq. (6.37), shows that Eq. (9.74) is the same as Eq. (9.67).

9.4 Diffusion equation

The inhomogeneous diffusion equation, Eq. (3.8), for a system in volume V bounded by surface S can be written in terms of a *diffusion operator*:

$$\mathcal{L}\psi \equiv \left(D\nabla^2 - \frac{\partial}{\partial t}\right)\psi(\boldsymbol{r},t) = -\rho(\boldsymbol{r},t), \tag{9.75}$$

where $\psi(\boldsymbol{r},t)$ often denotes a *concentration* (number of particles per volume) with $\rho(\boldsymbol{r},t)$ representing the number of particles created (or destroyed) per volume per time. Equation (9.75) is a parabolic PDE (see Appendix D) which has unique solutions for Dirichlet or Neumann boundary conditions on S, but not both, and for $\psi(\boldsymbol{r},t=t_0)$ specified throughout V, the *initial condition.*

In this section, we find the Green function for the diffusion operator. We seek the solution of

$$\left(D\nabla^2 - \frac{\partial}{\partial t}\right)G(\boldsymbol{r},\boldsymbol{r}',t,t') = \delta(\boldsymbol{r}-\boldsymbol{r}')\delta(t-t'), \tag{9.76}$$

subject to the appropriate boundary conditions, which are given in the following, Eq. (9.77). Peeking ahead to Eq. (9.86), we'll show how the boundary conditions on G allow the boundary conditions on ψ to be incorporated into the solution of Eq. (9.75). We'll also see how the boundary conditions lead to the reciprocity relation, Eq. (9.82), which captures a key feature of the behavior of physical systems in time.

9.4.1 Boundary conditions, causality, and reciprocity

Boundary conditions

$$\begin{aligned} G(\boldsymbol{r},\boldsymbol{r}',t,t') &= 0 && \text{Dirichlet condition on } \psi;\ \boldsymbol{r} \text{ on } S \\ \boldsymbol{\nabla}_r G(\boldsymbol{r},\boldsymbol{r}',t,t')\cdot\hat{\boldsymbol{n}} &= 0 && \text{Neumann condition on } \psi;\ \boldsymbol{r} \text{ on } S \\ G(\boldsymbol{r},\boldsymbol{r}',t,t') &= 0 && t<t' \quad \text{(causality).} \end{aligned} \tag{9.77}$$

causality

Note the "boundary condtion" in time, which is a statement of *causality*: The *effect* at time t cannot precede the *cause* at time t'; G is nonzero only for $t > t'$.

Reciprocity relation: $G(\boldsymbol{r},\boldsymbol{r}',t,t') = G(\boldsymbol{r}',\boldsymbol{r},-t',-t)$

We're going to make some substitutions here that rely on $\delta(t-t')$ being an even function (Section 4.5). Let $t \to -t$, $t' \to -t''$, and $\boldsymbol{r}' \to \boldsymbol{r}''$ in Eq. (9.76) with the result

$$\left(D\nabla^2 + \frac{\partial}{\partial t}\right) G(\boldsymbol{r}, \boldsymbol{r}'', -t, -t'') = \delta(\boldsymbol{r} - \boldsymbol{r}'')\delta(t - t''). \tag{9.78}$$

Multiply Eq. (9.78) by $G(\boldsymbol{r}, \boldsymbol{r}', t, t')$, multiply Eq. (9.76) by $G(\boldsymbol{r}, \boldsymbol{r}'', -t, -t'')$, and subtract:

$$\begin{aligned} &D[G(\boldsymbol{r}, \boldsymbol{r}', t, t')\nabla^2 G(\boldsymbol{r}, \boldsymbol{r}'', -t, -t'') - G(\boldsymbol{r}, \boldsymbol{r}'', -t, -t'')\nabla^2 G(\boldsymbol{r}, \boldsymbol{r}', t, t')] \\ &\quad + \frac{\partial}{\partial t}[G(\boldsymbol{r}, \boldsymbol{r}', t, t')G(\boldsymbol{r}, \boldsymbol{r}'', -t, -t'')] \\ &= \delta(\boldsymbol{r} - \boldsymbol{r}'')\delta(t - t'')G(\boldsymbol{r}, \boldsymbol{r}', t, t') - \delta(\boldsymbol{r} - \boldsymbol{r}')\delta(t - t')G(\boldsymbol{r}, \boldsymbol{r}'', -t, -t''). \end{aligned} \tag{9.79}$$

Integrate Eq. (9.79) over V and all time t:

$$\begin{aligned} &\int_{-\infty}^{\infty}\int_V \left\{D[G(\boldsymbol{r}, \boldsymbol{r}', t, t')\nabla^2 G(\boldsymbol{r}, \boldsymbol{r}'', -t, -t'') - G(\boldsymbol{r}, \boldsymbol{r}'', -t, -t'')\nabla^2 G(\boldsymbol{r}, \boldsymbol{r}', t, t')]\right. \\ &\qquad \left. + \frac{\partial}{\partial t}\{G(\boldsymbol{r}, \boldsymbol{r}', t, t')G(\boldsymbol{r}, \boldsymbol{r}'', -t, -t'')\}\right\} \mathrm{d}^3 r\, \mathrm{d}t \\ &\qquad = G(\boldsymbol{r}'', \boldsymbol{r}', t'', t') - G(\boldsymbol{r}', \boldsymbol{r}'', -t', -t'') \end{aligned} \tag{9.80}$$

Perform the volume integration on the terms in square brackets in Eq. (9.80) making use of Green's identity, Eq. (9.28), and perform the time integration on the time derivative. The left side of Eq. (9.80) is equivalent to:

$$\begin{aligned} &D\int_{-\infty}^{\infty}\int_S [G(\boldsymbol{r}, \boldsymbol{r}', t, t')\boldsymbol{\nabla} G(\boldsymbol{r}, \boldsymbol{r}'', -t, -t'') - G(\boldsymbol{r}, \boldsymbol{r}'', -t, -t'')\boldsymbol{\nabla} G(\boldsymbol{r}, \boldsymbol{r}', t, t')] \cdot \hat{\boldsymbol{n}}\, \mathrm{d}s\, \mathrm{d}t \\ &\qquad + \int_V \{G(\boldsymbol{r}, \boldsymbol{r}', t, t')G(\boldsymbol{r}, \boldsymbol{r}'', -t, -t'')\}\, \mathrm{d}^3 r \Big|_{t=-\infty}^{t=\infty}. \end{aligned} \tag{9.81}$$

The first integral in expression (9.81) vanishes by the boundary conditions on G, while the second integral vanishes because of the causality condition (show this). Thus, we have from Eq. (9.80),

$$G(\boldsymbol{r}, \boldsymbol{r}', t, t') = G(\boldsymbol{r}', \boldsymbol{r}, -t', -t). \tag{9.82}$$

The reciprocity relation, Eq. (9.82), expresses the invariance of the order of cause and effect with respect to the zero of time. Referring to Figure 9.3, a cause at $(\boldsymbol{r}', t')$ resulting in an effect at $(\boldsymbol{r}, t)$ is equivalent to a cause at $(\boldsymbol{r}, -t)$ resulting in the same effect at $(\boldsymbol{r}', -t')$. An *arrow of time* is implied by diffusion: Substances move in such a way as to disperse. The reciprocity relation shows that the Green function depends only on the *time difference* $(t - t')$. We can take that idea further. The Green function is *time translation invariant*: It is invariant under the shift $t \to t + a$, $t' \to t' + a$.

time translation invariance

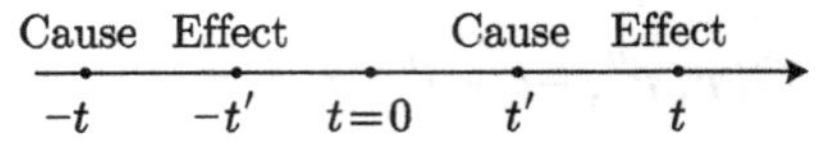

Figure 9.3 Arrow of time.

9.4.2 Solution to the diffusion equation

In Eq. (9.76), let $\boldsymbol{r} \leftrightarrow \boldsymbol{r}'$ and $t \leftrightarrow -t'$, and use Eq. (9.82). We find

$$\left(D\nabla^2_{r'} + \frac{\partial}{\partial t'}\right) G(\boldsymbol{r}, \boldsymbol{r}', t, t') = \delta(\boldsymbol{r} - \boldsymbol{r}')\delta(t - t'). \tag{9.83}$$

In Eq. (9.75), let $\boldsymbol{r} \rightarrow \boldsymbol{r}'$ and $t \rightarrow t'$; we have

$$\left(D\nabla^2_{r'} - \frac{\partial}{\partial t'}\right) \psi(\boldsymbol{r}', t') = -\rho(\boldsymbol{r}', t'). \tag{9.84}$$

Multiply Eq. (9.84) by $G(\boldsymbol{r}, \boldsymbol{r}', t, t')$, multiply Eq. (9.83) by $\psi(\boldsymbol{r}', t')$, and subtract:

$$\begin{aligned} &D[G(\boldsymbol{r}, \boldsymbol{r}', t, t')\nabla^2_{r'}\psi(\boldsymbol{r}', t') - \psi(\boldsymbol{r}', t')\nabla^2_{r'}G(\boldsymbol{r}, \boldsymbol{r}', t, t')] - \frac{\partial}{\partial t'}(G(\boldsymbol{r}, \boldsymbol{r}', t, t')\psi(\boldsymbol{r}', t')) \\ &= -G(\boldsymbol{r}, \boldsymbol{r}', t, t')\rho(\boldsymbol{r}', t') - \psi(\boldsymbol{r}', t')\delta(\boldsymbol{r} - \boldsymbol{r}')\delta(t - t'). \end{aligned} \tag{9.85}$$

We are, of course, going to integrate Eq. (9.85) with respect to d^3r' over V and with respect to $\mathrm{d}t'$ for[32] $t_0 \le t' \le t$, where the initial condition on ψ is given at t_0. We find:

$$\begin{aligned} \psi(\boldsymbol{r}, t) &= -\int_{t_0}^{t}\int_V G(\boldsymbol{r}, \boldsymbol{r}', t, t')\rho(\boldsymbol{r}', t')\,\mathrm{d}^3r'\,\mathrm{d}t' - \int_V G(\boldsymbol{r}, \boldsymbol{r}', t, t_0)\psi(\boldsymbol{r}', t_0)\,\mathrm{d}^3r' \\ &\quad - D\int_{t_0}^{t}\int_S [G(\boldsymbol{r}, \boldsymbol{r}', t, t')\boldsymbol{\nabla}_{r'}\psi(\boldsymbol{r}', t') - \psi(\boldsymbol{r}', t')\boldsymbol{\nabla}_{r'}G(\boldsymbol{r}, \boldsymbol{r}', t, t')]\cdot\hat{\boldsymbol{n}}\,\mathrm{d}s'\,\mathrm{d}t' \\ &\equiv F_\rho(\boldsymbol{r}, t) + F_\psi(\boldsymbol{r}, t) + F_B(\boldsymbol{r}, t), \end{aligned} \tag{9.86}$$

where $t > t_0$. There are three types of integrals in Eq. (9.86): $F_\rho(\boldsymbol{r}, t)$ represents the contribution of sources $\rho(\boldsymbol{r}, t)$ – the Green function "propagates" the effects of sources acting between t_0 and t; $F_\psi(\boldsymbol{r}, t)$ represents the evolution of the initial condition $\psi(\boldsymbol{r}, t_0)$; and $F_B(\boldsymbol{r}, t)$ represents the contributions of the boundary conditions. Parabolic equations require Dirichlet or Neumann boundary conditions, but not both (in addition to an initial condition). If ψ is to satisfy Dirichlet conditions, choose $G = 0$ so as to exclude any contribution of $\hat{\boldsymbol{n}}\cdot\boldsymbol{\nabla}\psi$ on S; if ψ is to satisfy Neumann conditions, choose $\hat{\boldsymbol{n}}\cdot\boldsymbol{\nabla}G = 0$ to rule out any contribution from ψ on S. The functions F_ψ and F_B satisfy the homogeneous diffusion equation (Exercise 9.12).

---

[32]Technically, the upper limit of the time integrals should be "t^+," where $t^+ \equiv t + \epsilon$. We should integrate to $t' = t + \epsilon$ and then let $\epsilon \to 0$, with $\epsilon > 0$, what's denoted $\epsilon \to 0^+$. In that way, the Green function vanishes at the upper limit, $G(\boldsymbol{r}, \boldsymbol{r}', t, t + \epsilon) = 0$. It's usually not necessary to display t^+ explicitly. For the lower limit, there's no problem with extending it to $-\infty$; the initial condition specified at $t = t_0$ is zero for all times $t < t_0$.

Does Eq. (9.86) reproduce the initial condition on ψ? What's the limit of Eq. (9.86) as $t \to t_0$? Clearly F_ρ and F_B vanish as $t \to t_0$. Is

$$\lim_{\epsilon \to 0^+} \int_V G(\boldsymbol{r}, \boldsymbol{r}', t_0 + \epsilon, t_0) \psi(\boldsymbol{r}', t_0) \, \mathrm{d}^3 r' = -\psi(\boldsymbol{r}, t_0) \quad ?$$

We now show the answer is yes. We can get a handle on the limiting property of the Green function by integrating Eq. (9.76) over t from $t' - \epsilon$ to $t' + \epsilon$:

$$\int_{t'-\epsilon}^{t'+\epsilon} \left(\nabla^2 - \frac{\partial}{\partial t} \right) G(\boldsymbol{r}, \boldsymbol{r}', t, t') \, \mathrm{d}t = \int_{t'-\epsilon}^{t'+\epsilon} \delta(\boldsymbol{r} - \boldsymbol{r}') \delta(t - t') \, \mathrm{d}t = \delta(\boldsymbol{r} - \boldsymbol{r}'). \tag{9.87}$$

The time integral over $\nabla^2 G$ on the left of Eq. (9.87) vanishes as $\epsilon \to 0$ because G is continuous in its arguments (see discussion in Section 9.1); the singularity on the right of Eq. (9.87) must be supplied by the time derivative of G. Thus,

$$\begin{aligned} -\int_{t'-\epsilon}^{t'+\epsilon} \mathrm{d}t \frac{\mathrm{d}}{\mathrm{d}t} G(\boldsymbol{r}, \boldsymbol{r}', t, t') &= -G(\boldsymbol{r}, \boldsymbol{r}', t, t')\Big|_{t=t'-\epsilon}^{t=t'+\epsilon} \\ &= -G(\boldsymbol{r}, \boldsymbol{r}', t' + \epsilon, t') = \delta(\boldsymbol{r} - \boldsymbol{r}'), \end{aligned}$$

where we've used $G = 0$ for $t < t'$ (causality). We therefore have[33]

$$\lim_{\epsilon \to 0^+} G(\boldsymbol{r}, \boldsymbol{r}', t_0 + \epsilon, t_0) = -\delta(\boldsymbol{r} - \boldsymbol{r}'). \tag{9.88}$$

9.4.3 Free-space Green function

For free space, $G_0(\boldsymbol{r}, \boldsymbol{r}', t, t')$ can only be a function[34] of the *differences* $\boldsymbol{r} - \boldsymbol{r}'$ and $t - t'$, so that Eq. (9.76) can be written

translational invariance in time and space

$$\left(D\nabla^2 - \frac{\partial}{\partial t} \right) G_0(\boldsymbol{r} - \boldsymbol{r}', t - t') = \delta(\boldsymbol{r} - \boldsymbol{r}') \delta(t - t').$$

Without loss of generality, we can set $\boldsymbol{r}' = 0$, $t' = 0$, with[35]

$$\left(D\nabla^2 - \frac{\partial}{\partial t} \right) G_0(\boldsymbol{r}, t) = \delta(\boldsymbol{r}) \delta(t). \tag{9.89}$$

[33]We should appreciate the singular nature of $G(\boldsymbol{r}, \boldsymbol{r}', t, t')$ in its time variables. In addition to the limiting property in Eq. (9.88), we have $\lim_{\epsilon \to 0^+} G(\boldsymbol{r}, \boldsymbol{r}', t, t + \epsilon) = 0$ from the causality condition.

[34]Time translation invariance was established through the reciprocity relation, Eq. (9.82), which if there are no pesky boundaries (free space) implies translational invariance in space as well.

[35]The singularity thus occurs at the origin of the space-time coordinate system, which can be placed anywhere we want in free space.

The causality condition is $G_0(\boldsymbol{r}, t<0)=0$. Equation (9.88) can then be written

$$\lim_{\epsilon\to 0^+} G_0(\boldsymbol{r},\epsilon) = -\delta(\boldsymbol{r}). \tag{9.90}$$

Equation (9.89) can be solved through Fourier transformation.[36] Let

$$G_0(\boldsymbol{r},t) \equiv \frac{1}{8\pi^3}\int_K \mathrm{e}^{\mathrm{i}\boldsymbol{k}\cdot\boldsymbol{r}} G_0(\boldsymbol{k},t)\,\mathrm{d}^3k, \tag{9.91}$$

where K denotes a space of three-dimensional vectors $\boldsymbol{k}$ (*reciprocal space*), much as V denotes the space of all position vectors $\boldsymbol{r}$. Equation (9.91) is a three-dimensional Fourier transform.[37] The vector $\boldsymbol{k}$ can be represented in Cartesian components $\boldsymbol{k} = k_x\hat{\boldsymbol{x}} + k_y\hat{\boldsymbol{y}} + k_z\hat{\boldsymbol{z}}$, just as the position vector $\boldsymbol{r} = x\hat{\boldsymbol{x}} + y\hat{\boldsymbol{y}} + z\hat{\boldsymbol{z}}$, so that $\boldsymbol{k}\cdot\boldsymbol{r} = k_x x + k_y y + k_z z$, with $\int_K \mathrm{e}^{i\boldsymbol{k}\cdot\boldsymbol{r}}\,\mathrm{d}^3k \equiv \iiint_{-\infty}^{\infty} \mathrm{e}^{\mathrm{i}k_x x}\mathrm{e}^{\mathrm{i}k_y y}\mathrm{e}^{\mathrm{i}k_z z}\,\mathrm{d}k_x\,\mathrm{d}k_y\,\mathrm{d}k_z$. We don't have to use Cartesian basis vectors; we break it down this way to gain familiarity. We're indulging in the typical physicist sloppiness of using the same symbol twice, once for $G_0(\boldsymbol{r})$ (the *real-space* Green function) and $G_0(\boldsymbol{k})$ (the *k-space* Green function). It should be clear from the context which is being used.

Because three-dimensional Fourier transforms may be unfamiliar, let's spend a moment on their properties. Equation (9.91) can be inverted with $G_0(\boldsymbol{k}) = \int_V \mathrm{e}^{-\mathrm{i}\boldsymbol{k}\cdot\boldsymbol{r}'} G_0(\boldsymbol{r}')\,\mathrm{d}^3r'$. Let's show that:

$$\begin{aligned}
\int_V \mathrm{e}^{-\mathrm{i}\boldsymbol{k}\cdot\boldsymbol{r}'} G_0(\boldsymbol{r}')\,\mathrm{d}^3r' &= \frac{1}{8\pi^3}\int_V \mathrm{e}^{-\mathrm{i}\boldsymbol{k}\cdot\boldsymbol{r}'}\int_K \mathrm{e}^{\mathrm{i}\boldsymbol{k}'\cdot\boldsymbol{r}'} G_0(\boldsymbol{k}')\,\mathrm{d}^3k'\,\mathrm{d}^3r' \\
&= \frac{1}{8\pi^3}\int_K\left(\int_V \mathrm{e}^{\mathrm{i}(\boldsymbol{k}'-\boldsymbol{k})\cdot\boldsymbol{r}'}\,\mathrm{d}^3r'\right) G_0(\boldsymbol{k}')\,\mathrm{d}^3k' \\
&= \frac{1}{8\pi^3}\int_K (8\pi^3\delta(\boldsymbol{k}-\boldsymbol{k}'))G_0(\boldsymbol{k}')\,\mathrm{d}^3k' = G_0(\boldsymbol{k}),
\end{aligned}$$

where we've used three copies of Eq. (4.33):

$$\begin{aligned}
\delta(\boldsymbol{k}-\boldsymbol{k}') &= \delta(k_x-k_x')\delta(k_y-k_y')\delta(k_z-k_z') \\
&= \frac{1}{(2\pi)^3}\int_{-\infty}^{\infty}\mathrm{e}^{\mathrm{i}(k_x-k_x')x}\,\mathrm{d}x\int_{-\infty}^{\infty}\mathrm{e}^{\mathrm{i}(k_y-k_y')y}\,\mathrm{d}y\int_{-\infty}^{\infty}\mathrm{e}^{\mathrm{i}(k_z-k_z')z}\,\mathrm{d}z \\
&= \frac{1}{8\pi^3}\int_V \mathrm{e}^{\mathrm{i}(\boldsymbol{k}-\boldsymbol{k}')\cdot\boldsymbol{r}}\,\mathrm{d}^3r.
\end{aligned}$$

---

[36] We were able to quickly derive G_0 for the Laplace and Helmholtz operators because the independent variables in the PDE for G_0 all refer to spatial variables. G_0 only depends on the distance, and thus, we're free to discard the angular part of the Laplacian, reducing the PDE for G_0 to an ODE. No such simplification occurs for the diffusion operator, where we have second-order spatial derivatives and a first-order time derivative. By Fourier transforming the spatial variables, the PDE for G_0 is reduced to a first-order ODE, Eq. (9.92). G_0 for the diffusion operator is considerably more complicated than that for the Laplace and Helmholtz operators.

[37] We treated one-dimensional Fourier transforms in Chapter 5.

Substituting Eq. (9.91) in Eq. (9.89), and using the integral representation of the three-dimensional delta function $\delta(\boldsymbol{r}) = (1/(8\pi^3))\int_K \mathrm{d}^3k \mathrm{e}^{\mathrm{i}\boldsymbol{k}\cdot\boldsymbol{r}}$, we find

$$\frac{1}{8\pi^3}\int_K \left[\left(-Dk^2 - \frac{\partial}{\partial t}\right) G_0(\boldsymbol{k}, t) - \delta(t)\right] \mathrm{e}^{\mathrm{i}\boldsymbol{k}\cdot\boldsymbol{r}}\, \mathrm{d}^3k = 0,$$

implying the first-order differential equation[38]

$$\left(\frac{\partial}{\partial t} + Dk^2\right) G_0(\boldsymbol{k}, t) = -\delta(t). \tag{9.92}$$

For $t > 0$, the solution to Eq. (9.92) is

$$G_0(\boldsymbol{k}, t) = A\mathrm{e}^{-Dk^2t} \qquad (t > 0). \tag{9.93}$$

By integrating Eq. (9.92) over the singularity in time, we have, in the usual way,

$$\int_{-\epsilon}^{\epsilon} \left(\frac{\partial}{\partial t} + Dk^2\right) G_0(\boldsymbol{k}, t)\, \mathrm{d}t = -1,$$

implying for $\epsilon \to 0^+$,

$$G_0(\boldsymbol{k}, \epsilon) - G_0(\boldsymbol{k}, -\epsilon) = -1. \tag{9.94}$$

But the causality condition requires $G_0(\boldsymbol{k}, -\epsilon) = 0$. Thus, we can take $A = -1$. The Fourier-transformed free-space Green function for the diffusion operator is thus

$$G_0(\boldsymbol{k}, t) = \begin{cases} 0 & t < 0 \\ -\mathrm{e}^{-Dk^2t} & t > 0. \end{cases} \tag{9.95}$$

To obtain the real-space Green function, we must evaluate the Fourier transform:

$$G_0(\boldsymbol{r}, t) = \frac{1}{8\pi^3}\int_K \mathrm{e}^{\mathrm{i}\boldsymbol{k}\cdot\boldsymbol{r}} G_0(\boldsymbol{k}, t)\, \mathrm{d}^3k.$$

We omit the details of evaluating the integral, and simply write down the result:

$$G_0(\boldsymbol{r}, \boldsymbol{r}', t, t') = \begin{cases} 0 & t < t' \\ -\frac{1}{[4\pi D(t-t')]^{3/2}} \exp\left[-\frac{|\boldsymbol{r}-\boldsymbol{r}'|^2}{4D(t-t')}\right] & t > t'. \end{cases} \tag{9.96}$$

Equation (9.96) can be written more compactly using the step function $\theta(t - t')$, Eq. (4.34):

$$G_0(\boldsymbol{r}, \boldsymbol{r}', t, t') = -\frac{\theta(t-t')}{[4\pi D(t-t')]^{3/2}} \exp\left[-\frac{|\boldsymbol{r}-\boldsymbol{r}'|^2}{4D(t-t')}\right]. \tag{9.97}$$

[38] There is thus an ODE to solve for each $\boldsymbol{k}$-vector.

We catch a break with G_0 for the diffusion operator: Its form is the same in any number of dimensions. In three dimensions, $|\boldsymbol{r}|^2 = \sum_{i=1}^{3} x_i^2$. Then,

$$\frac{1}{[4\pi Dt]^{3/2}} \mathrm{e}^{-r^2/(4Dt)} = \prod_{i=1}^{3} \left(\frac{1}{\sqrt{4\pi Dt}} \mathrm{e}^{-x_i^2/(4Dt)} \right).$$

G_0 in three dimensions is the product of three, one-dimensional Green functions.

Example. Consider an unbounded system in one dimension where initially the concentration of a substance is zero. Because of the initial condition, and letting the boundaries recede to infinity,[39] we need only consider the first integral in Eq. (9.86), which represents the contribution from sources. Let the source function be given by $\rho(x', t') = \psi_0 \delta(x')\delta(t')$, that is, an "injection" of strength ψ_0 at the origin at time $t = 0$. Using the one-dimensional free-space Green function, we have, using Eq. (9.86),

$$\begin{aligned}\psi(x,t) &= -\int_{-\infty}^{t}\int_{-\infty}^{\infty} \left(-\frac{\theta(t-t')}{\sqrt{4\pi D(t-t')}} \mathrm{e}^{-(x-x')^2/(4Dt)} \right) \psi_0 \delta(x')\delta(t') \, \mathrm{d}x' \, \mathrm{d}t' \\ &= \frac{\psi_0}{\sqrt{4\pi Dt}} \mathrm{e}^{-x^2/(4Dt)} \theta(t).\end{aligned}$$

The behavior of $\psi(x, t)$ is shown in Figure 9.4.

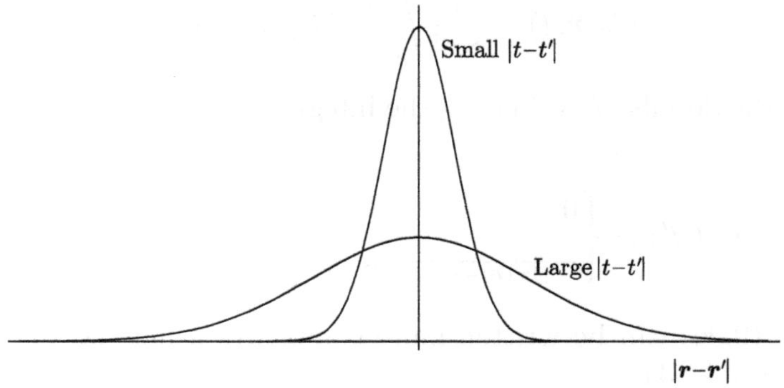

Figure 9.4 Diffusion of a localized source.

[39]The limits $x \to \infty$ and $t \to \infty$ do not commute in G_0. We expect there to be no contribution from infinitely far-removed boundaries. That expectation is met in Eq. (9.97) if we first let $\boldsymbol{r} \to \infty$, *then* let $t \to \infty$. With diffusion, the order of the limits matters: first let the spatial variable get large, then time.

Applying the diffusion operator to $G_0(\boldsymbol{r},\boldsymbol{r}',t,t')$ in the form of Eq. (9.97), it can be shown (Exercise 9.13) that

$$\left(D\nabla^2 - \frac{\partial}{\partial t}\right)G_0(\boldsymbol{r},\boldsymbol{r}',t,t') = \frac{1}{[4\pi D(t-t')]^{3/2}}\exp\left[-\frac{|\boldsymbol{r}-\boldsymbol{r}'|^2}{4D(t-t')}\right]\delta(t-t'). \tag{9.98}$$

The right side of Eq. (9.98) doesn't appear to be in the form of Eq. (9.76), yet it secretly is. For $t \neq t'$, the right side of Eq. (9.98) vanishes, the same as Eq. (9.76). The singularity occurs at $t = t'$, and there is the delta function $\delta(t-t')$ on the right of Eq. (9.98), the same as Eq. (9.76). We recover a delta function in the spatial variables from the behavior of the right side of Eq. (9.98) as $t' \to t$. We know from Eq. (9.88) that for $t = t' + \epsilon$, the Green function approaches $-\delta(\boldsymbol{r}-\boldsymbol{r}')$ in the limit $\epsilon \to 0^+$. From Exercise 3.1, the sequence of normalized functions $\delta_\epsilon(x) = (1/\sqrt{\pi\epsilon})\exp(-x^2/\epsilon)$ produce a one-dimensional delta function in the limit $\epsilon \to 0$ – the very form of the term on the right of Eq. (9.98). The same sequence applied in each of three dimensions produces a three-dimensional delta function. Thus, Eq. (9.98) is equivalent to Eq. (9.76). We have to "synthesize" $\delta(\boldsymbol{r}-\boldsymbol{r}')$ as occurring as $t \to t'$.

9.5 Wave equation

The inhomogeneous wave equation, Eq. (3.14), for a system in volume V bounded by surface S can be written in terms of a *wave operator*[40]: wave operator

$$\mathcal{L}\psi = \left(\nabla^2 - \frac{1}{c^2}\frac{\partial^2}{\partial t^2}\right)\psi(\boldsymbol{r},t) = -\alpha\rho(\boldsymbol{r},t). \tag{9.99}$$

The wave equation is a hyperbolic PDE (see Appendix D), the solutions of which require the specification on S of Dirichlet or Neumann boundary conditions, but not both, and initial conditions $\psi(\boldsymbol{r},t_0)$ and $\partial\psi(\boldsymbol{r},t)/\partial t|_{t=t_0}$ throughout V.

We seek the Green function associated with the wave operator, the solution of

$$\left(\nabla^2 - \frac{1}{c^2}\frac{\partial^2}{\partial t^2}\right)G(\boldsymbol{r},\boldsymbol{r}',t,t') = \delta(\boldsymbol{r}-\boldsymbol{r}')\delta(t-t') \tag{9.100}$$

subject to boundary conditions. The boundary conditions here are the same as those for the diffusion operator listed in Eq. (9.77), including the causality condition.

[40]The wave operator is sometimes called the *d'Alembertian* operator, denoted $\Box$ or $\Box^2$, a notation we don't use. The parameter α in Eq. (9.99) is a conversion factor between the dimension of the source term ρ, and the dimension of ψ, with the dimension of the product $\alpha\rho$ equal to that of ψ/L^2, where L^2 denotes the dimension of length squared.

Following the pattern of previous sections, we should establish the reciprocity relation. It turns out we've already done that task (Exercise 9.15). The Green function for the wave equation satisfies the same reciprocity relation as for the diffusion equation:

reciprocity relation for wave equation Green function

$$G(\boldsymbol{r}, \boldsymbol{r}', t, t') = G(\boldsymbol{r}', \boldsymbol{r}, -t', -t). \tag{9.82}$$

time translation invariance

The Green function for the wave operator is therefore invariant under time translations and depends on t and t' only through the difference $t - t'$ (Section 9.4). We note that the *homogeneous* wave equation is additionally *time-reversal invariant*: Equation (9.99) with no source function is invariant[41] under $t \to -t$. The causality condition, however, $G(\boldsymbol{r}, \boldsymbol{r}', t, t') = 0$ for $t < t'$, introduces a fundamental distinction between t and t', so that even though Eq. (9.100) appears to be time-reversal invariant (let $t \to -t$, $t' \to -t'$), its Green function solution is not.[42] The causality condition imposes an arrow of time: It breaks time-reversal symmetry.[43]

To find the solution of the inhomogeneous wave equation, follow the steps in Section 9.4.2. Let $\boldsymbol{r} \leftrightarrow \boldsymbol{r}'$ and $t \leftrightarrow -t'$ in Eq. (9.100) and make use of the reciprocity relation, Eq. (9.82). Call that equation A. Let $\boldsymbol{r} \to \boldsymbol{r}'$ and $t \to t'$ in Eq. (9.99); call that equation B. Multiply equation B by $G(\boldsymbol{r}, \boldsymbol{r}', t, t')$ and subtract from it equation A multiplied by $\psi(\boldsymbol{r}', t')$. Integrate with respect to d^3r' over V and integrate with respect to $\mathrm{d}t'$ from[44] t_0 to t. Make use of Green's identity, Eq. (9.28), in the volume integral involving the Laplacians, and the identity noted in Exercise 9.15 in the time integral over the time derivative. After algebra, we have an equation similar to Eq. (9.86): (for $t > t_0$)

$$\begin{aligned}
\psi(\boldsymbol{r}, t) &= -\alpha \int_{t_0}^{t} \int_V G(\boldsymbol{r}, \boldsymbol{r}', t, t') \rho(\boldsymbol{r}', t') \,\mathrm{d}^3r' \,\mathrm{d}t' \\
&\quad - \frac{1}{c^2} \int_V \left(G(\boldsymbol{r}, \boldsymbol{r}', t, t_0) \left.\frac{\partial \psi(\boldsymbol{r}', t')}{\partial t'}\right|_{t'=t_0} - \psi(\boldsymbol{r}', t_0) \left.\frac{\partial G(\boldsymbol{r}, \boldsymbol{r}', t, t')}{\partial t'}\right|_{t'=t_0} \right) \mathrm{d}^3r' \\
&\quad - \int_{t_0}^{t} \int_S [G(\boldsymbol{r}, \boldsymbol{r}', t, t' \boldsymbol{\nabla}_{r'} \psi(\boldsymbol{r}', t') - \psi(\boldsymbol{r}', t') \boldsymbol{\nabla}_{r'} G(\boldsymbol{r}, \boldsymbol{r}', t, t')] \cdot \hat{\boldsymbol{n}} \,\mathrm{d}s' \,\mathrm{d}t' \\
&\equiv F_\rho(\boldsymbol{r}, t) + F_\psi(\boldsymbol{r}, t) + F_B(\boldsymbol{r}, t).
\end{aligned} \tag{9.101}$$

The functions F_ψ and F_B (associated with the initial conditions and boundary conditions) satisfy the homogeneous wave equation, for the same reasons given in Exercise 9.12. Equation (9.101) reproduces the correct initial condition because of the result found in Exercise 9.17.

---

[41]The inhomogeneous wave equation, Eq. (9.99), is time-reversal invariant only if the source function satisfies $\rho(\boldsymbol{r}, t) = \rho(\boldsymbol{r}, -t)$.

[42]Even though a differential equation may display a symmetry (in this case time-reversal invariance of Eq. (9.100)), the solution need not have the same symmetry if the initial conditions or boundary conditions break the symmetry.

[43]There *are* theories in physics involving time-symmetric waves. Here one must conceive of signals that arrive from the future to influence events now.

[44]Perform the integration to $t' = t + \epsilon$ and let $\epsilon \to 0^+$. In that way, the Green function vanishes at the upper time limit by the causality condition.

Free-space Green function

For systems without boundaries at infinite distances, the Green function exhibits translational invariance with respect to time and space (Section 9.4). The differential equation for G_0 is then

$$\left(\nabla^2 - \frac{1}{c^2}\frac{\partial^2}{\partial t^2}\right) G_0(\boldsymbol{r}-\boldsymbol{r}', t-t') = \delta(\boldsymbol{r}-\boldsymbol{r}')\delta(t-t').$$

Without loss of generality, we can set $\boldsymbol{r}' = 0$ and $t' = 0$. Thus,

$$\left(\nabla^2 - \frac{1}{c^2}\frac{\partial^2}{\partial t^2}\right) G_0(\boldsymbol{r}, t) = \delta(\boldsymbol{r})\delta(t), \tag{9.102}$$

subject to the requirement $G_0(\boldsymbol{r}, t) = 0$ for $t < 0$. Equation (9.102) can be solved through Fourier transformation. Under the substitution in Eq. (9.91), introducing the k-space Green function $G_0(\boldsymbol{k}, t)$, we have, following the same steps as in Section 9.4, the second-order differential equation:

$$\left(\frac{1}{c^2}\frac{\partial^2}{\partial t^2} + k^2\right) G_0(\boldsymbol{k}, t) = -\delta(t). \tag{9.103}$$

For $t > 0$, the solution of Eq. (9.103) is

$$G_0(\boldsymbol{k}, t) = A\mathrm{e}^{ikct} + B\mathrm{e}^{-ikct}. \tag{9.104}$$

The constants A, B can be obtained from the matching conditions at $t = 0$. We require continuity in the time variable and a discontinuity in the time derivative (Section 9.1). Integrate Eq. (9.103) over t from $t = -\epsilon$ to $t = \epsilon$, with the results

$$\lim_{\epsilon\to 0^+} \left[G_0(\boldsymbol{k}, \epsilon) - G_0(\boldsymbol{k}, -\epsilon)\right] = 0 \qquad \lim_{\epsilon\to 0^+} \left.\frac{\partial G_0(\boldsymbol{k}, t)}{\partial t}\right|_{t=-\epsilon}^{t=\epsilon} = -c^2. \tag{9.105}$$

From the causality condition, however, $G_0(\boldsymbol{k}, -\epsilon) = 0$, implying that $B = -A$. The matching condition on the derivative, Eq. (9.105), implies $A = \mathrm{i}c/(2k)$. The k-space Green function is thus

$$G_0(\boldsymbol{k}, t) = -\frac{c}{k}\theta(t)\sin(kct). \tag{9.106}$$

We note that $G_0(\boldsymbol{k}, t)$ has the form of Eq. (9.106) in any number of spatial dimensions; only the nature (dimension) of the $\boldsymbol{k}$-vectors change, which impacts the form of $G_0(\boldsymbol{r}, t)$ through the Fourier transform.[45] To obtain

[45] It should be observed that the k-space Green function for the diffusion operator, Eq. (9.95), also has the same form for $\boldsymbol{k}$-vectors in any dimension. Because the Fourier transform of a Gaussian is a Gaussian, the free-space Green function $G_0(\boldsymbol{r}, t)$ for the diffusion operator has the same form in any number of spatial dimensions, Eq. (9.97). The form of $G_0(\boldsymbol{r}, t)$ for the wave operator is different in one, two, or three dimensions.

$G_0(\boldsymbol{r}, t)$ in three dimensions, substitute Eq. (9.106) in Eq. (9.91). We omit the details of evaluating that integral, and write down the result:

$$G_0(\boldsymbol{r}, \boldsymbol{r}', t, t') = -\frac{c}{4\pi|\boldsymbol{r} - \boldsymbol{r}'|}\theta(t - t')\delta(|\boldsymbol{r} - \boldsymbol{r}'| - c(t - t')). \tag{9.107}$$

Note that the reciprocity relation, Eq. (9.82), is satisfied.

Example. Consider an unbounded one-dimensional system, for which the quantity $\psi(x, t)$ (which satisfies the wave equation) has the value $\psi(x, t) = 0$ for $t \leq 0$. At time $t = 0$, a source is switched on at $x = 0$ and *stays* on, with $\rho(x, t) = \delta(x)\theta(t)$. Calculate $\psi(x, t)$ for $t > 0$. Because of the boundary conditions and the initial conditions, only the source term in Eq. (9.101) contributes:

$$\psi(x, t) = -\alpha \int_0^t \mathrm{d}t' \int_{-\infty}^{\infty} G(x, x', t, t')\delta(x')\theta(t')\mathrm{d}x' = -\alpha \int_0^t \mathrm{d}t' G_0(x, t - t'),$$

where we can use the free-space Green function G_0 (which has translational invariance in time as well as space). Let $\tau \equiv t - t'$; then

$$\psi(x, t) = -\alpha \int_0^t \mathrm{d}\tau G_0(x, \tau).$$

The free-space Green function associated with the one-dimensional wave operator is calculated in Exercise 9.18: $G_0(x, \tau) = -(c/2)\theta(\tau)\theta(c\tau - |x|)$. The integral we have to do is therefore as follows:

$$\psi(x, t) = \frac{\alpha c}{2} \int_0^t \theta(c\tau - |x|)\, \mathrm{d}\tau.$$

The integrand is nonzero only for $(|x|/c) < \tau < t$. Thus, as shown in Figure 9.5

$$\psi(x, t) = \frac{\alpha}{2}\theta(ct - |x|)(ct - |x|).$$

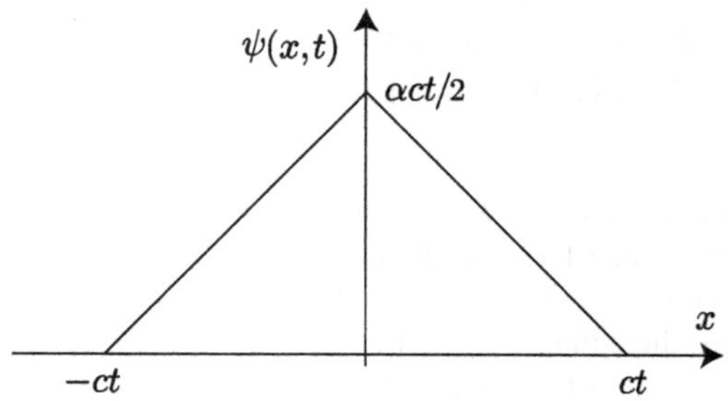

Figure 9.5 Propagation of $\psi(x, t)$ at speed c in one dimension.

9.6 The Kirchhoff integral theorem

Consider again the Poisson equation, Eq. (9.26). In Section 9.2, we sought the solution of this problem for sources contained in a region V bounded by a closed surface S, subject to prescribed boundary conditions. As we now discuss, this boundary value problem can be generalized in a way that finds use in various applications.

The idea is to use, in an equation such as Eq. (9.29), instead of the Green function G (that satisfies boundary conditions), the *free-space* Green function G_0. Repeating the steps that lead to Eq. (9.29), start with Green's identify, Eq. (9.28). Let $f = \psi(\boldsymbol{r})$ and $g = G_0(\boldsymbol{r}, \boldsymbol{r}')$, where $\psi(\boldsymbol{r})$ solves the Poisson equation ($\nabla^2\psi(\boldsymbol{r}) = -\alpha\rho(\boldsymbol{r})$), but where no attempt is made to make it satisfy boundary conditions other than to demand that $\psi(\boldsymbol{r}) \to 0$ as $r \to \infty$ (we know physically that the potential ψ due to localized sources must vanish at large distances), and $G_0(\boldsymbol{r}, \boldsymbol{r}')$ satisfies $\nabla_r^2 G_0(\boldsymbol{r}, \boldsymbol{r}') = \delta(\boldsymbol{r} - \boldsymbol{r}')$, but which also does not satisfy boundary conditions. With these substitutions,

$$\int_V [\psi(\boldsymbol{r}')\delta(\boldsymbol{r} - \boldsymbol{r}') + \alpha G_0(\boldsymbol{r}', \boldsymbol{r})\rho(\boldsymbol{r}')]\, \mathrm{d}^3r' =$$

$$\int_S [\psi(\boldsymbol{r}')\boldsymbol{\nabla}_{r'} G_0(\boldsymbol{r}', \boldsymbol{r}) - G_0(\boldsymbol{r}', \boldsymbol{r})\boldsymbol{\nabla}_{r'}\psi(\boldsymbol{r}')] \cdot \hat{\boldsymbol{n}}\, \mathrm{d}s'. \qquad (9.108)$$

The key point now is that V and S in Eq. (9.108) are *mathematical* constructs, and do not necessarily refer to anything physical. In particular, we could allow for the possibility that the observation point located by the vector $\boldsymbol{r}$ is not contained in V. Whether $\boldsymbol{r}$ does or does not lie in V determines whether the first integral on the left of Eq. (9.108) is or is not zero:

Kirchhoff integral representation

$$-\alpha\int_V G_0(\boldsymbol{r}', \boldsymbol{r})\rho(\boldsymbol{r}')\, \mathrm{d}^3r' + \int_S [\psi(\boldsymbol{r}')\boldsymbol{\nabla}_{r'} G_0(\boldsymbol{r}', \boldsymbol{r}) - G_0(\boldsymbol{r}', \boldsymbol{r})\boldsymbol{\nabla}_{r'}\psi(\boldsymbol{r}')] \cdot \hat{\boldsymbol{n}}\, \mathrm{d}s' \qquad (9.109)$$

$$= \begin{cases} \psi(\boldsymbol{r}) & \boldsymbol{r} \in V \\ 0 & \boldsymbol{r} \notin V. \end{cases}$$

Equation (9.109) is known as the *Kirchhoff integral representation*[46] of $\psi(\boldsymbol{r})$. It is not a *solution* of the Poisson equation because we do not in general know ψ and $\boldsymbol{\nabla}\psi \cdot \hat{\boldsymbol{n}}$ on S. Equation (9.109) *represents* $\psi(\boldsymbol{r})$ inside V in terms of its values and that of the normal derivative on S, and in terms of sources inside V. This equation is in the form of an *integral equation* (the unknown function ψ occurs under an integral sign, Chapter 10). The Kirchhoff representation comes into its own for the Helmholtz equation, where it forms the basis for a theory of diffraction.

[46]Also the Kirchhoff integral theorem.

Summary

The method of Green functions for solving inhomogeneous problems was introduced. The form of the Green function for ODEs and for each of the major types of PDEs were given. More detail could be given in each case. We didn't explore eigenfunction expansions for the Green functions associated with PDEs nor did we give examples involving every type of boundary condition for each PDE. Such an endeavor would rapidly turn into a "handbook" on Green functions, which is not our intention.

Exercises

9.1. Derive Eq. (9.6) as a consequence of Eq. (2.21).

9.2. Show that $C \sin kx + D \cos kx = 0$ at $x = l$ is equivalent to $E \sin k(x - l)$, where E is another constant.

9.3. Construct the generic form of the Green function for the operator $\mathcal{L}$ in Eq. (9.1) on the interval $0 \le x \le l$. Let there be a solution of $\mathcal{L}f = 0$ that vanishes at $x = 0$, call it $f_0(x)$, i.e. $f_0(0) = 0$. Likewise, let $f_1(x)$ be a linearly independent solution of $\mathcal{L}f = 0$ that vanishes at $x = l$, $f_1(l) = 0$. Use the matching conditions Eqs. (9.11) and (9.12) to show that

$$G(x, x') = \frac{1}{D} \begin{cases} f_0(x) f_1(x') & x < x' \\ f_0(x') f_1(x) & x > x', \end{cases}$$

where $D \equiv p(x')(f_0(x') f_1'(x') - f_0'(x') f_1(x'))$. Evaluate the quantity D for the example that starts in last example of Section 9.1.4. Does your answer agree with Eq. (9.14)?

9.4. Find the solution of the inhomogeneous differential equation $f''(x) = \phi(x)$ for $0 \le x \le 1$, where $\phi(x)$ is a known function, and where f satisfies the nonhomogeneous boundary conditions $f(0) = a$ and $f(1) = b$. Work this problem in two ways, first using Eq. (9.5), and then using Eq. (9.8). Do your answers agree? (They should.) First find the Green function satisfying homogeneous boundary conditions:

$$G(x, x') = \begin{cases} x(x' - 1) & x < x' \\ x'(x - 1) & x > x'. \end{cases}$$

9.5. Consider the operator $\mathcal{L} = \mathrm{d}^2/\mathrm{d}x^2$ for a system confined to $0 \le x \le l$.

(a) Show that the functions

$$\psi_n(x) \equiv \sqrt{\frac{2}{l}} \cos\left(\frac{n\pi x}{l}\right) \qquad n = 1, 2, 3, \ldots$$

are normalized eigenfunctions of L satisfying homogeneous Neumann boundary conditions at $x = 0$ and $x = l$, i.e. $\psi_n'(0) = \psi_n'(l) = 0$.

(b) Argue that this system admits an additional normalized eigenfunction

$$\psi_0(x) \equiv \sqrt{\frac{1}{l}}$$

meeting the same boundary conditions corresponding to eigenvalue zero.

(c) Show that $\psi_0(x)$ *must* be included for the eigenfunctions of this system to be a complete set. That is, show that

$$\sum_{n=0}^{\infty} \psi_n(x)\psi_n(x') = \delta(x - x').$$

Hint: $\cos x \cos y = \frac{1}{2}[\cos(x-y) + \cos(x+y)]$. Use Eqs. (5.13) and (5.16) which imply $1 + 2\sum_{n=1}^{\infty} \cos(n\alpha) = 2\pi\delta(\alpha)$. Don't forget that $\delta(\pi(x-x')/l) = (l/\pi)\delta(x-x')$ (Section 4.5).

9.6. Show for the Poisson equation, Eq. (9.26), that Neumann boundary conditions on ψ must satisfy a *self-consistency condition*: The normal derivative specified on the surface S bounding a volume V in which charges are contained must satisfy $\int_S \boldsymbol{\nabla}\psi \cdot \hat{\boldsymbol{n}}\, \mathrm{d}s = -\alpha \int_V \rho(\boldsymbol{r})\, \mathrm{d}^3 r$. Thus, the normal derivative cannot be arbitrarily specified; it must be self-consistent with the charge distribution. Hint: From the divergence theorem $\int_V \nabla^2\psi\, \mathrm{d}^3 r = \int_S \boldsymbol{\nabla}\psi \cdot \hat{\boldsymbol{n}}\, \mathrm{d}s$.

9.7. Derive the formulas in Eq. (9.45). Use the matching conditions, Eqs. (9.11) and (9.12).

9.8. (a) Show that the Green function as expressed in Eq. (9.47) can be written as Eq. (9.48). Hint: Use the generating function for P_l, Eq. (6.50).

(b) Show that the infinite series in Eq. (9.48) (corresponding to the function $F(\boldsymbol{r}, \boldsymbol{r}')$ in Eq. (9.36)) satisfies the Laplace equation. Hint: Show that $\nabla^2_{\theta,\phi} P_l(\hat{\boldsymbol{r}} \cdot \hat{\boldsymbol{r}}') = -l(l+1)P_l(\hat{\boldsymbol{r}} \cdot \hat{\boldsymbol{r}}')$.

9.9. Show, in three-dimensional space, that

$$(\nabla^2 + k^2)\left(\frac{\mathrm{e}^{\mathrm{i}k|\boldsymbol{r}-\boldsymbol{r}'|}}{|\boldsymbol{r}-\boldsymbol{r}'|}\right) = -4\pi \mathrm{e}^{\mathrm{i}k|\boldsymbol{r}-\boldsymbol{r}'|}\delta(\boldsymbol{r}-\boldsymbol{r}').$$

This result is not a "one liner." It relies on several vector identities, such as those in Appendix A. Don't forget that $\nabla^2 \equiv \boldsymbol{\nabla} \cdot \boldsymbol{\nabla}$.

9.10. Derive the constants A and E in Eq. (9.60) using the matching conditions on the Green function in Eq. (9.59). Hint: You'll need the Wronskian for Bessel and Neumann functions, Eq. (5.33).

9.11. Show that the summation in Eq. (9.64) satisfies the homogeneous Helmholtz equation. Show that

$$(\nabla^2 + k^2)J_n(kr)\cos n\theta = 0,$$

where ∇^2 is the Laplacian for circular polar coordinates.

9.12. Show that the functions F_ψ and F_B in Eq. (9.86) satisfy the homogeneous diffusion equation, i.e. $\mathcal{L}F_\psi = 0$ and $\mathcal{L}F_B = 0$. Hint: In calculating $\mathcal{L}F_\psi$, $t > t_0$, and in calculating $\mathcal{L}F_B$, $\boldsymbol{r}$ is not on the surface.

9.13. Derive Eq. (9.98). Show first that for constant α,

$$\nabla^2_r \mathrm{e}^{-\alpha|\boldsymbol{r}-\boldsymbol{r}'|^2} = (-6\alpha + 4\alpha^2|\boldsymbol{r}-\boldsymbol{r}'|^2)\mathrm{e}^{-\alpha|\boldsymbol{r}-\boldsymbol{r}'|^2}.$$

Note that $\boldsymbol{\nabla}_r(\boldsymbol{r} \cdot \boldsymbol{r}') = \boldsymbol{r}'$. The time derivatives are straightforward. Use Eq. (4.35), $\theta'(t) = \delta(t)$.

9.14. In the example given in last example Section 9.4.3, what is the dimension of the quantity ψ_0?

9.15. Derive the reciprocity relation, Eq. (9.82), for the Green function associated with the wave operator. Repeat the steps that begin in Section 9.4.1. Write down a fresh copy of Eq. (9.100) and let $\boldsymbol{r}' \to \boldsymbol{r}''$, $t \to -t$, and $t' \to -t''$. Multiply that equation by $G(\boldsymbol{r}, \boldsymbol{r}', t, t')$ and subtract from it Eq. (9.100) multiplied by $G(\boldsymbol{r}, \boldsymbol{r}'', -t, -t'')$. Note that for any two functions, $fg'' - gf'' = (\partial/\partial t)(fg' - gf')$. Integrate the equation that ensues over V, and over t for all times. Make use of Green's identity, Eq. (9.28), and the boundary conditions, Eq. (9.77).

9.16. (a) What are the dimensions of the Green function for the diffusion operator? What are the dimensions of the source function in Eq. (9.75)? What are the dimensions of ψ in Eq. (9.75)?

(b) What are the dimensions of the Green function for the wave operator? What would be the dimensions of the Green function for the wave operator in n spatial dimensions? A: $[G] = L^{2-n}T^{-1}$, where L, T denote the dimensions of length and time.

9.17. By integrating Eq. (9.100) from $t = t' - \epsilon$ to $t = t' + \epsilon$, show that

$$\lim_{\epsilon \to 0^+} \frac{\partial}{\partial t} G(\boldsymbol{r}, \boldsymbol{r}', t, t')\bigg|_{t=t'+\epsilon} = -c^2 \delta(\boldsymbol{r} - \boldsymbol{r}').$$

9.18. Show that the free-space Green function for the one-dimensional wave equation is

$$G_0(x, t) = -\frac{c}{2}\theta(t)\theta(ct - |x|). \tag{9.110}$$

Combine Eq. (9.106) with the one-dimensional Fourier transform,

$$G_0(x, t) = \frac{-c}{2\pi}\theta(t)\int_{-\infty}^{\infty} \frac{\sin(kct)}{k} \mathrm{e}^{\mathrm{i}kx}\, \mathrm{d}k.$$

Hint: $\sin x \cos y = \frac{1}{2}[\sin(x + y) + \sin(x - y)]$. Use the integral $\int_{-\infty}^{\infty} (\sin ax/x)\, \mathrm{d}x = \pi \mathrm{sgn}(a)$, where sgn denotes the *sign* function, $\mathrm{sgn}(a) = \pm 1$ for $a \gtrless 0$. You should find

$$G_0(x, t) = -\frac{c}{4}\theta(t)[\mathrm{sgn}(ct + x) + \mathrm{sgn}(ct - x)],$$

which is equivalent to Eq. (9.110). Note that for any finite time $t > 0$, $G_0(x, t) = 0$ for $|x| > ct$.

10

Integral equations

10.1 Introduction

A predominant theme of this book has been methods of solving partial differential equations (PDEs), for the simple reason that the fundamental equations of physics are often in the form of PDEs. Many key equations, however, occur as *integral equations.* One need only look at Eq. (9.3) involving Green functions, or the Cauchy integral formula, Eq. (8.21), as integral equations. It turns out (as we show) that *linear PDEs are equivalent to integral equations*[1] if the associated Green function exists. Differential equations are therefore equivalent to integral equations, and there are times when *it's advantageous to formulate problems in terms of integral equations.* In this section, we discuss why and how integral equations arise, before taking up their proper study in Section 10.2.

10.1.1 Equivalence of integral and differential equations

Consider Eq. (9.16),

$$\mathcal{L}f + \lambda w(x)f(x) = \phi(x), \tag{9.16}$$

where ϕ and w are known functions, and λ is a parameter. Assume that $\mathcal{L}$ possesses a Green function such that $\mathcal{L}_x G(x, x') = \delta(x - x')$. As is readily verified, the solution to Eq. (9.16) can be written

$$f(x) = -\lambda \int G(x, x')w(x')f(x')\,\mathrm{d}x' + \int G(x, x')\phi(x')\,\mathrm{d}x'. \tag{10.1}$$

[1]Recall that Maxwell's equations can be given in integral as well as differential form.

Mathematical Methods in Physics, Engineering, and Chemistry, First Edition.
Brett Borden and James Luscombe.

Finding the solution to the differential equation Eq. (9.16) is thus equivalent to finding the solution to the integral equation Eq. (10.1).[2]

10.1.2 Role of coordinate systems in capturing boundary data

We saw in Section 4.7 that linear and shift-invariant measurement systems are described by the convolution integral

$$m(x) \equiv \int_{-\infty}^{\infty} k(x-s) f(s)\, \mathrm{d}s, \tag{4.45}$$

where $f(x)$ is the "object" to be determined and $k(x)$ describes the behavior of the measurement system. The usual goal is to estimate $f(x)$ from the measurements $m(x)$ – and we'll return to this problem later. Our purpose in displaying Eq. (4.45) is that (as we explain) it exemplifies how the integral equation approach to the description of physical systems has advantages over that based on PDEs.

The separation-of-variables method developed in Section 3.2 is a great "vehicle" for introducing the special functions of mathematical physics. As a tool for solving realistic boundary value problems, however, it's of rather *limited utility* because it relies on the boundaries of systems having simple shapes (spherical, cylindrical, etc.). As noted in Section 3.2.1, the Helmholtz equation is separable in eleven different coordinate systems, and while that may sound like a large number, they each describe systems having a fair amount of symmetry. Consider radar-wave scattering from the surface of an aircraft, a problem described by the Helmholtz equation. The surfaces of most aircraft do not neatly coincide with the coordinate surfaces of any of the coordinate systems in which the Helmholtz equation is separable.

A common way to "get around" the separability issue is to recast the boundary value problem into a form in which the boundary values are included as integrals over the boundaries (rather than being enforced on a set of general solutions). In usual practice, these integral equations require that (approximate) numerical methods be applied to the evaluation of the integrals in question (e.g. integrals over the surface of an aircraft). But, it turns out, these numerical methods are often much more readily applied when the problem is expressed in integral-equation form, and in this form, they are often more amenable to such numerical methods. Integrals are "smoother" than derivatives and so are more stable.[3]

These two properties – automatic incorporation of boundary values and increased numerical stability in solution – mean that it is sometimes preferable to convert a differential equation into integral form. It is also

---

[2]Note, however, that the class of functions that can be integrated includes those which are discontinuous (nondifferentiable) and so possible solutions to integral equations are somewhat more general.

[3]Suppose $\varphi(x)$ is some estimate of a function – either by physical measurement or numerical approximation. Then, generally, this estimate will carry with it some error ϵ, and

sometimes possible to convert an integral equation into its corresponding differential equation and solve the resulting boundary value problem. More often, however, integral equations are best tackled "head-on."

Example. Quantum scattering

Quantum scattering is a typical problem in which the solution is expressed in terms of an integral. The time-independent Schrödinger equation for a particle of mass m and energy E, interacting with a potential energy environment $V(\boldsymbol{r})$, is

$$-\frac{\hbar^2}{2m}\nabla^2\psi(\boldsymbol{r}) + V(\boldsymbol{r})\psi(\boldsymbol{r}) = E\psi(\boldsymbol{r}),$$

where $V(\boldsymbol{r})$ is known and localized in the vicinity of the origin, i.e. $V(\boldsymbol{r}) \to 0$ outside some finite range. By writing $E = \hbar^2 k^2/(2m)$, the Schrödinger equation takes the form of the inhomogeneous Helmholtz equation (Section 9.3):

$$(\nabla^2 + k^2)\psi(\boldsymbol{r}) = \frac{2m}{\hbar^2}V(\boldsymbol{r})\psi(\boldsymbol{r}) \equiv U(\boldsymbol{r})\psi(\boldsymbol{r}). \tag{10.2}$$

The energy E is known – *this is not a bound-state problem* where we're solving for an energy eigenvalue. For *scattering problems*, we're looking for solutions of Eq. (10.2) that have the *form* for large r

$$\psi(\boldsymbol{r}) \underset{r\to\infty}{\to} \mathrm{e}^{\mathrm{i}\boldsymbol{k}_0\cdot\boldsymbol{r}} + f(\theta,\phi)\frac{\mathrm{e}^{\mathrm{i}kr}}{r}. \tag{10.3}$$

The first term in Eq. (10.3) represents an incident plane wave, with $\boldsymbol{k}_0$ the incident wavevector with magnitude $|\boldsymbol{k}_0| = k$ and $r \equiv |\boldsymbol{r}|$; because $V(\boldsymbol{r}) \to 0$ outside a finite range, we can treat the incident particle as a free particle. The second term in Eq. (10.3) represents an outgoing spherical wave (the scattered solution, caused by the interaction of the plane wave with $V(\boldsymbol{r})$), where $f(\theta,\phi)$ (what we *don't know*), is called the *scattering amplitude*. Equation (10.3) is a type of boundary condition; in scattering problems, we seek solutions where the *form* of $\psi(\boldsymbol{r})$ is prescribed.

$\Delta\varphi(x) = \varphi(x+\Delta x) - \varphi(x)$ will have an error $\Delta\epsilon$. The numerical estimate of the derivative of φ is then actually

$$\frac{\mathrm{d}\varphi}{\mathrm{d}x} \approx \frac{\Delta\varphi + \Delta\epsilon}{\Delta x} = \frac{\Delta\varphi}{\Delta x} + \frac{\Delta\epsilon}{\Delta x}$$

the last term of which becomes large as $\Delta x \to 0$, unless $\Delta\epsilon \to 0$ at least as fast as $\Delta x \to 0$ (which is unlikely for random estimation error). On the other hand,

$$\int \varphi(x)\,\mathrm{d}x \approx \sum_i \varphi(x_i)\Delta x + \sum_i \epsilon_i \Delta x$$

the last term of which is more likely to become small as $\Delta x \to 0$.

We know the free-space Green function for the Helmholtz operator, Eq. (9.66). The solution to Eq. (10.2) is then

$$\psi(\boldsymbol{r}) = \mathrm{e}^{\mathrm{i}\boldsymbol{k}_0\cdot\boldsymbol{r}} - \frac{1}{4\pi}\int \frac{\mathrm{e}^{\mathrm{i}k|\boldsymbol{r}-\boldsymbol{r}'|}}{|\boldsymbol{r}-\boldsymbol{r}'|}U(\boldsymbol{r}')\psi(\boldsymbol{r}')\,\mathrm{d}^3r', \tag{10.4}$$

where we've added a solution of the homogeneous Helmholtz equation, $\mathrm{e}^{\mathrm{i}\boldsymbol{k}_0\cdot\boldsymbol{r}}$ (see Exercise 10.1). Equation (10.4) solves the Schrödinger equation for $\psi(\boldsymbol{r})$ in the form of an integral equation. Of course, we can't do the integral until we know $\psi(\boldsymbol{r})$! We'll see that the form of Eq. (10.4) suggests a natural approximation scheme for computing $\psi(\boldsymbol{r})$.

We need to check that Eq. (10.4) satisfies the boundary condition, Eq. (10.3). Combining Eq. (9.71) with Eq. (10.4), we see that Eq. (10.4) is indeed in the form of Eq. (10.3) for large r, with

$$f(\theta,\phi) = -\frac{1}{4\pi}\int \mathrm{e}^{-\mathrm{i}\boldsymbol{k}'\cdot\boldsymbol{r}'}U(\boldsymbol{r}')\psi(\boldsymbol{r}')\,\mathrm{d}^3r', \tag{10.5}$$

where θ,ϕ are the spherical angles between $\hat{\boldsymbol{n}}$ and $\boldsymbol{k}_0$. The boundary condition is thus built into the integral equation, Eq. (10.4).

10.2 Classification of integral equations

Just as with differential equations, integral equations can involve multiple variables with integrals over surfaces, volumes, and arbitrary dimensional spaces. For simplicity, we consider integral equations in one variable only (the extension to higher dimensions being more or less straightforward, as we show in the following). Moreover, we limit our attention to *linear* integral equations which involve linear integral operators of the form given in Eq. (1.12), then Eq. (9.16) takes on a general form akin to Eq. (5.2), which was used to launch our discussion of ordinary differential equations (ODEs):

$$\beta y(x) - \lambda\int_a^b K(x,z)y(z) = f(x). \tag{10.6}$$

Here, λ and β are constant parameters and $f(x)$ is a known (source) function, included to define the *inhomogeneous* case. (When $f(x)=0$, the integral equation is *homogeneous.*) And, just as in the case of ODEs, the parameters can be used to help classify the integral equation: When $\beta = 0$, Eq. (10.6) is said to be a *first kind* equation; Whereas $\beta = 1$ for *second kind equations.* (The parameter λ will be discussed in the following.)

An important difference between integral equations and differential equations, however, is evident in the limits of integration terms a and b. Each limit can, generally, depend on x, but (possibly after a change of variables) the integral equation can be reduced to either the case $b = x$ which is called a *Volterra equation* or $b =$ constant – defining a *Fredholm equation* (these possibilities comprise the traditional[4] classifications): For example,

Fredholm and Volterra equations of the first and second kind

$$\int_a^x K(x,z)y(z)\,\mathrm{d}z = f(x), \tag{10.7}$$

is a *Volterra equation of the first kind,* and

$$y(x) - \lambda \int_a^b K(x,z)y(z)\,\mathrm{d}z = f(x), \tag{10.8}$$

is a *Fredholm equation of the second kind,* etc.

The function $K(x, z)$ is known as the *kernel* of the integral equation.[5] The goal in all cases is to determine the unknown function $y(x)$.

In the case that $a = -\infty$ and/or $b = \infty$, or if the kernel becomes infinite in the range of integration, the integral equation is said to be *singular.* With the limits of integration extended to infinity, first-kind equations are known as "integral transforms." (Recall, for example, the Fourier transform, Eq. (4.39).)

singular integral equations

The specific method used to obtain the solution $y(x)$ will depend on the form of f and K and, of course, the type of integral equation.

Example. The integral equation of quantum scattering theory, Eq. (10.4), is an inhomogeneous Fredholm equation of the second kind.

10.3 Neumann series

When the integral equation is of the second kind and $f(x) \neq 0$, then we can *formally* replace $y(z)$ in the integrand of Eq. (10.8) with the form

$$y(z) = f(z) + \lambda \int_a^b K(z,z')y(z')\,\mathrm{d}z'$$

to obtain

$$\begin{aligned} y(x) &= f(x) + \lambda \int_a^b K(x,z) \left(f(z) + \lambda \int_a^b K(z,z')y(z') \right) \mathrm{d}z\,\mathrm{d}z' \\ &= f(x) + \lambda \int_a^b K(x,z)f(z)\,\mathrm{d}z + \lambda^2 \int_a^b\int_a^b K(x,z)K(z,z')y(z')\,\mathrm{d}z\mathrm{d}z'. \end{aligned}$$

[4] Although the Volterra equations can be shown to be special cases of the (respective) Fredholm equations for kernels obeying $K(x, z) = 0$ if $x < z$, this is not the usual convention.

[5] The parameter λ could be absorbed into the kernel. In many problems, however, this parameter may take on different constant values depending on the situation at hand. Moreover, we will see that keeping the dependence on this parameter explicit will facilitate the follow-on theoretical discussion.

This formal substitution can be continued (repeatedly) in the obvious way, and we generally obtain (after n such substitutions)

$$y_n(x) = f(x) + \sum_{i=1}^{n} \lambda^i \int_a^b K_i(x,z) f(z)\, \mathrm{d}z \tag{10.9}$$

where $K_i(x,z)$ is defined recursively by

$$K_i(x,z) \equiv \int_a^b K(x,z') K_{i-1}(z',z)\, \mathrm{d}z',$$

where $K_0 \equiv 1$. In the limit as $n \to \infty$, Eq. (10.9) is an infinite series. When this series converges, the representation $y(x) = \lim_{n\to\infty} y_n(x)$ is the solution to the second kind integral equation. We write

Neumann series

$$y(x) = f(x) + \lambda \int_a^b R(x,z;\lambda) f(z)\, \mathrm{d}z$$

where

$$R(x,z;\lambda) \equiv \sum_{i=0}^{\infty} \lambda^i K_{i+1}(x,z) \tag{10.10}$$

resolvent kernel

is the *resolvent kernel.* Clearly, this series converges provided $|\lambda|$ is sufficiently small.

Example. Use the method of repeated substitution to solve the Volterra equation of the second kind

$$y(x) = x + \lambda \int_0^x (x-z) y(z)\, \mathrm{d}z.$$

Solution: The Neumann series starts with $y_0(x) = x$ from which we obtain

$$y_1(x) = x + \lambda \int_0^x (x-z) y_0(z)\, \mathrm{d}z = x + \lambda \int_0^x (x-z) z\, \mathrm{d}z = x + \frac{\lambda x^3}{3!}$$

$$y_2(x) = x + \lambda \int_0^x (x-z) y_1(z)\, \mathrm{d}z = x + \lambda \int_0^x (x-z)\left(z + \frac{\lambda z^3}{3!}\right) \mathrm{d}z$$

$$= x + \frac{\lambda}{3!} x^3 + \frac{\lambda^2}{5!} x^5.$$

Continuing in this way, we obtain

$$y_n(x) = x + \frac{\lambda}{3!} x^3 + \frac{\lambda^2}{5!} x^5 + \cdots + \frac{\lambda^n}{(2n+1)!} x^{2n+1}. \tag{10.11}$$

We note that this kernel is a difference kernel and so the integral equation could also have been solved using the Laplace transform.

When the original integral equation is first-kind, rather than second-kind, then this successive substitution method is useless. It is sometimes possible, however, to convert first-kind equations into second-kind equations in order to apply the Neumann series method.

Example. Consider the Volterra integral equation of the first kind

$$f(x) = \int_0^x K(x,z)y(z)\,\mathrm{d}z.$$

Assuming that the kernel is continuously differentiable when $z \leq x$, then we can differentiate to obtain (using Leibnitz' rule for differentiating integrals)

$$f'(x) = K(x,x)y(x) + \int_0^x \frac{\partial K(x,z)}{\partial x} y(z)\,\mathrm{d}z.$$

If $K(x,x)$ is never zero, then we can divide this equation by $K(x,x)$ to obtain

$$y(x) = \frac{f'(x)}{K(x,x)} - \int_0^x \frac{1}{K(x,x)} \frac{\partial K(x,z)}{\partial x} y(z)\,\mathrm{d}z$$

which is a second-kind equation.

10.4 Integral transform methods

There are two cases quite common in physics and engineering that merit special mention.

10.4.1 Difference kernels

Integrals of the form

$$\int_{-\infty}^{\infty} k(x-z)y(z)\,\mathrm{d}z \quad \text{or} \quad \int_0^x k(x-z)y(z)\,\mathrm{d}z \tag{10.12}$$

are said to have *difference* kernels. The first integral is a Fourier convolution while the second is a Laplace convolution. Each can be solved using the (respective) convolution theorem for the Fourier or Laplace transform.

difference kernel

We consider the case where the integral equation involves a Fourier convolution so that Eq. (10.8) becomes

$$y(x) = f(x) + \lambda \int_{-\infty}^{\infty} k(x-z)y(z)\,\mathrm{d}z.$$

The Fourier transform of both sides of this equation is

$$Y(\alpha) = F(\alpha) + \lambda K(\alpha)Y(\alpha) \tag{10.13}$$

where

$$Y(\alpha) = \int_{-\infty}^{\infty} y(x)e^{-i\alpha x}dx, \quad F(\alpha) = \int_{-\infty}^{\infty} f(x)e^{-i\alpha x}dx, \quad K(\alpha) = \int_{-\infty}^{\infty} k(x)e^{-i\alpha x}dx \tag{10.14}$$

are the Fourier transforms of y, f, and k (respectively), and we have applied the convolution theorem, Eq. (4.47).

Solving Eq. (10.13) for $Y(\alpha)$, we find

$$Y(\alpha) = \frac{F(\alpha)}{1 - \lambda K(\alpha)}$$

and so the solution to the integral equation can be obtained by inverse Fourier transform as

$$y(x) = \frac{1}{2\pi} \int_{-\infty}^{\infty} \frac{F(\alpha)e^{i\alpha x}}{1 - \lambda K(\alpha)} d\alpha.$$

(Had the convolution been of the Laplace type then we could have applied the Bromwich integral (Section 8.12) instead of the inverse Fourier transform.)[6]

10.4.2 Fourier kernels

When $K(x, z) = \exp(-ixz)$ and $a = -\infty$, $b = \infty$, the integral is in the form of a Fourier transform:

$$y(x) = f(x) + \lambda \int_{-\infty}^{\infty} y(z)e^{-ixz}\, dz = f(x) + \lambda Y(x) \tag{10.15}$$

where $Y(x)$ is defined in Eq. (10.14). (Note that $\alpha \to x$ in this case.) The Fourier transform of this equation is

$$Y(x) = F(x) + 2\pi\lambda y(-x) \tag{10.16}$$

where we have used

$$\begin{aligned}
\int_{-\infty}^{\infty} Y(z)e^{-ixz}\, dz &= \int_{-\infty}^{\infty} \left[\int_{-\infty}^{\infty} y(z')e^{-izz'}\, dz' \right] e^{-ixz}\, dz \\
&= \int_{-\infty}^{\infty} y(z') \left[\int_{-\infty}^{\infty} e^{-iz(x+z')}\, dz \right] dz' \\
&= 2\pi \int_{-\infty}^{\infty} y(z')\delta(x + z')\, dz' = 2\pi y(-x).
\end{aligned}$$

Substituting Eq. (10.16) into Eq. (10.15) yields

$$y(x) = f(x) + \lambda[F(x) + 2\pi\lambda y(-x)].$$

[6]We will see that this scheme doesn't work very well for type 1 equations with noise (since physical kernels will have K which vanish for some values of α) – something guaranteed as a result of Section 8.11.

Making the change of variables $x \to -x \implies y(-x) = f(-x) + \lambda(F(-x) + 2\pi\lambda y(x))$, we obtain

$$y(x) = f(x) + \lambda\{F(x) + 2\pi\lambda[f(-x) + \lambda(F(-x) + 2\pi\lambda y(x))]\}.$$

Finally, solving for $y(x)$ yields

$$y(x) = \frac{f(x) + \lambda F(x) + 2\pi\lambda^2(f(-x) + \lambda F(-x))}{1 - (2\pi)^2\lambda^4}.$$

Clearly, $\lambda \neq \pm 1/\sqrt{2\pi}$ or $\lambda \neq \pm \mathrm{i}/\sqrt{2\pi}$ (unless the numerator also vanishes).

Example. Solve

$$y(x) = \exp\left(\frac{-x^2}{2}\right) + \lambda \int_{-\infty}^{\infty} y(z)\mathrm{e}^{-\mathrm{i}xz}\, \mathrm{d}z.$$

Solution: The Fourier transform of $f(x) = \exp(-x^2/2)$ was found in Exercise 8.13 to be

$$F(x) = \sqrt{2\pi}\ \exp\left(\frac{-x^2}{2}\right) = \sqrt{2\pi}\ f(x).$$

Then, since $f(-x) = f(x)$, we obtain

$$y(x) = \frac{(1 + \sqrt{2\pi}\lambda)(1 + 2\pi\lambda^2)}{1 - (2\pi)^2\lambda^4} \exp\left(\frac{-x^2}{2}\right) = \frac{\exp(-x^2/2)}{1 - \sqrt{2\pi}\ \lambda}$$

provided $\lambda \neq 1/\sqrt{2\pi}$.

10.5 Separable kernels

Kernels of the form (margin: separable kernel)

$$K(x, z) = \sum_{i=1}^{N} \phi_i(x)\psi_i(z) \quad (N \text{ finite}) \tag{10.17}$$

are said to be *separable*. When the limits of integration are constants (Fredholm equations), a closed-form solution can be found in a straightforward manner: In the case of Eq. (10.8), we obtain

$$\begin{aligned} y(x) &= f(x) + \lambda \int_a^b K(x, z)y(z)\, \mathrm{d}z = f(x) + \lambda \sum_{i=1}^{n} \phi_i(x) \int_a^b \psi_i(z)y(z)\, \mathrm{d}z \\ &= f(x) + \lambda \sum_{i=1}^{n} c_i\phi_i(x) \text{ where } c_i \equiv \int_a^b \psi_i(z)y(z)\, \mathrm{d}z \text{ are constants.} \end{aligned}$$

The solution is therefore expressed in terms of the known functions $\phi_i(x)$ and the constants c_i, and the latter can be found by substitution: Writing $y(z) = f(z) + \lambda \sum_{j=1}^{n} c_j \phi_j(z)$, we find that

$$c_i = \int_a^b \psi_i(z) y(z) \, \mathrm{d}z = \int_a^b \psi_i(z) \left(f(z) + \lambda \sum_{j=1}^{n} c_j \phi_j(z) \right) \mathrm{d}z$$
$$= \int_a^b \psi_i(z) f(z) \, \mathrm{d}z + \lambda \sum_{j=1}^{n} c_j \int_a^b \psi_i(z) \phi_j(z) \, \mathrm{d}z, \quad (\text{for } i = 1, 2, \dots, n)$$

which are n equations in the n unknowns c_i.

Example. The integral equation

$$y(x) = x + \lambda \int_0^1 (xz^2 + x^2 z) y(z) \, \mathrm{d}z \tag{10.18}$$

has a separable kernel with $\phi_1(x) = x$, $\psi_1(z) = z^2$, $\phi_2(x) = x^2$, and $\psi_2(z) = z$. Equation (10.18) becomes

$$y(x) = x + \lambda(xc_1 + x^2 c_2) \text{ where } c_1 = \int_0^1 z^2 y(z) \mathrm{d}z \text{ and } c_2 = \int_0^1 z y(z) \mathrm{d}z.$$

Substituting $y(z) = z + \lambda(zc_1 + z^2 c_2)$ into these integrals (for c_1 and c_2) yields (after a little algebra)

$$\begin{pmatrix} 1 - \lambda/4 & -\lambda/5 \\ -\lambda/3 & 1 - \lambda/4 \end{pmatrix} \begin{pmatrix} c_1 \\ c_2 \end{pmatrix} = \begin{pmatrix} 1/4 \\ 1/3 \end{pmatrix}.$$

This system of equations will have a unique solution for c_1 and c_2 provided

$$\begin{vmatrix} 1 - \lambda/4 & -\lambda/5 \\ -\lambda/3 & 1 - \lambda/4 \end{vmatrix} = \frac{240 - 120\lambda - \lambda^2}{240} \neq 0$$

in which case

$$c_1 = \frac{60 + \lambda}{240 - 120\lambda - \lambda^2} \quad \text{and} \quad c_2 = \frac{80}{240 - 120\lambda - \lambda^2},$$

from which the solution is found as

$$y(x) = x + \lambda(xc_1 + x^2 c_2) = \frac{60(4 - \lambda)x + 80\lambda x^2}{240 - 120\lambda - \lambda^2}.$$

Example. Quantum scattering revisited.

The kernel in Eq. (10.4) is not separable, yet it can be expressed in separable form using the identity established in Eq. (9.67),

$$\frac{\exp(\mathrm{i}k|\boldsymbol{r}-\boldsymbol{r}'|)}{\mathrm{i}k|\boldsymbol{r}-\boldsymbol{r}'|} = 4\pi \sum_{l=0}^{\infty}\sum_{m=-l}^{l} j_l(kr_<)h_l^{(1)}(kr_>)Y_l^m(\theta,\phi)(Y_l^m(\theta',\phi'))^*. \tag{9.67}$$

Consider the case where the scattering potential in Eq. (10.4) is spherically symmetric, $U(\boldsymbol{r}) = U(r)$. In that case, the scattered wavefunction $\psi(\boldsymbol{r})$ is independent of the azimuth angle ϕ and can be expanded in the *form* of Eq. (7.58) (expansion of a plane wave),

$$\psi(\boldsymbol{r}) = \sum_{l=0}^{\infty}(2l+1)\mathrm{i}^l\psi_l(r)P_l(\cos\theta), \tag{10.19}$$

where $\psi_l(r)$ is an unknown expansion coefficient. Equation (10.19) is referred to as a *partial wave expansion.*[7] Using Eqs. (10.19) and (9.67), we find (you're asked to fill in the steps in Exercise 10.10)

$$\begin{aligned}\frac{1}{4\pi}\int \frac{\mathrm{e}^{\mathrm{i}k|\boldsymbol{r}-\boldsymbol{r}'|}}{|\boldsymbol{r}-\boldsymbol{r}'|}U(\boldsymbol{r}')\psi(\boldsymbol{r}')\,\mathrm{d}^3r' \\ = \mathrm{i}k\sum_{l=0}^{\infty}\mathrm{i}^l(2l+1)P_l(\cos\theta)\int_0^{\infty}(r')^2\,\mathrm{d}r' j_l(kr_<)h_l^{(1)}(kr_>)U(r')\psi_l(r').\end{aligned} \tag{10.20}$$

Using Eqs. (10.20) and (10.4), and making use of Eqs. (7.58) and (10.19), we find

$$\psi_l(r) = j_l(kr) - \mathrm{i}k\int_0^{\infty}(r')^2\,\mathrm{d}r' j_l(kr_<)h_l^{(1)}(kr_>)U(r')\psi_l(r'). \tag{10.21}$$

Thus, Eq. (10.4), which involves a three-dimensional integral, has been replaced (for a spherically symmetric scattering potential) with *an infinite set* of one-dimensional integral equations.

10.6 Self-adjoint kernels

For the linear integral operator first defined in Section 1.4,

$$\mathcal{K}y(x) \equiv \int_a^b K(x,z)y(z)\,\mathrm{d}z, \tag{1.12}$$

[7] See any textbook on quantum mechanics, for example Schiff [30, p. 117].

where we have changed the notation slightly to conform with our current discussion. Then, the general form of the linear integral equation, Eq. (10.6), can be written as

$$\beta y(x) - \lambda \mathcal{K} y(x) = f(x). \tag{10.22}$$

This form of the integral equations is the same general form as Eq. (2.1) (except with integral operators replacing differential operators), and we will see that many of the same solution methods will apply.[8]

We start by concentrating on second-kind equations which, as in the case of differential equations, we subdivide into homogeneous ($f(x) = 0$) and inhomogeneous ($f(x) \neq 0$) sub-cases. We assume that $K(x, z)$ is defined over the rectangle $a \leq x \leq b$, $a \leq z \leq b$. The homogeneous equation $y(x) = \lambda \mathcal{K} y(x)$ becomes

$$y(x) = \lambda \int_a^b K(x, z) y(z) \, dz. \tag{10.23}$$

This equation is an eigenequation and will possess nontrivial solutions only for characteristic values of λ. We denote these eigenvalues by λ_i and the corresponding eigenfunctions by $y_i(x)$ so that

$$y_i(x) = \lambda_i \int_a^b K(x, z) y_i(z) \, dz. \tag{10.24}$$

We see from Eq. (10.24) that if $y_i(x)$ is a solution then so are constant multiples of $y_i(x)$, and, consequently, we may consider each eigenfunction to be normalized.

Forming the inner product as

$$\langle y_j | y_i \rangle = \int_a^b y_j^*(x) y_i(x) \, dx = \lambda_i \int_a^b \int_a^b y_j^*(x) K(x, z) y_i(z) \, dz \, dx \tag{10.25}$$

we use $\langle y_j | y_i \rangle = \langle y_i | y_j \rangle^*$ to obtain

$$\begin{aligned} \lambda_i \int_a^b \int_a^b y_j^*(x) K(x, z) y_i(z) \, dz \, dx &= \left[\lambda_j \int_a^b \int_a^b y_i^*(x) K(x, z) y_j(z) \, dz \, dx \right]^* \\ &= \lambda_j^* \int_a^b \int_a^b y_i(x) K^*(x, z) y_j^*(z) \, dz \, dx. \end{aligned}$$

Hermitian kernel

Kernels which obey[9] $K(x, z) = K^*(z, x)$ are said to be Hermitian. We have seen that the eigenfunctions of Hermitian operators enjoy useful

[8]It might be useful to review Chapter 2 at this point.

[9]This situation defines the so-called "Hilbert–Schmidt" theory of integral equations.

properties, and we consider this case yet again. Switching the variables of integration in the last line we obtain, for Hermitian operators, the result

$$\lambda_i \int_a^b \int_a^b y_j^*(x) K(x,z) y_i(z)\ \mathrm{d}z\ \mathrm{d}x - \lambda_j^* \int_a^b \int_a^b y_j^*(x) K(x,z) y_i(z)\ \mathrm{d}z\ \mathrm{d}x = 0$$

$$\implies \quad (\lambda_i - \lambda_j^*) \int_a^b \int_a^b y_j^*(x) K(x,z) y_i(z)\ \mathrm{d}z\ \mathrm{d}x = 0$$

$$\text{or} \quad \frac{\lambda_i - \lambda_j^*}{\lambda_i} \langle y_j | y_i \rangle = 0. \quad \text{(using Eq. (10.25))}$$

(Note that $\lambda_i = 0$ is not allowed since it leads only to a trivial solution.)

We consider two cases:

1. When $i = j$, we see that

$$\frac{\lambda_i - \lambda_i^*}{\lambda_i} |y_i|^2 = 0 \quad \implies \quad \lambda_i = \lambda_i^*$$

and so the eigenvalues of Hermitian kernels are real-valued.

2. When $i \neq j$, we have

$$\frac{\lambda_i - \lambda_j}{\lambda_i} \langle y_j | y_i \rangle = 0 \quad \implies \quad \langle y_j | y_i \rangle = 0 \quad \text{when} \quad \lambda_i \neq \lambda_j$$

and so the eigenfunctions associated with distinct eigenvalues are orthogonal.

The eigenfunctions therefore form an orthonormal set.

And, as in Section 2.5 the eigenfunctions are complete. But there is a difference – now we must restrict ourselves to a more limited class of functions:

Any function $g(x)$ that can be generated as

$$g(x) = \int_a^b K(x,z) y(z)\ \mathrm{d}z$$

where $y(x)$ is some *continuous function and $K(x,z)$ is continuous and Hermitian, can be represented over $[a,b]$ by a linear combination of the eigenfunctions of the homogeneous integral equation, Eq.* (10.23) *with $k(x,z)$ as its kernel.*[10]

We now consider the full (inhomogeneous) integral equation with Hermitian kernel. Owing to the observation of the previous paragraph, we can express the solution $y(x)$ as

[10] This does not mean that *any* function defined on $[a,b]$ can be expanded in these eigenfunctions. This observation can have important consequences since $f(x)$ in Eq. (10.8) may not be representable in terms of the eigenfunctions of $k(x,z)$.

$$y(x) = f(x) + \sum_i a_i y_i(x)$$

(since $f(x)$ may not be expressible in terms of the eigenfunctions it is explicitly included). Substituting this representation into Eq. (10.8) yields

$$\begin{aligned} f(x) + \sum_i a_i y_i(x) &= f(x) + \lambda \int_a^b K(x,z) \left(f(z) + \sum_i a_i y_i(z) \right) \mathrm{d}z \\ &= f(x) + \lambda \int_a^b K(x,z) f(z) \, \mathrm{d}z + \lambda \sum_i a_i \int_a^b K(x,z) y_i(z) \, \mathrm{d}z \\ &= f(x) + \lambda \int_a^b K(x,z) f(z) \, \mathrm{d}z + \lambda \sum_i \frac{a_i y_i(x)}{\lambda_i} \end{aligned}$$

where we have used Eq. (10.24) in the last step. Forming the inner product of both sides with $y_j(x)$ results in

$$\sum_i a_i \langle y_j | y_i \rangle = \lambda \int_a^b \int_a^b y_j^*(x) K(x,z) f(z) \, \mathrm{d}z \, \mathrm{d}x + \lambda \sum_i \frac{a_i \langle y_j | y_i \rangle}{\lambda_i}. \tag{10.26}$$

The second term of this equation simplifies since $K(x,z)$ is Hermitian. We obtain

$$\begin{aligned} \int_a^b \int_a^b y_j^*(x) K(x,z) f(z) \, \mathrm{d}z \, \mathrm{d}x &= \int_a^b \int_a^b y_j^*(x) K^*(z,x) f(z) \, \mathrm{d}x \, \mathrm{d}z \\ &= \frac{1}{\lambda_j} \int_a^b y_j^*(z) f(z) \, \mathrm{d}z \end{aligned}$$

and, since the $y_i(x)$ are orthonormal, Eq. (10.26) reduces to

$$a_j = \frac{\lambda \langle y_j | f \rangle}{\lambda_j} + \frac{\lambda a_j}{\lambda_j} \quad \Longrightarrow \quad a_j = \frac{\lambda \langle y_j | f \rangle}{\lambda_j - \lambda}.$$

Therefore, the solution is

$$y(x) = f(x) + \lambda \sum_i \frac{\langle y_i | f \rangle}{\lambda_i - \lambda} \, y_i(x).$$

Example. Consider again the first example of Section 10.5

$$y(x) = x + \lambda \int_0^1 (xz^2 + x^2 z) y(z) \, \mathrm{d}z. \tag{10.18}$$

This kernel is real and symmetric in x and z and so is clearly Hermitian. We saw before that the corresponding homogeneous equation had eigensolutions of the form $y = xc_1 + x^2 c_2$ where

$$\begin{pmatrix} 1 - \lambda/4 & -\lambda/5 \\ -\lambda/3 & 1 - \lambda/4 \end{pmatrix} \begin{pmatrix} c_1 \\ c_2 \end{pmatrix} = \begin{pmatrix} 0 \\ 0 \end{pmatrix}$$

has nontrivial solutions only when $\lambda^2 + 120\lambda - 240 = 0$; that is, when $\lambda_\pm = -60 \pm 16\sqrt{15} \approx -60 \pm 62$. We have

$$c_2 = \frac{1 - \lambda_\pm/4}{-\lambda_\pm/5} c_1$$

(or, equivalently for eigenvalues, $c_2 = c_1(\lambda_\pm/3)/(1 - \lambda_\pm/4)$) and so the associated eigenfunctions are

$$\begin{aligned} y_1(x) &\approx A(x - 1.29x^2) \quad \text{for} \lambda_- \approx -122 \\ y_2(x) &\approx B(x + 1.25x^2) \quad \text{for} \lambda_+ \approx 2. \end{aligned}$$

Normalizing these functions over $(0, 1)$ finds

$$\int_0^1 A^2(x - 1.29x^2)^2 \, \mathrm{d}x = 1 \qquad \Longrightarrow \qquad A \approx 6.87$$

and

$$\int_0^1 B^2(x + 1.25x^2)^2 \, \mathrm{d}x = 1 \qquad \Longrightarrow \qquad B \approx 0.89$$

so our normalized eigenfunctions are $y_1(x) = 6.87(x - 1.29x^2)$ and $y_2(x) = 0.89(x + 1.25x^2)$. With $\langle y_1|x\rangle \approx 0.074$ and $\langle y_2|x\rangle \approx 0.575$, the solution for Eq. (10.18) is therefore

$$y(x) = x - \frac{0.074\lambda}{122 + \lambda} y_1(x) + \frac{0.575\lambda}{2 - \lambda} y_2(x).$$

Nothing prevents us, of course, from choosing any complete orthonormal basis in which to expand Eq. (10.8). When the kernel is Hermitian, the choice of eigenfunctions is especially convenient because we needn't consider the case of

$$\int_a^b K(x, z) y(z) \, \mathrm{d}z = 0 \tag{10.27}$$

and so the matrix with elements

$$K_{ij} \equiv \int_a^b \int_a^b y_i^* K(x, z) y_j(z) \, \mathrm{d}z \, \mathrm{d}x$$

will be diagonal.

The same cannot be said about first kind equations for which nontrivial solutions to the corresponding homogeneous equation (Eq. (10.27)) play an important role with, as we shall see, significant practical consequences – the difference between first and second kind equations is more than just formal.

10.7 Numerical approaches

matrix elements

By far the most common practical methods for dealing with integral equations are numerical (computer based). These methods first select a complete set of basis functions $\{\phi_n(x)\}$, $n = 1, 2, \ldots, \infty$, and start by representing the linear integral equation in terms of its matrix elements (with respect to this basis). First kind equations, for example, become

$$f(x) = \int_a^b K(x,z)y(z)\,\mathrm{d}z \qquad \Longrightarrow \qquad f_i = \sum_{j=1}^{\infty} K_{ij}y_j \quad i = 1, 2, \ldots, \infty$$

where

$$f_i \equiv \langle \phi_i(x)|f(x)\rangle, \;\; y(z) = \sum_{j=1}^{\infty} y_j\phi_j(z), \;\; \text{and} \;\; K_{ij} \equiv \langle \phi_i(x)|K(x,z)\phi_j(z)\rangle.$$

In this way, the integral equation is reduced to a matrix equation and the methods of Chapter 1 can be applied.

But, while easy to state in this form, the actual implementation of this scheme is fraught with potential peril, and a more careful analysis is often required. In this section, we present a brief overview of some of the pitfalls. We concentrate on first kind equations associated with linear measurement systems because they are both the most commonly seen in applied situations, and they also display most of the important points to consider.

10.7.1 Matrix form

First (and foremost) of the problems encountered in transferring an integral equation to a form appropriate for numerical analysis is the problem of the choice of basis. Computers can't deal with infinite sums and so the set $\{\phi_n(x)\}$ must be immediately truncated to a finite number of elements. Ideally, we would choose a basis in which the representation of $f(x)$ and $y(z)$ quickly decays. But these are two different functions, and there is no guarantee that the basis "best" suited to an abbreviated description of $f(x)$ will be well suited for a representation of $y(z)$. Moreover, measurements are discrete, finite, and defined by the measurement system (and often not the function being measured). A common choice is $\phi_n(x) = \delta(x - x_n)$, $n = 1, 2, \ldots, N$, where the x_n are uniformly distributed on the interval $[c, d]$ (though nonuniform sampling is also common). With this choice, our linear measurement model of Eq. (4.46) becomes

$$m(x_i) = \int_a^b K(x_i, s)f(s)\,\mathrm{d}s + n(x_i)\,, \quad c \le x_1 < x_2 < \cdots < x_N \le d, \tag{10.28}$$

where we have now included an additive term $n(x_i)$ which represents any error in the measurements and/or numerical representation error. That is, we write $m(x_i) = m_{\text{true}}(x_i) + n(x_i)$, where $m_{\text{true}}(x_i)$ is determined by the integral in Eq. (10.28). (This additive noise term will play an important role later on.) Here, the usual problem is to estimate $f(s)$ from the measurements $m_i \equiv m(x_i)$.

The object function $f(s)$, however, will not be limited by the discrete nature of the measurement system. The function $f(s)$ is generally of *infinite dimension;* that is, f belongs to Hilbert space (see Section 1.6). And, a moment's reflection should convince you that there's no compelling reason that the object function should be best represented in the same basis as the measurement function (or even that its representation should be limited to N terms). "Infinite" dimensions are still a problem for computers, of course, so we will simply assume that $f_j \equiv f(s_j)$ is represented at M points in the interval $[a, b]$ with $M \gg N$. Then, Eq. (10.28) can be written in matrix form as

$$\boldsymbol{m} = \boldsymbol{K}\boldsymbol{f} + \boldsymbol{n} \tag{10.29}$$

Here, $\boldsymbol{m} \in \mathbb{M} \subset \mathbb{C}^N$, where $\mathbb{M}$ denotes the measurement space, defined in the following, $\boldsymbol{f} \in \mathbb{O} \subset \mathbb{C}^M$ are vectors, with $\mathbb{O}$ the object space, and $\boldsymbol{K} \in \mathbb{C}^{N\times M}$ is a matrix, where $\mathbb{C}^{N\times M}$ is the space of all complex $N \times M$ matrices.

Equation (10.29) offers some reassurance since, after all, we sort of know how to work with matrices. But before we examine this equation further, lets make a few obligatory mathematical comments:

1. Equation (4.46) is an operator equation: It maps the function f to the measurements m. The main reason that we can convert this equation into the form of Eq. (10.29) is that the kernel K is *compact.*[11]

2. We need to bear in mind that the matrix $\boldsymbol{K}$ is $N \times M$ and that, in reality, M is infinitely large. Consequently, $\boldsymbol{K}$ is a "matrix" only in an abstract sense.

Ok. Now back to Eq. (10.29).

10.7.2 Measurement space

The fact that $\boldsymbol{K}$ is not generally a square matrix poses an immediate problem. If we denote *estimated values* of objects by "overlines" (i.e. $\overline{\boldsymbol{f}}$), then our basic goal is to get a good estimate $\overline{\boldsymbol{f}}$ of $\boldsymbol{f}$ from the measurements $\boldsymbol{m}$. We'd like to try something like

[11] A compact, or *completely continuous* operator $\mathcal{K}$ is one that maps an infinite, bounded set of vectors $\{\phi_n\}$, i.e. $\|\phi_n\| \leq C$, into a sequence $\{\mathcal{K}\phi_n\}$ that's bounded. We're glossing over points of mathematical finery here; suffice to say that compact operators on Hilbert space are those most directly analogous to operators on finite-dimensional vector spaces (which are always bounded – Section 1.4.5). A sufficient condition for $\mathcal{K}$ to be compact is that $Tr\mathcal{K}\mathcal{K}^{\dagger} < \infty$ [31, p. 198]. It's not *simply* that compact operators are bounded, it's in the sense we can say that the sets $\{\phi_n\}$ and $\{\mathcal{K}\phi_n\}$ form convergent sequences as $n \to \infty$.

$$\bar{\boldsymbol{f}} = \boldsymbol{K}^{-1}\boldsymbol{m},$$

but this is an issue – what exactly does the inverse of a nonsquare matrix look like?

One thing's for sure: when $M > N$ there are more unknowns than there are linearly independent equations and so the system cannot have a unique solution. The best we can hope for is a solution $\bar{\boldsymbol{f}}$ that is *consistent* with Eq. (10.29), and it's useful at this stage to make some geometric observations. Consider all the possible object functions $\boldsymbol{f} \in \mathbb{O} \subset \mathbb{C}^M$. The matrix $\boldsymbol{K}$ maps these functions to an N-dimensional subspace called the measurement space $\mathbb{M}$ according to the rule $\boldsymbol{K} : \boldsymbol{f} \mapsto \boldsymbol{K}\boldsymbol{f} \in \mathbb{M} \subset \mathbb{C}^N$. Alternatively, we can see that the measurement space is spanned by the columns of $\boldsymbol{K}$ (considered as N-dimensional vectors) – i.e. the measurement space consists of all vectors formed of linear combinations of

$$\boldsymbol{k}_j = \begin{pmatrix} K(x_1, s_j) \\ K(x_2, s_j) \\ \vdots \\ K(x_N, s_j) \end{pmatrix}, \quad j = 1, 2, \ldots, M$$

nullspace

The measurement space has dimension N. The vector $\boldsymbol{v} = \boldsymbol{f} - \boldsymbol{K}\boldsymbol{f}$ must live in a different space since it obviously can't have any component in measurement space. We say that the vector $\boldsymbol{v}$ lies in the *nullspace* $\mathbb{N} \subset \mathbb{C}^{M-N}$ of $\boldsymbol{K}$ (see Figure 10.1 and Section 1.4.4).

The whole nullspace thing is pretty important: This space is determined by the kernel $K(x, s)$ which, in turn, represents the measurement system. As such, nullspace contains all the things that *cannot be measured* – a vector $\boldsymbol{v} \in \mathbb{N}$ is "invisible" to the system. (This "invisibility" may represent an "out of band" frequency, the black-and-white picture of Uncle Ernie's blue eyes, or the sound of a tree falling in an unpopulated forest). As we will see, nullspace accounts for measurement artifacts and must be carefully considered.

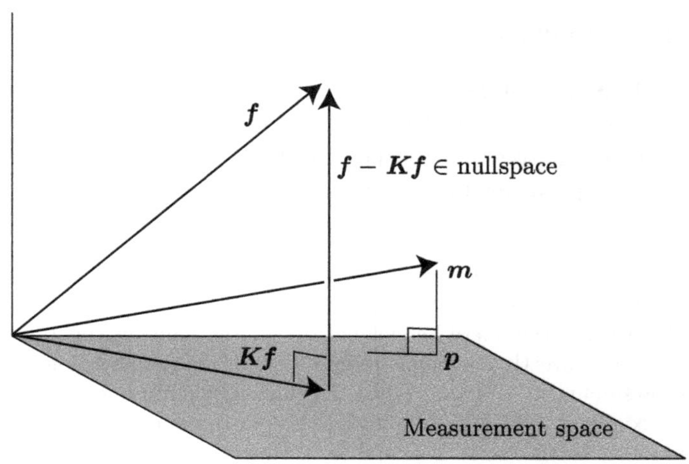

Figure 10.1 Measurement space.

Example. An example of a measurement artifact associated with nullspace.

Consider a movie camera filming a rotating wheel (one with spokes). The camera (usually) has a frame rate of 24 pictures per second. If the wheel rotates at the angular velocity of $24 \times 2\pi$ rad/s, then the camera will fail to measure any rotation at all. Moreover, if the N spokes are identical in appearance, then the wheel will appear to be stationary if its rotation rate is any integer multiple of $24 \times 2\pi/N$ – even when the stage-coach that it is attached to is clearly in motion. (This effect is familiar to fans of old movie westerns.) The problem here is that the camera will not correctly measure changes that happen faster that 1/24th of a second: it is a nullspace problem that must be addressed by a *sampling theorem.*

Example. The nullspace of a rectangular operator.

The vector

$$\begin{pmatrix} 2 \\ 2 \\ -1 \end{pmatrix}$$

lies in the nullspace of the matrix

$$\begin{pmatrix} 1 & 0 & 2 \\ 1 & 1 & 4 \end{pmatrix}$$

since

$$\begin{pmatrix} 1 & 0 & 2 \\ 1 & 1 & 4 \end{pmatrix} \begin{pmatrix} 2 \\ 2 \\ -1 \end{pmatrix} = \begin{pmatrix} 0 \\ 0 \end{pmatrix}$$

Example. The nullspace of a square operator.

The matrix

$$\boldsymbol{A} = \begin{pmatrix} 1 & 0 & 1 \\ 5 & 4 & 9 \\ 2 & 4 & 6 \end{pmatrix}$$

has its third column formed as the sum of the first two. It is easy to show that

$$\begin{pmatrix} 1 & 0 & 1 \\ 5 & 4 & 9 \\ 2 & 4 & 6 \end{pmatrix} \begin{pmatrix} b \\ b \\ -b \end{pmatrix} = \begin{pmatrix} 0 \\ 0 \\ 0 \end{pmatrix}$$

and so the nullspace of $\boldsymbol{A}$ is the line containing all points $x = b$, $y = b$, and $z = -b$.

Generally, the vector of measurements $\boldsymbol{m}$ in Eq. (10.29) *will not lie in the measurement space:* Because $\boldsymbol{m} = \boldsymbol{K f} + \boldsymbol{n}$ and $\boldsymbol{K f} \in \mathbb{M}$, we see that $\boldsymbol{m} \in \mathbb{M}$ only if $\boldsymbol{n} \in \mathbb{M}$ – and this will not usually be true (since $\boldsymbol{n}$ is independent of the system kernel, it will not usually be formed from the columns of $\boldsymbol{K}$). In addition, the object vector $\boldsymbol{f}$ will also have a component in measurement space and a component in nullspace. Since the null-space component of $\boldsymbol{f}$ cannot be determined from the measurements, our "best guess" at $\boldsymbol{f}$ will choose the estimate that minimizes the distance between $\boldsymbol{K f}$ and $\boldsymbol{m}$. That is, our estimate $\overline{\boldsymbol{f}}$ should minimize the norm $\|\boldsymbol{m} - \boldsymbol{K f}\|$ and we denote this as

$$\overline{\boldsymbol{f}} = \arg \min_{\text{all } \boldsymbol{f}} \|\boldsymbol{m} - \boldsymbol{K f}\| \tag{10.30}$$

We can use Figure 10.1 to obtain the solution to Eq. (10.30) by inspection. The vector $\boldsymbol{K f}$ must lie in $\mathbb{M}$, and its closest distance to $\boldsymbol{m}$ occurs when $\boldsymbol{K}\overline{\boldsymbol{f}} = \boldsymbol{p}$. The vector $\boldsymbol{m} - \boldsymbol{p} = \boldsymbol{m} - \boldsymbol{K}\overline{\boldsymbol{f}}$ is orthogonal to the measurement space $\mathbb{M}$ and so is perpendicular to all vectors in $\mathbb{M}$. Every vector in $\mathbb{M}$ is of the form $\boldsymbol{K y}$ (for some $\boldsymbol{y}$) and so the inner product between $\boldsymbol{K y}$ and $\boldsymbol{m} - \boldsymbol{K}\overline{\boldsymbol{f}}$ must vanish. The inner product of vectors $\boldsymbol{a}$ and $\boldsymbol{b}$ can be written in terms of the Hermitian-transpose operator as $\boldsymbol{a}^\dagger \boldsymbol{b}$ and so we have

$$(\boldsymbol{K y})^\dagger(\boldsymbol{m} - \boldsymbol{K}\overline{\boldsymbol{f}}) = \boldsymbol{y}^\dagger \boldsymbol{K}^\dagger(\boldsymbol{m} - \boldsymbol{K}\overline{\boldsymbol{f}}) = \boldsymbol{y}^\dagger(\boldsymbol{K}^\dagger \boldsymbol{m} - \boldsymbol{K}^\dagger \boldsymbol{K}\overline{\boldsymbol{f}}) = 0$$

This equation must hold for all $\boldsymbol{y}$ and so we conclude that $\overline{\boldsymbol{f}}$ must obey

$$\boldsymbol{K}^\dagger \boldsymbol{K}\overline{\boldsymbol{f}} = \boldsymbol{K}^\dagger \boldsymbol{m} \tag{10.31}$$

This matrix equation is really a set of equations – they are known as *the normal equations.*

10.7.3 The generalized inverse

We've seen in Section 4.7 that when the measurement model is in the form of a convolution, then Fourier analysis is a natural tool. Actual measurement systems will be naturally band-limited to $[c, d]$ and so the discrete and finite sampling can now be performed in the Fourier domain. Moreover, the Fourier transform matrix formed from the basis elements $\phi_n(x) = \mathrm{e}^{\mathrm{i}\alpha_n x}$ (refer to Eq. (4.21)) is unitary and so the normal equations (10.31) reduce to[12]

$$\overline{\boldsymbol{F}} = (\mathcal{F}\boldsymbol{K})^{-1}\boldsymbol{m}$$

where $\mathcal{F}\boldsymbol{K}$ denotes the Fourier transform of $\boldsymbol{K}$ (this is just the Fourier convolution theorem in the Fourier domain). In addition, since the convolution kernel $\boldsymbol{K}$ is *diagonal* in the Fourier domain (see Exercise 4.18),

[12] Although, as we will see, it is unwise to apply this equation without modification.

it's inverse is simply the reciprocal of its diagonal elements. Of course, here we've implicitly assumed that the dimension of the object function equals that of the data so that the system is not underdetermined.

But, often, the measurement model is not a convolution. And, almost always, the system is underdetermined. Luckily, a remarkable result from linear algebra – the so-called *singular value decomposition* – allows us to apply an analog to the Fourier transform which behaves in very much the same way [32, p. 142].

The matrix equation (10.31) (the normal equations) is a set of conditions that must be satisfied by the least-squares solution $\overline{\boldsymbol{f}}$ to Eq. (10.31) (with noisy measurements.) The matrix $\boldsymbol{K}^\dagger\boldsymbol{K} \in \mathbb{C}^{M\times M}$ is square, and, when it can be inverted, we can write

$$\overline{\boldsymbol{f}} = (\boldsymbol{K}^\dagger\boldsymbol{K})^{-1}\boldsymbol{K}^\dagger\boldsymbol{m} \tag{10.32}$$

The matrix $(\boldsymbol{K}^\dagger\boldsymbol{K})^{-1}\boldsymbol{K}^\dagger$ is known as the *Moore-Penrose pseudo-inverse* [32, p. 138] (or, sometimes, simply the *generalized inverse*) of $\boldsymbol{K}$. If we multiply both sides (matrix multiply on the left) of (10.32) by $\boldsymbol{K}$ we see that

pseudo-inverse

$$\boldsymbol{K}\overline{\boldsymbol{f}} = \boldsymbol{K}(\boldsymbol{K}^\dagger\boldsymbol{K})^{-1}\boldsymbol{K}^\dagger\boldsymbol{m} = \boldsymbol{p}$$

and so $\boldsymbol{K}(\boldsymbol{K}^\dagger\boldsymbol{K})^{-1}\boldsymbol{K}^\dagger$ is a *projection matrix* mapping $\boldsymbol{m}$ to its component in measurement space.

Eigen-analysis approach to least-squares

Equation (10.32) requires the inversion of the $M \times M$ matrix $\boldsymbol{K}^\dagger\boldsymbol{K}$, and, since this is fundamental to our ability to estimate the object $\boldsymbol{f}$ from the measurements $\boldsymbol{m}$, we will analyze $(\boldsymbol{K}^\dagger\boldsymbol{K})^{-1}\boldsymbol{K}^\dagger$ more carefully. Our aim here is not to develop fast algorithms for computing matrix inverses – that would be appropriate for an advanced course in applied linear algebra. Rather, we want to make some general observations about matrices of the form $(\boldsymbol{K}^\dagger\boldsymbol{K})^{-1}\boldsymbol{K}^\dagger$. We begin by recalling some results from elementary linear algebra (see Chapter 1) where we showed that:

1. $\boldsymbol{K}^\dagger\boldsymbol{K}$ is Hermitian since $(\boldsymbol{K}^\dagger\boldsymbol{K})^\dagger = \boldsymbol{K}^\dagger\boldsymbol{K}$. This $M \times M$ matrix will be diagonal in a basis formed from its orthonormal eigenvectors. That is (by the similarity transform),

$$\boldsymbol{K}^\dagger\boldsymbol{K} = \boldsymbol{S}\begin{pmatrix} \lambda_1 & 0 & \cdots & 0 \\ 0 & \lambda_2 & \cdots & 0 \\ \vdots & \vdots & \ddots & \vdots \\ 0 & 0 & \cdots & \lambda_M \end{pmatrix}\boldsymbol{S}^{-1} = \boldsymbol{S}\boldsymbol{\Lambda}\boldsymbol{S}^{-1}$$

   where the columns of $\boldsymbol{S}$ are formed from the eigenvectors $\{\hat{e}_m\}$, $m = 1, 2, \ldots, M$, of $\boldsymbol{K}^\dagger\boldsymbol{K}$ and the $\{\lambda_m\}$ are the corresponding eigenvalues.

2. $\boldsymbol{S}$ is unitary so that $\boldsymbol{S}^{-1} = \boldsymbol{S}^\dagger$.

3. Consequently, we can write

$$\boldsymbol{K}^\dagger \boldsymbol{K} = \sum_{m=1}^{M} \lambda_m \hat{\mathbf{e}}_m \hat{\mathbf{e}}_m^\dagger$$

(the *spectral theorem*) which is a representation of $\boldsymbol{K}^\dagger \boldsymbol{K}$ in terms of the matrices $\hat{\mathbf{e}}_m \hat{\mathbf{e}}_m^\dagger$.
The inverse of $\boldsymbol{K}^\dagger \boldsymbol{K}$ is

$$(\boldsymbol{K}^\dagger \boldsymbol{K})^{-1} = \boldsymbol{S} \begin{pmatrix} 1/\lambda_1 & 0 & \cdots & 0 \\ 0 & 1/\lambda_2 & \cdots & 0 \\ \vdots & \vdots & \ddots & \vdots \\ 0 & 0 & \cdots & 1/\lambda_M \end{pmatrix} \boldsymbol{S}^\dagger = \sum_{m=1}^{M} \frac{1}{\lambda_m} \hat{\mathbf{e}}_m \hat{\mathbf{e}}_m^\dagger \tag{10.33}$$

4. Since $\{\hat{\mathbf{e}}_i\}$ is an orthonormal basis of $\mathbb{O} \subset \mathbb{C}^M$, then any M dimensional vector $\boldsymbol{f}$ also has a representation in terms of the $\hat{\mathbf{e}}_i$:

$$\boldsymbol{f} = \sum_{i=1}^{M} (\hat{\mathbf{e}}_i^\dagger \boldsymbol{f}) \hat{\mathbf{e}}_i = \sum_{i=1}^{M} f_i \hat{\mathbf{e}}_i \tag{10.34}$$

where $f_i \equiv \hat{\mathbf{e}}_i^\dagger \boldsymbol{f}$ are the components of the vector $\boldsymbol{f}$ in this basis. Consequently,

$$\boldsymbol{x} = \boldsymbol{K}^\dagger \boldsymbol{K} \boldsymbol{f} = \sum_{i=1}^{M} \lambda_i \hat{\mathbf{e}}_i \hat{\mathbf{e}}_i^\dagger \boldsymbol{f} = \sum_{i=1}^{M} \lambda_i f_i \hat{\mathbf{e}}_i = \sum_{i=1}^{M} x_i \hat{\mathbf{e}}_i$$

with $x_i \equiv \lambda_i f_i$, is a representation of $\boldsymbol{x} = \boldsymbol{K}^\dagger \boldsymbol{K} \boldsymbol{f}$ in terms of the basis $\{\hat{\mathbf{e}}_i\}$.

So, what about an orthonormal basis for measurement space? Since measurement space contains vectors of the form $\boldsymbol{K}\boldsymbol{f}$, lets try making "basis vectors" of the form $\boldsymbol{K}\hat{\mathbf{e}}_i$. Some of the $\{\hat{\mathbf{e}}_j\}$ will live in nullspace, of course, and for them, we have $\boldsymbol{K}\hat{\mathbf{e}}_i = \mathbf{0}$. The other (nonzero) vectors will be orthogonal if $(\boldsymbol{K}\hat{\mathbf{e}}_i)^\dagger \boldsymbol{K}\hat{\mathbf{e}}_j \propto \delta_{ij}$ – that is, if

$$(\boldsymbol{K}\hat{\mathbf{e}}_i)^\dagger \boldsymbol{K}\hat{\mathbf{e}}_j = \hat{\mathbf{e}}_i^\dagger \boldsymbol{K}^\dagger \boldsymbol{K}\hat{\mathbf{e}}_j = \hat{\mathbf{e}}_i^\dagger (\boldsymbol{K}^\dagger \boldsymbol{K}\hat{\mathbf{e}}_j) \propto \delta_{ij}.$$

But

$$(\boldsymbol{K}\hat{\mathbf{e}}_i)^\dagger \boldsymbol{K}\hat{\mathbf{e}}_j = \hat{\mathbf{e}}_i^\dagger (\boldsymbol{K}^\dagger \boldsymbol{K}\hat{\mathbf{e}}_j) = \hat{\mathbf{e}}_i^\dagger (\lambda_j \hat{\mathbf{e}}_j) = \lambda_j \hat{\mathbf{e}}_i^\dagger \hat{\mathbf{e}}_j = \lambda_j \delta_{ij}$$

and so we see that if $\{\hat{\mathbf{e}}_j\}$ are the eigenvectors of $\boldsymbol{K}^\dagger \boldsymbol{K}$, then not only do they form an orthonormal basis for "object space" but the nonzero elements of the set $\{\boldsymbol{K}\hat{\mathbf{e}}_j\}$ also form an orthogonal basis (but not usually ortho*normal*) for measurement space.

We need an orthonormal basis for measurement space: define

$$\sigma_i \equiv \|\boldsymbol{K}\,\hat{\mathbf{e}}_i\| \quad \text{and} \quad \hat{\boldsymbol{\mu}}_i \equiv \begin{cases} \sigma_i^{-1}\boldsymbol{K}\,\hat{\mathbf{e}}_i & \text{if } \sigma_i \neq 0 \\ \boldsymbol{0} & \text{if } \sigma_i = 0 \end{cases} \tag{10.35}$$

The nonzero elements of the set $\{\hat{\boldsymbol{\mu}}_j\}$ will obviously have unit length and so will form an orthonormal basis of measurement space. The zero elements of $\{\hat{\boldsymbol{\mu}}_j\}$ will correspond to the $\{\hat{\mathbf{e}}_j\}$ in the nullspace of $\boldsymbol{K}$ – but our least squares solution $\overline{\boldsymbol{f}}$ to Eq. (10.31) throws those components out anyway.

So, now we can finally say something definitive about the $M \times N$ matrix $(\boldsymbol{K}^\dagger\boldsymbol{K})^{-1}\boldsymbol{K}^\dagger$. Let $\{\hat{\mathbf{e}}_j\}$ be the eigenvectors of $\boldsymbol{K}^\dagger\boldsymbol{K}$, and let $\boldsymbol{S}$ be the orthogonal matrix with columns equal to $\hat{\mathbf{e}}_j$ that diagonalizes $\boldsymbol{K}^\dagger\boldsymbol{K}$. If the eigenvalues λ_i are all nonzero, then the inversion of $\boldsymbol{K}^\dagger\boldsymbol{K}$ is easy and is given by (10.33). Then

$$(\boldsymbol{K}^\dagger\boldsymbol{K})^{-1}\boldsymbol{K}^\dagger = \boldsymbol{S}\boldsymbol{\Lambda}^{-1}\boldsymbol{S}^\dagger\boldsymbol{K}^\dagger = \boldsymbol{S}\boldsymbol{\Lambda}^{-1}(\boldsymbol{K}\boldsymbol{S})^\dagger = \boldsymbol{S}\boldsymbol{\Lambda}^{-1}\left(\boldsymbol{K}\begin{pmatrix} | & | & | & | \\ \boldsymbol{e}_1 & \boldsymbol{e}_2 & \cdots & \boldsymbol{e}_M \\ | & | & | & | \end{pmatrix}\right)^\dagger$$

$$= \boldsymbol{S}\boldsymbol{\Lambda}^{-1}\begin{pmatrix} | & | & | & | \\ \boldsymbol{K}\boldsymbol{e}_1 & \boldsymbol{K}\boldsymbol{e}_2 & \cdots & \boldsymbol{K}\boldsymbol{e}_M \\ | & | & | & | \end{pmatrix}^\dagger = \boldsymbol{S}\boldsymbol{\Lambda}^{-1}\begin{pmatrix} - & \sigma_1\hat{\boldsymbol{\mu}}_1^\dagger & - \\ - & \sigma_2\hat{\boldsymbol{\mu}}_2^\dagger & - \\ - & \vdots & - \\ - & \sigma_M\hat{\boldsymbol{\mu}}_M^\dagger & - \end{pmatrix}$$

$$= \boldsymbol{S}\begin{pmatrix} \sigma_1/\lambda_1 & 0 & \cdots & 0 \\ 0 & \sigma_2/\lambda_2 & \cdots & 0 \\ \vdots & \vdots & \ddots & \vdots \\ 0 & 0 & \cdots & \sigma_M/\lambda_M \end{pmatrix}\begin{pmatrix} - & \hat{\boldsymbol{\mu}}_1^\dagger & - \\ - & \hat{\boldsymbol{\mu}}_2^\dagger & - \\ - & \vdots & - \\ - & \hat{\boldsymbol{\mu}}_M^\dagger & - \end{pmatrix}$$

Of course, *some of the eigenvalues λ_i of $\boldsymbol{K}^\dagger\boldsymbol{K}$ will be zero* when $\boldsymbol{K}$ represents an underdetermined system. We worry about this situation next.

We've shown that the nonzero elements of $\{\hat{\boldsymbol{\mu}}_j\}$ form an orthonormal basis of measurement space. If we reorder the set $\{\hat{\boldsymbol{\mu}}_j\}$ so that it starts with the $\hat{\boldsymbol{\mu}}_j$ that corresponds to the largest value of σ_j and ends with the $\sigma_j = 0$ values, then all the zero elements $\hat{\boldsymbol{\mu}}_j = \boldsymbol{0}$ are at the end and we can truncate the $M \times N$ matrix

$$\begin{pmatrix} - & \hat{\boldsymbol{\mu}}_1^\dagger & - \\ - & \hat{\boldsymbol{\mu}}_2^\dagger & - \\ - & \vdots & - \\ - & \hat{\boldsymbol{\mu}}_M^\dagger & - \end{pmatrix}$$

to an $N \times N$ orthogonal matrix by dropping the last $M - N$ rows (which are all zero)

$$\boldsymbol{V}^\dagger = \begin{pmatrix} - & \hat{\boldsymbol{\mu}}_1^\dagger & - \\ - & \hat{\boldsymbol{\mu}}_2^\dagger & - \\ - & \vdots & - \\ - & \hat{\boldsymbol{\mu}}_N^\dagger & - \end{pmatrix}$$

Of course, in performing this reordering, we've also forced the zero values $\sigma_j = 0$ of the $M \times M$ diagonal matrix to be on the lower right:

$$\begin{pmatrix} \sigma_1/\lambda_1 & 0 & \cdots & 0 & 0 \cdots 0 \\ 0 & \sigma_2/\lambda_2 & \cdots & 0 & 0 \cdots 0 \\ \vdots & \vdots & \ddots & \vdots & \vdots \cdots \vdots \\ b_N & 0 & \cdots & \sigma_N/\lambda_N & 0 \cdots 0 \\ 0 & 0 & 0 & 0 & 0 \cdots 0 \\ \vdots & \vdots & \vdots & \vdots & \vdots \ddots \vdots \\ 0 & 0 & 0 & 0 & 0 \cdots 0 \end{pmatrix} = \left(\boldsymbol{D}_{M\times N} \,\middle|\, \boldsymbol{0}_{M\times(M-N)} \right)$$

where $\boldsymbol{D}$ is the $M \times N$ matrix with entries σ_i/λ_i along the main diagonal and zeros everywhere else.

singular value decomposition

We're basically done: We have expressed the matrix $(\boldsymbol{K}^\dagger \boldsymbol{K})^{-1}\boldsymbol{K}^\dagger$ in terms of its *singular value decomposition*[13] (or, SVD):

$$(\boldsymbol{K}^\dagger \boldsymbol{K})^{-1}\boldsymbol{K}^\dagger = \boldsymbol{S}\boldsymbol{D}\boldsymbol{V}^\dagger \tag{10.36}$$

Observe that the form of Eq. (10.36) has dropped all the terms for which $\sigma_j = 0$. (These terms are associated with the nullspace of $\boldsymbol{K}$.) The remaining σ_i satisfy

$$\sigma_i^2 = \|\boldsymbol{K}\boldsymbol{e}_i\|^2 = (\boldsymbol{K}\boldsymbol{e}_i)^\dagger \boldsymbol{K}\boldsymbol{e}_i = \boldsymbol{e}_i^\dagger \boldsymbol{K}^\dagger \boldsymbol{K}\boldsymbol{e}_i = \boldsymbol{e}_i^\dagger \lambda_i \boldsymbol{e}_i = \lambda_i \implies \sigma_i = \sqrt{\lambda_i}$$

and so $\boldsymbol{D}$ will have diagonal elements of the form $1/\sqrt{\lambda_i}$ *with the offending* $\lambda_j = 0$ *values eliminated from consideration.*

In developing Eq. (10.36), we've omitted more than a few details. A more careful derivation is possible, and it's a remarkable fact that *any* rectangular matrix $\boldsymbol{B} \in \mathbb{C}^{M\times N}$ can be written in the form

$$\boldsymbol{B} = \boldsymbol{S}\boldsymbol{D}\boldsymbol{V}^\dagger$$

where $\boldsymbol{S} \in \mathbb{C}^{M\times M}$ and $\boldsymbol{V} \in \mathbb{C}^{N\times N}$ are unitary and $\boldsymbol{D} \in \mathbb{C}^{M\times N}$ is diagonal.

Example. SVD based solution of an underdetermined system.

For the system $\boldsymbol{m} = \boldsymbol{K}\boldsymbol{f}$ with

$$\boldsymbol{K} = \begin{pmatrix} 1 & 1 & 1 \\ -1 & 0 & 1 \end{pmatrix}$$

[13] Because the SVD can be applied to general kernels, it is actually considered to be a generalization of Fourier analysis.

we can easily calculate

$$\boldsymbol{K}^\dagger \boldsymbol{K} = \begin{pmatrix} 2 & 1 & 0 \\ 1 & 1 & 1 \\ 0 & 1 & 2 \end{pmatrix} \quad \text{and} \quad \boldsymbol{K}\boldsymbol{K}^\dagger = \begin{pmatrix} 3 & 0 \\ 0 & 2 \end{pmatrix}$$

The eigenstructure of $\boldsymbol{K}^\dagger \boldsymbol{K}$ is

$$\lambda_1 = 2\,, \quad \boldsymbol{e}_1 = \frac{1}{\sqrt{2}} \begin{pmatrix} -1 \\ 0 \\ 1 \end{pmatrix}; \quad \lambda_2 = 3, \quad \boldsymbol{e}_2 = \frac{1}{\sqrt{3}} \begin{pmatrix} 1 \\ 1 \\ 1 \end{pmatrix};$$

$$\lambda_3 = 0, \quad \boldsymbol{e}_3 = \frac{1}{\sqrt{6}} \begin{pmatrix} 1 \\ -2 \\ 1 \end{pmatrix}$$

(Note that the eigenvector corresponding to $\lambda_3 = 0$ is in the nullspace of $\boldsymbol{K}^\dagger \boldsymbol{K}$.) The eigenstructure of $\boldsymbol{K}\boldsymbol{K}^\dagger$ is

$$\lambda_1 = 2\,, \quad \hat{\mathrm{e}}_1^\dagger = [0 \quad 1]\,; \quad \lambda_2 = 3\,, \quad \hat{\mathrm{e}}_2^\dagger = [1 \quad 0]$$

The SVD representation of the pseudoinverse is therefore

$$(\boldsymbol{K}^\dagger \boldsymbol{K})^{-1} \boldsymbol{K}^\dagger = \underbrace{\begin{pmatrix} \frac{-1}{\sqrt{2}} & \frac{1}{\sqrt{3}} & \frac{1}{\sqrt{6}} \\ 0 & \frac{1}{\sqrt{3}} & \frac{-2}{\sqrt{6}} \\ \frac{1}{\sqrt{2}} & \frac{1}{\sqrt{3}} & \frac{1}{\sqrt{6}} \end{pmatrix}}_{\boldsymbol{S}} \underbrace{\begin{pmatrix} \frac{1}{\sqrt{2}} & 0 \\ 0 & \frac{1}{\sqrt{3}} \\ 0 & 0 \end{pmatrix}}_{\boldsymbol{D}} \underbrace{\begin{pmatrix} 0 & 1 \\ 1 & 0 \end{pmatrix}}_{\boldsymbol{V}^\dagger} = \frac{1}{6} \begin{pmatrix} 2 & -3 \\ 2 & 0 \\ 2 & 3 \end{pmatrix}$$

and, for our estimate of $\boldsymbol{f}$, we have

$$\begin{pmatrix} \overline{f}_1 \\ \overline{f}_2 \\ \overline{f}_3 \end{pmatrix} = \frac{1}{6} \begin{pmatrix} 2 & -3 \\ 2 & 0 \\ 2 & 3 \end{pmatrix} \begin{pmatrix} m_1 \\ m_2 \end{pmatrix}$$

"Condition number"

The singular value decomposition is one of the most important results of linear algebra. This last subsection collected a lot of mathematical results – and it was worth the effort for two reasons:

1. Linear algebra methods are fundamental to modern data analysis;
2. Equation (10.36) allows us to make some very general observations about solutions to Eq. (10.32) – and, in turn, about the use of the measurements $\boldsymbol{m}$ to estimate objects $\boldsymbol{f}$.

Because $\boldsymbol{K}$ is bounded (it represents a compact operator), it's obvious that

$$\sigma_i = \sqrt{\lambda_i} = \|\boldsymbol{K}\,\boldsymbol{e}_i\|$$

is always less than some number. Let $\lambda_{\max}$ be the largest eigenvalue of $\boldsymbol{K}^\dagger \boldsymbol{K}$ and reorder the set of all the eigenvalues $\{\lambda_i\}$ from largest to smallest values. Any vector $\boldsymbol{f}$ in object space can be written in the form $\boldsymbol{f} = \sum_{i=1}^{M} f_i \boldsymbol{e}_i$, and since $\boldsymbol{K}$ is bounded,

$$\frac{\|\boldsymbol{K f}\|^2}{\|\boldsymbol{f}\|^2} = \frac{\sum\limits_{i=1}^{M} f_i^2 \lambda_i}{\sum\limits_{i=1}^{M} f_i^2} < \text{some finite number}$$

But M is generally as large as we want to let it be (in fact, it's really infinitely large) – so $\{\lambda_i\}$ must form a sequence that gets arbitrarily close to 0 as M gets bigger. This means that the elements $1/\sqrt{\lambda_i}$ of $\boldsymbol{D}$ will become arbitrarily *large* as we try to use higher dimensional data (for which $M \geq N$ becomes larger).

And now we can see why there's a problem: since the data $\boldsymbol{m}$ are contaminated by noise, then solutions of the form (10.32) will multiply that noise by an arbitrarily large number. Let $\boldsymbol{m} = \boldsymbol{m}_{\text{true}} + \boldsymbol{n}$ where $\boldsymbol{m}_{\text{true}}$ denotes the ideal (noise-free) measurement. Inserting Eq. (10.36) into Eq. (10.32) yields

$$\overline{\boldsymbol{f}} = \boldsymbol{S D V}^\dagger \boldsymbol{m}_{\text{true}} + \boldsymbol{S D V}^\dagger \boldsymbol{n}$$

and the last term will increase like $\|\boldsymbol{n}\|/\sqrt{\lambda_{\min}}$.

This is a general result and can be shown to be true for every compact operator. Moreover, collecting more data to fill-out the measurement space doesn't help: the problem of using measurements of the form (4.46) to estimate $f(s)$ is said to be ill-posed [14] and small errors in our measurements will generally lead to large errors in our estimates.

well-posed problems

[14] A problem is said to be *well posed* if solutions *exist*, are *unique*, and are *stable* [17, p. 227].

1. *Existence.* While physicists tend not to concern themselves with existence proofs, in this context, existence is related to boundary conditions that do not *overdetermine* the problem (too much information), as for the case of closed boundaries for hyperbolic PDEs.
2. *Uniqueness.* Uniqueness is related to boundary conditions that do not *underdetermine* the problem (not enough information). The upshot is that if one has found a solution of a PDE satisfying the boundary conditions, it is the correct solution, no matter how it was obtained.
3. *Stability.* Stability refers to the solution remaining stable to small changes in the source function, the boundary conditions, or the initial conditions. This requirement is essential for a solution to correspond with observable physical phenomena because such data (source or boundary) are known only imprecisely, within the margins of experimental error.

The *condition number* of the matrix $\boldsymbol{A}$ is given by the ratio of the largest eigenvalue to the smallest eigenvalue, i.e.

$$\kappa = \frac{\lambda_{\max}}{\lambda_{\min}}$$

The condition number is a measure of the stable solvability of the set of equations $\boldsymbol{Ax} = \boldsymbol{y}$: when κ is large, we say the problem is *ill-conditioned.*

condition number

Example. Consider the $N \times N$ matrix which is commonly used to approximate the second derivative of a function $-\,\mathrm{d}^2 f/\,\mathrm{d}x^2$ where $f(0) = f(1) = 0$ (the "finite difference matrix"):

$$\begin{pmatrix} 2 & -1 & 0 & 0 & \cdots & 0 \\ -1 & 2 & -1 & 0 & \cdots & 0 \\ 0 & -1 & 2 & -1 & \cdots & 0 \\ 0 & 0 & -1 & 2 & \cdots & 0 \\ \vdots & \vdots & \vdots & \vdots & \ddots & \vdots \\ 0 & 0 & 0 & 0 & -1 & 2 \end{pmatrix}$$

The largest eigenvalue of this matrix is approximately $\lambda_1 \approx 4$, and the smallest eigenvalue is $\lambda_N \approx \pi^2/N^2$. Consequently, the condition number is $\kappa \approx N^2/2$. This means that the better we attempt to estimate $-f''$ by increasing the number of unknowns, the harder it is to accurately compute the approximation.

Regularization (generalized noise filtering)

The main problem associated with noise contamination of the data has been shown to be related to the fact that the (nonzero) eigenvalues of $\boldsymbol{K}^\dagger \boldsymbol{K}$ get small as the dimensionality of the problem gets large: the error in estimating $\boldsymbol{f}$ grows like $\|\boldsymbol{n}\|/\sqrt{\lambda_{\min}}$.

There are actually many things we can do to mitigate this effect. Perhaps the easiest is to simply truncate the SVD representation: choose some small threshold value α and, if $\lambda_j < \alpha$, then replace the $1/\sqrt{\lambda_j}$ in $\boldsymbol{D}$ by 0. This method is called the *truncation filter.*

truncation filter

Another approach makes an *ad hoc* modification to the measurement equation: Rewrite the normal equations (10.31) in the form

$$(\boldsymbol{K}^\dagger \boldsymbol{K} + \alpha \boldsymbol{I})\overline{\boldsymbol{f}} = \boldsymbol{K}^\dagger \boldsymbol{m} \tag{10.37}$$

where α is some fixed scalar. Obviously, $\alpha = 0$ just returns the original normal equations and so it can be expected that small values of α don't really change the system description too much. Now, when we repeat the analysis of Section 10.7.3 we find that the eigenvalues $\tilde{\lambda}_i$ of the modified matrix $\boldsymbol{K}^\dagger \boldsymbol{K} + \alpha \boldsymbol{I}$ are

$$\tilde{\lambda}_i = \lambda_i + \alpha$$

where λ_i are the eigenvalues of $\boldsymbol{K}^\dagger \boldsymbol{K}$. We still have $\sigma_i = \sqrt{\lambda_i}$ and so the diagonal elements of $\boldsymbol{D}$ become

$$D_{ii} = \frac{\sqrt{\lambda_i}}{\lambda_i + \alpha}$$

When $\lambda_i \gg \alpha$, it's clear that $D_{ii} \approx 1/\sqrt{\lambda_i}$ (as before). When $\lambda_i \ll \alpha$, however, the $D_{ii} \approx \sqrt{\lambda_i}/\alpha$ will not become arbitrarily large, and the noise problem will be controlled. This method is called *Tikhonov regularization* [33].

Tikhonov regularization

Note that because $\|\boldsymbol{a} + \boldsymbol{b}\| \le \|\boldsymbol{a}\| + \|\boldsymbol{b}\|$ (for any vectors $\boldsymbol{a}$ and $\boldsymbol{b}$), Eq. (10.37) can be interpreted in a slightly different way. The least-squares solution $\overline{\boldsymbol{f}}$ to our regularized problem will now satisfy

$$\overline{f} = \arg \min_{\text{all } \boldsymbol{f}} \{ \|\boldsymbol{K}^\dagger \boldsymbol{K} \boldsymbol{f} - \boldsymbol{K}^\dagger \boldsymbol{m}\| + \alpha \|\boldsymbol{f}\| \}$$

I.e. we seek the estimate $\overline{\boldsymbol{f}}$ that minimizes the distance between $\boldsymbol{K}\boldsymbol{f}$ and $\boldsymbol{m}$ and also has the smallest "scaled length" $\alpha\|\boldsymbol{f}\|$. This interpretation has a nice geometric feel to it, and we can obviously consider extensions of the form

$$\overline{f} = \arg \min_{\text{all } \boldsymbol{f}} \{ \|\boldsymbol{K}^\dagger \boldsymbol{K} \boldsymbol{f} - \boldsymbol{K}^\dagger \boldsymbol{m}\| + \alpha \|\boldsymbol{f} - \boldsymbol{f}_0\| \}$$

where we can choose $\boldsymbol{f}_0$ to include any nullspace information that we might know *by other means*. In this way, we can refine the estimate to better fit our expectations.

In general, our regularized least-square scheme can be cast in the form

$$\overline{f} = \arg \min_{\text{all } \boldsymbol{f}} \{ \mathcal{J}_1(\boldsymbol{f}, \boldsymbol{m}) + \alpha \mathcal{J}_2(\boldsymbol{f}, \boldsymbol{f}_0) \} \tag{10.38}$$

where $\mathcal{J}_1(\boldsymbol{f}, \boldsymbol{m})$ is some measure of the distance between $\boldsymbol{f}$ and the measurements $\boldsymbol{m}$, and $\mathcal{J}_2(\boldsymbol{f}, \boldsymbol{f}_0)$ is some measure of the distance between $\boldsymbol{f}$ and our *expected* object $\boldsymbol{f}_0$.

Truncation filtering and Tikhonov regularization are actually quite powerful and useful methods for limiting the effects of noise. As presented, however, neither technique really offers a method for choosing a "best" threshold value α. And it's pretty obvious that a too-large α could seriously affect the ability to accurately estimate the object $\overline{\boldsymbol{f}}$. A too-small value won't be very useful in controlling noise.

Summary

- Many differential equations are equivalent to an integral equation form. The integral equation form is sometimes preferred – especially when the problem is multidimensional and the boundaries form a non separable system (requiring numerical solution).

- The traditional classification of integral equations (Volterra and Fredholm) is based on the limits of integration (variable or fixed, respectively). Equations can also be homogeneous (first kind) or inhomogeneous (second kind) – see Eq. (10.6). Homogeneous equations define an eigenproblem.
- Straightforward procedures exist for the solution of integral equations with a difference kernel (Eq. (10.12)) or a separable kernel (Eq. (10.17)). Second kind equations may also allow series solutions formed by repeated substitution
- Self-adjoint (Hermitian) kernels allow the integral equation to be addressed in the same manner as for equations with differential operators – Chapter 2. This approach is at the heart of Hilbert–Schmidt theory
- Numerical approaches must address the truncation of solutions to finite dimensional subspaces of $\mathbb{H}$. This approximation, in turn, requires that we carefully examine truncation error as well as measurement error. Moreover, dimension truncation naturally leads to the concepts of *measurement space*, *nullspace*, *generalized inverse* (required for nonsquare matrices), the *singular value decomposition*, and regularization of ill-posed problems.

Exercises

10.1. (a) Show that $e^{i\boldsymbol{k}_0\cdot\boldsymbol{r}}$ (a plane wave) is a solution of the homogeneous Helmholtz equation $(\nabla^2 + k^2)\psi = 0$ when $k_0^2 = k^2$. Hint: $\boldsymbol{k}_0 \cdot \boldsymbol{r} = k_0 r \cos\theta$.

(b) Show that e^{ikr}/r (no dot product in the exponential, a spherical wave) is a solution of $(\nabla^2 + k^2)\psi = 0$.

10.2. Volterra equations can often be turned into differential equations, as this and the next exercise show.

(a) Convert $f(x) = e^x + \int_0^x (x-y) f(y)\, dy$ into a differential equation. Use the Leibniz integral rule to show that f satisfies the differential equation

$$f''(x) = e^x + f(x).$$

(b) Show that $f(x)$ has the form, $f(x) = (a + bx)e^x + ce^{-x}$, where a, b, and c are constants. Evaluate the constants by showing that $f(x)$ satisfies the integral equation. A: $f(x) = \cosh x + \frac{1}{2}xe^x$.

10.3. Solve the integral equation

$$u(x) = x + \int_0^x xyu(y)\, dy \equiv x + xf(x).$$

(a) Develop a differential equation for the function $f(x) = \int_0^x yu(y)\, dy$. Show that

$$f'(x) = xu(x) = x^2 + x^2 f(x).$$

(b) Guess a solution in the form $f(x) = \exp(\alpha x^{\beta}) + \gamma$, and show that $f(x) = \exp(x^3/3) - 1$. Hence,

$$u(x) = x e^{x^3/3}.$$

10.4. Solve $u(x) = e^x + \lambda \int_0^1 xtu(t)\,dt$, where λ is a constant.
(a) What kind of integral equation is it? Is the kernel separable?
(b) What is $u(x)$?
(c) Is there a restriction on the value of λ?

10.5. Show that the solution of

$$\phi(x) = f(x) + \lambda \int_0^{\infty} \cos(\alpha x s)\phi(s)\,ds,$$

is

$$\phi(x) = \frac{f(x) + \lambda \int_0^{\infty} \cos(\alpha x s) f(s)\,ds}{1 - \lambda^2 \pi/(2\alpha)}.$$

Hint: Use Eq. (4.50).

10.6. Solve

$$\int_0^{\infty} f(t) e^{-st}\,dt = \frac{a}{a^2 + s^2}.$$

Hint: Consult Table 8.1.

10.7. Solve $y(x) = f(x) + \lambda \int_0^1 (1 - 2xz) y(z)\,dz$.

10.8. Use the method of repeated substitution to solve $y(x) = x + \lambda \int_0^1 xzy(z)\,dz$.

10.9. Use the Laplace transform to solve $y(x) = x + \lambda \int_0^x (x - z) y(z)\,dz$ and compare your result to the solution Eq. (10.11).

10.10. Fill in the steps leading to Eq. (10.20).

10.11. Derive Eq. (10.21).

10.12. Show for a spherically symmetric scattering potential $U(r)$, that the scattering amplitude in Eq. (10.5), $f(\theta, \phi)$, can be written

$$f(\theta) = \sum_{l=0}^{\infty} (2l + 1) f_l(k) P_l(\cos\theta),$$

where

$$f_l(k) = -\int_0^{\infty} (r')^2 j_l(kr') \psi_l(r')\,dr'.$$

Hint: Use Eq. (7.58).

10.13. The kernel of the integral equation $y(x) = \lambda \int_a^b K(x, z) y(z)\,dz$ has the form $K(x, z) = \sum_{n=0}^{\infty} \phi_n(x) g_n(z)$, where $\{\phi_n(x)\}$ form a complete orthonormal set of functions over the interval $[a, b]$.
(a) Show that the eigenvalues λ_i are given by $\|\boldsymbol{m} - \lambda^{-1}\boldsymbol{I}\| = 0$, where $\boldsymbol{M}$ is the matrix with elements $M_{ij} = \int_a^b g_i(z)\phi_j(z)\,dz$. If the corresponding solutions are $y_n(x) = \sum_{i=0}^{\infty} \alpha_i^{(n)} \phi_i(x)$, find an expression for $\alpha_i^{(n)}$.
(b) Obtain the eigenvalues and eigenfunctions over the interval $[0, 2\pi]$ if $K(x, z) = \sum_{i=1}^{\infty} \frac{1}{i} \cos(ix) \cos(iz)$.

10.14. The operator $\mathcal{M}$ is given by

$$\mathcal{M}f(x) = \int_{-\infty}^{\infty} M(x,y)f(y)\ dy$$

where $M(x,y) = 1$ inside the square $-a \leq x \leq a$, $-a \leq y \leq a$, and $M(x,y) = 0$ elsewhere. Find the eigenvalues of $\mathcal{M}$ and their corresponding eigenfunctions; show that the only eigenvalues are $\lambda_1 = 0$ and $\lambda_2 = 2a$. Find the general solution of

$$f(x) = g(x) + \lambda \int_{-\infty}^{\infty} M(x,y)f(y)\ dy.$$

11

Tensor analysis

We began this book with a discussion of vectors and vector spaces, as many physical phenomena have a vector nature. Quantities more general than vectors exist however – *tensors* – and are the subject of this chapter. Scalars and vectors turn out to be special cases of tensors. Tensors are classified by their *rank* (defined in Section 11.2.3); scalars are rank-zero tensors, and vectors are rank-one tensors. Vectors in three-dimensional space require the specification of three numbers; vector components relative to a basis in N dimensions require N numbers. A second-rank tensor in N-dimensional space requires the specification of N^2 numbers; pth-rank tensors require N^p numbers. Tensors are a powerful mathematical tool used in many areas of science and engineering.[1]

11.1 Once over lightly: A quick intro to tensors

Figure 11.1 shows a force $\delta \boldsymbol{F}$ applied[2] to a small region δs of a surface S, where it has unit normal vector $\hat{\boldsymbol{n}}$. The surface element, considered as a vector, has magnitude δs and direction $\hat{\boldsymbol{n}}$, with $\delta \boldsymbol{s} = \hat{\boldsymbol{n}} \delta s$. The components of $\hat{\boldsymbol{n}}$ in the orthonormal basis $\hat{\boldsymbol{x}}$, $\hat{\boldsymbol{y}}$, and $\hat{\boldsymbol{z}}$ are shown in the figure, with $\hat{\boldsymbol{n}} = n_x \hat{\boldsymbol{x}} + n_y \hat{\boldsymbol{y}} + n_z \hat{\boldsymbol{z}}$. Because $\hat{\boldsymbol{n}} \cdot \hat{\boldsymbol{n}} = 1$, we have $n_x^2 + n_y^2 + n_z^2 = 1$. Three pieces of information characterize $\delta \boldsymbol{s}$: its magnitude δs and the numbers n_x, n_y, and n_z, only two of which are independent because of the constraint $n_x^2 + n_y^2 + n_z^2 = 1$. In what follows, we will generally adopt the notation $\boldsymbol{e}_i$, $i = 1, 2, 3$, instead of $\hat{\boldsymbol{x}}$, $\hat{\boldsymbol{y}}$, $\hat{\boldsymbol{z}}$. With this

[1] Tensors occur in the theory of elasticity, fluid mechanics, electrodynamics, the theory of relativity, thermodynamics, classical and quantum mechanics, and the theory of quantum computation, to name just a few.

[2] The quantity $\delta \boldsymbol{F}$ is the net force acting at the surface, the difference in forces between that applied at the "top" side of the surface (the direction of $\hat{\boldsymbol{n}}$) and that applied to the "bottom" side.

Mathematical Methods in Physics, Engineering, and Chemistry, First Edition.
Brett Borden and James Luscombe.

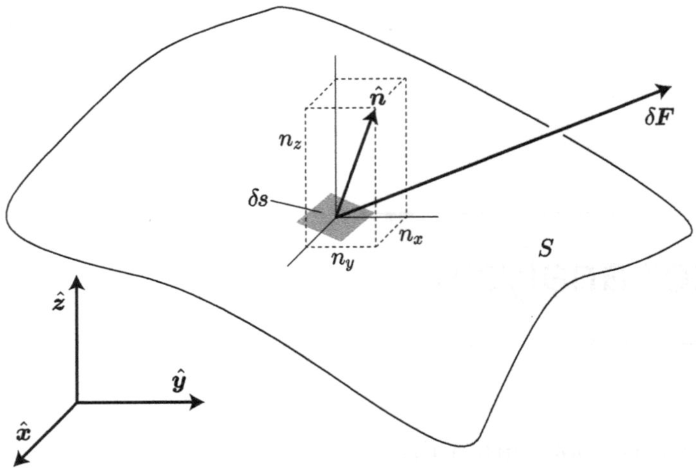

Figure 11.1 Force $\delta \boldsymbol{F}$ applied to surface S with local normal $\hat{\boldsymbol{n}}$.

notation, formulas can be expressed in a compact way, e.g, $\boldsymbol{e}_i \cdot \boldsymbol{e}_j = \delta_{ij}$ or $\hat{\boldsymbol{n}} = \sum_{i=1}^{3} n_i \boldsymbol{e}_i$.

In Figure 11.1 we have two vectors at the same location – $\delta \boldsymbol{F}$ and $\delta \boldsymbol{s}$ – which are not in general parallel. The concept of *pressure*, force per area, is a scalar quantity. *Stress* is a generalization of force per area where we keep track of the direction of the force and the direction of the local surface element. There are nine possible combinations of force per area when each is a vector,

$$T_{ij} \equiv \lim_{\delta s \to 0} \frac{\delta F_i}{\delta s_j}. \qquad (i, j = 1, 2, 3) \tag{11.1}$$

Equation (11.1) defines the nine elements $\{T_{ij}\}$ of the *stress tensor*, a relation involving the *ratio* of the components of two vectors, force and area. If we know the stress tensor, we can find the net force in the ith direction at the location characterized by $\delta \boldsymbol{s}$:

$$\delta F_i = \sum_{j=1}^{3} T_{ij} \delta s_j. \qquad (i = 1, 2, 3) \tag{11.2}$$

Equation (11.2) can be integrated to find the net force in the ith direction acting over the entire surface,

$$F_i = \int_S \sum_j T_{ij} \, \mathrm{d}s_j. \qquad (i = 1, 2, 3) \tag{11.3}$$

Example. The Maxwell stress tensor

In the theory of electromagnetism, if one knows the static electric and magnetic field vectors $\boldsymbol{E}$ and $\boldsymbol{B}$ at a surface, the *Maxwell stress tensor*, which in SI units have the elements[3]

[3]See for example Griffiths [34, p. 362].

$$T_{ij} \equiv \epsilon_0 \left(E_i E_j - \frac{1}{2} \delta_{ij} \boldsymbol{E} \cdot \boldsymbol{E} \right) + \frac{1}{\mu_0} \left(B_i B_j - \frac{1}{2} \delta_{ij} \boldsymbol{B} \cdot \boldsymbol{B} \right), \qquad (11.4)$$

allows one to calculate the forces on charges contained within a closed surface. The ith component of the total force is found by integrating the stress tensor over S:

$$F_i = \oint_S \sum_{j=1}^{3} T_{ij} \, \mathrm{d}s_j. \qquad (i = 1, 2, 3)$$

Tensors, therefore, are quantities associated with indices – two for second-rank tensors, but more for higher rank tensors. Some, but not all, tensors have a definite relation between their indices, and it's useful at this point to introduce the following definition.

Definition. *A* symmetric *second-rank tensor is one for which* $T_{ij} = T_{ji}$. *An* antisymmetric tensor *is one for which* $T_{ij} = -T_{ji}$.

The Maxwell stress tensor is an example of a symmetric tensor (see Eq. (11.4)). Tensors generally have no particular symmetry; neither symmetric nor antisymmetric. The stress tensor must be symmetric if angular momentum is to be conserved [35, p. 6].

Definition of tensor

Equation (11.1) defines the *components* of a tensor, but not what a tensor is, just as the components of a vector do not define a vector. *Tensors are generalizations of vectors.* Like a force field $\boldsymbol{F}(\boldsymbol{r})$, the stress tensor represents a physical stress field – it's a quantity, however, that requires more than three components. Equation (11.2) is the component form of $\delta \boldsymbol{F} = \boldsymbol{T} \delta \boldsymbol{s}$, a type of equation discussed at length in Chapter 1: Matrix algebra is well understood by now. Tensors, therefore, operate on vectors (like matrices), and if we want to extend the concept of matrix operations on vectors to more complex objects, we need to be careful with our bookkeeping – we can do this by reverting back to the vectors themselves (and not just focus on components). In terms of the basis vectors

$$\hat{\boldsymbol{x}} = \begin{pmatrix} 1 \\ 0 \\ 0 \end{pmatrix} \quad \hat{\boldsymbol{y}} = \begin{pmatrix} 0 \\ 1 \\ 0 \end{pmatrix} \quad \hat{\boldsymbol{z}} = \begin{pmatrix} 0 \\ 0 \\ 1 \end{pmatrix}$$

the vector $\delta \boldsymbol{s}$ becomes

$$\begin{pmatrix} \delta s_x \\ \delta s_y \\ \delta s_z \end{pmatrix} = \delta s_x \hat{\boldsymbol{x}} + \delta s_y \hat{\boldsymbol{y}} + \delta s_z \hat{\boldsymbol{z}}.$$

We can extend this approach to matrices by defining a basis of the form

$$\hat{\boldsymbol{x}}\hat{\boldsymbol{x}} = \begin{pmatrix} 1 & 0 & 0 \\ 0 & 0 & 0 \\ 0 & 0 & 0 \end{pmatrix} \quad \hat{\boldsymbol{x}}\hat{\boldsymbol{y}} = \begin{pmatrix} 0 & 1 & 0 \\ 0 & 0 & 0 \\ 0 & 0 & 0 \end{pmatrix} \quad \hat{\boldsymbol{x}}\hat{\boldsymbol{z}} = \begin{pmatrix} 0 & 0 & 1 \\ 0 & 0 & 0 \\ 0 & 0 & 0 \end{pmatrix}$$

$$\hat{\boldsymbol{y}}\hat{\boldsymbol{x}} = \begin{pmatrix} 0 & 0 & 0 \\ 1 & 0 & 0 \\ 0 & 0 & 0 \end{pmatrix} \quad \hat{\boldsymbol{y}}\hat{\boldsymbol{y}} = \begin{pmatrix} 0 & 0 & 0 \\ 0 & 1 & 0 \\ 0 & 0 & 0 \end{pmatrix} \quad \text{etc.} \tag{11.5}$$

so that

$$\begin{pmatrix} T_{xx} & T_{xy} & T_{xz} \\ T_{yx} & T_{yy} & T_{yz} \\ T_{zx} & T_{zy} & T_{zz} \end{pmatrix} = T_{xx}\hat{\boldsymbol{x}}\hat{\boldsymbol{x}} + T_{xy}\hat{\boldsymbol{x}}\hat{\boldsymbol{y}} + T_{xz}\hat{\boldsymbol{x}}\hat{\boldsymbol{z}} + T_{yx}\hat{\boldsymbol{y}}\hat{\boldsymbol{x}} + T_{yy}\hat{\boldsymbol{y}}\hat{\boldsymbol{y}}$$
$$+ T_{yz}\hat{\boldsymbol{y}}\hat{\boldsymbol{z}} + T_{zx}\hat{\boldsymbol{z}}\hat{\boldsymbol{x}} + T_{zy}\hat{\boldsymbol{z}}\hat{\boldsymbol{y}} + T_{zz}\hat{\boldsymbol{z}}\hat{\boldsymbol{z}}.$$

In terms of this representation, the matrix operation of $\boldsymbol{T}$ on $\delta\boldsymbol{s}$ *is enforced by introducing the dot product* $\boldsymbol{T}\cdot\delta\boldsymbol{s}$ with

$$\hat{\boldsymbol{x}}\hat{\boldsymbol{x}}\cdot\hat{\boldsymbol{x}} = \hat{\boldsymbol{x}}(\hat{\boldsymbol{x}}\cdot\hat{\boldsymbol{x}}) = \hat{\boldsymbol{x}}(1) = \hat{\boldsymbol{x}}\,, \quad \hat{\boldsymbol{x}}\hat{\boldsymbol{y}}\cdot\hat{\boldsymbol{x}} = \hat{\boldsymbol{x}}(\hat{\boldsymbol{y}}\cdot\hat{\boldsymbol{x}}) = \hat{\boldsymbol{x}}(0) = 0, \quad \text{etc.}$$

so that the usual result $\boldsymbol{T}\cdot\delta\boldsymbol{s} = (T_{xx}\delta s_x + T_{xy}\delta s_y + T_{xz}\delta s_z)\hat{\boldsymbol{x}} + (T_{yx}\delta s_x + T_{yy}\delta s_y + T_{yz}\delta s_z)\hat{\boldsymbol{y}} + (T_{zx}\delta s_x + T_{zy}\delta s_y + T_{zz}\delta s_z)\hat{\boldsymbol{z}}$ is recovered (show this!). This behavior explains the choice of notation which places basis vectors adjacent to each other.[4] And, since the operational behavior of a tensor is effected through use of the dot product, we see that a vector can also be considered a tensor.

The *advantage* of this approach is that it allows us to readily develop the algebra associated with a new class of objects: Tensors. Just as vectors (rank-one tensors) can be expressed in terms of the basis $\{\hat{\boldsymbol{x}}, \hat{\boldsymbol{y}}, \hat{\boldsymbol{z}}\}$ and matrices (rank-two tensors) can be expressed in terms of the basis $\{\hat{\boldsymbol{x}}\hat{\boldsymbol{x}}, \hat{\boldsymbol{x}}\hat{\boldsymbol{y}}, \hat{\boldsymbol{x}}\hat{\boldsymbol{z}}, \hat{\boldsymbol{y}}\hat{\boldsymbol{x}}, \hat{\boldsymbol{y}}\hat{\boldsymbol{y}}, \text{etc.}\}$, it's easy to consider bases consisting of elements of the form $\{\hat{\boldsymbol{x}}\hat{\boldsymbol{x}}\hat{\boldsymbol{x}}, \hat{\boldsymbol{x}}\hat{\boldsymbol{x}}\hat{\boldsymbol{y}}, \ldots\}$ (rank-three tensors) or more. *The result of a tensor operating on another tensor is effected by applying the dot product in the obvious way.*

The *disadvantage* is in the tedious manner by which higher rank tensors must be expressed (rank-two tensors expressed in terms of $\hat{\boldsymbol{x}}\hat{\boldsymbol{x}}$, $\hat{\boldsymbol{x}}\hat{\boldsymbol{y}}$, etc. are known as "dyads" and their use is sufficiently cumbersome that it's largely been abandoned in the modern literature). Nevertheless, this approach does form the foundation for the generalization of vectors and much of the rest of this chapter is devoted to developing rules for simplifying the algebra required for their application.

We therefore define a second-rank tensor:

$$\mathbf{T} \equiv \sum_{i=1}^{3}\sum_{j=1}^{3} T_{ij}\boldsymbol{e}_i\boldsymbol{e}_j. \tag{11.6}$$

[4]Note that the convention is $\hat{\boldsymbol{x}}\hat{\boldsymbol{y}}\cdot\hat{\boldsymbol{x}} = \hat{\boldsymbol{x}}(\hat{\boldsymbol{y}}\cdot\hat{\boldsymbol{x}})$ while $\hat{\boldsymbol{x}}\cdot\hat{\boldsymbol{x}}\hat{\boldsymbol{y}} = (\hat{\boldsymbol{x}}\cdot\hat{\boldsymbol{x}})\hat{\boldsymbol{y}}$, so order is important. See Exercise 11.1

This innocent looking expression raises a host of issues.

- We reserve a special type of symbol to denote tensors (of second and higher rank): boldface **Roman** font for tensors, $\mathbf{T}$, as distinguished from boldface ***italic*** font for vectors, $\boldsymbol{A}$. It's common practice to refer to symbols like T_{ij} as "tensors," but that's not correct. It's important to distinguish the tensor components T_{ij} from the tensor as a whole, $\mathbf{T}$, just as we distinguish a vector $\boldsymbol{A}$ from its components, A_i. At some point, we'll break training and start referring to tensor components as tensors (despite our admonition); continually referring to "the tensor whose components are T_{ij}" becomes cumbersome. The distinction between a tensor and its components should be kept in mind nonetheless.
- Tensors involve linear combinations of all possible products of "directions," dyadic products $\boldsymbol{e}_i\boldsymbol{e}_j$, which provide the sense in which tensors are operators. The *order* in which basis vectors are written in Eq. (11.6) is important; in general $T_{ij} \neq T_{ji}$. The notation $\boldsymbol{e}_i \bigotimes \boldsymbol{e}_j$ is sometimes used (instead of $\boldsymbol{e}_i\boldsymbol{e}_j$) to emphasize that there are ordered "slots" for the basis vectors, $\mathbf{1} \bigotimes \mathbf{2}$, one vector in the first slot, another in the second. The set of all *ordered* dyadic products $\boldsymbol{e}_i\boldsymbol{e}_j$ (where $\boldsymbol{e}_i\boldsymbol{e}_j \neq \boldsymbol{e}_j\boldsymbol{e}_i$) comprise an N^2-dimensional basis for second-rank tensors. Tensors form a vector space: For second-rank tensors $\mathbf{S}$ and $\mathbf{T}$ defined on the same basis set $\{\boldsymbol{e}_i\}$, one can add tensors $\mathbf{R} = \mathbf{S} + \mathbf{T}$ componentwise to produce new tensors, $R_{ij} = S_{ij} + T_{ij}$, and tensors can be multiplied by scalars, $\mathbf{R} = \lambda\mathbf{T}$, where $R_{ij} = \lambda T_{ij}$.
- Tensors comprise vector spaces because of their additive and scalar multiplicative properties. As we saw in Chapter 1, the next step in building up the full properties of vectors is to add a rule for multiplication, and we identified two types of vector multiplication: the inner product, which associates a scalar with a pair of vectors (Section 1.3), and the outer product which creates a higher dimensional object (Section 1.4.8). For spaces of arbitrary dimension, the outer product of N-dimensional vectors produces a second-rank tensor, defined on an N^2-dimensional space of ordered dyadic products.
- As we've noted, the juxtaposition of vectors $\boldsymbol{e}_i\boldsymbol{e}_j$ is intended as an operator: $(\boldsymbol{e}_i\boldsymbol{e}_j)\cdot\boldsymbol{f} = \boldsymbol{e}_i(\boldsymbol{e}_j\cdot\boldsymbol{f})$ for vector $\boldsymbol{f}$. In the dyadic product, therefore, the "leading" vector $\boldsymbol{e}_j$ in $\boldsymbol{e}_i\boldsymbol{e}_j$ plays the role of a linear functional (Section 1.2.5) – it forms the inner product with whatever vector the tensor is going to act on,[5] and one has to understand this from the context in which the tensor is used. In Dirac notation (Section 1.3.1), $\boldsymbol{e}_i\boldsymbol{e}_j$ would be written $|e_i\rangle\langle e_j|$ to emphasize the different roles played by the vectors. In the notation of Chapter 1, a second-rank tensor effects a mapping between elements of a vector space $\mathbb{V}_N$, $\mathbf{T}: \mathbb{V}_N \to \mathbb{V}_N$. We

[5]Second-rank tensors are sometimes denoted with double-headed arrows $\overleftrightarrow{\mathbf{T}}$, or two arrows $\overset{\rightrightarrows}{\mathbf{T}}$, to indicate that they act on vectors to produce vectors. We don't use such notations.

have motivated tensors through the example of the stress tensor appropriate for a three-dimensional space, but the idea of tensor-as-operator holds independently of the dimension of the space, N.

- Having offered a definition of second-rank tensor in Eq. (11.6), we note (already) that it is limited in scope – Eq. (11.6) makes sense only in an orthonormal basis, for reasons that we'll develop. Tensors defined in terms of orthonormal bases are referred to as *Cartesian tensors.* To work with tensors in general coordinate systems requires that we define another type of basis set, the *dual basis* (defined in the following, but denoted $\{\boldsymbol{e}^i\}$). For orthonormal bases, the distinction between the dual basis and the "regular" basis disappears (as we'll show). The extra layer of complication introduced in working with general tensors is worth the effort as it lends geometric insight into the meaning of tensors, and one doesn't have to unlearn certain habits that really only make sense for Cartesian tensors.

Example. Conductivity tensor

The microscopic form of Ohm's law is a relation $\boldsymbol{J} = \sigma \boldsymbol{E}$ between two vectors, the current density $\boldsymbol{J}$ and the applied electric field $\boldsymbol{E}$, where σ is a material-specific parameter, the *conductivity.* In many materials, the direction of the current is parallel to that of the applied field, and the conductivity is a simple proportionality between two vectors. There are other materials, however, *anisotropic* conductors, where the direction of $\boldsymbol{J}$ is not parallel to $\boldsymbol{E}$, where

$$J_i = \sum_j \sigma_{ij} E_j. \tag{11.7}$$

The quantities $\{\sigma_{ij}\}$ are elements of the *conductivity tensor*, $\boldsymbol{\sigma}$. We can either write Ohm's law in component form, as in Eq. (11.7), or as a vector relation

$$\boldsymbol{J} = \boldsymbol{\sigma} \cdot \boldsymbol{E}. \tag{11.8}$$

Equation (11.8) emphasizes the mathematical sense in which the conductivity tensor $\boldsymbol{\sigma}$ maps one vector, $\boldsymbol{E}$, the externally applied electric field, to the response of the material in terms of its current density, $\boldsymbol{J}$.

Example. Inertia tensor

A spherically symmetric, rigid object, rotating about a fixed axis of rotation described by unit vector $\hat{\boldsymbol{n}}$, at angular frequency ω, has angular momentum $\boldsymbol{L} = I\boldsymbol{\omega}$ and rotational kinetic energy $\frac{1}{2}I\omega^2$, where I is the *moment of inertia* of the object and the rotation vector $\boldsymbol{\omega} = \omega\hat{\boldsymbol{n}}$. For other rigid objects, the angular momentum is not aligned with the rotation axis, with $L_i = \sum_j I_{ij}\omega_j$, where $\{I_{ij}\}$ are elements of the *inertia tensor*, $\mathbf{I}$. As a vector relation, $\boldsymbol{L} = \mathbf{I} \cdot \boldsymbol{\omega}$. The inertia tensor maps the rotation vector $\boldsymbol{\omega}$, something set up by external forces, to the angular momentum vector $\boldsymbol{L}$, the response of the object to the imposed rotation. The rotational kinetic energy is expressed as $\frac{1}{2}\boldsymbol{\omega}\cdot\mathbf{I}\cdot\boldsymbol{\omega}$.

Example. Stress tensor in a fluid

Referring to Eq. (3.7), the balance equation for the momentum of a fluid in a volume V, having mass density ρ, flowing with velocity $\boldsymbol{v}$, is, where S is a surface bounding V,

$$\underbrace{\frac{\mathrm{d}}{\mathrm{d}t}\int_V \rho\boldsymbol{v}\,\mathrm{d}^3r}_{\text{rate of change of momentum in } V} = \underbrace{-\oint_S \rho\boldsymbol{v}\boldsymbol{v}\cdot\mathrm{d}\boldsymbol{s}}_{\text{momentum flux through } S} + \underbrace{\int_V \rho\boldsymbol{F}\,\mathrm{d}^3r}_{\text{momentum produced by external forces}} + \underbrace{\oint_S \mathbf{T}\cdot\mathrm{d}\boldsymbol{s}}_{\text{momentum produced by short-range internal forces at } S}, \tag{11.9}$$

where we've used dyadic notation: $\boldsymbol{v}\boldsymbol{v}$. The first integral on the right accounts for the momentum convected through S, where $\rho\boldsymbol{v}$ is the momentum density (momentum per volume). The other two integrals represent *sources* of momentum production: $\boldsymbol{F}$ is an external force (per mass) that couples to the particles within V (such as the gravitational field), and $\mathbf{T}$, the stress tensor, represents the effect of short-range internal forces acting at S. In a fluid, there is always a normal component of the surface force, the pressure P, with $\mathbf{T} = -P\mathbf{I}$, where I is the[6] *unit dyadic*, $\mathbf{I} = \sum_{i=1}^{3}\boldsymbol{e}_i\boldsymbol{e}_i$. In terms of components, $T_{ij} = -P\delta_{ij}$. There are additional terms to the stress tensor of a fluid due to the effects of viscosity [36, p. 45], but we omit them here. Viscosity (internal friction) involves a transfer of momentum from points where the velocity is large to those where it is small.

Where to now?

In a narrow sense, we're done. We've introduced one of the primary purposes in which tensors are used, as objects that relate one vector to another. Tensors are a mathematical tool for incorporating the effects of anisotropy in the description of physical systems. There are *many* such relationships involving tensors for anisotropic systems.[7] To name just two, the susceptibility tensor, a relation between the electric polarization $\boldsymbol{P}$ of a dielectric material and the electric field, $P_i = \sum_j \chi_{ij}E_j$, and the permeability tensor, a relation between the magnetic vectors $\boldsymbol{B}$ and $\boldsymbol{H}$, $B_i = \sum_j \mu_{ij}H_j$.

There's much more to say about tensors however. There are higher rank tensors, alluded to previously. The elastic energy E of a deformed crystal is written in terms of tensors [35, p. 37]

[6] The unit dyadic $\mathbf{I} = \sum_i \boldsymbol{e}_i\boldsymbol{e}_i$ is the identity operator (sometimes written $\mathbf{I} = \hat{\boldsymbol{x}}\hat{\boldsymbol{x}} + \hat{\boldsymbol{y}}\hat{\boldsymbol{y}} + \hat{\boldsymbol{z}}\hat{\boldsymbol{z}}$): $\mathbf{I}\cdot\boldsymbol{f} = \sum_i \boldsymbol{e}_i(\boldsymbol{e}_i\cdot\boldsymbol{f}) = \sum_i \boldsymbol{e}_i f_i = \boldsymbol{f}$, where in an orthonormal basis $f_i = \boldsymbol{e}_i\cdot\boldsymbol{f}$.

[7] See for example Nye [37].

$$E = \frac{1}{2}\sum_{i=1}^{3}\sum_{j=1}^{3}\sum_{l=1}^{3}\sum_{m=1}^{3}\lambda_{iklm}u_{ik}u_{lm}. \tag{11.10}$$

The quantities $\{u_{ik}\}$ are the elements of the *strain tensor* (a second-rank tensor), with

$$u_{ik} \equiv \frac{1}{2}\left(\frac{\partial u_i}{\partial x_k} + \frac{\partial u_k}{\partial x_i}\right),$$

where $u_i \equiv x'_i - x_i$ are the elements of the *displacement vector* characterizing the deformation of an object due to applied forces, where x'_i are the components of the position vector $\boldsymbol{r}'$ of material points that, prior to the imposition of forces, were at locations specified by the position vector $\boldsymbol{r}$. What a mouthful, just to define the terms in Eq. (11.10), when the intention is to give an example of a higher rank tensor.[8] The quantities $\{\lambda_{iklm}\}$ are the elements of the *elastic modulus tensor*, a tensor of rank four. The point of introducing Eq. (11.10) is to note that a fourth-rank tensor (elastic modulus tensor) provides a relation between *two* second-rank tensors (strain tensor) and a scalar, the energy. It can be shown that the components of the stress tensor are related to those of the strain tensor through the elastic modulus tensor (a generalization of Hooke's law):

$$T_{ik} = \sum_{l=1}^{3}\sum_{m=1}^{3}\lambda_{iklm}u_{lm}. \tag{11.11}$$

Equation (11.11) shows that a fourth-rank tensor also provides a relation between second-rank tensors, generalizing expressions such as Eq. (11.7), in which second-rank tensors provide relations between vectors. Tensors therefore wear many hats, as operators that effect mappings between tensors and scalars, and between tensors. There's clearly much to learn about the workings of tensors. We develop the properties of tensors in a systematic manner in the remainder of this chapter.

Perhaps you've noticed (or not) a potentially troubling issue that tensors are defined in terms of vector components (such as the stress and strain tensors), which are basis dependent. If we change the basis, how do the elements of tensors change? We'll see that the question of whether a set of functions $\{T_{ij}\}$ represents a tensor depends on how they transform under a change of basis. Tensors have *invariance* properties that are independent of basis (as we'll show). Because tensors involve entities and properties that are independent of the choice of basis, they form an ideal tool for the study of natural laws – a second broad reason to study tensors (the effects of anisotropy being the first). Laws of physics, when expressed as relations between tensors, if true in one coordinate system, are true in all coordinate systems.[9] We return to this point in Section 11.2.6.

---

[8] Tensor rank is defined in Section 11.2.3.

[9] Tensors feature prominently in the general theory of relativity.

11.2 Transformation properties

Do just any set of N^2 quantities constitute the elements of a second-rank tensor? The answer to that question, "No," lies in a detailed examination of how tensors transform under a change of basis – the goal of this section. The elements of a tensor must possess an internal consistency such that their behavior under coordinate transformations is fairly narrowly constrained. The material in this section may seem dry and technical, yet it's very important for learning about tensors.

11.2.1 The two types of vector: Contravariant and covariant

Before immersing ourselves in the transformation properties of tensors, we observe that there are two *types* of vector – loosely, *forces* and *fluxes.* The velocity of a particle, for example, is the limit $\mathrm{d}\boldsymbol{r}/\mathrm{d}t$; velocity, therefore, has the geometric *character* of a line element, $\mathrm{d}\boldsymbol{r}$, and we'll explain what we mean by the character of a vector. Other vectors are specified by *gradients*, such as the heat current density $\boldsymbol{J} = -\kappa \boldsymbol{\nabla} T$ (Section 3.1). Gradients are orthogonal to the *level sets* of functions[10]; flux-related vectors therefore have the character of oriented surface elements, $\mathrm{d}\boldsymbol{s}$. Line elements and surface elements behave differently under coordinate transformations (as we'll show), which is what we mean by their vector character. The two types of vector are referred to mathematically as *contravariant* and *covariant.* The distinction between them is moot as long as we restrict ourselves to orthogonal coordinate systems. It might be supposed that a *mathematical* distinction is not essential for physics. Anisotropy, however, is one reason to adopt nonorthogonal coordinate systems. At a more fundamental level, the general theory of relativity establishes that *geometry is physical* – the metric tensor (see Section 11.4), which *defines* a geometry, is determined by physical conditions, and, for example, represents gravitational potentials. The mathematics of general relativity is the province of *differential geometry*, a topic outside the intended scope of this book. Nevertheless, there are sufficiently many applications involving general coordinate systems to warrant the treatment of contra- and covariant vectors and their generalizations as tensors. The material in this chapter is a springboard for more advanced work.

[10]The *level set* of a function f is the locus of points for which f takes on a constant value c, $f(x, y, z) = c$, a familiar concept having different names in physical applications. In electrostatics, we refer to *equipotential* surfaces as those for which the scalar potential has a constant value. In thermodynamics, we refer to *isotherms* as the set of all thermodynamic variables for which the temperature of a system has a fixed value. See also Section A.1.2.

11.2.2 Coordinate transformations

We're going to adopt the traditional notation of tensor analysis, and that is to denote coordinates with superscripts,[11] x^i, and basis vectors with subscripts, $\boldsymbol{e}_i$. Thus, $\boldsymbol{V} = \sum_{i=1}^{3} V^i \boldsymbol{e}_i$. The reason for this change in notation will become apparent shortly. Denoting vector components with superscripts takes some getting used to, but experience shows you'll pick it up pretty fast. When there's a possibility of confusion, we'll denote the square of a quantity A as $(A)^2$ to avoid mistaking it with the vector component A^2; contrary to what you might think, problems of this sort do not occur often. In what follows, we'll assume a space of N dimensions, where N is left unspecified.[12]

We will encounter expressions involving numerous sums over various indices, and writing out the summation symbols becomes cumbersome (such as in Eq. (11.10)). The *Einstein summation convention* is to interpret repeated raised and lowered indices as implying a sum. Thus, we can abbreviate $\sum_{i=1}^{N} A^i \boldsymbol{e}_i \equiv A^i \boldsymbol{e}_i$, where the range of the summation index is understood from context. Of course the index i is a dummy index having no absolute meaning. The expressions $A^i \boldsymbol{e}_i = A^j \boldsymbol{e}_j = A^k \boldsymbol{e}_k$ are equivalent and imply the same sum. Remember: The rule is that *repeated* upper and lower indices imply a summation. Terms such as $A^i \boldsymbol{e}_j$ do not imply a sum. We will gradually work in the summation convention to gain practice with it but after a point, we will use it without comment. In the rare instances, we wish to refer to the general term $A^i \boldsymbol{e}_i$ without implying a summation, we will explicitly note that.

A point P in an N-dimensional space has N independent coordinates $\{x^i\}_{i=1}^{N}$. Assume we can find N independent, single-valued analytic functions of the coordinates x^i,

$$y^i = y^i(x^1, \ldots, x^N). \qquad i = 1, 2, \ldots, N \tag{11.12}$$

A set of functions is *independent* if the determinant of the matrix of partial derivatives $\partial y^i / \partial x^j$ (the *Jacobian matrix*) is not identically zero. In that case, the functions $\{y^i\}$ provide *another set of coordinates*, a new set of numbers, to attach to the *same points in space* (also labeled by

[11] The use of subscripts and superscripts is not as arbitrary as it might first appear, and the way the two types of indices are used in calculations is quite logical and consistent. Subscripts and superscripts have been used in tensor analysis for 100 years. We quote from a 1927 book [38, p. 1]: "Recent advances in the theory of differential invariants and the wide use of this theory in physical investigations have brought about a rather general acceptance of a particular type of notation, the essential feature of which is the systematic use of subscripts and superscripts."

[12] We of course live in a three-dimensional world of height, width, and depth. The theory of relativity adds time as another dimension; general relativity models spacetime as a four-dimensional geometry. In bosonic string theory, spacetime is 26-dimensional, while in supersymmetric string theory spacetime is 10-dimensional.

x^i), and hence are a *transformation of coordinates.* By assumption (non vanishing of the Jacobian determinant), Eq. (11.12) is invertible[13]:

$$x^j = x^j(y^1, \ldots, y^N). \qquad j = 1, 2, \ldots, N \tag{11.13}$$

Equations (11.12) and (11.13) are *passive* coordinate transformations, where new coordinates are assigned to the same points of space.[14] *Points* are thus invariant under coordinate transformations, which do nothing to the points themselves. Any *set* of points is therefore invariant, as well as any point function.[15] A *scalar field* is a function $\phi(\boldsymbol{r})$ that assigns a scalar to every point of space. The *value* of scalar fields is invariant under passive coordinate transformations. Thus, for scalar field ϕ,

$$\phi(x^i) = \phi(x^i(y^j)) \equiv \phi'(y^j). \tag{11.14}$$

Equation (11.14) indicates that the *form* of the function of the transformed coordinates may change, ϕ', but not its value $\phi'(y^j)$. Scalars are *invariant* under coordinate transformations. Scalars, however, do not exhaust the possible types of invariants under coordinate transformations: *Invariants other than scalars exist,* as we'll see.

Let point Q in an infinitesimal neighborhood of P have coordinates $x^i + \mathrm{d}x^i$ (see Figure 11.2). The points (P, Q) define the infinitesimal displacement vector $\mathrm{d}\boldsymbol{r} \equiv \overrightarrow{PQ}$ with components $\mathrm{d}x^i$. In a new coordinate system, determined by the transformation in Eq. (11.12), the same points P and Q have coordinates y^i and $y^i + \mathrm{d}y^i$. In this coordinate system, the *same vector* $\mathrm{d}\boldsymbol{r}$ has components $\mathrm{d}y^i$. The components of $\mathrm{d}\boldsymbol{r}$ in the two coordinate systems are related by calculus

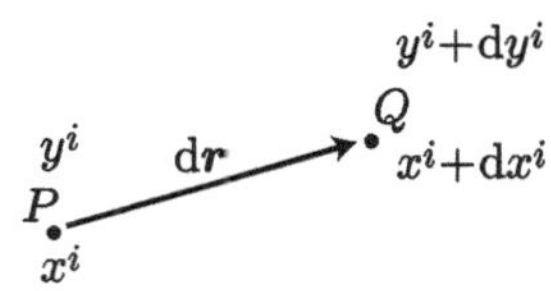

Figure 11.2 Points P and Q infinitesimally separated, described in two coordinate systems.

$$\mathrm{d}y^i = \sum_{j=1}^{N} \left.\frac{\partial y^i}{\partial x^j}\right|_P \mathrm{d}x^j \equiv \sum_{j=1}^{N} A^i_j \,\mathrm{d}x^j = A^i_j \,\mathrm{d}x^j, \quad \text{(summation convention)} \tag{11.15}$$

where the partial derivatives

$$A^i_j \equiv \left.\frac{\partial y^i}{\partial x^j}\right|_P, \tag{11.16}$$

evaluated at P, comprise the elements of the Jacobian matrix. The derivatives exist through the analyticity of the transformation equations, Eq. (11.12). We're denoting (in this chapter only) the elements of a matrix

[13] We impose the minimal requirement that the functions $y^i(x^j)$ in Eq. (11.12) are continuous, together with their first partial derivatives. These conditions suffice for the functions $x^i(y^j)$ in Eq. (11.13) to be single-valued and to have continuous first partial derivatives.

[14] *Active* coordinate transformations (which we don't use in this book) assign coordinates to *different points* of the *same* coordinate system.

[15] We're making a distinction between points and coordinates. The temperature at a point doesn't care what coordinates you attach to the point.

as A^i_j, where the superscript i labels the rows and the subscript j labels columns. The quantities A^i_j are *constant* for Q in an infinitesimal neighborhood of P. Equation (11.15) then represents a *locally linear* transformation, even though the transformation equations, Eq. (11.12) are not necessarily linear.

Example. Let the transformation equations be linear,

$$y^i = c^i + \sum_{j=1}^{N} A^i_j x^j = c^i + A^i_j x^j, \qquad \text{(summation convention)} \qquad (11.17)$$

where the c^i are constants, as are the matrix elements A^i_j. If $c^i = 0$, the transformation is said to be *homogeneous*. A homogeneous, linear transformation was given in Section 1.4.10. Equation (1.47) specifies how the coordinates (x, y) for a point in a plane change upon a rigid rotation of the coordinate axes through the angle φ:

$$\begin{pmatrix} x' \\ y' \end{pmatrix} = \begin{pmatrix} \cos\varphi & \sin\varphi \\ -\sin\varphi & \cos\varphi \end{pmatrix} \begin{pmatrix} x \\ y \end{pmatrix},$$

see Figure 1.5. Renaming $(x, y) = (x^1, x^2)$, Eq. (1.47) is written in the notation of this chapter

$$\begin{pmatrix} (x^1)' \\ (x^2)' \end{pmatrix} = \begin{pmatrix} A^1_1 & A^1_2 \\ A^2_1 & A^2_2 \end{pmatrix} \begin{pmatrix} x^1 \\ x^2 \end{pmatrix}.$$

A widely used notational device that simplifies tensor equations, one which we'll adopt, is the *Schouten index convention* [39, p. 2]. Instead of coming up with a different symbol for each new coordinate system (y, x', $\overline{x}$, etc.), choose x to represent coordinates once and for all. The coordinates in *different coordinate systems* are distinguished by primes attached to *indices*. Thus, Eq. (11.15) is written $dx^{i'} = A^{i'}_j \, dx^j$ (summation convention), with $A^{i'}_j \equiv \partial x^{i'}/\partial x^j|_P$; Eq. (11.12) is written $x^{i'} = x^{i'}(x^j)$.

Example. The transformation between Cartesian coordinates in a plane (x, y) and plane polar coordinates (r, θ) is invertible.

$$r = \sqrt{x^2 + y^2} \qquad x = r\cos\theta$$
$$\theta = \tan^{-1}(y/x) \qquad y = r\sin\theta.$$

The elements of the Jacobian matrix are found from Eq. (11.16), with

$$[A^{i'}_j] = \begin{pmatrix} \partial r/\partial x & \partial r/\partial y \\ \partial \theta/\partial x & \partial \theta/\partial y \end{pmatrix} = \begin{pmatrix} x/r & y/r \\ -y/r^2 & x/r^2 \end{pmatrix}.$$

The Jacobian determinant $J = r^{-1}$; the transformation is invertible except at $r = 0$.

A set of differentials in one coordinate system thus determines a set of differentials in another coordinate system, Eq. (11.15). The transformation inverse to Eq. (11.15) is

$$\mathrm{d}x^i = \sum_{j'=1}^{N} \left.\frac{\partial x^i}{\partial x^{j'}}\right|_P \mathrm{d}x^{j'} \equiv A^i_{j'}\,\mathrm{d}x^{j'}. \qquad i = 1, \ldots, N \tag{11.18}$$

Compare the placement of indices in Eq. (11.18) with that in Eq. (11.15). Combining Eqs. (11.18) and (11.15), $\mathrm{d}x^i = A^i_{j'}\,\mathrm{d}x^{j'} = A^i_{j'}A^{j'}_k\,\mathrm{d}x^k$, or

$$(\delta^i_k - A^i_{j'}A^{j'}_k)\,\mathrm{d}x^k = 0, \tag{11.19}$$

where we've written the Kronecker symbol in a new way,

$$\delta^i_k = \begin{cases} 1 & \text{if} i = k \\ 0 & \text{if} i \neq k \end{cases} \qquad (i, j = 1, 2, \ldots, N).$$

Because the $\mathrm{d}x^k$ in Eq. (11.19) can be independently varied, the *matrices* $[A^i_{j'}]$ and $[A^{j'}_k]$ are inversely related[16]

$$A^i_{j'}A^{j'}_k = \delta^i_k. \qquad \text{(summation convention)} \tag{11.20}$$

Equation (11.20) is simply the chain rule: $\sum_{j'}(\partial x^i/\partial x^{j'})(\partial x^{j'}/\partial x^k) = \partial x^i/\partial x^k = \delta^i_k$.

---

[16] In other notational schemes, one must come up with different symbols for the Jacobian matrix and its inverse; e.g. U and $\overline{U}$, or U and U^{-1}. In either case, one has to remember which matrix applies to which transformation. In the Schouten method there is one symbol with two types of indices, primed and unprimed. Other schemes use the same symbol for the Jacobian matrix and its inverse, but with two ways of writing the indices, ${A^i}_j$ and ${A_j}^i$. Not only does one have to remember which applies to which transformation, but such a scheme quickly becomes unintelligible to students at the back of the room.

11.2.3 Contravariant vectors and tensors

Let's return to the idea introduced in Section 11.2.2, that the vector $\mathrm{d}\boldsymbol{r}$ has coordinates $\mathrm{d}x^i$ in one coordinate system, and $\mathrm{d}x^{j'}$ in another. We have yet to consider the basis vectors in the two coordinate systems, $\{\boldsymbol{e}_i\}$ and $\{\boldsymbol{e}_{j'}\}$. Equations (11.15) and (11.18) indicate how the *components* of $\mathrm{d}\boldsymbol{r}$ transform between coordinate systems. How do the basis vectors transform? We can use the fact that it's the same vector to equate[17]

$$\mathrm{d}\boldsymbol{r} = \mathrm{d}x^i \boldsymbol{e}_i = \mathrm{d}x^{j'} \boldsymbol{e}_{j'}. \tag{11.21}$$

By combining Eq. (11.15) with Eq. (11.21), we have

$$\mathrm{d}x^i \boldsymbol{e}_i = A^i_{j'}\, \mathrm{d}x^{j'} \boldsymbol{e}_i = \mathrm{d}x^{j'} \boldsymbol{e}_{j'} \;\Rightarrow\; \mathrm{d}x^{j'}(\boldsymbol{e}_{j'} - A^i_{j'} \boldsymbol{e}_i) = 0.$$

Because the quantities $\mathrm{d}x^{j'}$ can be independently varied, we have the transformation relation for basis vectors

$$\boldsymbol{e}_{j'} = A^i_{j'} \boldsymbol{e}_i. \qquad j' = 1, \ldots, N \tag{11.22}$$

Of course, we can swap primes and unprimes to arrive at the inverse of Eq. (11.22),

$$\boldsymbol{e}_i = A^{j'}_i \boldsymbol{e}_{j'}. \qquad i = 1, \ldots, N \tag{11.23}$$

We can easily confirm that Eq. (11.23) is the inverse of Eq. (11.22): $\boldsymbol{e}_i = A^{j'}_i \boldsymbol{e}_{j'} = A^{j'}_i A^k_{j'} \boldsymbol{e}_k = \delta^k_i \boldsymbol{e}_k = \boldsymbol{e}_i$, where we've used Eq. (11.20). Compare Eq. (11.22) with Eq. (11.15) ($\mathrm{d}x^{i'} = A^{i'}_j\, \mathrm{d}x^j$): The components of $\mathrm{d}\boldsymbol{r}$ transform *inversely* (or *contra*) to how the basis vectors transform.

The apparent maze of indices with primes and unprimes, which seems daunting at first, can be "psyched out" fairly quickly. Take an equation like Eq. (11.22). You want the transformation equation for basis vectors in the "new" coordinate system, $\boldsymbol{e}_{j'}$, in terms of the "old," $\boldsymbol{e}_i$. Line up the index j' on the left side of the equation (which is "downstairs," a subscript), with a similarly placed index on the elements of the transformation matrix, $A^i_{j'}$. The remaining index i on $A^i_{j'}$, a superscript, must, in the summation convention, be matched with a subscript on the basis vector $\boldsymbol{e}_i$. Respect the placement of indices on both sides of the equation, and it's straightforward to simply write down expressions like Eq. (11.22). The same applies to the transformation of vector components, as in Eq. (11.18): The index i, a superscript on the left side of the equation, lines up with the superscript on the right side of the equation. Some terminology: An index like i in Eq. (11.18) or j' in Eq. (11.22) is said to be a *free index* – it ranges over the values $1, \ldots, N$. In understanding tensor equations, it helps to first identify the free indices. The other indices, such as i in Eq. (11.22) or j' in Eq. (11.18), are dummy indices, which in the older

free index

[17] Note that we have chosen once and for all to represent basis vectors with the symbol $\boldsymbol{e}$.

literature were referred to as *umbral* indices – they're in effect not visible externally, outside the inner workings of the equation.

How do the components of vectors *other* than $\mathrm{d}\boldsymbol{r}$ transform under a change of basis? The question is easy to answer now that we know how the basis vectors transform. In Eqs. (11.15) and (11.18), calculus was used to specify how the *components* $\mathrm{d}x^i$ of $\mathrm{d}\boldsymbol{r} = \mathrm{d}x^i \boldsymbol{e}_i$ transform. We could establish how the basis vectors transform by requiring that $\mathrm{d}\boldsymbol{r}$ be the same when represented in two basis sets, $\{\boldsymbol{e}_i\}$ and $\{\boldsymbol{e}_{j'}\}$: $\mathrm{d}\boldsymbol{r} = \mathrm{d}x^i \boldsymbol{e}_i = \mathrm{d}x^{j'} \boldsymbol{e}_{j'}$, Eq. (11.21). The fact that $\mathrm{d}\boldsymbol{r}$ is the same when expressed in different bases implies it has *absolute meaning*, with an existence independent of coordinate system. A quantity having this property is said to be a *geometric object*, an invariant that's not simply a number.[18] *The infinitesimal displacement vector* $\mathrm{d}\boldsymbol{r}$ *is the prototype of a class of geometric objects referred to as* contravariant vectors.[19]

contravariant vector

A vector $\boldsymbol{T} \equiv T^i \boldsymbol{e}_i$ has contravariant components if they transform as $T^{i'} = A^{i'}_j T^j$, i.e. as in Eq. (11.15). In that way, when we use Eq. (11.22) for the transformation of the basis vectors, $\boldsymbol{T}$ is a geometric object: $\boldsymbol{T} = T^j \boldsymbol{e}_j = T^j A^{i'}_j \boldsymbol{e}_{i'} = T^{i'} \boldsymbol{e}_{i'} = \boldsymbol{T}'$. Any set of N quantities $\{T^i\}$ that transform like the components of $\mathrm{d}\boldsymbol{r}$,

$$T^{k'} = A^{k'}_j T^j = \frac{\partial x^{k'}}{\partial x^j} T^j, \qquad k' = 1, \ldots, N \tag{11.24}$$

are said to be contravariant components of a vector. The contravariant components of a vector transform inversely ("contra") to the transformation of basis vectors.

Any set of N^2 quantities $\{T^{ij}\}$ that transform as

$$T^{i'j'} = A^{i'}_k A^{j'}_l T^{kl} = \frac{\partial x^{i'}}{\partial x^k} \frac{\partial x^{j'}}{\partial x^l} T^{kl} \tag{11.25}$$

are said to be the contravariant components of a second-rank tensor. If the quantities $\{T^{ij}\}$ transform as in Eq. (11.25), then $\mathbf{T} \equiv T^{ij} \boldsymbol{e}_i \boldsymbol{e}_j$ (note the change in notation from Eq. (11.6)) is a geometric object:

$$\begin{aligned}\mathbf{T}' &= T^{i'j'} \boldsymbol{e}_{i'} \boldsymbol{e}_{j'} = A^{i'}_k A^{j'}_l T^{kl} A^m_{i'} \boldsymbol{e}_m A^n_{j'} \boldsymbol{e}_n = (A^{i'}_k A^m_{i'})(A^{j'}_l A^n_{j'}) T^{kl} \boldsymbol{e}_m \boldsymbol{e}_n \\ &= \delta^m_k \delta^n_l T^{kl} \boldsymbol{e}_m \boldsymbol{e}_n = T^{mn} \boldsymbol{e}_m \boldsymbol{e}_n = \mathbf{T},\end{aligned}$$

where we've used Eqs. (11.25), (11.22), and (11.20). The contravariant components of a tensor of rank r are a set of N^r objects $\{T^{k_1 k_2 \cdots k_r}\}$ that transform as

$$T^{k'_1 k'_2 \cdots k'_r} = A^{k'_1}_{l_1} A^{k'_2}_{l_2} \cdots A^{k'_r}_{l_r} T^{l_1 l_2 \cdots l_r}.$$

The *rank* of a tensor is therefore the number of indices it has, or, equivalently, the number of terms $A^{i'}_j$ required for its transformation.

tensor rank

[18]Physically, the electric field vector $\boldsymbol{E}$ exists independently of the coordinate system we use to describe it.

[19]By a contravariant vector, we mean a vector with contravariant components. The same terminology applies to contravariant tensors.

Thus, vectors are rank-one tensors (a single factor of $A^{i'}_j$ in Eq. (11.24)), and scalars are rank-zero tensors because their transformation property, Eq. (11.14), involves no transformation matrix. An rth-rank tensor, considered as a geometric object is defined as $\mathbf{T} \equiv T^{k_1 \cdots k_r} \boldsymbol{e}_{k_1} \cdots \boldsymbol{e}_{k_r}$, where we've introduced an r-fold, ordered "dyadic-like" product $\boldsymbol{e}_{k_1} \cdots \boldsymbol{e}_{k_r}$.

Equation (11.25) is often given as a *definition* of second-rank tensor, i.e. a tensor is anything that transforms as a tensor. In the same way, Eq. (11.24) is sometimes used to define a vector.[20] We have motivated the definition of tensor ($\mathbf{T} = T^{ij} \boldsymbol{e}_i \boldsymbol{e}_j$) from the idea that, like a vector, a tensor is a physical quantity that retains its identity however the coordinate system is transformed. The transformation properties of tensors, Eq. (11.25), then follow from that idea, as we've shown.

Example. Building tensors from vectors

One way to produce tensors is through products of vectors. Suppose one has vectors $\boldsymbol{X}$ and $\boldsymbol{Y}$ for which their components satisfy the transformation property Eq. (11.24), i.e. $X^{i'} = A^{i'}_k X^k$ and $Y^{j'} = A^{j'}_l Y^l$. Then, the products of vector components $T^{ij} \equiv X^i Y^j$ comprise the contravariant elements of a second-rank tensor because they automatically transform according to Eq. (11.25): $T^{i'j'} = A^{i'}_k A^{j'}_l T^{kl}$. The tensor $\mathbf{T}$ produced this way is often denoted $\mathbf{T} = \boldsymbol{X} \otimes \boldsymbol{Y}$. Not every tensor can be constructed from a product of vectors – the Maxwell stress tensor, Eq. (11.4), is an example.

The coordinate basis

How do partial derivatives transform under a coordinate transformation? Using the chain rule, for coordinates $\{x^{i'}\}$ and $\{x^j\}$,

$$\frac{\partial}{\partial x^{i'}} = \sum_j \frac{\partial x^j}{\partial x^{i'}} \frac{\partial}{\partial x^j} \equiv A^j_{i'} \frac{\partial}{\partial x^j}. \tag{11.26}$$

Compare Eq. (11.26) with Eq. (11.22): Partial derivatives transform in the same manner as the vectors $\boldsymbol{e}_i$ – they "covary" with the basis vectors. One might wonder, on the basis of this comparison, whether basis vectors $\boldsymbol{e}_i$ *are* partial derivatives. Is it feasible to conjecture such a correspondence, $\boldsymbol{e}_i \leftrightarrow \partial/\partial x^i$? As we now show, the answer is yes.

Figure 11.3 shows a three-dimensional coordinate system with point P at the intersection of three *coordinate curves*,[21] (u, v, w). For nearby

---

[20] Of course, that begs the question: What *is* a vector? We began this book with a consideration of that question, where the traditional view of vector was generalized to an element of an inner-product space. One *could* define vector components to be quantities that transform as in Eq. (11.24).

[21] In three dimensions, if two of the coordinates are held fixed and the other coordinate is allowed to vary, the resulting curve is a coordinate curve. Holding one coordinate fixed, with the other two allowed to vary, produces a *coordinate surface*.

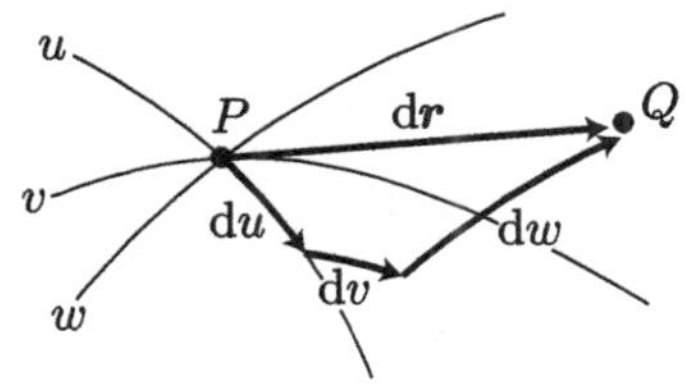

Figure 11.3 (uvw) coordinate system. The tangents to coordinate curves define a basis.

point Q, define the vector $\mathrm{d}\boldsymbol{r} = \vec{PQ}$. Point P is located relative to the origin (not shown) by vector $\boldsymbol{r}$, with Q located by vector $\boldsymbol{r} + \mathrm{d}\boldsymbol{r}$. The location of P is specified by the three coordinates (uvw), and hence $\boldsymbol{r}$ is a function of (uvw). To first order in small quantities,

$$\mathrm{d}\boldsymbol{r} = \frac{\partial \boldsymbol{r}}{\partial u}\,\mathrm{d}u + \frac{\partial \boldsymbol{r}}{\partial v}\,\mathrm{d}v + \frac{\partial \boldsymbol{r}}{\partial w}\,\mathrm{d}w, \tag{11.27}$$

where the derivatives with respect to coordinates are evaluated at P. The derivatives

$$\boldsymbol{e}_u \equiv \left.\frac{\partial \boldsymbol{r}}{\partial u}\right|_P \qquad \boldsymbol{e}_v \equiv \left.\frac{\partial \boldsymbol{r}}{\partial v}\right|_P \qquad \boldsymbol{e}_w \equiv \frac{\partial \boldsymbol{r}}{\partial w} \tag{11.28}$$

thus form a local basis – an arbitrary vector $\mathrm{d}\boldsymbol{r}$ in the neighborhood of P can be expressed as a linear combination of them – and they're each tangent to a coordinate curve. A local set of basis vectors is determined by the local structure of the coordinate system, the *coordinate basis.* Any set of N linearly independent vectors that spans a space comprises a basis. Not every basis is a coordinate basis, but, given a coordinate system, one can always find a basis, the coordinate basis.

Example. Coordinate basis for spherical coordinates

The position vector in spherical coordinates can be written

$$\boldsymbol{r} = r\sin\theta\cos\phi\hat{\boldsymbol{x}} + r\sin\theta\sin\phi\hat{\boldsymbol{y}} + r\cos\theta\hat{\boldsymbol{z}}.$$

Applying Eq. (11.28), we have the vectors of the coordinate basis

$$\begin{aligned}
\boldsymbol{e}_r &= \frac{\partial \boldsymbol{r}}{\partial r} = \sin\theta\cos\phi\hat{\boldsymbol{x}} + \sin\theta\sin\phi\hat{\boldsymbol{y}} + \cos\theta\hat{\boldsymbol{z}} \\
\boldsymbol{e}_\theta &= \frac{\partial \boldsymbol{r}}{\partial \theta} = r\cos\theta\cos\phi\hat{\boldsymbol{x}} + r\cos\theta\sin\phi\hat{\boldsymbol{y}} - r\sin\theta\hat{\boldsymbol{z}} \\
\boldsymbol{e}_\phi &= \frac{\partial \boldsymbol{r}}{\partial \phi} = -r\sin\theta\sin\phi\hat{\boldsymbol{x}} + r\sin\theta\cos\phi\hat{\boldsymbol{y}},
\end{aligned} \tag{11.29}$$

which are tangential to the coordinate curves (see Figure A.6).

The norms of these vectors are $||\boldsymbol{e}_r|| = 1$, $||\boldsymbol{e}_\theta|| = r$, and $||\boldsymbol{e}_\phi|| = r\sin\theta$ (show this). *The magnitude and direction of the basis vectors are not constant* and vary throughout the coordinate system. We have the unit vectors,

$$\hat{\boldsymbol{e}}_r = \sin\theta\cos\phi\hat{\boldsymbol{x}} + \sin\theta\sin\phi\hat{\boldsymbol{y}} + \cos\theta\hat{\boldsymbol{z}}$$
$$\hat{\boldsymbol{e}}_\theta = \cos\theta\cos\phi\hat{\boldsymbol{x}} + \cos\theta\sin\phi\hat{\boldsymbol{y}} - \sin\theta\hat{\boldsymbol{z}}$$
$$\hat{\boldsymbol{e}}_\phi = -\sin\phi\hat{\boldsymbol{x}} + \cos\phi\hat{\boldsymbol{y}}.$$

By definition, $\boldsymbol{e}_r = \hat{\boldsymbol{e}}_r$, $\boldsymbol{e}_\theta = r\hat{\boldsymbol{e}}_\theta$, and $\boldsymbol{e}_\phi = r\sin\theta\hat{\boldsymbol{e}}_\phi$.

11.2.4 Covariant vectors and tensors

Contravariant vectors are those whose "provenance" can be traced to the prototype, $\mathrm{d}\boldsymbol{r}$. Velocity, $\boldsymbol{v} = \mathrm{d}\boldsymbol{r}/\,\mathrm{d}t$, is a contravariant vector, as are the acceleration and force vectors, $\boldsymbol{a} = \mathrm{d}\boldsymbol{v}/\,\mathrm{d}t$ and $\boldsymbol{F} = m\boldsymbol{a}$. Contravariant vector components V^i transform inversely to how the basis vectors $\boldsymbol{e}_i$ transform (Eqs. (11.24) and (11.22)), which is what makes geometric objects possible, $V^i\boldsymbol{e}_i$ = basis-independent invariant. There is another class of vectors, however, (and their generalizations as tensors) – *covariant vectors* – whose components transform in the *same way* as the basis vectors $\boldsymbol{e}_i$.

Consider the gradient of a scalar function, $\boldsymbol{\nabla}\phi = (\partial\phi/\partial x)\hat{\boldsymbol{x}} + (\partial\phi/\partial y)\hat{\boldsymbol{y}} + (\partial\phi/\partial z)\hat{\boldsymbol{z}}$. Its components transform as

$$\frac{\partial\phi'}{\partial x^{j'}} = \frac{\partial x^i}{\partial x^{j'}}\frac{\partial\phi}{\partial x^i} \equiv A^i_{j'}\frac{\partial\phi}{\partial x^i}, \tag{11.30}$$

where we've used Eq. (11.14). The gradient has the geometric property of being orthogonal to the level sets of functions, and it's physical, e.g. the electric field is the gradient of the scalar potential. We want the gradient to be a geometric object.

That outcome requires the use of a different set of basis vectors. Geometric invariance of contravariant vectors is achieved because vector components A^i and basis vectors $\boldsymbol{e}_i$ transform inversely to each other under coordinate transformations. The components of the gradient, $\partial\phi/\partial x^i$, transform in the same way as the vectors $\boldsymbol{e}_i$. We require, therefore, a set of basis vectors that transform inversely to the vectors $\boldsymbol{e}_i$.

The *dual basis* (or the *reciprocal basis*) is a set of vectors associated with the original basis $\{\boldsymbol{e}_i\}$, denoted with superscripts $\{\boldsymbol{e}^i\}$, defined such that[22]:

$$\boldsymbol{e}^i\cdot\boldsymbol{e}_j \equiv \delta^i_j. \qquad (i,j = 1, \dots, N) \tag{11.31}$$

[22]The dot product in Eq. (11.31) is notional and should not be taken literally; it's used to *indicate* orthonormality. A better notation would be to write Eq. (11.31) as $\boldsymbol{e}^i(\boldsymbol{e}_j) = \delta^i_j$ to indicate that $\boldsymbol{e}^i$ is an operator that acts on $\boldsymbol{e}_j$ in such a way as to produce δ^i_j. In a first exposure to the subject, we'll use the dot product in Eq. (11.31) as it's less abstract. However, we start using the more general notation in Section 11.8.

Equation (11.31) holds for every element of the basis set $\{\boldsymbol{e}_j\}_{j=1}^{N}$, and because there is only one vector that satisfies $\boldsymbol{e}^j \cdot \boldsymbol{e}_j = 1$ (no sum), namely $i = j$, the index i is restricted to the values $i = 1, \ldots, N$. The number of elements in the sets $\{\boldsymbol{e}_i\}$ and $\{\boldsymbol{e}^j\}$ are therefore the same.[23]

It remains to be shown that the vectors defined by Eq. (11.31) transform as intended. Assume that the dual basis vectors transform under a coordinate transformation as a linear combination $\boldsymbol{e}^{j'} = B^{j'}_k \boldsymbol{e}^k$, where the quantities $B^{j'}_k$ are an unknown set of expansion coefficients. We require that the dual basis vectors $\{\boldsymbol{e}^{j'}\}$ (associated with a transformed coordinate system) obey the definition Eq. (11.31) in that coordinate system.[24] Thus,

$$\boldsymbol{e}^{j'} \cdot \boldsymbol{e}_{i'} = B^{j'}_k \boldsymbol{e}^k \cdot A^j_{i'} \boldsymbol{e}_j = B^{j'}_k A^j_{i'} \delta^k_j = B^{j'}_k A^k_{i'} = \delta^{j'}_{i'},$$

where we've used Eqs. (11.22) and (11.31). The set of coefficients $\{B^{j'}_k\}$, considered as a matrix, is therefore the inverse of the matrix $[A^k_{i'}]$. The inverse of $[A^k_{i'}]$ is, however, from Eq. (11.20), $[A^{j'}_k]$, and because the inverse of a matrix is unique, $B^{j'}_k = A^{j'}_k$. Thus, the dual basis vectors have the desired property of transforming inversely to how the vectors $\boldsymbol{e}_i$ transform,

$$\boldsymbol{e}^{i'} = A^{i'}_j \boldsymbol{e}^j. \tag{11.32}$$

Compare Eq. (11.32) with Eq. (11.22).

The gradient $\boldsymbol{\nabla}\phi \equiv (\partial\phi/\partial x^i)\boldsymbol{e}^i$ is therefore a geometric object:

$$(\boldsymbol{\nabla}\phi)' = \frac{\partial\phi'}{\partial x^{i'}}\boldsymbol{e}^{i'} = A^j_{i'}\frac{\partial\phi}{\partial x^j}A^{i'}_k\boldsymbol{e}^k = \frac{\partial\phi}{\partial x^j}\boldsymbol{e}^k\delta^j_k = \frac{\partial\phi}{\partial x^k}\boldsymbol{e}^k = \boldsymbol{\nabla}\phi,$$

where we've used Eqs. (11.30), (11.32), and (11.20). Just as the infinitesimal displacement vector $\mathrm{d}\boldsymbol{r}$ is the prototype contravariant vector, *the gradient* $\boldsymbol{\nabla}\phi$ *is the prototype of a class of geometric objects called covariant vectors.* A quantity $\boldsymbol{T} \equiv T_m\boldsymbol{e}^m$ has covariant components if they transform as in Eq. (11.26), $T_{n'} = A^m_{n'}T_m$, so that $\boldsymbol{T}' = T_{m'}\boldsymbol{e}^{m'} = A^n_{m'}T_nA^{m'}_j\boldsymbol{e}^j = T_n\boldsymbol{e}^j\delta^n_j = \boldsymbol{T}$, where we've used Eqs. (11.32) and (11.20). Any set of N quantities $\{T_i\}$ that transform like covariant vector

$$T_{j'} = A^i_{j'}T_i \tag{11.33}$$

are the covariant components of a vector – they "co vary" with the basis vectors[25] $\boldsymbol{e}_i$.

[23] The two sets $\{\boldsymbol{e}_i\}$ and $\{\boldsymbol{e}^j\}$ are isomorphic. Given the isomorphism, it's natural to pair $\boldsymbol{e}^j \leftrightarrow \boldsymbol{e}_j$.

[24] The dual basis is associated with a given coordinate system because the vectors of the coordinate basis are tangent to coordinate curves (see Section 11.2.3), and the dual basis is associated with the original basis $\{\boldsymbol{e}_i\}$.

[25] A useful mnemonic for the placement of indices is "co goes below."

A set of N^2 objects $\{T_{mn}\}$ that transform like

$$T_{m'n'} = A^k_{m'} A^l_{n'} T_{kl} \tag{11.34}$$

are covariant components of a second-rank tensor. A second-rank tensor $\mathbf{T}$ with covariant components is defined as $\mathbf{T} \equiv T_{mn} \boldsymbol{e}^m \boldsymbol{e}^n$; $\mathbf{T}$ is independent of basis:

$$\begin{aligned}\mathbf{T}' &= T_{i'j'} \boldsymbol{e}^{i'} \boldsymbol{e}^{j'} = A^k_{i'} A^l_{j'} T_{kl} A^{i'}_m A^{j'}_n \boldsymbol{e}^m \boldsymbol{e}^n = (A^k_{i'} A^{i'}_m)(A^l_{j'} A^{j'}_n) T_{kl} \boldsymbol{e}^m \boldsymbol{e}^n \\ &= \delta^k_m \delta^l_n T_{kl} \boldsymbol{e}^m \boldsymbol{e}^n = T_{mn} \boldsymbol{e}^m \boldsymbol{e}^n = \mathbf{T},\end{aligned}$$

where we've used Eqs. (11.34), (11.32), and (11.20). A set of N^r objects $\{T_{k_1 \cdots k_r}\}$ that transforms like $T_{k'_1 \cdots k'_r} = A^{m_1}_{k'_1} \cdots A^{m_r}_{k'_r} T_{m_1 \cdots m_r}$ are the covariant components of an rth-rank tensor $\mathbf{T} = T_{k_1 \cdots k_r} \boldsymbol{e}^{k_1} \cdots \boldsymbol{e}^{k_r}$.

We can use the dual basis to provide another interpretation of the Jacobian matrix. Start with Eq. (11.21) and take the inner product with $\boldsymbol{e}^k$:

$$\boldsymbol{e}^k \cdot \mathrm{d}\boldsymbol{r} = \mathrm{d}x^i \boldsymbol{e}^k \cdot \boldsymbol{e}_i = \mathrm{d}x^i \delta^k_i = \mathrm{d}x^k = \mathrm{d}x^{j'} \boldsymbol{e}^k \cdot \boldsymbol{e}_{j'} = A^{j'}_l \, \mathrm{d}x^l \boldsymbol{e}^k \cdot \boldsymbol{e}_{j'}, \tag{11.35}$$

where we've used Eqs. (11.31) and (11.16). Equation (11.35) indicates that

$$A^{j'}_l \boldsymbol{e}^k \cdot \boldsymbol{e}_{j'} \, \mathrm{d}x^l = \mathrm{d}x^k. \tag{11.36}$$

By a familiar line of reasoning, Eq. (11.36) implies that $A^{j'}_l \boldsymbol{e}^k \cdot \boldsymbol{e}_{j'} = \delta^k_l$, in turn implying that $[\boldsymbol{e}^k \cdot \boldsymbol{e}_{j'}]$ is the inverse of the matrix $[A^{j'}_l]$ (because the inverse of a matrix is unique). Referring to Eq. (11.20), we identify

$$\boldsymbol{e}^k \cdot \boldsymbol{e}_{j'} = A^k_{j'} = \frac{\partial x^k}{\partial x^{j'}}. \tag{11.37}$$

Jacobian matrices therefore connect basis vectors in different coordinate systems (as well as coordinate differentials).[26] By repeating the argument, one can show that

$$\boldsymbol{e}^{k'} \cdot \boldsymbol{e}_i = A^{k'}_i = \frac{\partial x^{k'}}{\partial x^i}, \tag{11.38}$$

that is, swap primes and unprimes in Eq. (11.37).

[26]Because of the prime on the index, there is (hopefully) no chance of confusing $A^k_{j'} = \boldsymbol{e}^k \cdot \boldsymbol{e}_{j'}$ (from Eq. (11.37)) with $\boldsymbol{e}^k \cdot \boldsymbol{e}_j = \delta^k_j$ (from Eq. (11.31)). In the Schouten index scheme, the symbol A^j_k (without any primes) is *defined* as the Kronecker delta, $A^j_k \equiv \delta^j_k$.

11.2.5 Mixed tensors

There is a third class of tensors (besides contravariant and covariant) – *mixed tensors* – that are expressed in terms of both types of basis vectors. For example, a set of N^3 objects $\{T^i{}_{jk}\}$ that transform as

$$T^{i'}{}_{j'k'} = A^{i'}_p A^l_{j'} A^m_{k'} T^p{}_{lm} \tag{11.39}$$

are the components of a mixed third-rank tensor with one contravariant and two covariant indices. Notationally, the upper and lower indices are set apart: $T^i{}_{jk}$ instead of T^i_{jk}. It's *good hygiene* in writing the components of mixed tensors *not* to put superscripts aligned above subscripts. Adopting this convention helps avoid mistakes; we don't have the tools yet to see what could go wrong with the innocent looking quantity T^i_{jk}, but we will.

The components of a mixed tensor of type (p,q), with p contravariant indices and q covariant indices are any set of $N^{(p+q)}$ objects $\{T^{i_1\cdots i_p}{}_{j_1\cdots j_q}\}$ that transform as

$$T^{k'_1\cdots k'_p}{}_{m'_1\cdots m'_q} = A^{k'_1}_{t_1}\cdots A^{k'_p}_{t_p} A^{s_1}_{m'_1}\cdots A^{s_q}_{m'_q} T^{t_1\cdots t_p}{}_{s_1\cdots s_q}. \tag{11.40}$$

The tensor $\mathbf{T}$ of type (p,q) is $\mathbf{T} = T^{k_1\cdots k_p}{}_{m_1\cdots m_q}\boldsymbol{e}_{k_1}\cdots\boldsymbol{e}_{k_p}\boldsymbol{e}^{m_1}\cdots\boldsymbol{e}^{m_q}$. In this notation, a pth-rank tensor with contravariant components is a type $(p,0)$ tensor, and a qth-rank tensor with covariant components is a type $(0,q)$ tensor. We've attempted not to refer to "contravariant tensors" and "covariant tensors" (although it's unavoidable in usage) – there are only tensors that have contravariant, covariant, or mixed components. After we develop a few more rules about how to manipulate tensors, we'll see that the same tensor $\mathbf{T}$ can be expressed in any sets of basis vectors: $\mathbf{T} = T_{ij}\boldsymbol{e}^i\boldsymbol{e}^j = T^i_j\boldsymbol{e}_i\boldsymbol{e}^j = T_i{}^j\boldsymbol{e}^i\boldsymbol{e}_j = T^{ij}\boldsymbol{e}_i\boldsymbol{e}_j$.

Is the Kronecker symbol δ^i_j the component of a mixed tensor (as the notation would suggest)? How does it transform? Using Eq. (11.40),

$$(\delta^i_j)' = A^{i'}_k A^m_{j'}\delta^k_m = A^{i'}_k A^k_{j'} = \delta^{i'}_{j'},$$

where we've used Eq. (11.20). The transformation of δ^i_j, $(\delta^i_j)'$, has the same values of δ^i_j in the new coordinate system. The Kronecker delta is a *constant tensor*, one that has the same value in all coordinate systems.

11.2.6 Covariant equations

If an equation involving tensors is true in one coordinate system, it's true in all coordinate systems. Suppose we have a relation between tensors, $A_{ij} = B_{ij}$. Write this equation as $D_{ij} = 0$, where $D_{ij} \equiv A_{ij} - B_{ij}$. If $D_{ij} = 0$ in one coordinate system, then $D_{i'j'} = 0$ in *any* coordinate system because the tensor transformation properties are linear and homogeneous, $D_{i'j'} = A^k_{i'}A^l_{j'}D_{kl}$. Thus, $A_{i'j'} = B_{i'j'}$ in all coordinate systems. While the individual *components* A_{ij}, B_{ij} transform between coordinate systems, the *form* of the equation is the same for all coordinate systems.

Tensor equations are called *covariant* because their form covaries with transformations between coordinate systems. This idea is referred to as the *principle of covariance.* Covariant equations are *form invariant* – the mathematical form of the equation remains the same under coordinate transformations.[27] This idea is fundamental to physics: For physical laws to be the same in all coordinate systems, they must be formulated in a covariant manner, which is why it's so important to be able to establish whether a given set of mathematical objects constitute a tensor.

It's not out of place to quote from Einstein's 1916 article that established the general theory of relativity. A sizable portion of that article is devoted to tensors in a section, "Mathematical Aids to the Formation of Generally Covariant Equations." Based on what we've covered in this chapter, you should be able to follow what he says [40, p. 121]:[28]

> Let certain things ("tensors") be defined with respect to any system of coordinates by a number of functions of the coordinates, called the "components" of the tensor. There are then certain rules by which these components can be calculated for a new system of coordinates, if they are known for the original system of coordinates, and if the transformation connecting the two systems is known. The things hereafter called tensors are further characterized by the fact that the equations of transformation for their components are linear and homogeneous. Accordingly, all the components in the new system vanish, if they all vanish in the original system. If, therefore, a law of nature is expressed by equating all the components of a tensor to zero, it is generally covariant. By examining the laws of the formation of tensors, we acquire the means of formulating generally covariant laws.

11.3 Contraction and the quotient theorem

inner product

Consider vectors $\boldsymbol{T}$ and $\boldsymbol{U}$, one with covariant components and the other with contravariant. The inner product $\boldsymbol{T} \cdot \boldsymbol{U}$ when expressed in terms of components is

[27] Note that we've just introduced a *second* use of the word "covariant," distinct from covariant and contravariant transformation properties, which can be a source of confusion. An equation, for example, expressed among contravariant tensor elements, $T^{ij} = U^{ij}$, is a covariant equation.

[28] On a historical note, observe that Einstein placed quotes around the word tensor – tensors were not part of the lexicon of physics at that time. Something similar happened in the development of quantum mechanics, where Heisenberg in his initial theory of quantum mechanics (what would become known as *matrix mechanics*) was unaware of the existence of matrices and had to be instructed in them from his thesis advisor, Max Born. Today, all students of science and engineering must learn matrix algebra, with knowledge of tensors a standard requirement for many applications.

$$\boldsymbol{T} \cdot \boldsymbol{U} = T_i \boldsymbol{e}^i \cdot U^j \boldsymbol{e}_j = T_i U^j \delta^i_j = T_i U^i, \qquad (11.41)$$

where we've used Eq. (11.31). The construct $T_i U^i$ is a scalar under coordinate transformations:

$$(\boldsymbol{T} \cdot \boldsymbol{U})' \equiv T_{i'} U^{i'} = A^k_{i'} A^{i'}_j T_k U^j = \delta^k_j T_k U^j = T_j U^j = \boldsymbol{T} \cdot \boldsymbol{U}, \qquad (11.42)$$

where we've used Eqs. (11.24), (11.33), and (11.20). Thus, *if we know the inner product in one coordinate system, we know its value in any coordinate system* – a highly useful feature that allows one to evaluate the inner product in whatever coordinate system permits the simplest calculation.

The inner product is an instance of a more general process known as *tensor contraction.* When a contravariant (upper) index is set equal to a covariant (lower) index, and summed over, it reduces a type (p, q) tensor to one of type $(p-1, q-1)$, i.e. *it lowers the tensor rank by two.* Consider $T^i{}_j \equiv U^i V_j$, the components of a second-rank tensor formed from the product of vector components U^i and V_j. If we set $j = i$, and sum over i, $T^i{}_i \equiv U^i V_i$, we form a scalar, Eq. (11.42). More generally, $T^{ii_1 \cdots i_{p-1}}{}_{ij_1 \cdots j_{q-1}}$ indicates, for a type (p, q) tensor, the summation ("contraction") of a contravariant and covariant index to produce a type $(p-1, q-1)$ tensor. Consider the contraction of a third-rank tensor (with components) $T^{ij}{}_k \equiv U^{ij} V_k$, formed from the product of components of a second-rank, contravariant tensor U^{ij} with the components of a covariant vector, V_k. If we set $j = k$ and sum, we lower the rank by two to form a contravariant vector (first-rank tensor) $T^i \equiv U^{ik} V_k$. To prove that T^i is the component of a vector, we must show that it transforms as one:[29]

$$\begin{aligned} T^{i'} &= U^{i'k'} V_{k'} = A^{i'}_l A^{k'}_m U^{lm} A^n_{k'} V_n = A^{i'}_l (A^{k'}_m A^n_{k'}) U^{lm} V_n \\ &= A^{i'}_l \delta^n_m U^{lm} V_n = A^{i'}_l U^{ln} V_n = A^{i'}_l T^l, \end{aligned}$$

where we've used Eq. (11.25), (11.33), and (11.20).

The inverse process, of forming the components of higher-rank tensors, from products of the components of lower-rank tensors is called the outer product, which increases the rank. The product of the components of a tensor of type (r, s) with the components of a tensor of type (p, q) form the components of a new tensor of type $(r+p, s+q)$. outer product

The direct test for whether a set of mathematical objects form tensor components is to verify that they transform appropriately. There is, however, an indirect method for checking the tensorial character of a set of quantities known loosely as the *quotient theorem*, which says that in an equation $UV = T$, if V and T are known to be elements of a tensor, then

[29] At some point, in any course involving tensor analysis, the phrase "index gymnastics" will be uttered.

U is also a tensor element. With the quotient theorem, we use known tensors to ascertain the tensor character of putative tensors.

Suppose $\{T_r\}$ is a set of quantities that we wish to test for its tensor character. Let $\{X^r\}$ be the components of a contravariant vector. If the sum $T_r X^r$ is an invariant, then by the quotient theorem, the quantities T_r form the elements of a covariant vector. From the given invariance, we have $T_r X^r = T_{s'} X^{s'}$. We can use the known transformation properties of X^r, Eq. (11.24), to write $T_r X^r = T_{s'} A_j^{s'} X^j$. Rewrite this equation: $(T_j - T_{s'} A_j^{s'}) X^j = 0$. Because the X^j are arbitrary, the terms in parentheses must vanish, establishing $T_j = A_j^{s'} T_{s'}$ as the elements of a covariant vector, Eq. (11.33).

Let's look at a more challenging example. Suppose we run into an equation,

$$T^{mn}{}_{kl} = U^m S^n{}_{kl}, \tag{11.43}$$

where it's known that $\mathbf{T}$ is a type $(2,2)$ tensor and $\boldsymbol{U}$ is a vector. By the quotient theorem, we're entitled to conclude that $\mathbf{S}$ is a tensor of type $(1,2)$. To show this, introduce contravariant vector components $\{x^i\}$ and covariant vector components $\{y_r\}$. Multiply Eq. (11.43) by $y_m y_n x^k x^l$ and contract,

$$T^{mn}{}_{kl} y_m y_n x^k x^l = U^m S^n_{kl} y_m y_n x^k x^l = (U^m y_m) S^n_{kl} y_n x^k x^l. \tag{11.44}$$

We take this step so that the left side of Eq. (11.44) is a scalar (we've contracted all indices); $U^m y_m$ is also a scalar. We've therefore established that $S^n_{kl} y_n x^k x^l$ is a scalar, and thus $S^{n'}_{k'l'} y_{n'} x^{k'} x^{l'} = S^n_{kl} y_n x^k x^l$. Now use the transformation properties of x^i and y_m, implying that $S^{n'}_{k'l'} A^i_{n'} y_i A^{k'}_j x^j A^{l'}_m x^m = S^i{}_{jm} y_i x^j x^m$. Because x^i and y_j are arbitrary, $S^{n'}_{k'l'} A^i_{n'} A^{k'}_j A^{l'}_m = S^i_{jm}$, establishing $\mathbf{S}$ as a type $(1,2)$ tensor.

11.4 The metric tensor

We began this book by generalizing vectors in two ways – as elements of vector spaces (Section 1.2) and in the use of the inner product, a rule which associates a scalar to a pair of vectors (Section 1.3). In this chapter, vectors have been generalized to tensors, which, in addition to representing physical quantities such as the elastic stress field, includes their role as operators that act on vectors and other tensors.

In tensor analysis, the inner product is specified by another tensor, the *metric tensor* $\mathbf{g}$, a bilinear operator[30] that assigns to a pair of vectors $\boldsymbol{u}$, $\boldsymbol{v}$, a scalar $\mathbf{g}(\boldsymbol{u}, \boldsymbol{v})$. Because of its linearity properties, for $\boldsymbol{u} = u^i \boldsymbol{e}_i$ and $\boldsymbol{v} = v^j \boldsymbol{e}_j$,

$$\mathbf{g}(\boldsymbol{u}, \boldsymbol{v}) = u^i v^j \mathbf{g}(\boldsymbol{e}_i, \boldsymbol{e}_j) \equiv g_{ij} u^i v^j. \tag{11.45}$$

[30] A *bilinear* operator is linear in each argument: $\mathbf{g}(\alpha_1 \boldsymbol{u}_1 + \alpha_2 \boldsymbol{u}_2, \boldsymbol{v}) = \alpha_1 \mathbf{g}(\boldsymbol{u}_1, \boldsymbol{v}) + \alpha_2 \mathbf{g}(\boldsymbol{u}_2, \boldsymbol{v})$, for scalars α_1, α_2, and $\mathbf{g}(\boldsymbol{u}, \alpha_1 \boldsymbol{v}_1 + \alpha_2 \boldsymbol{v}_2) = \alpha_1 \mathbf{g}(\boldsymbol{u}, \boldsymbol{v}_1) + \alpha_2 \mathbf{g}(\boldsymbol{u}, \boldsymbol{v}_2)$.

The action of $\mathbf{g}$ on a pair of vectors is specified by its values on pairs of basis vectors,[31] $g(\boldsymbol{e}_i, \boldsymbol{e}_j)$, symbolized g_{ij}. The simplest way to specify the action of $\mathbf{g}$ is through the dot product,[32]

$$g_{ij} \equiv \boldsymbol{e}_i \cdot \boldsymbol{e}_j. \tag{11.46}$$

The tensor elements g_{ij} are always symmetric,[33] $g_{ij} = g_{ji}$. A primary use of the metric tensor is to provide the distance "squared" between two points infinitesimally separated by the vector $\mathrm{d}\boldsymbol{r} = \mathrm{d}x^i \boldsymbol{e}_i$, the arc length $(\mathrm{d}l)^2$:

$$(\mathrm{d}l)^2 = \mathbf{g}(\mathrm{d}\boldsymbol{r},\, \mathrm{d}\boldsymbol{r}) = g_{ij}\, \mathrm{d}x^i\, \mathrm{d}x^j. \tag{11.47}$$

The arc length of a curve can then be found from the metric tensor, $\int \mathrm{d}l = \int \sqrt{g_{ij}\, \mathrm{d}x^i\, \mathrm{d}x^j}$. It also provides a way to calculate the angle θ between two vectors, $\boldsymbol{u}$, $\boldsymbol{v}$, $\cos\theta = \mathbf{g}(\boldsymbol{u}, \boldsymbol{v})/(\|\boldsymbol{u}\| \cdot \|\boldsymbol{v}\|)$, where $\|\boldsymbol{u}\| = \sqrt{\mathbf{g}(\boldsymbol{u}, \boldsymbol{u})}$. In general, the metric tensor *defines* a geometry – each geometry has its own metric tensor.

For example, in the spherical coordinate system, we have from Eq. (11.46), using the coordinate basis given by Eq. (11.29),

$$[g_{ij}] = \begin{pmatrix} 1 & 0 & 0 \\ 0 & r^2 & 0 \\ 0 & 0 & r^2 \sin^2\theta \end{pmatrix}. \tag{11.48}$$

The matrix $[g_{ij}]$ is diagonal in this case because the spherical coordinate system utilizes an orthogonal set of basis vectors; in general, the metric tensor is a full matrix. The magnitude and direction of basis vectors are not constant and vary throughout the coordinate system (noted in Section 11.2.3). In general $\mathrm{d}\boldsymbol{r}$ has the form

$$\mathrm{d}\boldsymbol{r} = \sum_i (\text{coordinate differential})^i \times (\text{basis vector})_i = \sum_i \mathrm{d}x^i \boldsymbol{e}_i.$$

The metric tensor carries the information about distances between nearby points; *coordinates* do not necessarily have the dimension of length, such as angular variables.[34] Combining Eqs. (11.47) and (11.48), we have the expression for arc length:

$$(\mathrm{d}l)^2 = (\mathrm{d}r)^2 + r^2(\mathrm{d}\theta)^2 + r^2 \sin^2\theta(\mathrm{d}\phi)^2.$$

---

[31] This generalizes what we found in Eq. (1.1), where the value of an operator acting on a single vector is determined by its value acting on basis vectors.

[32] We've defined the *covariant* elements of $\mathbf{g}$. We'll shortly introduce g^{ij}, its contravariant elements.

[33] In advanced treatments, $\mathbf{g}$ is *defined* as a symmetric, bilinear operator.

[34] Check out the street addresses in your neighborhood. There may not be any relation between the numerical assignment of addresses and the distance between houses.

In Euclidean geometry, $(\mathrm{d}l)^2 = g_{ij}\,\mathrm{d}x^i\,\mathrm{d}x^j > 0$ *for any sign of the coordinate differentials* $\mathrm{d}x^i$. A metric is said to be *positive definite*[35] if $(\mathrm{d}l)^2 > 0$ for all $\mathrm{d}x^i$, unless $\mathrm{d}x^i = 0$. Said differently, the distance between two points in Euclidean geometry vanishes only if the points are coincident. In the theory of relativity, the geometry of spacetime features an *indefinite* metric where $(\mathrm{d}l)^2$ can be either positive, negative, or zero.

11.5 Raising and lowering indices

We have at our disposal two basis sets, the coordinate basis, and the dual basis, and a given vector $\boldsymbol{A}$ can be expressed in either:

$$\boldsymbol{A} = A^i\boldsymbol{e}_i = A_k\boldsymbol{e}^k. \tag{11.49}$$

There must be a connection between the components A^i and A_k (of the same vector). Take the inner product of both sides of Eq. (11.49) with $\boldsymbol{e}_j$,

$$\boldsymbol{e}_j\cdot\boldsymbol{A} = A^i\boldsymbol{e}_i\cdot\boldsymbol{e}_j = A^i g_{ji} = A_k\boldsymbol{e}_j\cdot\boldsymbol{e}^k \equiv A_k\boldsymbol{e}^k\cdot\boldsymbol{e}_j = A_k\delta^k_j = A_j, \tag{11.50}$$

where we've used Eqs. (11.46) and (11.31). Note the step taken in Eq. (11.50),

$$\boldsymbol{e}_j\cdot\boldsymbol{e}^k = \boldsymbol{e}^k\cdot\boldsymbol{e}_j = \delta^k_j, \tag{11.51}$$

which is standard in tensor analysis.[36] Thus, from Eq. (11.50),

$$A_j = g_{ji}A^i. \tag{11.52}$$

Equation (11.52) is an instance of *lowering an index.* The metric tensor connects the components of a vector in the dual basis, A_j, with its components in the coordinate basis.[37] Now take the inner product between Eq. (11.49) and $\boldsymbol{e}^j$;

$$\boldsymbol{e}^j\cdot\boldsymbol{A} = A^i\boldsymbol{e}^j\cdot\boldsymbol{e}_i = A^i\delta^j_i = A^j = A_k\boldsymbol{e}^k\cdot\boldsymbol{e}^j, \tag{11.53}$$

---

[35] A test for positive definiteness is provided by *Sylvester's criterion* that various determinants (principal minors) associated with $[g_{ij}]$ all be positive [41, p. 52].

[36] In Section 1.2.5, we noted that linear functionals, operators that map vectors into scalars, form a vector space of their own, the dual space, denoted V_N^*. The dual basis vectors $\boldsymbol{e}^k$ perform that role; see Eq. (11.31). For Eq. (11.51) to make sense, $\boldsymbol{e}_j$ must be an element of the vector space *dual* to the space that $\boldsymbol{e}^k$ belongs to. In advanced treatments, it's shown that the "dual to the dual space" is isomorphic to the original vector space, V_N, and we can simply take $\boldsymbol{e}_j\cdot\boldsymbol{e}^k = \boldsymbol{e}^k\cdot\boldsymbol{e}_j$ in Eq. (11.51). See, for example [42, p. 77]. This step allows one to preserve for tensors the symmetry of the inner product one has for elementary vectors, $\boldsymbol{A}\cdot\boldsymbol{B} = \boldsymbol{B}\cdot\boldsymbol{A}$.

[37] Note from Eq. (11.52) that *all* vector components in the coordinate basis are required to obtain the components in the dual basis.

where we've used Eq. (11.31). We define the *contravariant* elements of the metric tensor as (compare with Eq. (11.46))

$$g^{ij} \equiv \boldsymbol{e}^i \cdot \boldsymbol{e}^j, \tag{11.54}$$

where $g^{ij} = g^{ji}$. Combining Eqs. (11.54) and (11.53),

$$A^j = A_k g^{kj}. \tag{11.55}$$

Equation (11.55) is an instance of *raising an index.* The tensors g_{ij} and g^{ij} are termed the *fundamental tensors* because of the essential role they play in subsequent considerations.

Is there a relation between g_{ij} and g^{ij}? Combining Eq. (11.55), the raising of an index, $A^i = g^{ij}A_j$, with Eq. (11.52), the lowering of an index, $A_j = g_{jk}A^k$,

$$A^i = g^{ij}A_j = g^{ij}g_{jk}A^k. \tag{11.56}$$

Equation (11.56) is equivalent to

$$(\delta^i_k - g^{ij}g_{jk})A^k = 0. \tag{11.57}$$

But because the $\{A^k\}$ are arbitrary,

$$g^{ij}g_{jk} = \delta^i_k. \tag{11.58}$$

The two types of metric tensors are inverses of each other. Using Eq. (11.48), we have for the spherical coordinate system[38]

$$[g^{ij}] = \begin{pmatrix} 1 & 0 & \\ 0 & 1/r^2 & 0 \\ 0 & 0 & 1/(r^2 \sin^2\theta) \end{pmatrix}. \tag{11.59}$$

We can now explain the convention, noted in Section 11.2.5, of writing mixed tensors so that contravariant indices are not placed over covariant indices. Consider a tensor C^{ijk}. Suppose (for whatever reason) we wish to lower the middle index. By writing $g_{jl}C^{ilk} = C^{i\ k}_{\ j}$, we know where j "comes from." Had we written $g_{jl}C^{ilk} = C^{ik}_j$, where would j "go" if later we decide to raise the index? The order of indices matters. It's not our intention to present detailed calculations where one engages in various acts of index gymnastics, but the need for such acts exists, and errors can be incurred by being sloppy about the placement of indices.

Raising and lowering indices applies to basis vectors as well. Expand a dual basis vector in the coordinate basis, $\boldsymbol{e}^i = c^{ij}\boldsymbol{e}_j$, where the c^{ij} are unknown expansion coefficients. Take the inner product with $\boldsymbol{e}^k$: $\boldsymbol{e}^k \cdot \boldsymbol{e}^i \equiv$

[38]The form of $[g^{ij}]$ in Eq. (11.59) is so simple because the form of $[g_{kl}]$ is likewise simple for this coordinate system. For general metric tensors, with matrix elements $[g_{kl}]$, one has to find the matrix inverse to obtain $[g^{ij}]$.

$g^{ik} = c^{ij}\boldsymbol{e}^k \cdot \boldsymbol{e}_j = c^{ij}\delta_j^k = c^{ik}$, where we've used Eqs. (11.54) and (11.31). Hence, $c^{ik} = g^{ik}$, and we have, just like Eq. (11.55),

$$\boldsymbol{e}^i = g^{ij}\boldsymbol{e}_j. \tag{11.60}$$

Thus, dual basis vectors are a linear combination of *all* the coordinate basis vectors through the metric tensor, g^{ij}. We can either accept the definition of g^{ij} in Eq. (11.54) as the dot product of dual basis vectors, *or* we can take Eq. (11.60) as defining the dual basis, where g^{ij} is obtained as the inverse of the metric tensor g_{ij}; see Eq. (11.58). By repeating the argument, it's simple to show that $\boldsymbol{e}_i = g_{ij}\boldsymbol{e}^j$.

Example. Use Eq. (11.60) to find the dual basis vectors associated with the spherical coordinate system. Using the inverse metric tensor in Eq. (11.59),

$$\begin{pmatrix} \boldsymbol{e}^r \\ \boldsymbol{e}^\theta \\ \boldsymbol{e}^\phi \end{pmatrix} = \begin{pmatrix} 1 & 0 & 0 \\ 0 & 1/r^2 & 0 \\ 0 & 0 & 1/(r^2 \sin^2\theta) \end{pmatrix} \begin{pmatrix} \boldsymbol{e}_r \\ \boldsymbol{e}_\theta \\ \boldsymbol{e}_\phi \end{pmatrix}.$$

Thus, $\boldsymbol{e}^r = \boldsymbol{e}_r$, $\boldsymbol{e}^\theta = \boldsymbol{e}_\theta/r^2$, and $\boldsymbol{e}^\phi = \boldsymbol{e}_\phi/(r^2\sin^2\theta)$. One can check that $\boldsymbol{e}^r \cdot \boldsymbol{e}_r = 1$, $\boldsymbol{e}^\theta \cdot \boldsymbol{e}_\theta = 1$, and $\boldsymbol{e}^\phi \cdot \boldsymbol{e}_\phi = 1$ – the basis vectors $\boldsymbol{e}_\theta$ and $\boldsymbol{e}_\phi$ are *not* unit vectors; see the last example of Section 11.2.3

Using these relations, we have the identity for the sum of dyadic products:

$$\boldsymbol{e}_k\boldsymbol{e}^k = g_{kj}\boldsymbol{e}^j g^{kl}\boldsymbol{e}_l = \delta_j^l \boldsymbol{e}^j\boldsymbol{e}_l = \boldsymbol{e}^l\boldsymbol{e}_l, \tag{11.61}$$

where we've used Eq. (11.58). The unit dyadic or identity operator can be written as

$$\mathbf{I} = \boldsymbol{e}^i\boldsymbol{e}_i = \boldsymbol{e}_i\boldsymbol{e}^i. \tag{11.62}$$

Let $\mathbf{I}$ act on a vector defined first with respect to the dual basis, and then with respect to the coordinate basis,

$$\begin{aligned} \mathbf{I}\cdot\boldsymbol{A} &= \boldsymbol{e}^i\boldsymbol{e}_i \cdot (A_j\boldsymbol{e}^j) = \boldsymbol{e}^i A_j(\boldsymbol{e}_i \cdot \boldsymbol{e}^j) = \boldsymbol{e}^i A_j \delta_i^j = \boldsymbol{e}^i A_i = \boldsymbol{A} \\ &= \boldsymbol{e}^i\boldsymbol{e}_i \cdot (A^j\boldsymbol{e}_j) = \boldsymbol{e}^i A^j(\boldsymbol{e}_i \cdot \boldsymbol{e}_j) = \boldsymbol{e}^i A^j g_{ji} = \boldsymbol{e}^i A_i = \boldsymbol{A}, \end{aligned}$$

where we've used Eq. (11.51) in the first line and Eq. (11.52) in the second line to lower an index. Using the identity operator, we can derive the basis transformation equations, Eqs. (11.22) and (11.32), in another way:

$$\boldsymbol{e}_{i'} = \mathbf{I}\cdot\boldsymbol{e}_{i'} = \boldsymbol{e}_j\boldsymbol{e}^j \cdot \boldsymbol{e}_{i'} = \boldsymbol{e}_j A^j_{i'} \quad \boldsymbol{e}^{i'} = \mathbf{I}\cdot\boldsymbol{e}^{i'} = \boldsymbol{e}^j\boldsymbol{e}_j \cdot \boldsymbol{e}^{i'} = \boldsymbol{e}^j\boldsymbol{e}^{i'} \cdot \boldsymbol{e}_j = A^{i'}_j \boldsymbol{e}^j,$$

where we've used Eqs. (11.37) and (11.38).

11.6 Geometric properties of covariant vectors

We've now largely met the players in tensor analysis: scalars, contravariant, and covariant vectors, and their generalizations as tensors, including mixed tensors.[39] Contravariant vectors share the attributes of the displacement vector, but what are covariant vectors? Their components transform like those of the gradient vector, which doesn't immediately convey a picture of what they are. One might question the relevance of covariant vectors if one has arrived this far in a scientific education without previously encountering them.[40] Can we provide a geometric interpretation of covariant vectors?

Covariant vectors behave geometrically as families of parallel planes. Figure 11.4 shows a plane and three vectors, $\boldsymbol{r}_0$, $\boldsymbol{r}$, and $\boldsymbol{w}$, where $\boldsymbol{w}$ is orthogonal to the plane. The vector $\boldsymbol{r}_0$ locates a fixed point on the plane having coordinates (x_0^1, x_0^2, x_0^3) and $\boldsymbol{r}$ locates an arbitrary point on the plane with coordinates (x^1, x^2, x^3); $\boldsymbol{r} - \boldsymbol{r}_0$ thus lies in the plane. The vector $\boldsymbol{w}$ is therefore orthogonal to *all vectors lying in the plane*, $\boldsymbol{w} \cdot (\boldsymbol{r} - \boldsymbol{r}_0) = 0$. By the quotient theorem, $\boldsymbol{w}$ must be a covariant vector (because $\boldsymbol{r} - \boldsymbol{r}_0$ is a contravariant vector). Let's write the components of $\boldsymbol{w}$ as w_i. The coordinates $\{x^j\}$ of all points in the plane therefore satisfy the *equation of a plane*, $w_i x^i = d$, where $d \equiv \boldsymbol{w} \cdot \boldsymbol{r}_0$ is a constant. The *intercepts* $p^i = d/w_i$ of a plane – where it crosses through the coordinate axes – are found from the equation of the plane by setting all other coordinates $x^j = 0$ for $j \neq i$. The coordinates of the points on *another* plane, parallel to the first (i.e. orthogonal to the same vector $\boldsymbol{w}$), satisfy another equation of plane, $w_i x^i = d'$, where $d' \neq d$ is another constant. The intercepts of the parallel plane are given by $q^i = d'/w_i$. Subtracting these equations, we find $w_i = (d' - d)/(q^i - p^i)$. The components of $\boldsymbol{w}$ are therefore inversely related to the distance between the intercepts of a pair of parallel planes. The direction of $\boldsymbol{w}$ is orthogonal to both planes, with its magnitude inversely proportional to the interplanar separation. This

equation of a plane

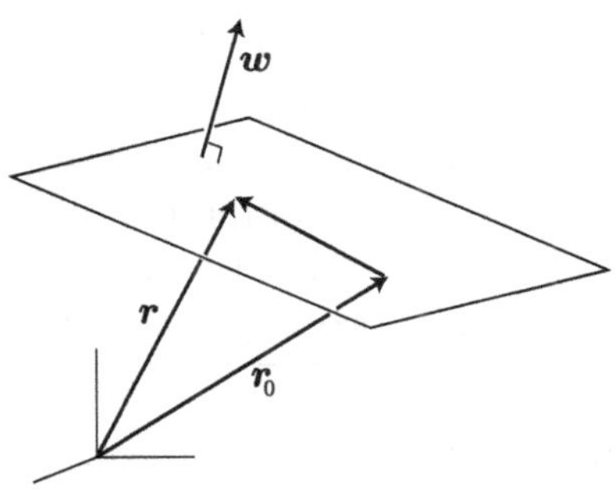

Figure 11.4 Vector $\boldsymbol{w}$ is perpendicular to all vectors $\boldsymbol{r} - \boldsymbol{r}_0$ lying in the plane.

[39]There is a fourth, and final concept, introduced in the next section – tensor densities.

[40]Covariant vectors are likely unfamiliar because scientific calculations usually employ orthogonal coordinate systems where the distinction between contravariant and covariant disappears.

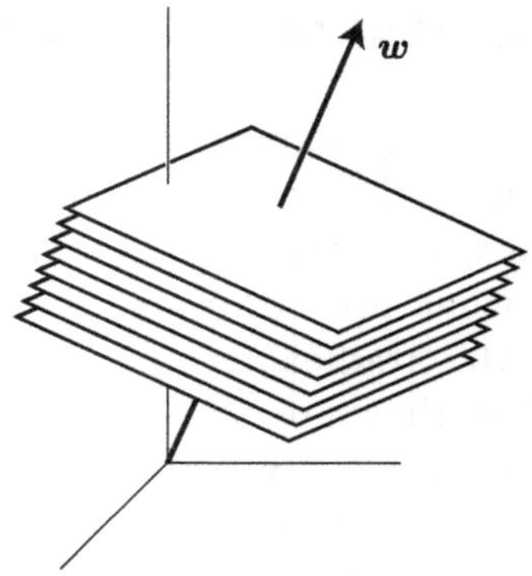

Figure 11.5 Covariant vectors represent families of equally spaced, parallel planes, with magnitude inversely proportional to the distance between surfaces.

construction implies that $\boldsymbol{w}$ is not orthogonal to just two parallel planes, but to a *family* of equally spaced, parallel planes, shown in Figure 11.5. Oriented stacks of parallel planes can be added to oriented stacks to produce new oriented stacks [43] – covariant vectors form a vector space.

Example. The wavevector $\boldsymbol{k}$ is a covariant vector that can be visualized as in Figure 11.5. Referring to Figure 7.3, the construct $\boldsymbol{k} \cdot \boldsymbol{r} =$ constant specifies a plane of constant phase orthogonal to the wavevector (plane waves). The magnitude of $\boldsymbol{k}$, $k = 2\pi/\lambda$, where λ is the wavelength. Consecutive planes of equal phase get closer together as the magnitude of $\boldsymbol{k}$ increases. In quantum mechanics, the de Broglie relation for the momentum of a particle, $\boldsymbol{p} = \hbar\boldsymbol{k}$, indicates that momentum here is naturally considered a covariant vector, $p_i = \hbar k_i$. In other applications, momentum is represented as a contravariant vector, $\boldsymbol{p} = m\boldsymbol{v}$, with $p^i = mv^i$. The two are related by $p_i = g_{ij}p^j$.

Example. The electric field $\boldsymbol{E}$ is a geometric object. It has a natural representation as a covariant vector $E_i = -\nabla_i\phi$ from its role as the gradient of the electrostatic potential function $\phi(\boldsymbol{r})$. The magnitude of $\boldsymbol{E}$ is inversely proportional to the distance between equipotential surfaces, the hallmark of covariant vectors. The electric field is also represented as a contravariant vector from its relation to the Newtonian equation of motion $E^i = (m/q)a^i$, where a^i is a component of the acceleration vector. The two quantities are related through the metric tensor $E^i = g^{ij}E_j$. It's the same vector, represented in two ways: $\boldsymbol{E} = E^i\boldsymbol{e}_i = E_j\boldsymbol{e}^j$.

Physically, the connection between gradients and covariant vectors is apparent – their magnitudes are inversely proportional to the distance between surfaces. Mathematically, consider the change in a scalar field ϕ over a displacement $\mathrm{d}\boldsymbol{r} = \mathrm{d}x^i\boldsymbol{e}_i$, with

$$\mathrm{d}\phi = \frac{\partial\phi}{\partial x^i}\,\mathrm{d}x^i. \qquad (11.63)$$

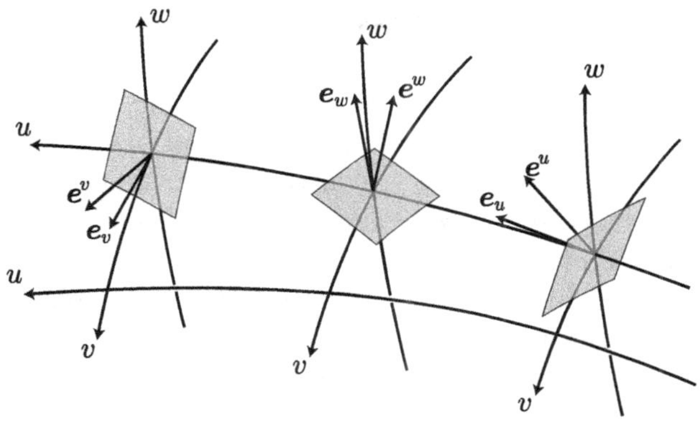

Figure 11.6 Vectors of the coordinate basis $\{e_i\}$ are tangent to coordinate curves; vectors of the dual basis $\{e^k\}$ are orthogonal to coordinate surfaces.

We can represent $\mathrm{d}\phi$ (a scalar) as an inner product between $\mathrm{d}\boldsymbol{r}$ and a new vector (the gradient) such that $\mathrm{d}\phi = \boldsymbol{\nabla}\,\phi \cdot \mathrm{d}\boldsymbol{r}$. If $\mathrm{d}\boldsymbol{r}$ lies within the level set of ϕ, $\mathrm{d}\phi = 0$, implying that $\boldsymbol{\nabla}\,\phi$ is orthogonal to level sets of ϕ. By the quotient theorem, $\boldsymbol{\nabla}\,\phi$ is a covariant vector, which we can represent in the dual basis, $\boldsymbol{\nabla} = \boldsymbol{e}^i \nabla_i$,

$$\mathrm{d}\phi = \boldsymbol{\nabla}\,\phi \cdot \mathrm{d}\boldsymbol{r} = (\nabla_i \phi)\boldsymbol{e}^i \cdot (\mathrm{d}x^j \boldsymbol{e}_j) = (\nabla_i \phi)\,\mathrm{d}x^j \delta^i_j = (\nabla_i \phi)\,\mathrm{d}x^i. \tag{11.64}$$

Comparing Eqs. (11.64) and (11.63), we identify $\nabla_i \equiv \partial/\partial x^i$. Note the location of the indices: *A derivative with respect to a contravariant component, x^i, is a covariant vector component,* ∇_i. To show that ∇_i is the component of a covariant vector is simple: see Eq. (11.26), $\nabla_{i'} = A^j_{i'} \nabla_j$.

Gradients normal to the coordinate surfaces of a (u, v, w) coordinate system provide a basis for covariant vectors,

$$\boldsymbol{e}^u \equiv \boldsymbol{\nabla}\, u \qquad \boldsymbol{e}^v \equiv \boldsymbol{\nabla}\, v \qquad \boldsymbol{e}^w \equiv \boldsymbol{\nabla}\, w. \tag{11.65}$$

A coordinate surface is the surface that results by holding one of the coordinates fixed. A sphere, for example, results by holding the radial coordinate fixed and letting the coordinates θ, ϕ vary; the sphere is the coordinate surface associated with the radial coordinate.[41] Figure 11.6 illustrates the distinction between coordinate basis vectors $\boldsymbol{e}_i$ (tangents to coordinate curves) and the dual basis vectors $\boldsymbol{e}^k$, orthogonal to coordinate surfaces. The vectors in Eq. (11.65) are dual to the basis vectors $\boldsymbol{e}_i$ in the sense of Eq. (11.31). If we assume that (u, v, w) are functions of an underlying Cartesian coordinate system $(x, y, z) \equiv x^i$ and if we let (u, v, w) be labeled u^i, then

$$\boldsymbol{e}^i \cdot \boldsymbol{e}_j = \boldsymbol{\nabla}\, u^i \cdot \frac{\partial \boldsymbol{r}}{\partial u^j} = \frac{\partial u^i}{\partial x^k}\frac{\partial x^k}{\partial u^j} = \frac{\partial u^i}{\partial u^j} = \delta^i_j. \tag{11.66}$$

[41] For a sphere, the unit vector $\hat{\boldsymbol{r}}$ is both tangent to the radial coordinate and orthogonal to the radial coordinate surface.

The gradient naturally forms a covariant vector, $\boldsymbol{\nabla} = \nabla_i \boldsymbol{e}^i$. Because geometric objects are independent of basis, however, we *could* have considered it a contravariant vector, $\boldsymbol{\nabla} \equiv \nabla^i \boldsymbol{e}_i$. Contravariant vector components can be found from covariant components by raising the index, $\nabla^i = g^{ij}\nabla_j$. For the gradient as a contravariant vector, the change in a scalar function $\mathrm{d}\phi = \boldsymbol{\nabla}\phi \cdot \mathrm{d}\boldsymbol{r}$ would require us to express $\mathrm{d}\boldsymbol{r} = \boldsymbol{e}^i\, \mathrm{d}x_i$ as a covariant vector, with the result that $\mathrm{d}\phi = (\nabla^i \phi)\, \mathrm{d}x_i$, in which case we would conclude that $\nabla^i = \partial/\partial x_i$; note the placement of the indices – *the derivative with respect to a covariant component* x_i *is a contravariant component* ∇^i. The quantity ∇^i, being the contravariant component of a vector, must transform as one:

$$\nabla^{i'} = \frac{\partial}{\partial x_{i'}} = \frac{\partial x_k}{\partial x_{i'}} \frac{\partial}{\partial x_k} = \left(\frac{\partial x_k}{\partial x_{i'}}\right) \nabla^k. \tag{11.67}$$

From Eq. (11.24), however, Eq. (11.67) should read $\nabla^i = A^i_{k'} \nabla^{k'}$. Comparing Eqs. (11.67) and (11.24), we conclude that

$$A^{i'}_k \equiv \frac{\partial x^{i'}}{\partial x^k} = \frac{\partial x_k}{\partial x_{i'}}. \tag{11.68}$$

Note the placement of indices.

Example. Let's reconsider the definition of the stress tensor in Eq. (11.1). We now know that oriented surface elements $\mathrm{d}\boldsymbol{s} = \mathrm{d}s\hat{\boldsymbol{n}}$ are covariant vectors, orthogonal to surfaces. The normal vector $\hat{\boldsymbol{n}}$ is thus a covariant vector, so that $\mathrm{d}\boldsymbol{s} = \mathrm{d}s_i \boldsymbol{e}^i$. The elements of the stress tensor should thus be indicated, not as T_{ij} in Eq. (11.1), but as T^{ij}:

$$T^{ij} \equiv \lim_{\delta s \to 0} \frac{\delta F^i}{\delta s_j}.$$

The analog of Eq. (11.2) is therefore

$$\delta F^i = T^{ij} \delta s_j. \tag{11.69}$$

11.7 Relative tensors

There is one more concept used in modeling physical phenomena, in addition to scalars and contra- and covariant vectors, and that is *density*, either scalar, vector, or tensor. Consider the integral of a scalar field, $\int \phi(x)\, \mathrm{d}^n x$. Is the *integral* a scalar? Not in general. While $\phi'(\boldsymbol{r}') = \phi(\boldsymbol{r})$ under coordinate transformations, we have to take into account the transformation of the volume element. Under a change of variables $x^i = x^i(x^{j'})$, the volume element of a multiple integral transforms as

$$\mathrm{d}^n x \equiv \mathrm{d}x^1 \cdots \mathrm{d}x^n = J\, \mathrm{d}x^{1'} \cdots \mathrm{d}x^{n'} \equiv J\, \mathrm{d}^n x' \qquad (11.70)$$

(so that the integral transforms as $\int \phi(x^i)\, \mathrm{d}^n x = \int \phi(x^i(x^{j'}))J\, \mathrm{d}^n x' \equiv \int \phi'(x^{j'})J\, \mathrm{d}^n x'$), where "the Jacobian" J is the determinant of the Jacobian matrix $A^i_{j'}$,

$$J = \det[A^i_{j'}] = \begin{vmatrix} \frac{\partial x^1}{\partial x^{1'}} & \cdots & \frac{\partial x^1}{\partial x^{n'}} \\ \vdots & & \vdots \\ \frac{\partial x^n}{\partial x^{1'}} & \cdots & \frac{\partial x^n}{\partial x^{n'}} \end{vmatrix}. \qquad (11.71)$$

Note that we *don't take the absolute value of the Jacobian determinant.*[42]

Relative tensors of weight w have components that transform according to the rules we've developed (such as Eq. (11.40)), with the additional requirement of the Jacobian raised to an integer power,[43] w:

relative tensors

$$T^{k'_1 \cdots k'_p}{}_{m'_1 \cdots m'_q} = J^w A^{k'_1}_{t_1} \cdots A^{k'_p}_{t_p} A^{s_1}_{m'_1} \cdots A^{s_q}_{m'_q} T^{t_1 \cdots t_p}{}_{s_1 \cdots s_q}. \qquad \text{(relative tensor)} \qquad (11.72)$$

Linear combinations of tensors *of the same weight* produce new tensors with that weight. Products of tensors of weights w_1 and w_2 produce tensors of weight $w_1 + w_2$. Contractions of relative tensors do not change the weight. *A tensor equation must be among tensors of the same weight.* We require that tensor equations valid in one coordinate system be valid in all others; this property would be lost in an equation among tensors of different weights. Relative tensors with $w = \pm 1$ occur frequently and are called *tensor densities*. Tensors that transform with $w = 0$ are called *absolute tensors*.

absolute tensor

The covariant metric tensor is an absolute tensor: From Eq. (11.34),

$$g_{i'j'} = A^l_{i'} A^m_{j'} g_{lm}. \qquad (11.73)$$

The *determinant* of the metric tensor, however, is a relative scalar of weight[44] $w = 2$. Let g denote the determinant of the *covariant* metric tensor (a fairly standard notation). Applying the product rule for determinants to Eq. (11.73),

$$g' = J^2 g, \qquad (11.74)$$

[42]We allow for the possibility of transformations with $J < 0$ (by not taking the absolute value of the determinant). Transformations with $J < 0$ allow us to further classify vectors and tensors as *pseudotensors* (Section 11.11) as those that transform as tensors when $J > 0$, but for $J < 0$ transform with an additional change of sign.

[43]Beware: Relative tensors are also defined with w replaced by $-w$. See Weinberg [44, p. 99]. We have adopted a definition that leads to $w = +2$ for the determinant of the covariant elements of the metric tensor.

[44]The same is true of the determinant of any covariant second-rank absolute tensor.

where we've used Eq. (11.71). Thus, *the sign of g is an absolute quantity, invariant under coordinate transformations.* Equation (11.74) then provides an alternate expression for the Jacobian, one that separates the contributions from the coordinate systems it connects: $J = \sqrt{g'/g}$. Using Eq. (11.74), $\sqrt{|g'|} = J\sqrt{|g|}$ and thus $\sqrt{|g|}$ is a scalar density (transforms with $w = 1$).

From Eq. (11.70), therefore, we have the *invariant volume element*

$$\sqrt{|g'|}\, \mathrm{d}y^1 \cdots \mathrm{d}y^n = \sqrt{|g|}\, \mathrm{d}x^1 \cdots \mathrm{d}x^n. \tag{11.75}$$

Note how Eq. (11.75) has a net weight of $w = 0$: $\sqrt{|g|}\, \mathrm{d}^n x$ is an absolute scalar. (Under $x \to y$, $\mathrm{d}^n y = J^{-1}\, \mathrm{d}^n x$ from Eq. (11.70).) Thus, the integral of a scalar field $\int \phi\, \mathrm{d}^n x$ is not invariant, but $\int \phi\sqrt{|g|}\, \mathrm{d}^n x$ is. Upon substituting $J = \sqrt{g'/g}$ in Eq. (11.72), we find that

$$(|g'|)^{-w/2} T^{k'_1 \cdots k'_p}{}_{m'_1 \cdots m'_q} = A^{k'_1}_{t_1} \cdots A^{k'_p}_{t_p} A^{s_1}_{m'_1} \cdots A^{s_q}_{m'_q} ((|g|)^{-w/2} T^{t_1 \cdots t_p}{}_{s_1 \cdots s_q}).$$

For a tensor $\mathbf{T}$ of weight w, $(|g|)^{-w/2}\mathbf{T}$ transforms as an absolute tensor. Conversely, an absolute tensor $\mathbf{U}$ when multiplied by $(|g|)^{w/2}$ becomes a tensor of weight w. In particular, $\sqrt{|g|}\mathbf{U}$ is a tensor density.

Example. The determinant of the metric tensor for spherical coordinates (see Eq. (11.48)) is $g = r^4 \sin^2\theta$, and thus $\sqrt{g} = r^2 \sin\theta$. The invariant volume element is thus (from Eq. (11.75)) the product of $\sqrt{g}$ with the product of the coordinate differentials, $\mathrm{d}r\, \mathrm{d}\theta\, \mathrm{d}\phi$, the usual volume element in spherical coordinates, $r^2 \sin\theta\, \mathrm{d}r\, \mathrm{d}\theta\, \mathrm{d}\phi$.

Example. The Lorentz transformation relates the coordinates assigned to an *event* (point in spacetime) in two inertial reference frames in uniform relative motion. For motion along a common x-axis with relative speed v, the Lorentz transformation is

$$\begin{pmatrix} ct' \\ x' \\ y' \\ z' \end{pmatrix} = \begin{pmatrix} \gamma & -\beta\gamma & 0 \\ -\beta\gamma & \gamma & 0 & 0 \\ 0 & 0 & 1 & 0 \\ 0 & 0 & 0 & 1 \end{pmatrix} \begin{pmatrix} ct \\ x \\ y \\ z \end{pmatrix}, \tag{11.76}$$

where c is the speed of light, $\beta \equiv v/c$, and $\gamma \equiv 1/\sqrt{1-\beta^2}$ is the Lorentz factor. The Jacobian of the Lorentz transformation[45] is unity (under quite general conditions), as can be verified in the special case of Eq.

[45]You should make the connection that the Lorentz transformation in special relativity plays the role of the Jacobian matrix in tensor analysis, $x^{i'} = L^{i'}_j x^j$, where $L^{i'}_j$ is given in Eq. (11.76).

(11.76). Because $J = 1$, the four-dimensional spacetime volume element is invariant,

$$\mathrm{d}t'\ \mathrm{d}x'\ \mathrm{d}y'\ \mathrm{d}z' = \mathrm{d}t\ \mathrm{d}x\ \mathrm{d}y\ \mathrm{d}z. \tag{11.77}$$

Equation (11.77) is used in high-energy physics in the analysis of scattering data.

11.8 Tensors as operators

Having examined the transformation properties of tensors, let's return to their role as operators, mentioned in Section 11.1. For that purpose, we introduce a special notation. Let $\boldsymbol{v}$ denote contravariant vectors; we'll denote a *collection* of k contravariant vectors as $(\boldsymbol{v}_1, \ldots, \boldsymbol{v}_k)$ (so the indices refer to individual vectors and not vector components). Likewise we'll denote covariant vectors with $\boldsymbol{\omega}$, with a collection of l covariant vectors as $(\boldsymbol{\omega}^1, \ldots, \boldsymbol{\omega}^l)$.

Consider a type $(0, 1)$ tensor $\mathbf{T}$, which we'll now show acts as a covariant vector. Let's start to write $\mathbf{T}(\boldsymbol{v})$ instead of $\mathbf{T} \cdot \boldsymbol{v}$:

$$\mathbf{T}(\boldsymbol{v}) = \mathbf{T}(v^i \boldsymbol{e}_i) = v^i \mathbf{T}(\boldsymbol{e}_i) = v^i T_j \boldsymbol{e}^j \cdot \boldsymbol{e}_i = v^i T_j \delta_i^j = v^i T_i, \tag{11.78}$$

where we've used that a tensor is a linear operator, that $\boldsymbol{v}$ has a representation in the basis set $\{\boldsymbol{e}_i\}$, with $\boldsymbol{v} = v^i \boldsymbol{e}_i$, and that as a type $(0, 1)$ tensor (see Section 11.2.5) $\mathbf{T}$ has the representation in the associated dual basis set, $\{\boldsymbol{e}^j\}$, with $\mathbf{T} = T_j \boldsymbol{e}^j$. From Eq. (11.78), we identify $T_i = \mathbf{T}(\boldsymbol{e}_i)$, i.e. the component T_i of $\mathbf{T}$ is the action of $\mathbf{T}$ on a basis vector, $T_i = \mathbf{T}(\boldsymbol{e}_i)$. From Eq. (11.78), we see that $\mathbf{T}$ operates on a contravariant vector and returns a scalar, i.e. $\mathbf{T}$ is a linear functional.

Now let $\mathbf{T}$ denote a type $(1, 0)$ tensor, which we'll show is a contravariant vector:

$$\mathbf{T}(\boldsymbol{\omega}) = \omega_j \mathbf{T}(\boldsymbol{e}^j) = \omega_j T^i \boldsymbol{e}_i \cdot \boldsymbol{e}^j = \omega_j T^i \delta_i^j = \omega_j T^j, \tag{11.79}$$

where we've used linearity, that $\boldsymbol{\omega} = \omega_j \boldsymbol{e}^j$, and Eq. (11.51). From Eq. (11.79), we identify $T^j = \mathbf{T}(\boldsymbol{e}^j)$, i.e. the component T^j of $\mathbf{T}$ is the action of $\mathbf{T}$ on an element of the dual basis. A type $(1, 0)$ tensor therefore operates on a covariant vector and returns a scalar.

We defined the metric tensor $\mathbf{g}$ as a *bilinear operator* (Section 11.4), with $\mathbf{g}(\boldsymbol{v}_1, \boldsymbol{v}_2) = v_1^i v_2^j \mathbf{g}(\boldsymbol{e}_i, \boldsymbol{e}_j) = v_1^i v_2^j g_{ij}$, so that $g_{ij} = \mathbf{g}(\boldsymbol{e}_i, \boldsymbol{e}_j)$. That is, $\mathbf{g}$ acts on a pair of contravariant vectors and returns a scalar. The metric tensor is a type $(0, 2)$ tensor, $\mathbf{g} = g_{ij} \boldsymbol{e}^i \boldsymbol{e}^j$. Let's work through the "mechanics" of how it functions as such, using the linearity of dyadic products:

$$\begin{aligned}\mathbf{g}(\boldsymbol{v}_1, \boldsymbol{v}_2) &= g_{ij} \boldsymbol{e}^i \boldsymbol{e}^j \cdot (v_1^k \boldsymbol{e}_k)(v_2^m \boldsymbol{e}_m) \equiv g_{ij} v_1^k v_2^m (\boldsymbol{e}^i \cdot \boldsymbol{e}_k)(\boldsymbol{e}^j \cdot \boldsymbol{e}_m) \\ &= g_{ij} v_1^k v_2^m \delta_k^i \delta_m^j = g_{ij} v_1^i v_2^j.\end{aligned} \tag{11.80}$$

The bilinear action of a dyadic product on a pair of vectors is defined as

$$\boldsymbol{e}^i\boldsymbol{e}^j(\boldsymbol{e}_k, \boldsymbol{e}_m) = (\boldsymbol{e}^i \cdot \boldsymbol{e}_k)(\boldsymbol{e}^j \cdot \boldsymbol{e}_m) \equiv \boldsymbol{e}^i(\boldsymbol{e}_k)\boldsymbol{e}^j(\boldsymbol{e}_m) = \delta^i_k\delta^j_m. \tag{11.81}$$

Equation (11.81), when expressed in terms of vectors, not basis vectors, defines a bilinear function $\boldsymbol{\omega}^1\boldsymbol{\omega}^2$ that acts on ordered pairs of contravariant vectors such that

$$\boldsymbol{\omega}^1\boldsymbol{\omega}^2(\boldsymbol{v}_1, \boldsymbol{v}_2) \equiv \boldsymbol{\omega}^1(\boldsymbol{v}_1)\boldsymbol{\omega}^2(\boldsymbol{v}_2). \tag{11.82}$$

Second-rank tensors of type $(2, 0)$ are also bilinear functions that act on pairs of covariant vectors and return a scalar,

$$\begin{aligned}\mathbf{T}(\boldsymbol{\omega}^1, \boldsymbol{\omega}^2) &= \omega^1_i\omega^2_j\mathbf{T}(\boldsymbol{e}^i, \boldsymbol{e}^j) = \omega^1_i\omega^2_j T^{kl}\boldsymbol{e}_k\boldsymbol{e}_l(\boldsymbol{e}^i, \boldsymbol{e}^j)\\ &= \omega^1_i\omega^2_j T^{kl}\boldsymbol{e}_k(\boldsymbol{e}^i)\boldsymbol{e}_l(\boldsymbol{e}^j) = \omega^1_i\omega^2_j T^{kl}\delta^i_k\delta^j_l = \omega^1_i\omega^2_j T^{ij}. \end{aligned}\tag{11.83}$$

Once again, we identify $T^{ij} = \mathbf{T}(\boldsymbol{e}^i, \boldsymbol{e}^j)$. Second-rank tensors of type $(1, 1)$ act similarly as bilinear operators:

$$\begin{aligned}\mathbf{T}(\boldsymbol{\omega}, \boldsymbol{v}) &= \omega_i v^j\mathbf{T}(\boldsymbol{e}^i, \boldsymbol{e}_j) = \omega_i v^j T^k_l\boldsymbol{e}_k\boldsymbol{e}^l(\boldsymbol{e}^i, \boldsymbol{e}_j) = \omega_i v^j T^k_l\boldsymbol{e}_k(\boldsymbol{e}^i)\boldsymbol{e}^l(\boldsymbol{e}_j)\\ &= \omega_i v^j T^k_l\delta^i_k\delta^l_j = \omega_i v^j T^i_j. \end{aligned}\tag{11.84}$$

We identify $T^i_j = \mathbf{T}(\boldsymbol{e}^i, \boldsymbol{e}_j)$.

We certainly see the pattern now. To complete this idea, a function of several variables is said to be *multilinear* if it's linear in each argument,

$$\begin{aligned}\mathbf{T}(\boldsymbol{v}_1, \ldots, \alpha\boldsymbol{v}_i + \beta\boldsymbol{v}'_i, \ldots, \boldsymbol{v}_r) &= \alpha\mathbf{T}(\boldsymbol{v}_1, \ldots, \boldsymbol{v}_i, \ldots, \boldsymbol{v}_r)\\ &\quad + \beta\mathbf{T}(\boldsymbol{v}_1, \ldots, \boldsymbol{v}'_i, \ldots, \boldsymbol{v}_r),\end{aligned}$$

for $1 \le i \le r$ and α, β scalars. A type (k, l) tensor is a multilinear operator that acts on k covariant vectors and l contravariant vectors and returns a scalar,

$$\mathbf{T}(\boldsymbol{\omega}^1, \ldots, \boldsymbol{\omega}^k, \boldsymbol{v}_1, \ldots, \boldsymbol{v}_l) = \omega^1_{i_1}\cdots\omega^k_{i_k}v^{m_1}_1\cdots v^{m_l}_l T^{i_1\cdots i_k}{}_{m_1\cdots m_l},$$

where

$$T^{i_1\cdots i_k}{}_{m_1\cdots m_l} = \mathbf{T}(\boldsymbol{e}^{i_1}, \ldots, \boldsymbol{e}^{i_k}, \boldsymbol{e}_{m_1}, \ldots, \boldsymbol{e}_{m_l}).$$

Tensors are thus mappings between ordered products of vectors and scalars. Tensors can also map tensors to tensors, however. A type $(1, 2)$ tensor, $\mathbf{T} = T^i{}_{jk}\boldsymbol{e}_i\boldsymbol{e}^j\boldsymbol{e}^k$, when acting on $(\boldsymbol{\omega}, \boldsymbol{v}_1, \boldsymbol{v}_2)$ produces a number. Consider, however, the action of $\mathbf{T}$ acting just on a covariant vector $\boldsymbol{\omega}$, symbolized $\mathbf{T}(\boldsymbol{\omega}, \cdot, \cdot)$:

$$\begin{aligned}\mathbf{T}(\boldsymbol{\omega}, \cdot, \cdot) &= T^i{}_{jk}\boldsymbol{e}_i\boldsymbol{e}^j\boldsymbol{e}^k(\boldsymbol{\omega}) = (T^i{}_{jk}\boldsymbol{\omega}(\boldsymbol{e}_i))\boldsymbol{e}^j\boldsymbol{e}^k\\ &= (T^i{}_{jk}\omega_i)\boldsymbol{e}^j\boldsymbol{e}^k \equiv B_{jk}\boldsymbol{e}^j\boldsymbol{e}^k \equiv \mathbf{B},\end{aligned}$$

where $\mathbf{B}$ is a type $(0,2)$ tensor. Note that we've used Eq. (11.51): $\boldsymbol{e}_i(\boldsymbol{\omega}) = \boldsymbol{\omega}(\boldsymbol{e}_i)$. The conclusion that $\mathbf{T}(\boldsymbol{\omega},\cdot,\cdot) = \mathbf{B}$ is really just an instance of a tensor equation written in component form: $T^i{}_{jk}\omega_i = B_{jk}$. The same tensor $\mathbf{T}$ acting on a pair of contravariant vectors produces another contravariant vector, $\mathbf{T}(\cdot,\boldsymbol{v}_1,\boldsymbol{v}_2) = \boldsymbol{v}$:

$$\begin{aligned}\mathbf{T}(\cdot,\boldsymbol{v}_1,\boldsymbol{v}_2) &= T^i{}_{jk}\boldsymbol{e}_i\boldsymbol{e}^j\boldsymbol{e}^k(\boldsymbol{v}_1,\boldsymbol{v}_2) = (T^i{}_{jk}\boldsymbol{e}^j(\boldsymbol{v}_1)\boldsymbol{e}^k(\boldsymbol{v}_2))\boldsymbol{e}_i \\ &= (T^i{}_{jk}v_1^j v_2^k)\boldsymbol{e}_i \equiv v^i\boldsymbol{e}_i.\end{aligned}$$

This equation in component form is $T^i{}_{jk}v_1^j v_2^k = v^i$.

Example. The stress tensor is a type $(2,0)$ tensor: It maps a pair covariant vectors to a scalar. It also maps a single covariant vector to a contravariant vector:

$$\delta\boldsymbol{F} = \delta F^i\boldsymbol{e}_i = \mathbf{T}(\delta\boldsymbol{s}) = T^{ij}\boldsymbol{e}_i\boldsymbol{e}_j(\delta s_k\boldsymbol{e}^k) = T^{ij}\boldsymbol{e}_i\delta s_k\delta_j^k = (T^{ij}\delta s_j)\boldsymbol{e}_i,$$

which is consistent with Eq. (11.69).

That tensors can wear different hats is exemplified by the metric tensor, a type $(0,2)$ symmetric, bilinear function that takes a pair of vectors (in either order) and produces the same number, $\mathbf{g}(\boldsymbol{v}_1,\boldsymbol{v}_2) = g_{ij}v_1^i v_2^j$, where $g_{ij} = g_{ji}$. Let $\mathbf{g}$ act on a single vector, however:

$$\mathbf{g}(\boldsymbol{v},\cdot) = g_{ij}\boldsymbol{e}^i\boldsymbol{e}^j(v^k\boldsymbol{e}_k) = g_{ij}\boldsymbol{e}^i v^k\delta_k^j = g_{ij}\boldsymbol{e}^i v^j = v_i\boldsymbol{e}^i \equiv \boldsymbol{\omega},$$

where we've lowered an index. Thus, $\mathbf{g}$ maps a vector $\boldsymbol{v}$ to a dual vector, $\boldsymbol{\omega}$: $\mathbf{g}(\boldsymbol{v},\cdot) = \boldsymbol{\omega}$; $\mathbf{g}$ effects a mapping between a vector space and its dual space: $\mathbf{g}: V_N \to V_N^*$, which provides a natural accounting for the lowering of an index, $g_{ij}v^j = v_i$. What about *raising* indices, can $\mathbf{g}$ act on a dual vector, $\mathbf{g}(\boldsymbol{\omega},\cdot)$? We know that V_N and V_N^* are isomorphic. The mapping $\mathbf{g}: V_N \to V_N^*$ must therefore be invertible: There must exist a unique mapping $\mathbf{g}^{-1}: V_N^* \to V_N$ such that

$$\begin{aligned}\boldsymbol{v} = \mathbf{g}^{-1}(\mathbf{g}(\boldsymbol{v})) &= \mathbf{g}^{-1}(v_j\boldsymbol{e}^j) = v_j\mathbf{g}^{-1}(\boldsymbol{e}^j) = v_j[(\mathbf{g}^{-1})^{kl}\boldsymbol{e}_k\boldsymbol{e}_l]\boldsymbol{e}^j \\ &= v_j(\mathbf{g}^{-1})^{kl}\boldsymbol{e}_k\delta_l^j = v_j(\mathbf{g}^{-1})^{kj}\boldsymbol{e}_k \equiv v^k\boldsymbol{e}_k.\end{aligned}$$

Thus, $(\mathbf{g}^{-1})^{kj}v_j \equiv g^{kj}v_j = v^k$. It's customary to omit the inverse symbol, where it's understood that the contravariant tensor ($\mathbf{g}$ with upper indices) is the inverse of $\mathbf{g}$ with lower indices. We could have, in asking whether $\mathbf{g}$ can act on a dual vector, considered $\mathbf{g}(\boldsymbol{\omega}) = g_{ij}\boldsymbol{e}^i\boldsymbol{e}^j(\omega_k\boldsymbol{e}^k)$, and simply defined $\boldsymbol{e}^j(\boldsymbol{e}^k) = g^{jk}$, which we symbolized as $\boldsymbol{e}^j\cdot\boldsymbol{e}^k$ in Eq. (11.54). The lowering of an index on g^{jk} produces: $g^j{}_k = g_{kl}g^{jl} = (g^{-1})^{jl}g_{kl} = \delta_k^j$, which we symbolized as $\boldsymbol{e}^j\cdot\boldsymbol{e}_k$ in Eq. (11.31).

11.9 Symmetric and antisymmetric tensors

A tensor whose values (components) remain unchanged when two of its covariant or two of its contravariant arguments are transposed, is said to be *symmetric in these two arguments.* For example, if

$$\mathbf{T}(\boldsymbol{\omega}^1, \ldots, \boldsymbol{\omega}^k, \boldsymbol{v}_1, \ldots, \boldsymbol{v}_i, \ldots, \boldsymbol{v}_j, \ldots, \boldsymbol{v}_l)$$
$$= \mathbf{T}(\boldsymbol{\omega}^1, \ldots, \boldsymbol{\omega}^k, \boldsymbol{v}_1, \ldots, \boldsymbol{v}_j, \ldots, \boldsymbol{v}_i, \ldots, \boldsymbol{v}_l),$$

then $\mathbf{T}$ is *symmetric* in contravariant arguments i and j. A tensor whose values *change sign* when two of its covariant or two of its contravariant arguments are transposed is *antisymmetric* (or *alternating* or *skew-symmetric*) in these arguments. Transposing covariant *and* contravariant arguments makes no sense, as they are defined with respect to different basis sets; *symmetry and antisymmetry are with respect to covariant or contravariant arguments only.*[46] The symmetry or antisymmetry of a tensor is a geometric property, independent of basis transformations. A tensor antisymmetric (symmetric) in *all* of its arguments is said to be *totally antisymmetric* (totally symmetric).

permutation

We now introduce a widely used notation for signifying symmetric and antisymmetric tensors; it's use can sometimes save a lot of writing. First, define a *permutation* π of the integers $(1, \ldots, n)$ as a one-to-one mapping of the set onto itself with the values $\pi(1), \ldots, \pi(n)$. The permutation $\pi : 1234 \to 4132$ has values $\pi(1) = 4$, $\pi(2) = 1$, $\pi(3) = 3$, and $\pi(4) = 2$. There are $n!$ permutations of n objects. The *symmetric part of a tensor* (with respect to indices $i_1 \cdots i_n$) is defined with (parentheses around the indices)

$$T_{(i_1 \cdots i_n)} \equiv \frac{1}{n!} \sum_{\pi} T_{i_{\pi(1)} \cdots i_{\pi(n)}},$$

where the sum is over the $n!$ permutations of $(1 \cdots n)$. The *antisymmetric part* (with respect to $i_1 \cdots i_n$) is defined with [square brackets around the indices]

$$T_{[i_1 \cdots i_n]} \equiv \frac{1}{n!} \sum_{\pi} \delta_\pi T_{i_{\pi(1)} \cdots i_{\pi(n)}},$$

where δ_π, the *parity* [also called the *signum* (sign)] of the permutation is $+1$ for even permutations and -1 for odd permutations.[47] An even (odd) permutation is a permutation obtained through an even (odd) number of pairwise interchanges of the numbers, starting from the reference sequence

[46]Thus, one cannot speak of an antisymmetric mixed tensor. One must first convert a mixed tensor to either a contra- or covariant tensor, e.g. $T^i{}_j \to T^{ij} = g^{jk} T^i{}_k$, and then apply symmetry operations.

[47]The parity of a permutation is unique. There are many ways of realizing a given permutation of the reference sequence through pairwise exchanges, but all ways require either an even or an odd number of exchanges [3, p. 47].

$(1 \cdots n)$. The sequence 2413 is an odd permutation of 1234. A second-rank tensor can always be decomposed into symmetric and antisymmetric parts, $T_{ij} = T_{(ij)} + T_{[ij]}$, but the same is not true of high-rank tensors. For example, $T_{ijk} \neq T_{(ijk)} + T_{[ijk]}$. The notation can apply to any groups of indices. For example, $T^{(ij)k}{}_{[lm]} \equiv \frac{1}{4}(T^{ijk}{}_{lm} + T^{jik}{}_{lm} - T^{ijk}{}_{ml} - T^{jik}{}_{ml})$ denotes a tensor symmetric in its first two contravariant indices and antisymmetric in its covariant indices.

11.10 The Levi-Civita tensor

Levi-Civita symbol

The *Levi-Civita symbol* (also known as the *permutation symbol* or the *totally antisymmetric symbol*) for an N-dimensional space is defined such that

$$\varepsilon_{i_1 \cdots i_N} \equiv \varepsilon^{i_1 \cdots i_N} = \begin{cases} +1 & \text{if } i_1 \cdots i_N \text{ is an even permutation of} 1 \cdots N, \\ -1 & \text{if } i_1 \cdots i_N \text{ is an odd permutation of} 1 \cdots N, \\ 0 & \text{if two or more indices are equal.} \end{cases} \tag{11.85}$$

We previously made use of the Levi-Civita symbol in Eq. (1.36); here we formally distinguish contra- and covariant versions. *Permutation symbols have N indices* (for an N-dimensional space). There are N indices, each of which can take on N values; the symbol therefore has N^N possible values, of which $N!$ are nonzero[48] – there are $N!$ distinct arrangements (permutations) of N indices. For $N = 3$, out of the $3^3 = 27$ possible values of ε_{ijk}, only $3! = 6$ are nonzero, with $\varepsilon_{123} = \varepsilon_{231} = \varepsilon_{312} = 1$ and $\varepsilon_{132} = \varepsilon_{213} = \varepsilon_{321} = -1$.

The Levi-Civita symbol is used to produce *antisymmetric* linear combinations, where the terms in linear combinations alternate in sign as terms are combined. We saw its use in Section 1.4.9 in defining determinants. Here we develop a more general definition of determinant based on the two types of symbol in Eq. (11.85).

Theorem 11.10.1. *The determinant of an $N \times N$ matrix $\boldsymbol{A}$ with elements A^i_j can be defined in terms of either of the Levi-Civita symbols in Eq. (11.85):*

$$\varepsilon_{i_1 \cdots i_N} A^{i_1}_{j_1} \cdots A^{i_N}_{j_N} = \varepsilon_{j_1 \cdots j_N} \det \boldsymbol{A} \tag{11.86}$$

$$\varepsilon^{i_1 \cdots i_N} A^{j_1}_{i_1} \cdots A^{j_N}_{i_N} = \varepsilon^{j_1 \cdots j_N} \det \boldsymbol{A}. \tag{11.87}$$

Proof. To show Eq. (11.86), define a function D_A of matrix elements, $D_A(j_1, \ldots, j_N) \equiv \varepsilon_{i_1 \cdots i_N} A^{i_1}_{j_1} \cdots A^{i_N}_{j_N}$. D_A involves an alternating sum of $N!$, N-tuple products of matrix elements (there are $N!$ ways to choose

[48] By Stirling's approximation, $N! \sim N^N \mathrm{e}^{-N}$ and thus $N! < N^N$.

$i_1 \cdots i_N$ such that they are all distinct). The terms within N-tuples each come from a different row of the matrix ($i_1 \cdots i_N$ are distinct) and, as we'll see, a different column. The function $D_A(j_1, \ldots, j_N)$ is antisymmetric in column indices j. To see that, interchange the upper *indices* $i_k \leftrightarrow i_l$ of two matrix elements in an N-tuple, then restore the *matrix elements* back to their previous positions with respect to the order of the indices in $\varepsilon_{i_1 \cdots i_N}$, and now interchange $i_k \leftrightarrow i_l$ in $\varepsilon_{i_1 \cdots i_N}$; $D_A(j_1, \cdots, j_N)$ is antisymmetric under $j_k \leftrightarrow j_l$. Hence, $D_A(j_1, \ldots, j_N)$ vanishes if any two indices are equal. The quantity $D_A(j_1, \ldots, j_N)$ has the properties of the determinant of $\boldsymbol{A}$, and in fact equals $\det \boldsymbol{A}$ ($-\det \boldsymbol{A}$) when $j_1 \cdots j_N$ is an even (odd) permutation of $1 \cdots N$. The determinant can be considered a function of matrix columns, $\det(\boldsymbol{C}_1, \ldots, \boldsymbol{C}_N)$ ($\boldsymbol{C}_j \equiv A^i_j$, $i = 1, \cdots, N$), such that it's multilinear and antisymmetric, and has the value unity for the identity matrix. Equation (11.87) follows by an analogous argument involving rows. □

Equations (11.86) and (11.87) indicate what we already know, that the determinant changes sign upon the interchange of columns or rows. They generalize the customary definitions of determinant: $\det \boldsymbol{A} = \varepsilon_{i_1 \cdots i_N} A^{i_1}_1 \cdots A^{i_N}_N = \varepsilon^{i_1 \cdots i_N} A^1_{i_N} \cdots A^N_{i_N}$ [45, p. 30], such as we gave in Section 1.4.9. These results allow us to prove the product rule of determinants:

$$\begin{aligned}\det \boldsymbol{A} \det \boldsymbol{B} &= \det \boldsymbol{A} \varepsilon_{\beta_1 \cdots \beta_N} B^{\beta_1}_1 \cdots B^{\beta_N}_N = \varepsilon_{\alpha_1 \cdots \alpha_N} A^{\alpha_1}_{\beta_1} \cdots A^{\alpha_N}_{\beta_N} B^{\beta_1}_1 \cdots B^{\beta_N}_N \\ &= \varepsilon_{\alpha_1 \cdots \alpha_N} (A^{\alpha_1}_{\beta_1} B^{\beta_1}_1) \cdots (A^{\alpha_N}_{\beta_N} B^{\beta_N}_N) \equiv \varepsilon_{\alpha_1 \cdots \alpha_N} C^{\alpha_1}_1 \cdots C^{\alpha_N}_N = \det \boldsymbol{C}.\end{aligned}$$

Equations (11.86) and (11.87) hold for any $N \times N$ matrix. They allow us to establish the transformation properties of the Levi-Civita symbols. By applying Eq. (11.86) to the Jacobian matrix, we have (see Eq. (11.71)) $\varepsilon_{\alpha_1 \cdots \alpha_n} A^{\alpha_1}_{\beta'_1} \cdots A^{\alpha_n}_{\beta'_n} = \varepsilon_{\beta'_1 \cdots \beta'_n} J$. The permutation symbol thus transforms as a relative tensor:

$$\varepsilon_{\beta'_1 \cdots \beta'_n} = J^{-1} A^{\alpha_1}_{\beta'_1} \cdots A^{\alpha_n}_{\beta'_n} \varepsilon_{\alpha_1 \cdots \alpha_n}. \tag{11.88}$$

Comparing with Eq. (11.72), $\varepsilon_{\alpha_1 \cdots \alpha_n}$ *is a covariant tensor with weight* $w = -1$. Using Eq. (11.87), it follows that

$$\varepsilon^{\gamma'_1 \cdots \gamma'_n} = J A^{\gamma'_1}_{\beta_1} \cdots A^{\gamma'_n}_{\beta_n} \varepsilon^{\beta_1 \cdots \beta_n}; \tag{11.89}$$

$\varepsilon^{i_1 \cdots i_n}$ *is a contravariant tensor of weight* $w = +1$. Combining Eqs. (11.88) and (11.86) as applied to the Jacobian matrix, $(\varepsilon_{\beta_1 \cdots \beta_n})' = J^{-1} A^{\alpha_1}_{\beta'_1} \cdots A^{\alpha_n}_{\beta'_n} \varepsilon_{\alpha_1 \cdots \alpha_n} = J^{-1} J \varepsilon_{\beta'_1 \cdots \beta'_n}$; thus $(\varepsilon_{\beta_1 \cdots \beta_n})' = \varepsilon_{\beta'_1 \cdots \beta'_n}$, the transformed permutation symbol has the value of $\varepsilon_{\beta_1 \cdots \beta_n}$ in the original coordinate system. *The permutation symbol* $\varepsilon_{\beta_1 \cdots \beta_n}$ *is a constant tensor*; ditto for $\varepsilon^{\beta_1 \cdots \beta_n}$.

By the rules established in Section 11.7, we define an absolute covariant tensor by multiplying $\varepsilon_{i_1 \cdots i_n}$ with $\sqrt{|g|}$. We define the covariant *Levi-Civita tensor* as

Levi-Civita tensor

$$\epsilon_{i_1\cdots i_n} \equiv \sqrt{|g|}\varepsilon_{i_1\cdots i_n}. \tag{11.90}$$

Note the change in notation from $\varepsilon_{i_1\cdots i_n}$ (permutation symbol, tensor density), to $\epsilon_{i_1\cdots i_n}$, absolute tensor. The contravariant Levi-Civita tensor follows from raising indices on $\epsilon_{i_1\cdots i_n}$. From Eq. (11.90),

$$\epsilon^{j_1\cdots j_n} \equiv g^{j_1 i_1}\cdots g^{j_n i_n}\epsilon_{i_1\cdots i_n} = \sqrt{|g|}g^{j_1 i_1}\cdots g^{j_n i_n}\varepsilon_{i_1\cdots i_n} = \sqrt{|g|}\det[g^{ij}]\varepsilon^{j_1\cdots j_n},$$

where in the last equality, we have the determinant of the *contravariant* metric tensor. Using Eq. (11.58), $\det[g^{ij}] = g^{-1}$, the inverse of g (determinant of the covariant metric tensor). Writing $g = \text{sgn}(g)|g|$ (where sgn(g) is the *sign* of g), we have for the contravariant tensor

$$\epsilon^{j_1\cdots j_n} = \frac{\text{sgn}(g)}{\sqrt{|g|}}\varepsilon^{j_1\cdots j_n}. \tag{11.91}$$

In Euclidean geometries, with positive definite metric tensors, $\text{sgn}(g) = 1$. In the special theory of relativity, however, $\text{sgn}(g) = -1$. Beware.

We now define a special tensor, the *generalized Kronecker delta*, that will allow us to derive an important property of the Levi-Civita symbols. Let

generalized Kronecker delta

$$\Delta^{i_1\cdots i_N}_{j_1\cdots j_N} \equiv \begin{cases} +1 & \text{if } i_1\cdots i_N \text{ is an even permutation of } j_1\cdots j_N, \\ -1 & \text{if } i_1\cdots i_N \text{ is an odd permutation of } j_1\cdots j_N, \\ 0 & \text{if } i_1\cdots i_N \text{ is not a permutation of } j_1\cdots j_N \text{ or } i_1\cdots i_N \text{ are not all distinct.} \end{cases}$$

The generalized Kronecker delta is a type (N, N) tensor,[49] antisymmetric in all contravariant indices and all covariant indices.[50] To understand what's implied here, consider the case of $N = 2$, where each of the indices can only take the values $(1, 2)$. By definition, $\Delta^{12}_{12} = +1$, $\Delta^{21}_{12} = -1$, $\Delta^{21}_{21} = +1$, and $\Delta^{12}_{21} = -1$; all other cases are zero.[51] We could "build" the values of this tensor out of products of ordinary Kronecker deltas,

$$\Delta^{ij}_{lm} = \delta^i_l\delta^j_m - \delta^j_l\delta^i_m = \begin{vmatrix} \delta^i_l & \delta^i_m \\ \delta^j_l & \delta^j_m \end{vmatrix}. \tag{11.92}$$

The first set of delta functions in Eq. (11.92) covers the case where the indices (i, j) are the same as (l, m) (no permutation, but certainly a case of an even permutation), while the second group of delta functions covers

---

[49]Let $T^{s_1\cdots s_N}$ be the components of a type $(N, 0)$ tensor. Then, $\Delta^{i_1\cdots i_N}_{j_1\cdots j_N}T^{j_1\cdots j_N}$ generates an alternating sum of $N!$ terms, $T^{i_1 i_2\cdots i_N} - T^{i_2 i_1\cdots i_N} + \cdots$, i.e. a linear combination of type $(N, 0)$ tensors, and hence a type $(N, 0)$ tensor. By the quotient theorem, $\Delta^{i_1\cdots i_N}_{j_1\cdots j_N}$ is a type (N, N) tensor.

[50]We've defined $\Delta^{i_1\cdots i_N}_{j_1\cdots j_N}$ in terms of permutations of the upper indices relative to a reference sequence of lower indices, but it can be defined the other way around, as involving permutations of the lower indices relative to a reference sequence of upper indices.

[51]In general there are N^{2N} possible values of this rank-$2N$ tensor, of which $(N!)^2$ are nonzero.

the case of (i, j) being an odd permutation of (l, m). Because $N = 2$, there are no other distinct permutations of the upper indices. The association with a determinant is not accidental – the determinant is antisymmetric in rows and columns, the same as for the upper and lower indices of $\Delta^{i_1\cdots i_N}_{j_1\cdots j_N}$. A determinant of Kronecker delta functions has the very properties we require of $\Delta^{i_1\cdots i_N}_{j_1\cdots j_N}$ (upper indices totally antisymmetric, which must be a permutation of the lower indices and has the values ± 1):

$$\Delta^{\alpha_1\cdots\alpha_N}_{\beta_1\cdots\beta_N} = \begin{vmatrix} \delta^{\alpha_1}_{\beta_1} & \cdots & \delta^{\alpha_1}_{\beta_N} \\ \vdots & & \vdots \\ \delta^{\alpha_N}_{\beta_1} & \cdots & \delta^{\alpha_N}_{\beta_N} \end{vmatrix}. \tag{11.93}$$

Another way to realize the generalized Kronecker delta (antisymmetric in all upper and lower indices having the nonzero values of ± 1) is from the outer products of Levi-Civita tensors, with $\Delta^{i_1\cdots i_N}_{j_1\cdots j_N} = \lambda \epsilon^{i_1\cdots i_N}\epsilon_{j_1\cdots j_N} = \lambda \text{sgn}(g)\varepsilon^{i_1\cdots i_N}\varepsilon_{j_1\cdots j_N}$, where λ is a proportionality factor. The proportionality constant can be evaluated with any set of indices for which both sides of the formula are nonzero. For the upper indices exactly the same as the lower indices, consistency requires that $\lambda = \text{sgn}(g)$. Combining with Eq. (11.93), we arrive at the useful result,

$$\text{sgn}(g)\epsilon^{i_1\cdots i_n}\epsilon_{j_1\cdots j_n} = \varepsilon^{i_1\cdots i_n}\varepsilon_{j_1\cdots j_n} = \begin{vmatrix} \delta^{i_1}_{j_1} & \cdots & \delta^{i_1}_{j_n} \\ \vdots & & \vdots \\ \delta^{i_n}_{j_1} & \cdots & \delta^{i_n}_{j_n} \end{vmatrix}. \tag{11.94}$$

Example. For the case of $N = 3$, use Eq. (11.94) to evaluate $\varepsilon^{ijk}\varepsilon_{kmn}$. First, evaluate the general term, then do the contraction. From Eq. (11.94),

$$\varepsilon^{ijk}\varepsilon_{lmn} = \begin{vmatrix} \delta^i_l & \delta^i_m & \delta^i_n \\ \delta^j_l & \delta^j_m & \delta^j_n \\ \delta^k_l & \delta^k_m & \delta^k_n \end{vmatrix} = \delta^i_l(\delta^j_m\delta^k_n - \delta^j_n\delta^k_m)$$

$$+ \delta^j_l(\delta^i_n\delta^k_m - \delta^i_m\delta^k_n) + \delta^k_l(\delta^i_m\delta^j_n - \delta^i_n\delta^j_m).$$

Setting $l = k$ and summing over k, we find that

$$\varepsilon^{ijk}\varepsilon_{kmn} = \delta^i_m\delta^j_n - \delta^i_n\delta^j_m. \tag{11.95}$$

To arrive at this result requires evaluating terms like $\delta^i_k\delta^k_m = \delta^i_m$ and $\delta^k_k = 3$.

11.11 Pseudotensors

In a right-handed Cartesian coordinate system (see Figure 11.7), associated with the unit vectors $\hat{\boldsymbol{x}}$ and $\hat{\boldsymbol{y}}$ is the third, $\hat{\boldsymbol{z}} = \hat{\boldsymbol{x}} \times \hat{\boldsymbol{y}}$. The cross product is antisymmetric in its operands ($\boldsymbol{A} \times \boldsymbol{B} = -\boldsymbol{B} \times \boldsymbol{A}$), and there

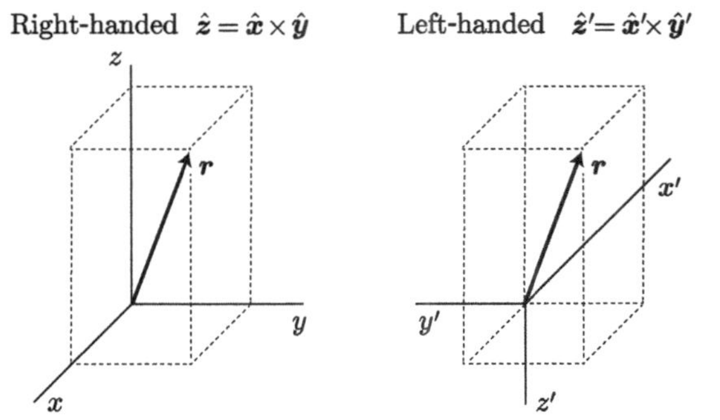

Figure 11.7 Right-handed and left-handed coordinate systems. The vector $\boldsymbol{r}$ has negative components in the inverted coordinate system, the hallmark of a polar vector.

is a definite (but conventional) role assigned to the vectors participating in the cross product, $\mathbf{Vector}_1 \times \mathbf{Vector}_2$, where $\mathbf{Vector}_1$ and $\mathbf{Vector}_2$ occupy definite "slots" in this operation. The direction of $\boldsymbol{A} \times \boldsymbol{B}$ is that of the thumb on a human right hand as $\boldsymbol{A}$ is "crossed" into $\boldsymbol{B}$. An *inversion* of the coordinate axes in a right-handed system produces a *left-handed coordinate system*, with $\hat{z}' = \hat{x}' \times \hat{y}'$ given by the direction of the thumb on the left hand, where $\hat{x}' = -\hat{x}$, $\hat{y}' = -\hat{y}$, $\hat{z}' = -\hat{z}$. Under an inversion, the coordinates of $\boldsymbol{r}$ in the transformed system are the *negative* of their values in the original system, $(x', y', z') = (-x, -y, -z)$. Vectors with components that transform under inversion like the components of $\boldsymbol{r}$ are called *polar vectors*. Note that $\boldsymbol{r}$ in Figure 11.7 is the *same vector* before and after the inversion, in keeping with the requirement that a vector is a geometric object that maintains its identity under a change of basis.

polar vector

Are all vectors polar? Yes (on semantic grounds – only vectors are vectors) and no because nonpolar vector-like quantities exist. Figure 11.8 shows the cross product $\boldsymbol{C} = \boldsymbol{A} \times \boldsymbol{B}$ between polar vectors. Under inversion, the components of $\boldsymbol{A}$ and $\boldsymbol{B}$ are negative relative to the inverted coordinate axes, yet those of $\boldsymbol{C}' = \boldsymbol{A}' \times \boldsymbol{B}'$ are *positive*.[52] The vector-like quantity $\boldsymbol{C}$ therefore does not preserve its identity under the transformation. *The cross product of polar vectors is a fundamentally different kind of object*: It's not a polar vector and *should not be called a vector*.[53] The cross product is a different kind of vector, one derived from two other vectors.[54] Vectors with components that do not change sign under inversion are called *axial vectors* or *pseudovectors*.

axial vector

[52]In the original coordinate system, $C_x = A_y B_z - B_y A_z$. In the inverted coordinate system, $C'_x = A'_y B'_z - B'_y A'_z = A_y B_z - B_y A_z = C_x$. If $C_x > 0$ in the original system, $C'_x > 0$ in the inverted system. The components of $\boldsymbol{A}$, $\boldsymbol{B}$ change sign under inversion, yet those of $\boldsymbol{C}$ do not. $\boldsymbol{C}$ is not a vector in the same sense that $\boldsymbol{A}$, $\boldsymbol{B}$ are.

[53]The position vector is the prototype vector; anything called vector must behave like the prototype.

[54]Quantities like the cross product, made up from *antisymmetric* linear combinations of the products of vector components, as in $C_x = A_y B_z - B_y A_z$, are termed *bivectors*. Under the interchange of *labels* $x \leftrightarrow y$, $C_x \to -C_x$.

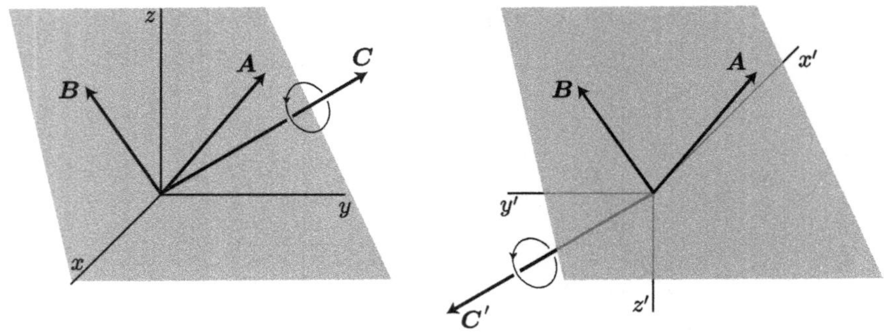

Figure 11.8 The vector cross product is a pseudovector.

pseudotensor

Pseudotensors transform as tensors *when the Jacobian of the transformation is positive*, but transform with an additional change of sign[55] for $J < 0$. The term pseudotensor is unfortunate, implying *false* tensor; perhaps "half-tensor" or "demi-tensor" would be better. For $J > 0$, pseudotensors have all the invariance properties expected of tensors. The Levi-Civita symbol is a pseudotensor *in spaces of an odd number of dimensions* (see Eq. (11.88)). The Jacobian matrix for an inversion of a three-dimensional Cartesian system has $(-1, -1, -1)$ on its diagonal (zeros otherwise), and hence, $\det J = -1$. It's the extra change of sign from J in Eq. (11.88) that keeps ε_{ijk} a constant tensor, i.e. it does *not* change sign under an inversion. A third-rank tensor made up from three covariant vectors $T_{ijk} = U_i V_j W_k$ (not a pseudotensor) *would* change sign under inversion, but not ε_{ijk}. The components of the cross product $\boldsymbol{C} \equiv \boldsymbol{F} \times \boldsymbol{G}$ can be written $C_i = \varepsilon_{ijk} F^j G^k$. It's readily shown using Eq. (11.88), for $\boldsymbol{F}$, $\boldsymbol{G}$ contravariant vectors, that C_i transforms as $C_{i'} = J^{-1} A^j_{i'} C_j$. The cross product is therefore a pseudovector.

Polar vectors are related to the prototype vector, $\boldsymbol{r}$: velocity, $\boldsymbol{v} = \dot{\boldsymbol{r}}$; acceleration, $\boldsymbol{a} = \dot{\boldsymbol{v}} = \ddot{\boldsymbol{r}}$; force, $\boldsymbol{F} = m\boldsymbol{a}$; and electric field, $\boldsymbol{E} = \boldsymbol{F}/q$. Axial vectors are associated with a cross product: torque, $\boldsymbol{\tau} = \boldsymbol{r} \times \boldsymbol{F}$; angular velocity, $\boldsymbol{v} = \boldsymbol{\omega} \times \boldsymbol{r}$; angular momentum, $\boldsymbol{L} = \boldsymbol{r} \times \boldsymbol{p}$; magnetic field, $\boldsymbol{F} = q\boldsymbol{v} \times \boldsymbol{B}$. Generally, we have the rules:

- polarvector $\times$ polarvector = pseudovector
- pseudovector $\times$ pseudovector = pseudovector
- polarvector $\times$ pseudovector = polarvector.

What about the inner product? A *pseudoscalar* results from the inner product between a polar vector and an axial vector, such as between electric and magnetic field vectors,[56] $\boldsymbol{E} \cdot \boldsymbol{B}$. *A pseudoscalar reverses sign*

[55]Hence the admonition in Section 11.7 not to take the absolute value of the Jacobian determinant.

[56]In "building" the Lagrangian for the electromagnetic field, $\boldsymbol{E} \cdot \boldsymbol{B}$ is ruled out because it's a pseudoscalar.

under inversion. For $\boldsymbol{A}$, $\boldsymbol{B}$, and $\boldsymbol{C}$ polar vectors, $\boldsymbol{A}\cdot(\boldsymbol{B}\times\boldsymbol{C})$ is a pseudoscalar.

Why should pseudotensors concern us? They provide another way to classify the equations of physics. Do the equations of physics depend on our choice of coordinate system? (Let's hope not!) Pseudovectors depend on a *handedness convention*, which is arbitrary. The cross product associates a vector with a plane, and there are two sides to a plane. Does the universe care which hand we use? If not, then *a valid equation of physics cannot equate polar and axial vectors.* Faraday's law relates two *axial* vectors, $\boldsymbol{\nabla}\times\boldsymbol{E} = -\partial\boldsymbol{B}/\partial t$. Note that the "curl of the curl" does not change the vector character of what it operates on, as in $\boldsymbol{\nabla}\times\boldsymbol{\nabla}\times\boldsymbol{E} = -(1/c^2)\partial^2\boldsymbol{E}/\partial t^2$ and $\boldsymbol{\nabla}\times\boldsymbol{\nabla}\times\boldsymbol{B} = -(1/c^2)\partial^2\boldsymbol{B}/\partial t^2$.

The cross product as a way of multiplying vectors *only applies in three dimensions.* Only in three dimensions can we associate a *vector* with a rotation (such as the direction of your thumb); higher dimensional spaces require a generalization of the cross product. Rotations should be seen, not as rotations about an axis (our picture in three dimensions), but as mappings affecting *pairs* of coordinate axes. In N dimensions, there are "N-choose-2"$= N(N-1)/2$ pairs of coordinate axes; rotations are therefore described by a quantity having $N(N-1)/2$ components, which is *not a vector in the same space*, unless $N=3$. $N=3$ is the nontrivial solution of $N = N(N-1)/2$. Rotations in a four-dimensional space require a quantity having six components, either a vector in *six* dimensions or as an antisymmetric second-rank tensor.

The generalization of the cross product to spaces of dimension $N>3$ is known as the *wedge product* (not studied in this book), denoted $\boldsymbol{A}\wedge\boldsymbol{B}$. The wedge product of vectors does not produce another vector but an antisymmetric tensor (unless $N=3$). By antisymmetric combinations, note from the orbital angular momentum $\boldsymbol{L}=\boldsymbol{r}\times\boldsymbol{p}$, that $L_x = yp_z - zp_y$ changes sign under the interchange of labels y and z, $L_x\to -L_x$. In higher dimensional spaces, angular momentum is an antisymmetric tensor, with components $L_{ij} = x_ip_j - x_jp_i$. Even in three dimensions, the angular momentum $\boldsymbol{L}$ isn't a vector (in the same sense that $\boldsymbol{r}$ is a vector), but a pseudovector.

11.12 Covariant differentiation of tensors

Is the derivative of a tensor a tensor? How would you answer such a question? Hopefully you're asking, "Does it transform like one?" Consider $\partial T^r/\partial x^b \equiv F^r_b$. We've written F^r_b in tensor notation, but is it a tensor? How does it transform? Differentiate Eq. (11.24), the transformation equation for contravariant vectors:

$$\frac{\partial T^{l'}}{\partial x^{a'}} = \frac{\partial x^b}{\partial x^{a'}}\frac{\partial}{\partial x^b}(A^{l'}_r T^r) = A^b_{a'}A^{l'}_r\frac{\partial T^r}{\partial x^b} + A^b_{a'}T^r\frac{\partial A^{l'}_r}{\partial x^b}. \tag{11.96}$$

Partial derivatives of tensors are not tensors

Equation (11.96) is not in the form of homogeneous transformations satisfied by tensors. *Derivatives of tensors are not tensors.* Only if the second term on the right of Eq. (11.96) is identically zero is the derivative of a tensor a tensor. If $\partial A^l_r/\partial x_b \equiv 0$ in Eq. (11.96), then we'd have that the partial derivative of a type $(1,0)$ tensor is a type $(1,1)$ tensor. For tensors of arbitrary rank, if not for those inhomogeneous terms, the derivative of a type (k,l) tensor would be a type $(k,l+1)$ tensor. The terms in question, $\partial A^{l'}_r/\partial x^b$, represent the *second derivatives* of the coordinate transformation equations. Under *strictly linear* coordinate transformations these terms vanish.[57]

This development is a setback. Physics is chock-full of equations involving derivatives, and we want tensors to help us write equations in covariant form. We must find a *generalized derivative* such that derivatives of tensors transform as tensors under general coordinate transformations, yet which reduces to the partial derivative for linear transformations. As an aside, geometries that can be "covered" by coordinate systems connected by linear transformations are such that the metric tensor consists of a collection of constants. What we're describing is a *flat* versus a curved geometry. Geometries that are globally curved (such as the surface of a sphere) can *locally* be treated as flat (such as a small region of the surface of Earth). At any point of a curved geometry, local coordinate systems can be found such that the metric tensor is constant within a sufficiently small neighborhood of that point. The generalized derivative we seek, the *covariant derivative*, is such that the derivative of the metric tensor is zero at a point.[58] As we'll see, *the metric tensor plays a significant role in formulating the covariant derivative.*

The covariant derivative

We adopt conventional notation and denote the covariant derivative (yet to be found) with respect to the ith coordinate as ∇_i (not to be confused with the gradient operator, $\boldsymbol{\nabla}$).[59] It's also conventional to denote the partial derivative $\partial/\partial x^i$ as ∂_i, a custom we now adopt. The covariant derivative is defined by a set of requirements.[60] Because scalar fields transform without the transformation matrix A^i_j (see Eq. (11.14)), and because it's the derivative of A^i_j that causes all the trouble in defining derivatives

[57] The Lorentz transformation of special relativity is linear, and thus *within the confines of special relativity, derivatives of tensors are tensors.* This is not the case for more general coordinate transformations, such as encountered in the general theory of relativity.

[58] The metric tensor is not a constant globally, but only in a sufficiently small neighborhood of a point.

[59] What we will write as $\nabla_i T^j$ is also written $T^j_{;i}$ – with a semicolon – another notational scheme that generally cannot be deciphered by students at the back of the room.

[60] We saw in Chapter 1 a similar situation where vector spaces and the inner product are defined by a set of requirements. The covariant derivative is any construct meeting the requirements.

of tensors, one simple requirement is that ∇_i be the same as the partial derivative when acting on scalar fields, $\nabla_i \phi = \partial_i \phi$. Partial derivatives have two familiar properties that we abstract for ∇_i – derivatives are linear operators and they obey the product rule. The covariant derivative is defined as an operator that satisfies five requirements [46, p. 31]:

1. ∇ is linear, $\nabla(\alpha \mathbf{S} + \beta \mathbf{T}) = \alpha \nabla \mathbf{S} + \beta \nabla \mathbf{T}$ (for tensors $\mathbf{S}, \mathbf{T}$ and scalars α, β);
2. ∇ satisfies the product rule, $\nabla(\mathbf{S} \bigotimes \mathbf{T}) = (\nabla \mathbf{S}) \bigotimes \mathbf{T} + \mathbf{S} \bigotimes (\nabla \mathbf{T})$;
3. ∇ commutes with contractions, $\nabla_i (T^j_{\ jk}) = (\nabla \mathbf{T})^{\ \ j}_{ijk}$;
4. For any scalar field ϕ, $\nabla_i \phi = \partial_i \phi$;
5. For any scalar field ϕ, $\nabla_i \nabla_j \phi = \nabla_j \nabla_i \phi$. (no torsion property)

The first two properties would be expected of any derivative. The third property indicates that $\nabla_i \mathbf{T}$ is a new tensor of higher rank – adds an additional covariant index – and as such is a generalization of the gradient. The last requirement imposes a level of smoothness; we'll comment on the no-torsion property at a later point.

connection coefficients

These requirements are satisfied by defining the covariant derivative of a vector component T^j as

$$\nabla_i T^j \equiv \partial_i T^j + \Gamma^j_{ik} T^k, \tag{11.97}$$

for a suitably defined set of *connection coefficients* $\{\Gamma^j_{ik}\}$. That is, the covariant derivative is the partial derivative, plus "something else," whatever is required to make it transform like a tensor. Note from Eq. (11.97) that the covariant derivative of T^j requires *all other* components, T^k. The linearity requirement is clearly met by the form of Eq. (11.97). Equation (11.97) defines the covariant derivative *only* for contravariant vector components. We'll see shortly how the product rule is satisfied and how the covariant derivative of covariant vector components is defined.

What do we require of the connection coefficients $\{\Gamma^j_{ik}\}$? If the inhomogeneous term in Eq. (11.96) were absent, the derivative of T^j, a type $(1, 0)$ tensor, would be a type $(1, 1)$ tensor. Thus, we want $\nabla_i T^j$ to transform as a type $(1, 1)$ tensor, $\nabla_{m'} T^{n'} = A^m_{m'} A^{n'}_n \nabla_m T^n$. From Eq. (11.97),

$$\begin{aligned} \nabla_{m'} T^{n'} \equiv \partial_{m'} T^{n'} + \Gamma^{n'}_{m'a'} T^{a'} &\overset{\text{what we want}}{\overset{\downarrow}{=}} A^m_{m'} A^{n'}_n \nabla_m T^n \\ &\equiv A^m_{m'} A^{n'}_n (\partial_m T^n + \Gamma^n_{ma} T^n). \end{aligned} \tag{11.98}$$

Use the transformation properties of $\partial_{m'}$ and $T^{n'}$ in Eq. (11.98) to effect a cancelation:

$$\Gamma^{n'}_{m'a'} T^{a'} \overset{\text{what we want}}{\overset{\downarrow}{=}} A^m_{m'} A^{n'}_n \Gamma^n_{ma} T^a - A^b_{m'} T^a \partial_b A^{n'}_a .$$

Substitute into the left of this equation, the transformation property $T^{a'} = A^{a'}_{a} T^a$:

$$(A^{a'}_{a}\Gamma^{n'}_{m'a'} - A^{m}_{m'}A^{n'}_{n}\Gamma^{n}_{ma} + A^{b}_{m'}\partial_b A^{n'}_{a})T^a \overset{\text{what we want}}{\overset{\downarrow}{=}} 0.$$

Because T^a is arbitrary, we require for each a that

$$A^{a'}_{a}\Gamma^{n'}_{m'a'} - A^{m}_{m'}A^{n'}_{n}\Gamma^{n}_{ma} + A^{b}_{m'}\partial_b A^{n'}_{a} \overset{\text{what we want}}{\overset{\downarrow}{=}} 0.$$

Note that a is a *free index*. Multiply by $A^{a}_{b'}$ and contract over a. For $\nabla_i T^j$ to transform as a type $(1, 1)$ tensor, the connection coefficients must transform as

$$\Gamma^{n'}_{m'b'} \overset{\text{what we want}}{\overset{\downarrow}{=}} A^{m}_{m'}A^{n'}_{n}A^{a}_{b'}\Gamma^{n}_{ma} - A^{a}_{b'}A^{b}_{m'}\partial_b A^{n'}_{a}. \tag{11.99}$$

Any set of N^3 quantities $\{\Gamma^{n}_{mb}\}$ that transform as in Eq. (11.99) is a valid set of connection coefficients. Note that they do not constitute a tensor! The right side of Eq. (11.97) consists of a sum of two nontensorial objects, such that their sum is a tensor.

Example. Show explicitly that $\nabla_m T^n$ transforms as a type $(1, 1)$ tensor. From Eqs. (11.97), (11.96), and (11.99), (and using orthogonality, $A^{a}_{a'}A^{a'}_{b} = \delta^a_b$)

$$\begin{aligned}
\nabla_{m'}T^{n'} &= \partial_{m'}T^{n'} + \Gamma^{n'}_{m'a'}T^{a'} \\
&= A^{m}_{m'}A^{n'}_{n}\partial_m T^n + A^{b}_{m'}(\partial_b A^{n'}_{r})T^r + (A^{n'}_{n}A^{m}_{m'}A^{a}_{a'}\Gamma^{n}_{ma} - A^{r}_{a'}A^{b}_{m'}(\partial_b A^{n'}_{r}))A^{a'}_{s}T^s \\
&= A^{m}_{m'}A^{n'}_{n}(\partial_m T^n + A^{a}_{a'}A^{a'}_{s}\Gamma^{n}_{ma}T^s) + A^{b}_{m'}(\partial_b A^{n'}_{r})(T^r - A^{r}_{a'}A^{a'}_{s}T^s) \\
&= A^{m}_{m'}A^{n'}_{n}(\partial_m T^n + \Gamma^{n}_{ma}T^a) = A^{m}_{m'}A^{n'}_{n}\nabla_m T^n.
\end{aligned}$$

Covariant derivative of covariant vectors

Equation (11.97) defines the covariant derivative of a contravariant vector component. What is the covariant derivative of a covariant vector component? Here's where the product rule plays a part. Use the fact that for a scalar field ϕ, $\nabla_i\phi = \partial_i\phi$. Build a scalar from the inner product between a covariant and a contravariant vector, $\phi \equiv \omega_n T^n$. Thus, $\nabla_i(\omega_n T^n) = \partial_i(\omega_n T^n)$, which, using Eq. (11.97), and the product rule is equivalent to $T^n(\nabla_i\omega_n - \partial_i\omega_n + \Gamma^{a}_{in}\omega_a) = 0$ (show this). Because T^n is arbitrary, we have the covariant derivative of a covariant vector component:

$$\nabla_i\omega_n = \partial_i\omega_n - \Gamma^{a}_{in}\omega_a. \tag{11.100}$$

Note that the covariant derivative of ω_n requires all components ω_a. Through the systematic use of Eqs. (11.97) and (11.100), rules such as the following can be derived

$$\nabla_i(T^j\omega_k) = \partial_i(T^j\omega_k) + \Gamma^j_{ib}\omega_k T^b - \Gamma^a_{ik}\omega_a T^j$$
$$\nabla_i(\omega_j\omega_k) = \partial_i(\omega_j\omega_k) - \Gamma^a_{ik}\omega_j\omega_a - \Gamma^a_{ij}\omega_k\omega_a. \tag{11.101}$$

In general, there is a + sign for every contravariant index and a − sign for every covariant index. The second result in Eq. (11.101) implies that

$$\nabla_i g_{ab} = \partial_i g_{ab} - \Gamma^k_{ib} g_{ak} - \Gamma^k_{ia} g_{bk}. \tag{11.102}$$

The action of the covariant derivative is determined once the connection coefficients have been specified, as Eqs. (11.97), (11.100), and (11.101) show. Beside their transformation property, however (Eq. (11.99)), the connection coefficients are unconstrained and must be *imposed* through some other consideration. The message here is that it's up to us to *choose* the connection coefficients. Every allowed set of connection coefficients determines a type of covariant derivative.

The torsion tensor

The connection coefficients are not tensors. Remarkably, however, the antisymmetric combination

$$T^i_{jk} \equiv \Gamma^i_{jk} - \Gamma^i_{kj} \tag{11.103}$$

does transform like a tensor. It's straightforward to show that T^i_{jk} transforms as a tensor:

$$\Gamma^{n'}_{m'b'} - \Gamma^{n'}_{b'm'} = A^m_{m'} A^{n'}_n A^a_{b'} (\Gamma^n_{ma} - \Gamma^n_{am}). \tag{11.104}$$

The quantities $\{T^i_{jk}\}$ comprise the *torsion tensor.* The name arises because $T^i_{jk} \neq 0$ implies that a "twist" exists at every point of the geometry (we don't show this in this book). It's standard practice to stipulate that the torsion tensor is zero,[61] which is achieved by requiring the connection coefficients to be *symmetric* in the lower indices, $\Gamma^i_{jk} = \Gamma^i_{kj}$. The no-torsion requirement constrains our choices of possible connection coefficients.

Metric compatibility and the Christoffel symbols

We seem to know a lot about the covariant derivative, *except for what it is*! What we don't have is an explicit expression for the connection

[61] There are physical reasons behind the no-torsion requirement. For one, the *equivalence principle* holds that a small enough region of spacetime is flat, i.e. free of gravity. A nonzero torsion tensor would imply that even at a microscopic scale, there is a "twist" built into spacetime. Unless there is a physical *source* of that twist, the torsion tensor should be set to zero.

coefficients. We can find one through the following derivation. Using Eq. (11.102) it can be shown that,

$$\nabla_i g_{ab} + \nabla_a g_{ib} - \nabla_b g_{ia} = \partial_i g_{ab} + \partial_a g_{ib} - \partial_b g_{ia} \tag{11.105}$$
$$+ g_{ik}(\Gamma^k_{ab} - \Gamma^k_{ba}) + g_{ak}(\Gamma^k_{bi} - \Gamma^k_{ib}) - g_{bk}(\Gamma^k_{ia} + \Gamma^k_{ai}).$$

Equation (11.105) simplifies when we invoke the symmetry of the connection coefficients, $\Gamma^k_{ij} = \Gamma^k_{ji}$. As previously mentioned, a flat geometry (or a locally flat geometry) has a metric tensor that's constant over some region, so that we require[62] $\nabla_i g_{ab} = 0$, a condition known as *metric compatibility*. With these assumptions, Eq. (11.105) implies that

metric compatibility

$$2g_{bk}\Gamma^k_{ia} = \partial_i g_{ab} + \partial_a g_{ib} - \partial_b g_{ia}.$$

Noting that b is a free index, multiply by g^{bl} and contract over b. We find:

$$\Gamma^l_{ia} = \frac{1}{2} g^{lb}(\partial_i g_{ab} + \partial_a g_{ib} - \partial_b g_{ia}). \tag{11.106}$$

The connection coefficients specified by Eq. (11.106) are the *Christoffel symbols*. Note the symmetry in the lower indices. For an N-dimensional space, there would in general be N^3 connection coefficients. Because of symmetry, however, the number of independent Christoffel symbols is reduced to $N^2(N+1)/2$. It can be shown (laboriously) that they transform as required by Eq. (11.99). The Christoffel symbols are *unique*[63] – there is one and only one set of torsion-free connection coefficients satisfying $\nabla_i g_{jk} = 0$.

Christoffel symbols

The Christoffel symbols are not tensors (even though we use tensor notation, which is why they're called symbols) – they don't transform as tensors, Eq. (11.99). Another way to see that they're not tensors is to consider the Cartesian coordinate system, for which $\Gamma^i_{jk} = 0$. If the Γ^i_{jk} *were* components of a tensor, we know that a tensor equation, if true in one coordinate system holds in all coordinate systems, and thus, we would have $\Gamma^i_{jk} = 0$ in every coordinate system.

Example. Evaluate the Christoffel symbols in plane polar coordinates. Use

$$[g_{ij}] = \begin{pmatrix} 1 & 0 \\ 0 & r^2 \end{pmatrix} \qquad [g^{ij}] = \begin{pmatrix} 1 & 0 \\ 0 & r^{-2} \end{pmatrix}.$$

Of the 2^3 possible symbols in this coordinate system, only three are nonzero:

---

[62]The covariant derivative of a constant tensor is zero. From its defining requirements, in particular that it commutes with contractions, $\nabla_i \delta^a_b = \nabla_i(\delta^a_c \delta^c_b) = \delta^a_c \nabla_i \delta^c_b + \delta^c_b \nabla_i \delta^a_c = 2\nabla_i \delta^a_b$. Thus, $\nabla_i \delta^a_b = 0$.

[63]The uniqueness of the Christoffel symbols is known as the Fundamental Theorem of Riemannian Geometry, a result that will not be proven here, although it's not difficult to show.

$$\Gamma^\theta_{r\theta} = \Gamma^\theta_{\theta r} = \frac{1}{r} \qquad \Gamma^r_{\theta\theta} = -r \tag{11.107}$$

(and thus $\Gamma^r_{rr} = \Gamma^r_{\theta r} = \Gamma^r_{r\theta} = \Gamma^\theta_{rr} = \Gamma^\theta_{\theta\theta} = 0$). Let's derive these results using Eq. (11.106),

$$\begin{aligned}
\Gamma^r_{r\theta} =& \Gamma^r_{\theta r} = \frac{1}{2}g^{rr}\left[\frac{\partial g_{r\theta}}{\partial r} + \frac{\partial g_{rr}}{\partial \theta} - \frac{\partial g_{r\theta}}{\partial r}\right] + \frac{1}{2}g^{r\theta}\left[\frac{\partial g_{\theta\theta}}{\partial r} + \frac{\partial g_{\theta r}}{\partial \theta} - \frac{\partial g_{r\theta}}{\partial \theta}\right] = 0 \\
\Gamma^r_{\theta\theta} =& \frac{1}{2}g^{rr}\left[\frac{\partial g_{r\theta}}{\partial \theta} + \frac{\partial g_{r\theta}}{\partial \theta} - \frac{\partial g_{\theta\theta}}{\partial r}\right] + \frac{1}{2}g^{r\theta}\left[\frac{\partial g_{\theta\theta}}{\partial \theta} + \frac{\partial g_{\theta\theta}}{\partial \theta} - \frac{\partial g_{\theta\theta}}{\partial \theta}\right] \\
=& -\frac{1}{2}g^{rr}\frac{\partial g_{\theta\theta}}{\partial r} = -\frac{1}{2}(1)(2r) = -r.
\end{aligned}$$

Special cases

While derivatives of tensors are not tensors, certain combinations of derivatives *do* transform as tensors. This is especially so for the curl and the divergence, two fundamental quantities in the description of vector fields (Helmholtz's theorem, Appendix A).

Covariant curl

Using Eq. (11.100), the antisymmetric combination

$$\nabla_i \omega_j - \nabla_j \omega_i = \partial_i \omega_j - \partial_j \omega_i, \tag{11.108}$$

a type $(0, 2)$ tensor, generates an expression for the curl valid for *arbitrary* coordinate systems.[64] Equation (11.108) applies *only* to covariant vectors, however – it does *not* generalize to the curl of contravariant vectors. Using Eq. (11.97),

$$\nabla_i T^j - \nabla_j T^i = \partial_i T^j - \partial_j T^i + (\Gamma^j_{ik} - \Gamma^i_{jk})T^k. \tag{11.109}$$

Covariant divergence

The covariant divergence, defined as $\nabla_k T^k$, generalizes the divergence in orthogonal coordinate systems and is given by the following expression:

$$\nabla_k T^k = \frac{1}{\sqrt{g}}\partial_i(\sqrt{g}T^i), \tag{11.110}$$

where g is the determinant of the metric tensor.

---

[64]For $N = 3$, the three independent components of the antisymmetric tensor $T_{ij} \equiv \partial_i \omega_j - \partial_j \omega_i$ can be placed in one-to-one correspondence with the three components of the cross product (see Exercise 11.19). For $N > 3$, the cross product does not exist *as an* N*-dimensional vector* but does survive as the antisymmetric tensor T_{ij}.

To derive Eq. (11.110), we require a formula for the derivative of a functional determinant, the determinant of a square matrix with elements that are differentiable functions. Equation (1.39), the expansion in minors, is just what we need. The determinant of an $N \times N$ matrix M with elements M_{ij} (we temporarily revert to the usual indexing scheme for matrices), denoted m, is, in terms of its expansion in minors,

$$m \equiv \det M = \sum_{j=1}^{N} M_{ij}(-1)^{i+j} A_{ij}, \qquad (i = \text{any row}) \qquad (11.111)$$

where A_{ij} is the (i, j) minor, the determinant of the $(N-1) \times (N-1)$ matrix obtained by eliminating the ith row and jth column of M. The expansion in minors facilitates taking the derivative because the minor does not depend on M_{ij}. From Eq. (11.111),

$$\frac{\partial m}{\partial M_{ij}} = (-1)^{i+j} A_{ij}. \qquad (11.112)$$

From Eq. (1.42), the matrix inverse has elements $(M^{-1})_{ij} = m^{-1}(-1)^{i+j} A_{ji}$. Thus, $A_{ij} = (-1)^{i+j} m (M^{-1})_{ji}$ (note the placement of indices), and hence, from Eq. (11.112),

$$\frac{\partial m}{\partial M_{ij}} = m(M^{-1})_{ji}. \qquad (11.113)$$

The total differential of a determinant, the expression we seek, is therefore

$$\mathrm{d}m \equiv \sum_{i=1}^{n} \sum_{j=1}^{n} \frac{\partial m}{\partial M_{ij}} \, \mathrm{d}M_{ij} = m \sum_{i=1}^{n} \sum_{j=1}^{n} (M^{-1})_{ji} \, \mathrm{d}M_{ij}. \qquad (11.114)$$

We can now compute the covariant divergence. From Eq. (11.97),

$$\nabla_k T^k = \partial_k T^k + \Gamma^k_{kl} T^l. \qquad (11.115)$$

We find, using Eq. (11.106),

$$\Gamma^k_{kl} = \frac{1}{2} g^{ka} \partial_l g_{ak}, \qquad \text{(true for the Christoffel symbols)} \qquad (11.116)$$

where the second and third terms in the expression of the Christoffel symbol cancel (through index substitution). Apply Eq. (11.114) to the determinant of the metric tensor: $\mathrm{d}g = g g^{ji} \, \mathrm{d}g_{ij}$, where we've used $(g^{-1})_{ji} \equiv g^{ji}$ (back to the summation convention). The total differential of any element g_{ak} is $\mathrm{d}g_{ak} = (\partial_l g_{ak}) \, \mathrm{d}x^l$, and likewise for the determinant, $\mathrm{d}g = (\partial_l g) \, \mathrm{d}x^l$. Thus, using Eq. (11.114),

$$\frac{\partial g}{\partial x^l} = g g^{ka} \frac{\partial g_{ak}}{\partial x^l}.$$

Combining with Eq. (11.116), we have the useful result

$$\Gamma^k_{kl} = \frac{1}{2g}\frac{\partial g}{\partial x^l} = \frac{\partial}{\partial x^l}\ln\sqrt{g}. \tag{11.117}$$

Combine Eqs. (11.117) and (11.115) and we're done; we have Eq. (11.110).

Example. The covariant divergence in spherical coordinates is, from Eq. (11.110), and using $g = r^4 \sin^2\theta$,

$$\begin{aligned}\frac{1}{\sqrt{g}}\frac{\partial}{\partial x^i}(\sqrt{g}A^i) &= \frac{1}{r^2\sin\theta}\left[\frac{\partial}{\partial r}(r^2\sin\theta A^r) + \frac{\partial}{\partial\theta}(r^2\sin\theta A^\theta) + \frac{\partial}{\partial\phi}(r^2\sin\theta A^\phi)\right]\\ &= \frac{1}{r^2}\frac{\partial}{\partial r}(r^2A^r) + \frac{1}{\sin\theta}\frac{\partial}{\partial\theta}(\sin\theta A^\theta) + \frac{\partial A^\phi}{\partial\phi} = \boldsymbol{\nabla}\cdot\boldsymbol{A},\end{aligned}$$

the usual expression for the divergence in spherical coordinates.

What does Γ^i_{jk} tell us?

Under a local change of coordinates $x^i \to x^i + \mathrm{d}x^i$, we expect there to be, in general, an associated change in basis vectors $\boldsymbol{e}_j \to \boldsymbol{e}_j + \mathrm{d}\boldsymbol{e}_j$. Because the coordinate basis vectors are tangent to coordinate curves, as the coordinate grid changes, so do the basis vectors. For small changes, $\mathrm{d}\boldsymbol{e}_j$ can be expressed as a linear combination of the basis vectors $\{\boldsymbol{e}_i\}$, where the expansion coefficients are proportional to the coordinate changes $\{\mathrm{d}x^k\}$:

$$\mathrm{d}\boldsymbol{e}_j = (\mathrm{d}x^k\gamma^i_{kj})\boldsymbol{e}_i. \tag{11.118}$$

That way, as $\mathrm{d}x^k \to 0$, $\mathrm{d}\boldsymbol{e}_j \to 0$. Two covariant vectors are related in general by a second-rank mixed tensor, as in $A_i = F^j_i B_j$. Here, the proportionality factor is itself proportional to the change in coordinates, $F^j_i = \mathrm{d}x^k\gamma^j_{ik}$; hence the three-index symbol. It's not an accident we've chosen the symbol γ in Eq. (11.118) – we now show that $\gamma^i_{jk} = \Gamma^i_{jk}$.

Assume the total differential of $\mathrm{d}\boldsymbol{e}_j$ can be written

$$\mathrm{d}\boldsymbol{e}_j = \partial_k\boldsymbol{e}_j\,\mathrm{d}x^k. \tag{11.119}$$

Equating Eqs. (11.118) and (11.119), we have the rate of change of $\boldsymbol{e}_j$ along the coordinate curve x^k is given by

$$\partial_k\boldsymbol{e}_j = \gamma^i_{kj}\boldsymbol{e}_i, \tag{11.120}$$

which we note involves *all* vectors of the basis.

We first show that Γ^i_{jk} is symmetric. In a coordinate basis, $\boldsymbol{e}_i = \partial_i$ (Section 11.2.3), and partial derivatives commute, $\partial_k\partial_j = \partial_j\partial_k$, implying that $\partial_k\boldsymbol{e}_j = \partial_j\boldsymbol{e}_k$. From Eq. (11.120), therefore, $\gamma^i_{kj}\boldsymbol{e}_i = \gamma^i_{jk}\boldsymbol{e}_i$, and thus γ^i_{kj} is symmetric. We now show that $\gamma^i_{kj} = \Gamma^i_{kj}$. Consider, from Eq. (11.46), that the total derivative of an element of the metric tensor, $\mathrm{d}g_{ij} = \boldsymbol{e}_i \cdot \mathrm{d}\boldsymbol{e}_j + \mathrm{d}\boldsymbol{e}_i \cdot \boldsymbol{e}_j$, and thus, using Eq. (11.118),

$$\begin{aligned}\mathrm{d}g_{ij} &= \boldsymbol{e}_i \cdot \mathrm{d}\boldsymbol{e}_j + \mathrm{d}\boldsymbol{e}_i \cdot \boldsymbol{e}_j = \boldsymbol{e}_i \cdot (\mathrm{d}x^k \gamma^m_{kj})\boldsymbol{e}_m + (\mathrm{d}x^k \gamma^m_{ki})\boldsymbol{e}_m \cdot \boldsymbol{e}_j \\ &= (g_{im}\gamma^m_{kj} + g_{mj}\gamma^m_{ki})\,\mathrm{d}x^k \equiv \frac{\partial g_{ij}}{\partial x^k}\,\mathrm{d}x^k. \end{aligned} \tag{11.121}$$

From Eq. (11.102), because $\nabla_k g_{ij} = 0$, the Christoffel symbols have the property that

$$\partial_k g_{ij} = \Gamma^l_{ki} g_{lj} + \Gamma^l_{kj} g_{il}. \tag{11.122}$$

Subtracting Eqs. (11.121) and (11.122),

$$0 = (\gamma^m_{kj} - \Gamma^m_{kj})g_{im} + (\gamma^m_{ki} - \Gamma^m_{ki})g_{jm}. \tag{11.123}$$

With an appeal to the uniqueness of the Christoffel symbols, we have that $\gamma^m_{kj} = \Gamma^m_{kj}$.

Combining Eqs. (11.118) and (11.120) with $\gamma^m_{kj} = \Gamma^m_{kj}$,

$$\mathrm{d}\boldsymbol{e}_j = \Gamma^i_{jk}\,\mathrm{d}x^k \boldsymbol{e}_i = \frac{\partial \boldsymbol{e}_j}{\partial x^k}\,\mathrm{d}x^k. \tag{11.124}$$

By making use of Eq. (11.31), we have therefore

$$\Gamma^i_{jk} = \boldsymbol{e}^i \cdot (\partial_j \boldsymbol{e}_k). \tag{11.125}$$

Thus, we can interpret the Christoffel symbols: They describe how basis vectors change along coordinate curves; Γ^i_{jk} is the rate of change of $\boldsymbol{e}_k$ along coordinate curve associated with x^j, projected onto the coordinate surface associated with $\boldsymbol{e}^i$.

Example. In polar coordinates $\Gamma^r_{\theta\theta} = -r$, Eq. (11.107). The rate of change of $\boldsymbol{e}_\theta = r\hat{\boldsymbol{\theta}}$ in the θ direction is $\mathrm{d}\boldsymbol{e}_\theta = -r\hat{\boldsymbol{r}}\,\mathrm{d}\theta$ and thus, $\Gamma^r_{\theta\theta} = -r$ (projection onto the r direction).

Example. Illustrate Eq. (11.124) in polar coordinates using the Christoffel symbols:

$$\begin{aligned}\mathrm{d}\boldsymbol{e}_r &= \Gamma^k_{rj}\boldsymbol{e}_k\,\mathrm{d}x^j = \Gamma^r_{rj}\boldsymbol{e}_r\,\mathrm{d}x^j + \Gamma^\theta_{rj}\boldsymbol{e}_\theta\,\mathrm{d}x^j = \Gamma^r_{rr}\boldsymbol{e}_r\,\mathrm{d}r + \Gamma^r_{r\theta}\boldsymbol{e}_r\,\mathrm{d}\theta \\ &\quad + \Gamma^\theta_{rr}\boldsymbol{e}_\theta\,\mathrm{d}r + \Gamma^\theta_{r\theta}\boldsymbol{e}_\theta\,\mathrm{d}\theta = \Gamma^\theta_{r\theta}\boldsymbol{e}_\theta\,\mathrm{d}\theta = \frac{1}{r}\boldsymbol{e}_\theta\,\mathrm{d}\theta = \frac{1}{r}(r\hat{\boldsymbol{\theta}})\,\mathrm{d}\theta = \hat{\boldsymbol{\theta}}\,\mathrm{d}\theta,\end{aligned}$$

where we've used Eq. (11.107). For the other basis vector we have, keeping only the nonzero terms,

$$\mathrm{d}\boldsymbol{e}_\theta = \Gamma^r_{\theta\theta}\boldsymbol{e}_r \,\mathrm{d}\theta + \Gamma^\theta_{\theta r}\boldsymbol{e}_\theta \,\mathrm{d}r = -r\boldsymbol{e}_r \,\mathrm{d}\theta + \frac{1}{r}\boldsymbol{e}_\theta \,\mathrm{d}r = -r\hat{\boldsymbol{r}} \,\mathrm{d}\theta + \hat{\boldsymbol{\theta}} \,\mathrm{d}r, \tag{11.126}$$

where we've used $\boldsymbol{e}_r = \hat{\boldsymbol{r}}$ and $\boldsymbol{e}_\theta = r\hat{\boldsymbol{\theta}}$. Using $\mathrm{d}\boldsymbol{e}_\theta = \mathrm{d}(r\hat{\boldsymbol{\theta}}) = r\,\mathrm{d}\hat{\boldsymbol{\theta}} + \hat{\boldsymbol{\theta}}\,\mathrm{d}r$, we see that Eq. (11.126) is the same as $\mathrm{d}\hat{\boldsymbol{\theta}} = -\mathrm{d}\theta\hat{\boldsymbol{r}}$, a familiar result.

Equation (11.125) is often given as the *definition* of the connection coefficients; Eq. (11.106) (the expression for the Christoffel symbols) is then obtained when metric compatibility is imposed. We have inverted that order, postulating the connection coefficient in Eq. (11.97), arriving at Eq. (11.106) through metric compatibility, and then motivating Eq. (11.125) as reasonable.

Summary

This chapter provided a first look at tensors, but we've just scratched the surface. The next step would be to see how tensors are used in analytical mechanics, relativistic mechanics, and the mechanics of deformable media. Tensors are difficult to define standing on one foot – abstract objects with invariant properties that are independent of the coordinate system used to describe them. Tensors are represented in a given coordinate system by a set of functions, components, just as vectors are determined in a given coordinate system by a set of components, and tensors are generalizations of vectors. Two classes of invariants exist (besides scalars), contravariant and covariant, defined by the properties under coordinate transformations of line element vectors $\mathrm{d}\boldsymbol{r}$ and the gradients of scalar fields, $\boldsymbol{\nabla}\,\phi$. One way to define tensors, therefore (the traditional approach in physics), is through their transformation properties, what we covered in Section 11.2. Studying *just* the transformation properties, however, leaves one with the lifeless view that tensors consist of quantities that transform according to certain rules. Another view of tensors, one that's not circular as in a tensor is anything that acts like a tensor, is to consider their role as operators. Just as vectors can be seen as operators, as the inner product between vectors $\boldsymbol{A}\cdot\boldsymbol{B}$ produces a scalar, one cannot fully appreciate what tensors are without consideration of their roles as operators, covered in Section 11.8. Tensors are multilinear generalizations of vectors as geometric objects (retain their identity under coordinate transformations) that map products of vectors to scalars, and other tensors to tensors. Just this one sentence requires the rest of the chapter to understand!

Exercises

11.1. The matrix algebra of Chapter 1 was largely developed to describe the action of a linear operator on a vector. When vectors are themselves considered as tensors, however, the "operator" quality requires additional structure to adjust to the matrix characterization – that's what dyads try to avoid. Show that the matrix form of dyads (Eq. (11.5)), when operating on vectors, is consistent with the interpretation

$$\boldsymbol{e}_i\boldsymbol{e}_j \cdot \boldsymbol{e}_k = \boldsymbol{e}_i(\boldsymbol{e}_j \cdot \boldsymbol{e}_k) = \boldsymbol{e}_i\delta_{jk}$$

Show, however, that the dyadic interpretation

$$\boldsymbol{e}_i \cdot \boldsymbol{e}_j\boldsymbol{e}_k = (\boldsymbol{e}_i \cdot \boldsymbol{e}_j)\boldsymbol{e}_k = \delta_{ij}\boldsymbol{e}_k$$

requires that the matrix representation of $\boldsymbol{e}_i$ be replaced by its *transpose*.

11.2. Show that $(a_{rst} + a_{str} + a_{srt})x^r x^s x^t = 3a_{rst}x^r x^s x^t$.

11.3. Show that a symmetric second-rank tensor in N dimensions has $N(N+1)/2$ independent components.

11.4. What is the metric tensor for cylindrical coordinates, (ρ, ϕ, z)? Hint: The line element is $\mathrm{d}\boldsymbol{l} = \mathrm{d}\rho\hat{\boldsymbol{\rho}} + \rho\,\mathrm{d}\phi\hat{\boldsymbol{\phi}} + \mathrm{d}z\hat{\boldsymbol{z}}$.

11.5. Simplify the expression $g_{ij}A^iB^j - A^kB_k$.

11.6. Show in polar coordinates that if $\mathrm{d}x^i = \begin{pmatrix} \mathrm{d}r \\ \mathrm{d}\theta \end{pmatrix}$, then $\mathrm{d}x_i = \begin{pmatrix} \mathrm{d}r & r^2\,\mathrm{d}\theta \end{pmatrix}$.

11.7. If $[g_{ij}] = \begin{pmatrix} A & 1 \\ 1 & 0 \end{pmatrix}$, what is $[g^{ij}]$?

11.8. In an N-dimensional space, evaluate $\delta_i^k\delta_k^i$. (Answer: N. Why?)

11.9. Let $\mathbf{F}$ be a second-rank antisymmetric covariant tensor with respect to the basis $\boldsymbol{e}^j$, i.e. $F_{ij} = -F_{ji}$. Show that the contravariant components F^{ij} are also antisymmetric. Show that $\mathbf{F}$ is antisymmetric in any coordinate system, i.e. show $F^{i'j'} = -F^{j'i'}$ and $F_{m'n'} = -F_{n'm'}$. Antisymmetry is thus an intrinsic (coordinate-independent) property.

11.10. Show that $\partial_i\omega_j - \partial_j\omega_i$ transforms as a tensor. Hint: Dummy indices.

11.11. Derive Eq. (11.104). Hint: Dummy indices are dummy indices. Make liberal substitutions of indices to show Eq. (11.104).

11.12. Show that Eq. (11.105) is an identity.

11.13. For an isotropic elastic medium under dynamic stress, at time t the displacement u_i and stress tensor p_{ij} satisfy

$$p_{ij} = c_{ijkl}\left(\frac{\partial u_k}{\partial x_l} + \frac{\partial u_l}{\partial x_k}\right) \quad \text{and} \quad \frac{\partial p_{ij}}{\partial x_j} = \rho\frac{\partial^2 u_i}{\partial t^2}$$

where

$$c_{ijkl} = \lambda\delta_{ij}\delta_{kl} + \eta\delta_{ik}\delta_{jl} + \nu\delta_{il}\delta_{jk}$$

with λ, η, ν, and ρ are constants. Show that both $\boldsymbol{\nabla}\cdot\boldsymbol{u}$ and $\boldsymbol{\nabla}\times\boldsymbol{u}$ satisfy wave equations and find the corresponding wave speeds.

11.14. In a general coordinate system u^i, $i = 1, 2, 3$, in three-dimensional Euclidean space, a volume element is given by

$$\mathrm{d}V = |\boldsymbol{e}_1\,\mathrm{d}u^1 \cdot (\boldsymbol{e}_2\,\mathrm{d}u^2 \times \boldsymbol{e}_3\,\mathrm{d}u^3)|$$

Show that an alternate form for this expression, written in terms of the determinant g of the metric tensor, is given by

$$dV = \sqrt{g}\, du^1\, du^2\, du^3$$

Show that, under a general coordinate transformation to a new coordinate system u'^i, the volume element remains unchanged (i.e. show that it is a scalar quantity).

11.15. By writing down the expression for the square of the infinitesimal arc length $(dl)^2$ in spherical coordinates, find the components g_{ij} of the metric tensor in this coordinate system. Hence, using

$$\nabla \cdot \boldsymbol{v} = \frac{1}{\sqrt{g}} \frac{\partial}{\partial u^j}(\sqrt{g} v^j)$$

find the expression for the divergence of a vector field $\boldsymbol{v}$ in spherical coordinates. Calculate the Christoffel symbols Γ^i_{jk} in this coordinate system. Use either Eq. (11.106) or Eq. (11.125).

11.16. Evaluate the sum $\varepsilon^{ijk}\varepsilon_{jkn}$. Hint: Use Eq. (11.95), and watch the order of the indices. A: $2\delta^i_n$.

11.17. The contraction of an antisymmetric tensor A_{ij} with a symmetric tensor S^{kl} produces zero; the following is a proof. Explain every step in this derivation:

$$A_{ij}S^{ij} = -A_{ji}S^{ij} = -A_{ji}S^{ji} = -A_{ij}S^{ij}.$$

The last step follows by changing the names of dummy indices.

11.18. The Levi-Civita symbol allows one to derive the traditional identities of vector calculus in a quick and efficient manner. The ith component of the curl of a covariant vector $\boldsymbol{A}$ (in three-dimensional space) is written, for arbitrary coordinate systems, as $(\nabla \times \boldsymbol{A})_i = \varepsilon_i^{jk}\partial_j A_k$, where $\varepsilon_i^{jk} = g^{jl}g^{km}\varepsilon_{ilm}$.

(a) Show that ε_i^{jk} is antisymmetric in the upper two indices.

(b) The divergence of the curl would be written, in general tensor notation,

$$\nabla \cdot (\nabla \times \boldsymbol{A}) = \partial^i(\nabla \times \boldsymbol{A})_i = g^{im}\partial_m \varepsilon_i^{jk}\partial_j A_k = \varepsilon^{mjk}\partial_m\partial_j A_k \tag{11.127}$$

Show that the right side of Eq. (11.127) is identically zero. Hint: The terms $\partial_m\partial_j$ are symmetric under $m \leftrightarrow j$, but ε^{mjk} is not.

(c) For the same reason, explain why $(\nabla \times \nabla\phi)_i = \varepsilon_i^{jk}\partial_j\partial_k\phi = 0$.

(d) Derive the *BAC-CAB* rule using the Levi-Civita symbol. Explain how the following accomplishes that task:

$$(\boldsymbol{A} \times \boldsymbol{B} \times \boldsymbol{C})_i = \varepsilon_{ijk}A^j(\boldsymbol{B} \times \boldsymbol{C})^k = \varepsilon_{ijk}\varepsilon^{klm}A^j B_l C_m = ?$$

Hint: Use Eq. (11.95).

11.19. Second-rank, antisymmetric tensors have zero for their diagonal elements, $T^{ii} = 0$ (because $T^{ii} = -T^{ii}$), and therefore, such tensors have $\frac{1}{2}N(N-1)$ independent components (show this). For $N = 3$, the three independent elements of an antisymmetric tensor T^{jk} can be associated with the elements of a vector (actually, pseudovector) P_i through the relation

$$P_i = \frac{1}{2}\varepsilon_{ijk}T^{jk}. \tag{11.128}$$

Show that Eq. (11.128) implies the correspondence

$$P_1 = T^{23} \qquad P_2 = T^{31} \qquad P_3 = T^{12}.$$

It's through this correspondence that we can associate the components of the angular momentum vector with the elements of a second-rank antisymmetric tensor, the angular momentum tensor, $T^{ij} = r^i p^j - r^j p^i$. The association goes the other way: If the components of a vector P_i are known, their association with the elements of an antisymmetric tensor is established through the relation

$$T^{lm} = \varepsilon^{lmi}P_i. \tag{11.129}$$

Show that Eq. (11.129) when combined with Eq. (11.128) generates an identity, $P_i = P_i$. Use the result of Exercise 11.16.

A

Vector calculus

Many physical quantities are "measured" by how much they change (or don't change) as they are observed as time progresses, or position is altered, or both. The important thing here is "rate of change" and that's what differential calculus is all about. For elementary functions $f(x)$ of a single variable, this mathematical tool naturally leads to an ordinary differential equation based description of the world which specifies how the function changes as x either increases or decreases. But when the function $f(x, y, z)$ depends on several variables (as with physical fields), we must consider how the change occurs in various directions. These directionally dependent rates of change are inherently expressed as the components of a multidimensional object – the behavior of which is the subject of this appendix.

A.1 Scalar fields

Scalar-valued fields associate some number with a position in space (or, space and time if the scalar field is time-varying). Examples include temperature distribution in the atmosphere and sound intensity in an auditorium.

A.1.1 The directional derivative

For scalar-valued fields $f(x, y, z)$, we know that the partial derivatives $\partial f/\partial x$, $\partial f/\partial y$, and $\partial f/\partial z$ describe the rate of change of $f(x, y, z)$ in the directions of the coordinate axes. But more generally useful is an understanding of the rate of change of $f(x, y, z)$ in arbitrary directions.

Consider the unit vector $\hat{\boldsymbol{u}} = u_x\hat{\boldsymbol{x}} + u_y\hat{\boldsymbol{y}} + u_z\hat{\boldsymbol{z}}$ which defines some (prespecified) direction. The line going through (x, y, z) in the direction $\hat{\boldsymbol{u}}$ has parametric representation $(x + tu_x, y + tu_y, z + tu_z)$, where $t \in \mathbb{R}$.

Mathematical Methods in Physics, Engineering, and Chemistry, First Edition.
Brett Borden and James Luscombe.

The *directional derivative* of $f(x, y, z)$ in the direction $\hat{\boldsymbol{u}}$ is denoted by $D_{\hat{\boldsymbol{u}}} f$ and is given by

$$\begin{aligned} D_{\hat{\boldsymbol{u}}} f(x, y, z) &= \lim_{t \to 0} \frac{f(x + tu_x, y + tu_y, z + tu_z) - f(x, y, z)}{t} \\ &= \frac{\partial f}{\partial x} u_x + \frac{\partial f}{\partial y} u_y + \frac{\partial f}{\partial z} u_z \end{aligned} \tag{A.1}$$

A.1.2 The gradient

It is convenient (in fact, *especially* convenient) to write Eq. (A.1) in the form

$$D_{\hat{\boldsymbol{u}}} f(x, y, z) = \hat{\boldsymbol{u}} \cdot \boldsymbol{\nabla} f(x, y, z)$$

where

definition of the gradient of a scalar field

$$\boldsymbol{\nabla} f(x, y, z) \equiv \frac{\partial f}{\partial x} \hat{\boldsymbol{x}} + \frac{\partial f}{\partial y} \hat{\boldsymbol{y}} + \frac{\partial f}{\partial z} \hat{\boldsymbol{z}}$$

is known as the *gradient of* $f(x, y, z)$ (also denoted as **grad** f). The gradient of a scalar-valued field is a vector-valued field so that a differential description of a scalar field requires a three-fold increase in information.

Note that since $\hat{\boldsymbol{u}}$ is a unit vector[1]:

- The maximum value of $D_{\hat{\boldsymbol{u}}} f(x, y, z)$ occurs in the direction of $\boldsymbol{\nabla} f(x, y, z)$;
- This maximum value is $|\boldsymbol{\nabla} f(x, y, z)|$.

As we will see, it is also convenient to define a vector differential *operator* (the "grad operator") by

$$\boldsymbol{\nabla} \equiv \hat{\boldsymbol{x}} \frac{\partial}{\partial x} + \hat{\boldsymbol{y}} \frac{\partial}{\partial y} + \hat{\boldsymbol{z}} \frac{\partial}{\partial z} \tag{A.2}$$

which operates on functions "to the right."

[1]It's important to note that the direction vector $\hat{\boldsymbol{u}}$ and $\boldsymbol{\nabla} f$ lie in the *domain* of the function – for example, the direction of $\boldsymbol{\nabla} g(x, y)$ is the direction in the x-y plane that leads to the greatest rate of change of $z = g(x, y)$. In contrast, the *constraint equation* $f(x, y, z) = C$ can be thought of as defining a surface in three-dimensions. If we define the *function* $w = f(x, y, z)$, then the gradient $\boldsymbol{\nabla} w$ will have a direction in three-dimensional space. Setting $w = f(x, y, z) = C$ means that $\boldsymbol{\nabla} w = \boldsymbol{\nabla} C = \boldsymbol{0}$ and so the function w does not change as we vary (x, y, z) *on the surface defined by* $f(x, y, z) = C$. Any change in the function $w = f(x, y, z)$ must therefore occur in the direction normal to the surface $f(x, y, z) = C$ and so $\boldsymbol{\nabla} f(x, y, z)$ will be normal to the surface defined by $f(x, y, z) = C$.

A.2 Vector fields

In distinction with scalar fields, vector-valued fields associate a magnitude *and a direction* with positions (and sometimes, time) in space. Examples include atmospheric wind conditions and – very important in physics – force fields (such as gravity).

Since such vector-valued fields are of the form

$$\boldsymbol{F}(x,y,z) = F_x(x,y,z)\hat{\boldsymbol{x}} + F_y(x,y,z)\hat{\boldsymbol{y}} + F_z(x,y,z)\hat{\boldsymbol{z}}$$

where $F_x(x,y,z)$, $F_y(x,y,z)$, and $F_z(x,y,z)$ are *three* (generally unrelated) scalar-valued functions. It follows that, as far as the differential rate-of-change description of vector-valued fields is concerned, we need to deal with three times more information than with scalar fields: instead of tracking only $\partial f/\partial x$, $\partial f/\partial y$, and $\partial f/\partial z$, we need also to know $\partial F_x/\partial x$, $\partial F_y/\partial x$, $\partial F_x/\partial y$, etc. These derivatives form the components of the three gradients $\boldsymbol{\nabla} F_x$, $\boldsymbol{\nabla} F_y$, and $\boldsymbol{\nabla} F_z$ which can be used to form the rows of the 3×3 *Jacobian* matrix of $\boldsymbol{F}$:

$$[\boldsymbol{\nabla} F_x \quad \boldsymbol{\nabla} F_y \quad \boldsymbol{\nabla} F_z]^T = \begin{bmatrix} \partial F_x/\partial x & \partial F_x/\partial y & \partial F_x/\partial z \\ \partial F_y/\partial x & \partial F_y/\partial y & \partial F_y/\partial z \\ \partial F_z/\partial x & \partial F_z/\partial y & \partial F_z/\partial z \end{bmatrix} \tag{A.3}$$

The "problem" now is that in order to track changes in the field, we need to deal with nine different scalar fields.

To get a handle on this increased amount of information, physicists typically attempt to reduce the dimensional complexity of the problem by appealing to matrix invariants and symmetries (in much the same way as examining the scalar magnitude of a multidimensional vector: cf. Section 1.4).

A.2.1 Divergence

An obvious matrix characterization is based on the trace of the Jacobian matrix (A.3) (see Section 1.4.11). This *scalar field* is called the "divergence of $\boldsymbol{F}$" and is denoted

definition of the divergence of a vector field

$$\textbf{div}\ \boldsymbol{F} \equiv \frac{\partial F_x}{\partial x} + \frac{\partial F_y}{\partial y} + \frac{\partial F_z}{\partial z} \tag{A.4}$$

In terms of the grad operator (A.2), the divergence can also be written as

$$\textbf{div}\ \boldsymbol{F} = \boldsymbol{\nabla} \cdot \boldsymbol{F}$$

(which is the more common modern notation).

As we shall see in Section A.4.2, this characteristic property of the matrix (A.3) is very important in physics.

A.2.2 Curl

In Section 1.5, we saw that any *real symmetric* matrix can be diagonalized by representing it in a system of properly rotated coordinate axes. Consequently, when the Jacobian matrix (A.3) is symmetric, the nine descriptive elements of the matrix (A.3) can be reduced to three elements in an "appropriate" coordinate system (known in physics as "principal coordinates"). Moreover, since the trace of a matrix is an invariant under such transformations, it follows that **div** $\boldsymbol{F}$ will be a uniquely important (scalar) characterizer of the vector field $\boldsymbol{F}$ – *but only when the matrix* (A.3) *is symmetric.* The question becomes, then: How symmetric is the Jacobian matrix?

To answer this question, physicists examine the *difference* between the off-diagonal elements: That is, they examine the values of $\partial F_z/\partial y - \partial F_y/\partial z$, $\partial F_x/\partial z - \partial F_z/\partial x$, and $\partial F_y/\partial x - \partial F_x/\partial y$ (and how much they differ from zero). These off-diagonal differences are three distinct pieces of information, and it is useful to represent them in the form of a vector field and give it a name:

definition of the curl of a vector field

$$\mathbf{curl}\ \boldsymbol{F} \equiv \begin{bmatrix} \partial F_z/\partial y - \partial F_y/\partial z \\ \partial F_x/\partial z - \partial F_z/\partial x \\ \partial F_y/\partial x - \partial F_x/\partial y \end{bmatrix} = \left(\frac{\partial F_z}{\partial y} - \frac{\partial F_y}{\partial z}\right)\hat{\boldsymbol{x}} + \left(\frac{\partial F_x}{\partial z} - \frac{\partial F_z}{\partial x}\right)\hat{\boldsymbol{y}} + \left(\frac{\partial F_y}{\partial x} - \frac{\partial F_x}{\partial y}\right)\hat{\boldsymbol{z}} \tag{A.5}$$

In terms of the grad operator (A.2), the curl can also be written as

$$\mathbf{curl}\ \boldsymbol{F} = \boldsymbol{\nabla} \times \boldsymbol{F}$$

A.2.3 The Laplacian

As an example of the usefulness of this "symmetry" analysis of the Jacobian matrix, note that when we are dealing with a *scalar* field f, the (vector field) gradient of f is

$$\boldsymbol{F} = \mathbf{grad}\ f = \frac{\partial f}{\partial x}\hat{\boldsymbol{x}} + \frac{\partial f}{\partial y}\hat{\boldsymbol{y}} + \frac{\partial f}{\partial z}\hat{\boldsymbol{z}}$$

and, owing to the fact that

$$\frac{\partial^2 f}{\partial x \partial y} = \frac{\partial^2 f}{\partial y \partial x} \quad \Longrightarrow \quad \frac{\partial F_y}{\partial x} = \frac{\partial F_x}{\partial y}, \qquad \text{etc.,}$$

the Jacobian matrix for the vector field $\boldsymbol{\nabla} f$ is guaranteed to be symmetric. Consequently, **curl (grad** f**)** $= \mathbf{0}$ and so **div (grad** f**)** is *automatically* a uniquely important characterization of the scalar field $f(x, y, z)$.

The operator **div** (**grad** f) $= \boldsymbol{\nabla} \cdot \boldsymbol{\nabla} f$ is known as the "Laplacian of f" and is denoted by the special symbol[2]

definition of the Laplacian of a scalar field

$$\boldsymbol{\nabla} \cdot \boldsymbol{\nabla} f \equiv \nabla^2 f$$

A.2.4 Vector operator formulae

It's almost always easier to work with vector notation than it is to work with individual vector components (when possible). In order to do this, a variety of (provable by direct application of the appropriate definitions) relations are often tabulated. Common among them are:

1. $\boldsymbol{\nabla}(\phi + \psi) = \boldsymbol{\nabla}\phi + \boldsymbol{\nabla}\psi$
2. $\boldsymbol{\nabla} \cdot (\boldsymbol{A} + \boldsymbol{B}) = \boldsymbol{\nabla} \cdot \boldsymbol{A} + \boldsymbol{\nabla} \cdot \boldsymbol{B}$
3. $\boldsymbol{\nabla} \times (\boldsymbol{A} + \boldsymbol{B}) = \boldsymbol{\nabla} \times \boldsymbol{A} + \boldsymbol{\nabla} \times \boldsymbol{B}$
4. $\boldsymbol{\nabla}(\phi\psi) = \phi \boldsymbol{\nabla}\psi + \psi \boldsymbol{\nabla}\phi$
5. $\boldsymbol{\nabla}(\boldsymbol{A} \cdot \boldsymbol{B}) = \boldsymbol{A} \times (\boldsymbol{\nabla} \times \boldsymbol{B}) + \boldsymbol{B} \times (\boldsymbol{\nabla} \times \boldsymbol{A}) + (\boldsymbol{A} \cdot \boldsymbol{\nabla})\boldsymbol{B} + (\boldsymbol{B} \cdot \boldsymbol{\nabla})\boldsymbol{A}$
6. $\boldsymbol{\nabla} \cdot (\phi \boldsymbol{A}) = \phi \boldsymbol{\nabla} \cdot \boldsymbol{A} + \boldsymbol{A} \cdot \boldsymbol{\nabla}\phi$
7. $\boldsymbol{\nabla} \cdot (\boldsymbol{A} \times \boldsymbol{B}) = \boldsymbol{B} \cdot (\boldsymbol{\nabla} \times \boldsymbol{A}) - \boldsymbol{A} \cdot (\boldsymbol{\nabla} \times \boldsymbol{B})$
8. $\boldsymbol{\nabla} \times (\phi \boldsymbol{A}) = \boldsymbol{\nabla}\phi \times \boldsymbol{A} + \phi \boldsymbol{\nabla} \times \boldsymbol{A}$
9. $\boldsymbol{\nabla} \times (\boldsymbol{A} \times \boldsymbol{B}) = \boldsymbol{A}(\boldsymbol{\nabla} \cdot \boldsymbol{B}) - \boldsymbol{B}(\boldsymbol{\nabla} \cdot \boldsymbol{A}) + (\boldsymbol{B} \cdot \boldsymbol{\nabla})\boldsymbol{A} - (\boldsymbol{A} \cdot \boldsymbol{\nabla})\boldsymbol{B}$
10. $\boldsymbol{\nabla} \times \boldsymbol{\nabla}\phi = \boldsymbol{0}$
11. $\boldsymbol{\nabla} \cdot (\boldsymbol{\nabla} \times \boldsymbol{A}) = 0$
12. $\boldsymbol{\nabla} \cdot \boldsymbol{\nabla}\phi = \boldsymbol{\nabla}^2\phi$
13. $\boldsymbol{\nabla} \times (\boldsymbol{\nabla} \times \boldsymbol{A}) = \boldsymbol{\nabla}(\boldsymbol{\nabla} \cdot \boldsymbol{A}) - \boldsymbol{\nabla}^2 \boldsymbol{A}$

If $\boldsymbol{r} = x\hat{\boldsymbol{x}} + y\hat{\boldsymbol{y}} + z\hat{\boldsymbol{z}}$ is the position vector and $r = |\boldsymbol{r}|$, then it can be shown that

1. $\boldsymbol{\nabla} r = \hat{\boldsymbol{r}}$
2. $\boldsymbol{\nabla} \cdot \boldsymbol{r} = 3$
3. $\boldsymbol{\nabla} \times \boldsymbol{r} = \boldsymbol{0}$
4. $\boldsymbol{\nabla}\left(\frac{1}{r}\right) = -\frac{\hat{\boldsymbol{r}}}{r^2}$
5. $\boldsymbol{\nabla} \cdot \left(\frac{\hat{\boldsymbol{r}}}{r^2}\right) = -\overset{2}{\boldsymbol{\nabla}}\left(\frac{1}{r}\right) = 4\pi\delta(r)$

where $\delta(r)$ is the delta function.

[2]Note that, despite this conventional notation, ∇^2 is a linear operator

A.3 Integration

A.3.1 Line integrals

A curve C is said to be "smooth" if it has a parametric representation $x = f(t)$, $y = g(t)$, and $z = h(t)$ such that f', g', and h' are continuous and not simultaneously 0. From elementary calculus, we know that when C is smooth, the differential element of length is

$$\mathrm{d}l = \sqrt{(f'(t))^2 + (g'(t))^2 + (h'(t))^2}\,\mathrm{d}t$$

and the line integral of $\phi(x, y, z)$ along a smooth curve C parameterized by t from $t = t_0$ to $t = t_1$ can be calculated by

$$\int_C \phi(x, y, z)\,\mathrm{d}l = \int_{t_0}^{t_1} \phi(f(t), g(t), h(t))\sqrt{(f'(t))^2 + (g'(t))^2 + (h'(t))^2}\,\mathrm{d}t$$

This definition can be used to describe an important type of line integral that is common in Physics applications. If $\boldsymbol{F}(x, y, z)$ denotes the force acting on an object at the point (x, y, z) as it is moved along a curve C, then the work done by the force along C is defined to be

$$W = \int_C \boldsymbol{F} \cdot \hat{\boldsymbol{l}}\,\mathrm{d}l$$

where $\hat{\boldsymbol{l}}$ denotes the unit tangent vector to C. Evidently, only the component of $\boldsymbol{F}$ in the direction of motion can contribute to W. If $\boldsymbol{r}(t) = f(t)\hat{\boldsymbol{x}} + g(t)\hat{\boldsymbol{y}} + h(t)\hat{\boldsymbol{z}}$ is a space curve parameterized by the scalar variable t, then the differential along the curve is

$$\mathrm{d}\boldsymbol{r} = \frac{\mathrm{d}\boldsymbol{r}}{\mathrm{d}t}\,\mathrm{d}t = \left(\frac{\mathrm{d}f}{\mathrm{d}t}\hat{\boldsymbol{x}} + \frac{\mathrm{d}g}{\mathrm{d}t}\hat{\boldsymbol{y}} + \frac{\mathrm{d}h}{\mathrm{d}t}\hat{\boldsymbol{z}}\right)\,\mathrm{d}t$$

The magnitude of this infinitesimal distance is $|\mathrm{d}\boldsymbol{r}| = \sqrt{\mathrm{d}\boldsymbol{r} \cdot \mathrm{d}\boldsymbol{r}} = \sqrt{\boldsymbol{r}' \cdot \boldsymbol{r}'}\,\mathrm{d}t$, and if the curve is smooth, then the arc length $l(t)$ of the curve from $\boldsymbol{r}(t_0)$ to $\boldsymbol{r}(t_1)$ is

$$l(t_1) = \int_{t_0}^{t_1} \sqrt{\boldsymbol{r}\prime(t) \cdot \boldsymbol{r}\prime(t)}\,\mathrm{d}t = \int_{t_0}^{t_1} \sqrt{\left[\frac{\mathrm{d}f(t)}{\mathrm{d}t}\right]^2 + \left[\frac{\mathrm{d}g(t)}{\mathrm{d}t}\right]^2 + \left[\frac{\mathrm{d}h(t)}{\mathrm{d}t}\right]^2}\,\mathrm{d}t$$

Note that $l'(t) = |\boldsymbol{r}'(t)|$. Consequently, $\mathrm{d}\boldsymbol{r} = \mathrm{d}x\hat{\boldsymbol{x}} + \mathrm{d}y\hat{\boldsymbol{y}} + \mathrm{d}z\hat{\boldsymbol{z}} = \hat{\boldsymbol{l}}\,\mathrm{d}l$ and so we can write the work integral as

Work performed by a force $\boldsymbol{F}$ in motion along a path C

$$W = \int_C \boldsymbol{F} \cdot \mathrm{d}\boldsymbol{r}$$

A.3.2 Surface integrals

Suppose $z = f(x, y)$ defines a smooth surface above the xy plane. Then, $\boldsymbol{r}(x, y) = x\hat{\boldsymbol{x}} + y\hat{\boldsymbol{y}} + f(x, y)\hat{\boldsymbol{z}}$ will be a point on that surface. Let $x = u(t)$ and $y = v(t)$ denote any curve (parameterized by t) in the xy-plane. If $\boldsymbol{r}(x, y)$ is a one-to-one mapping from (x, y) to some point $\boldsymbol{r}(x, y)$ in space, then $\boldsymbol{r}(u(t), v(t))$ will be a space curve (constrained to the surface). The tangent to that curve will be given by

$$\frac{\mathrm{d}\boldsymbol{r}(x, y)}{\mathrm{d}t} = \frac{\partial \boldsymbol{r}}{\partial x}\frac{\mathrm{d}x}{\mathrm{d}t} + \frac{\partial \boldsymbol{r}}{\partial y}\frac{\mathrm{d}y}{\mathrm{d}t}$$

The special curves $x =$ constant and $y =$ constant (in the xy-plane) are called "coordinate curves" (see Section A.5 for a more complete discussion). For these curves, we have

$$\frac{\mathrm{d}\boldsymbol{r}(x = \text{constant}, y)}{\mathrm{d}t} = \frac{\partial \boldsymbol{r}}{\partial y}\frac{\mathrm{d}y}{\mathrm{d}t} \quad \text{and} \quad \frac{\mathrm{d}\boldsymbol{r}(x, y = \text{constant})}{\mathrm{d}t} = \frac{\partial \boldsymbol{r}}{\partial x}\frac{\mathrm{d}x}{\mathrm{d}t}$$

These are both vectors tangent to the surface (scaled by $\mathrm{d}y/\mathrm{d}t$ and $\mathrm{d}x/\mathrm{d}t$, respectively), and, consequently, a vector normal to the surface is

$$\boldsymbol{n} = \frac{\partial \boldsymbol{r}}{\partial x} \times \frac{\partial \boldsymbol{r}}{\partial y}$$

An infinitesimal displacement of $\boldsymbol{r}$ on the surface is given by

$$\mathrm{d}\boldsymbol{r} = \frac{\partial \boldsymbol{r}}{\partial x}\,\mathrm{d}x + \frac{\partial \boldsymbol{r}}{\partial y}\,\mathrm{d}y$$

The area element on the surface is the infinitesimal parallelogram whose sides are the coordinate curves. This area element has magnitude[3]

$$\mathrm{d}s = \left|\frac{\partial \boldsymbol{r}}{\partial x}\,\mathrm{d}x \times \frac{\partial \boldsymbol{r}}{\partial y}\,\mathrm{d}y\right| = |\boldsymbol{n}|\,\mathrm{d}x\,\mathrm{d}y$$

($\boldsymbol{n}$, as defined here, is not generally a unit vector). Consequently, the area of the surface is

$$A = \iint_S \mathrm{d}s = \iint_{\mathcal{R}} |\boldsymbol{n}|\,\mathrm{d}x\,\mathrm{d}y$$

where $\mathcal{R}$ denotes the (regular) projection of S onto the x-y plane.

[3]Recall that the the quantity $\boldsymbol{A} \times \boldsymbol{B}$ is directed normal to the plane containing $\boldsymbol{A}$ and $\boldsymbol{B}$ with magnitude equal to the area of the parallelogram with sides $\boldsymbol{A}$ and $\boldsymbol{B}$.

More generally, if $\boldsymbol{r} = x\hat{\boldsymbol{x}} + y\hat{\boldsymbol{y}} + f(x,y)\hat{\boldsymbol{z}}$ is a point on the surface, then

$$\mathrm{d}s = \left|\frac{\partial \boldsymbol{r}}{\partial x}\,\mathrm{d}x \times \frac{\partial \boldsymbol{r}}{\partial y}\,\mathrm{d}y\right| = \sqrt{(f_x)^2 + (f_y)^2 + 1}\;\mathrm{d}x\,\mathrm{d}y$$

and, consequently, we can evaluate the "surface integral of a function $g(x,y,z)$" by using the formula

$$\iint_S g(x,y,z)\,\mathrm{d}s = \iint_{\mathcal{R}} g(x,y,f(x,y))\sqrt{(f_x(x,y))^2 + (f_y(x,y))^2 + 1}\;\mathrm{d}a$$

Flux of a vector field through a surface

Of special importance in physics is the surface integral formed when $g(x,y,z) = \boldsymbol{F}\cdot\hat{\boldsymbol{n}}$. This quantity is the local normal component to the surface of a vector field $\boldsymbol{F}$ and

$$\iint_S \boldsymbol{F}\cdot\hat{\boldsymbol{n}}\;\mathrm{d}s$$

is known as the flux of $\boldsymbol{F}$ through (or over) the surface S. The unit normal vector to the surface S follows as before and we obtain

$$\hat{\boldsymbol{n}} = \frac{1}{\left|\frac{\partial \boldsymbol{r}}{\partial x}\times\frac{\partial \boldsymbol{r}}{\partial y}\right|}\,\frac{\partial \boldsymbol{r}}{\partial x}\times\frac{\partial \boldsymbol{r}}{\partial y} = \frac{1}{\sqrt{(f_x)^2 + (f_y)^2 + 1}}(-f_x\hat{\boldsymbol{x}} - f_y\hat{\boldsymbol{y}} + \hat{\boldsymbol{z}})$$

A.4 Important integral theorems in vector calculus

A.4.1 Green's theorem in the plane

Green's theorem. *Let C be a piecewise smooth simple closed curve and let $\mathcal{R}$ be the region consisting of C and its interior. If M and N are functions that are continuous and have continuous first partial derivatives throughout an open region $\mathcal{D}$ containing $\mathcal{R}$, then*

Green's theorem in the plane

$$\oint_C (M\;\mathrm{d}x + N\;\mathrm{d}y) = \iint_{\mathcal{R}}\left(\frac{\partial N}{\partial x} - \frac{\partial M}{\partial y}\right)\,\mathrm{d}a$$

(The notation $\oint$ is used to denote the line integral along the closed path C once in the positive (counterclockwise) direction.)

Proof. We will only give a partial proof. Consider a region of convex shape so that it can be split up in either of two ways. Referring to the Figure A.1a, we have

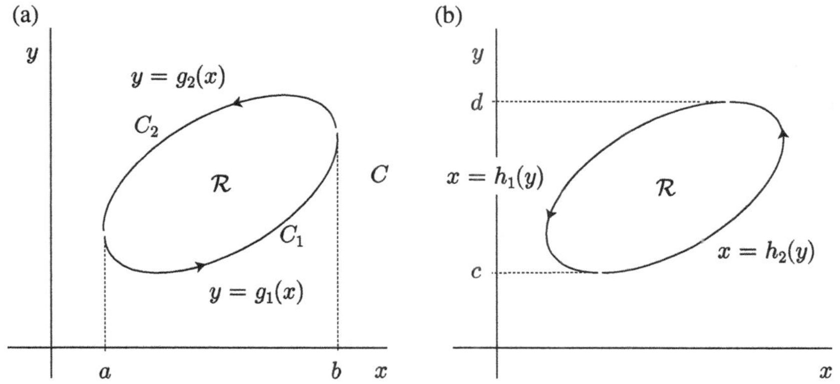

Figure A.1 Different order of integral iteration: (a) first x and then y; (b) first y and then x.

$$\begin{aligned}
\oint_C M \,\mathrm{d}x &= \int_{C_1} M(x,y)\,\mathrm{d}x + \int_{C_2} M(x,y)\,\mathrm{d}x \\
&= \int_a^b M(x,g_1(x))\,\mathrm{d}x + \int_b^a M(x,g_2(x))\,\mathrm{d}x \\
&= \int_a^b M(x,g_1(x))\,\mathrm{d}x - \int_a^b M(x,g_2(x))\,\mathrm{d}x \\
&= \int_a^b [M(x,g_1(x)) - M(x,g_2(x))]\,\mathrm{d}x
\end{aligned}$$

But also,

$$\begin{aligned}
\iint_{\mathcal{R}} \frac{\partial M}{\partial y}\,\mathrm{d}a &= \int_a^b \int_{g_1(x)}^{g_2(x)} \frac{\partial M}{\partial y}\,\mathrm{d}y\,\mathrm{d}x = \int_a^b M(x,y)\Big|_{g_1(x)}^{g_2(x)}\,\mathrm{d}x \\
&= \int_a^b [M(x,g_2(x)) - M(x,g_1(x))]\,\mathrm{d}x
\end{aligned}$$

so

$$\oint_C M\,\mathrm{d}x = -\iint_{\mathcal{R}} \frac{\partial M}{\partial y}\,\mathrm{d}a$$

In a similar way, the Figure A.1b can be used to show that

$$\oint_C N\,\mathrm{d}y = \iint_{\mathcal{R}} \frac{\partial N}{\partial x}\,\mathrm{d}a$$

and therefore Green's theorem is valid for regions that are convex. □

More general regions (*connected*, in the sense that any two points in the region $\mathcal{R}$ can be joined by a curve contained within the region), but not necessarily convex, also satisfy Green's theorem since any such region can be subdivided (as in Figure A.2) in such a way that each sub-region is convex. Note that the boundary integral contributions from adjacent sub-regions will cancel due to opposition in the direction traversed. Consequently, only the boundary contribution from the original region $\mathcal{R}$ need be credited.

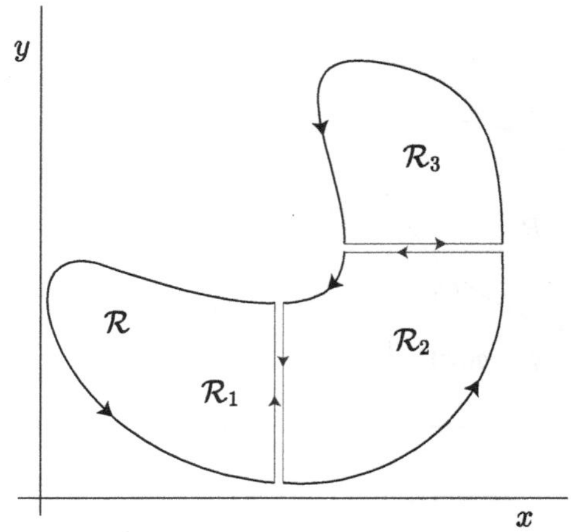

Figure A.2 Green's theorem for more general regions.

A.4.2 The divergence theorem

The divergence theorem. *Let $\mathcal{Q}$ be a region in three dimensions bounded by a closed surface S, and let $\hat{\boldsymbol{n}}$ denote the unit outward normal to S at (x, y, z). If $\boldsymbol{F}$ is a vector function that has continuous partial derivatives in $\mathcal{Q}$, then*

Gauss' theorem

$$\iint_S \boldsymbol{F} \cdot \hat{\boldsymbol{n}} \, \mathrm{d}s = \iiint_{\mathcal{Q}} \nabla \cdot \boldsymbol{F} \, \mathrm{d}V$$

The theorem is also sometimes called "Gauss' Theorem." (Note that this last equation is sometimes invoked to justify the interpretation of divergence, $\boldsymbol{\nabla} \cdot \boldsymbol{F}$, as the "flux per unit volume.")

The proof of Gauss' theorem is similar to that of Green's theorem in the plane (except with line and surface integrals being replaced with surface and volume integrals, respectively). This straightforward calculation is left to the reader.

A.4.3 Stokes' theorem

Green's theorem in the plane is

$$\oint_C (M \, \mathrm{d}x + N \, \mathrm{d}y) = \iint_{\mathcal{R}} \left(\frac{\partial N}{\partial x} - \frac{\partial M}{\partial y} \right) \mathrm{d}a$$

If $\boldsymbol{F}(x, y) = M(x, y)\hat{\boldsymbol{x}} + N(x, y)\hat{\boldsymbol{y}} + P(x, y)\hat{\boldsymbol{z}}$, then

$$(\boldsymbol{\nabla} \times \boldsymbol{F}) \cdot \hat{\boldsymbol{z}} = \frac{\partial N}{\partial x} - \frac{\partial M}{\partial y}$$

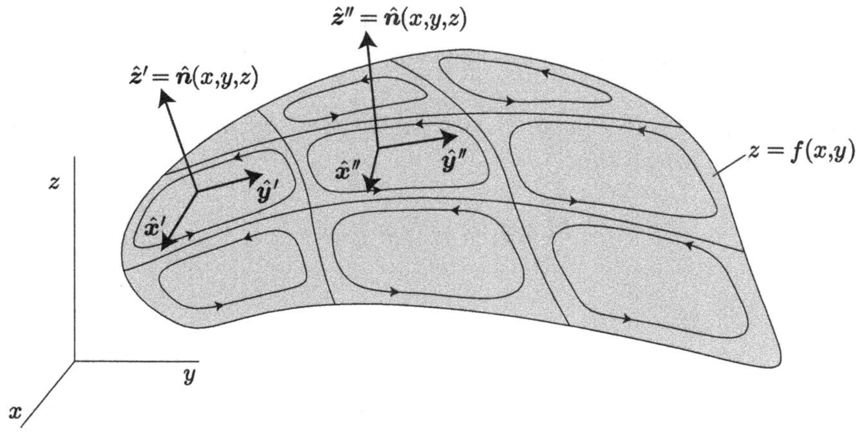

Figure A.3 Green's theorem for surfaces.

Consider the tangent vector to C as given by

$$\hat{\boldsymbol{l}} = \frac{\mathrm{d}x}{\mathrm{d}l}\hat{\boldsymbol{x}} + \frac{\mathrm{d}y}{\mathrm{d}l}\hat{\boldsymbol{y}} + \frac{\mathrm{d}z}{\mathrm{d}l}\hat{\boldsymbol{z}}$$

then Green's theorem in the plane can also be written as[4]

$$\oint_C \boldsymbol{F} \cdot \hat{\boldsymbol{l}}\, \mathrm{d}l = \iint_{\mathcal{R}} (\boldsymbol{\nabla} \times \boldsymbol{F}) \cdot \hat{\boldsymbol{z}}\, \mathrm{d}a$$

With the exception of the factor of $\hat{\boldsymbol{z}}$, this form is coordinate-free (i.e. it is expressed in terms of vectors rather than in terms of coordinates). The vector $\hat{\boldsymbol{z}}$, of course, is just a unit normal to the xy-plane, and it can be shown that the generalization of Green's theorem in the plane to an arbitrary surface S with normal $\hat{\boldsymbol{n}}$ bounded by a closed curve C is given by (see Figure A.3).

Stokes' theorem.

Stokes' theorem

$$\oint_C \boldsymbol{F} \cdot \hat{\boldsymbol{l}}\, \mathrm{d}l = \iint_S (\boldsymbol{\nabla} \times \boldsymbol{F}) \cdot \hat{\boldsymbol{n}}\, \mathrm{d}s$$

(Note that this last equation is sometimes invoked to justify the interpretation of curl, $(\boldsymbol{\nabla} \times \boldsymbol{F}) \cdot \hat{\boldsymbol{n}}$, as the "circulation per unit area.")

A.4.4 Conservative fields

Generally, the work done by $\boldsymbol{F}$ will depend on the particular curve C along which the force acts (this is not surprising, since only the component of $\boldsymbol{F}$ tangent to C actually contributes to the work). If $\boldsymbol{F}$ is a *conservative force,* however, then the work done in going from (x_1, y_1, z_1) to

conservative fields

[4]The integral $\oint_C \boldsymbol{F} \cdot \hat{\boldsymbol{l}}\, \mathrm{d}l$ is known as the circulation of $\boldsymbol{F}$ around C (and, when $\boldsymbol{F}$ is a force field, is equal to the work performed along the path C). This identity provides an interpretation of interpretation of $\boldsymbol{\nabla} \times \boldsymbol{F}$ since the $\hat{\boldsymbol{z}}$ component of $\boldsymbol{\nabla} \times \boldsymbol{F}$ is, evidently, the circulation per unit area of $\boldsymbol{F}$. Moreover, it is clear that if $\boldsymbol{\nabla} \times \boldsymbol{F}$ has zero $\hat{\boldsymbol{z}}$ component throughout $\mathcal{R}$, then $\boldsymbol{F}$ is conservative in $\mathcal{R}$.

(x_2, y_2, z_2) will be independent of the path taken. There is a theorem that is sometimes used to define a conservative force:

Theorem A.4.1. *If $\boldsymbol{F}(x, y, z)$ can be written as the gradient of some function $\phi(x, y, z)$ so that*

$$\boldsymbol{F}(x, y, z) = \nabla\phi(x, y, z)$$

then

$$\int_C \boldsymbol{F} \cdot \, \mathrm{d}\boldsymbol{r}$$

is independent of path. The function $\phi(x, y, z)$ in this case is called the potential function.

Rather than formally proving this theorem, let's just observe that if

$$\boldsymbol{F}(x, y, z) = \nabla\phi(x, y, z) = \frac{\partial\phi}{\partial x}\hat{\boldsymbol{x}} + \frac{\partial\phi}{\partial y}\hat{\boldsymbol{y}} + \frac{\partial\phi}{\partial z}\hat{\boldsymbol{z}}$$

then

$$\int_C \boldsymbol{F} \cdot \, \mathrm{d}\boldsymbol{r} = \int_C (F_x(x, y, z)\, \mathrm{d}x + F_y(x, y, z)\, \mathrm{d}y + F_z(x, y, z)\, \mathrm{d}z)$$

If C is parameterized by $x = u(t)$, $y = v(t)$, and $z = w(t)$ with (x_1, y_1, z_1) corresponding to t_1 and (x_2, y_2, z_2) corresponding to t_2 then

$$\begin{aligned}
\int_C \boldsymbol{F} \cdot \, \mathrm{d}\boldsymbol{r} &= \int_{t_1}^{t_2} (F_x(u(t), v(t), w(t))u'(t)\, \mathrm{d}t + F_y(u(t), v(t), w(t))v'(t)\, \mathrm{d}t \\
&\quad + F_z(u(t), v(t), w(t))w'(t)\, \mathrm{d}t) \\
&= \int_{t_1}^{t_2} (F_x(u(t), v(t), w(t))u'(t) + F_y(u(t), v(t), w(t))v'(t) \\
&\quad + F_z(u(t), v(t), w(t))w'(t))\, \mathrm{d}t \\
&= \int_{t_1}^{t_2} \frac{\mathrm{d}}{\mathrm{d}t}(\phi(u(t), v(t), w(t)))\, \mathrm{d}t
\end{aligned}$$

where the last line follows by the chain rule (in reverse). Then the fundamental theorem of calculus allows us to do the last integral and obtain

$$\begin{aligned}
\int_C \boldsymbol{F} \cdot \, \mathrm{d}\boldsymbol{r} &= \int_{t_1}^{t_2} \frac{\mathrm{d}}{\mathrm{d}t}(\phi(u(t), v(t), w(t)))\, \mathrm{d}t = \phi(u(t_2), v(t_2), w(t_2)) \\
&\quad - \phi(u(t_1), v(t_1), w(t_1)) = \phi(x_2, y_2, z_2) - \phi(x_1, y_1, z_1)
\end{aligned}$$

But this answer depends only on the endpoints of the curve (and not on the curve itself).

The theorem (together with Stokes' theorem) allows us to generalize the description of independence of path (for conservative fields). We have the following

Theorem A.4.2. *If $\boldsymbol{F}(x, y, z)$ has continuous first partial derivatives throughout a simply connected region, then the following conditions are equivalent:*

1. *$\boldsymbol{F}$ is conservative*
2. *$\int_C \boldsymbol{F} \cdot \mathrm{d}\boldsymbol{r}$ is independent of path*
3. *$\oint_C \boldsymbol{F} \cdot \mathrm{d}\boldsymbol{r} = 0$ for every simple closed curve C*
4. *$\boldsymbol{\nabla} \times \boldsymbol{F} = \boldsymbol{0}$*

A.4.5 The Helmholtz theorem

From Section A.2.4, we have the identity

$$\nabla^2 \boldsymbol{V} = \boldsymbol{\nabla}(\boldsymbol{\nabla} \cdot \boldsymbol{V}) - \boldsymbol{\nabla} \times (\boldsymbol{\nabla} \times \boldsymbol{V})$$

Define

$$\boldsymbol{V}(\boldsymbol{r}) = -\frac{1}{4\pi} \iiint_Q \frac{\boldsymbol{F}(\boldsymbol{r}')}{|\boldsymbol{r} - \boldsymbol{r}'|} \, \mathrm{d}^3 r'$$

The Laplacian of the i-th component of $\boldsymbol{F}$ can be computed using the last two identities of Section A.2.4:

$$\begin{aligned}
\nabla^2 V_i(\boldsymbol{r}) = \boldsymbol{\nabla} \cdot \boldsymbol{\nabla} V_i(\boldsymbol{r}) &= -\frac{1}{4\pi} \iiint_Q F_i(\boldsymbol{r}') \nabla_r^2 \frac{1}{|\boldsymbol{r} - \boldsymbol{r}'|} \, \mathrm{d}^3 r' \\
&= -\frac{1}{4\pi} \iiint_Q F_i(\boldsymbol{r}')(-4\pi\delta(\boldsymbol{r} - \boldsymbol{r}')) \, \mathrm{d}^3 r' \\
&= F_i(\boldsymbol{r})
\end{aligned}$$

where ∇_r^2 operates only on the $\boldsymbol{r}$ and not on the $\boldsymbol{r}'$ dependence of $1/||\boldsymbol{r} - \boldsymbol{r}'||$, and we have taken Q to be all space in the last integral. Consequently, we have the following result:

Helmholtz's theorem. *For any well-behaved vector field $\boldsymbol{F}$ whose divergence and curl decay to 0 faster than $1/r^2$ as $r \to \infty$, we can write*

$$\boldsymbol{F} = -\boldsymbol{\nabla}\phi + \boldsymbol{\nabla} \times \boldsymbol{A}$$

where $\phi \equiv -\boldsymbol{\nabla} \cdot \boldsymbol{V}$ is known as the scalar potential *and $\boldsymbol{A} \equiv -\boldsymbol{\nabla} \times \boldsymbol{V}$ is known as the* vector potential.

This theorem is sometimes called "the fundamental theorem of vector analysis."

A straightforward corollary of Helmholtz's theorem is

Corollary. *If* $|\boldsymbol{F}| \to 0$ *faster than* $1/r$ *as* $r \to \infty$ *then*[5]

$$\begin{aligned} \boldsymbol{F}(\boldsymbol{r}) = \boldsymbol{\nabla}\left(-\frac{1}{4\pi}\iiint \frac{\boldsymbol{\nabla}_{r'} \cdot \boldsymbol{F}(\boldsymbol{r}')}{|\boldsymbol{r}-\boldsymbol{r}'|}\,\mathrm{d}^3r'\right) \\ +\boldsymbol{\nabla}\times\left(\frac{1}{4\pi}\iiint \frac{\boldsymbol{\nabla}_{r'} \times \boldsymbol{F}(\boldsymbol{r}')}{|\boldsymbol{r}-\boldsymbol{r}'|}\,\mathrm{d}^3r'\right) \end{aligned}$$

where the integrals are over all space.

This result makes our earlier discussions of **div** $\boldsymbol{F}$ and **curl** $\boldsymbol{F}$ all the more relevant. We already knew that **curl** $\boldsymbol{F} = \boldsymbol{0}$ means that **div** $\boldsymbol{F}$ is a uniquely important scalar characterization of $\boldsymbol{F}$ (it means that $\boldsymbol{F} = -\boldsymbol{\nabla}\phi$). This corollary goes a bit further and says that when **curl** $\boldsymbol{F} = \boldsymbol{0}$, the scalar field **div** $\boldsymbol{F}$ is enough to determine $\boldsymbol{F}$. Moreover, when **curl** $\boldsymbol{F} \neq \boldsymbol{0}$, then the two potentials will reduce the nine-elements of the derivative matrix (Eq. (A.3)) to the four scalar fields that really matter.

A.5 Coordinate systems

It is often convenient to use coordinate systems other than the rectangular coordinates because of the types of symmetry that are associated with certain field distributions. Of the other possible coordinate systems, the two most commonly used are cylindrical and spherical polar coordinates. It is, of course, possible to directly calculate the gradient, divergence, and curl, in cylindrical and spherical coordinates, but it is more useful to derive expressions for these operators in generalized curvilinear coordinates, so that they may be applied to any coordinate system.

A.5.1 Orthogonal curvilinear coordinates

Consider the equation

$$f(x, y, z) = u$$

where u is a constant. This equation denotes a family of surfaces in space, such that each surface is characterized by a particular value of the parameter u. For example, if $f(x, y, z) = x$, then $x = u$ denotes the planar surface parallel to the yz-plane in rectangular coordinates, each surface having a different value of x.

[5]Since we have extended Q to all space in our derivation, we see that the divergence and curl of $\boldsymbol{F}$ must decay faster than $1/r^2$ as $r \to \infty$ in order that the integrals remain finite. This requirement, in turn, demands that $\boldsymbol{F}$ decays faster than $1/r$.

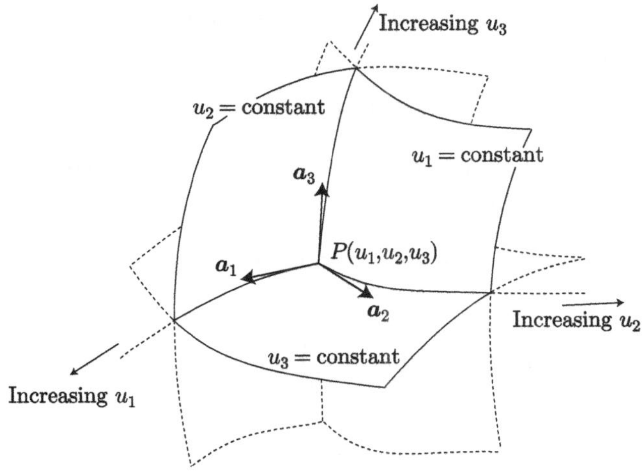

Figure A.4 Generalized curvilinear coordinates.

Next consider the three equations

$$f_1(x, y, z) = u_1$$
$$f_2(x, y, z) = u_2$$
$$f_3(x, y, z) = u_3$$

which are chosen so that *the three families of surfaces are mutually perpendicular,* or orthogonal. A point P in space can then be defined as the intersection of three of these surfaces, one of each family. P is completely defined if the values of u_1, u_2, and u_3 are stated. The variables u_1, u_2, and u_3 are called the *curvilinear coordinates* of P (see Figure A.4).

The intersection of the surfaces u_2 = constant and u_3 = constant (called the u_2-surface and the u_3-surface) is, in general, a curve that is normal to the u_1-surface. This curve is the "u_1-axis"; and similarly for the u_2- and u_3-axes. In general, the directions of these axes are variable. It is only in the Cartesian coordinate system that the directions are fixed throughout space.

A.5.2 Unit vectors

Suppose that dl_1 is an element of length *perpendicular* to the surface u_1 = constant at the point $P(u_1, u_2, u_3)$. The element dl_1 is the distance between the surfaces u_1 and $u_1 + du_1$ and is related to du_1 by the equation

$$dl_1 = h_1\, du_1$$

where h_1 is, in general, a function of the coordinates u_1, u_2, and u_3 and is called the *scale factor.* Similarly,

$$dl_2 = h_2\, du_2 \quad \text{and} \quad dl_3 = h_3\, du_3$$

(In the Cartesian system, h_1, h_2, and h_3 are all unity.)

The differential displacement of the point P resulting from arbitrary increments $\mathrm{d}u_1$, $\mathrm{d}u_2$, and $\mathrm{d}u_3$ in the coordinates is given by

$$\mathrm{d}\boldsymbol{r} = \hat{\boldsymbol{a}}_1\,\mathrm{d}l_1 + \hat{\boldsymbol{a}}_2\,\mathrm{d}l_2 + \hat{\boldsymbol{a}}_3\,\mathrm{d}l_3$$

where $\hat{\boldsymbol{a}}_1$, $\hat{\boldsymbol{a}}_2$, $\hat{\boldsymbol{a}}_3$ are unit vectors in the directions of the u_1, u_2, u_3 axes (i.e. normal to the u_1, u_2, u_3 planes), *at the point* P. Thus,

$$\mathrm{d}\boldsymbol{r} = \hat{\boldsymbol{a}}_1 h_1\,\mathrm{d}u_1 + \hat{\boldsymbol{a}}_2 h_2\,\mathrm{d}u_2 + \hat{\boldsymbol{a}}_3 h_3\,\mathrm{d}u_3$$

But

$$\mathrm{d}\boldsymbol{r} = \frac{\partial \boldsymbol{r}}{\partial u_1}\,\mathrm{d}u_1 + \frac{\partial \boldsymbol{r}}{\partial u_2}\,\mathrm{d}u_2 + \frac{\partial \boldsymbol{r}}{\partial u_3}\,\mathrm{d}u_3$$

which implies

$$\hat{\boldsymbol{a}}_k = \frac{1}{h_k}\frac{\partial \boldsymbol{r}}{\partial u_k} \qquad k = 1, 2, 3$$

and so

$$h_k\hat{\boldsymbol{a}}_k = \frac{\partial \boldsymbol{r}}{\partial u_k} \qquad \Longrightarrow \qquad h_k = \left|\frac{\partial \boldsymbol{r}}{\partial u_k}\right|.$$

In Cartesian coordinates, $\boldsymbol{r} = x\hat{\boldsymbol{x}} + y\hat{\boldsymbol{y}} + z\hat{\boldsymbol{z}}$

$$\Longrightarrow \qquad h_k = \sqrt{\left(\frac{\partial x}{\partial u_k}\right)^2 + \left(\frac{\partial y}{\partial u_k}\right)^2 + \left(\frac{\partial z}{\partial u_k}\right)^2}$$

The same result can be obtained differently as follows:

$$\hat{\boldsymbol{a}}_k = \frac{\text{any vector that is tangential to the} u_k\text{-axis at } P}{\text{magnitude of the vector}}$$

But $\partial \boldsymbol{r}/\partial u_k$ is the vector tangent to u_k at P, so

$$\hat{\boldsymbol{a}}_k = \frac{1}{\left|\frac{\partial \boldsymbol{r}}{\partial u_k}\right|}\frac{\partial \boldsymbol{r}}{\partial u_k}$$

Since $h_k = |\partial \boldsymbol{r}/\partial u_k|$, it follows that

$$h_k = \sqrt{\left(\frac{\partial x}{\partial u_k}\right)^2 + \left(\frac{\partial y}{\partial u_k}\right)^2 + \left(\frac{\partial z}{\partial u_k}\right)^2} \quad \text{and} \quad \hat{\boldsymbol{a}}_k = \frac{1}{h_k}\frac{\partial \boldsymbol{r}}{\partial u_k}.$$

A.5.3 Differential displacement

The differential vector displacement is

$$\mathrm{d}\boldsymbol{r} = \sum_k \hat{\boldsymbol{a}}_k h_k\,\mathrm{d}u_k$$

The magnitude of the differential displacement is

$$\sqrt{\mathrm{d}\boldsymbol{r}\cdot\mathrm{d}\boldsymbol{r}} = \sqrt{\sum_k h_k^2\,\mathrm{d}u_u^2}$$

A.5.4 Differential surface and volume elements

$$\mathrm{d}s_1 = \mathrm{d}l_2\,\mathrm{d}l_3 = h_2h_3\,\mathrm{d}u_2\,\mathrm{d}u_3$$
$$\mathrm{d}s_2 = \mathrm{d}l_3\,\mathrm{d}l_1 = h_3h_1\,\mathrm{d}u_3\,\mathrm{d}u_1$$
$$\mathrm{d}s_3 = \mathrm{d}l_1\,\mathrm{d}l_2 = h_1h_2\,\mathrm{d}u_1\,\mathrm{d}u_2$$

and $\mathrm{d}V = \mathrm{d}l_1\,\mathrm{d}l_2\,\mathrm{d}l_3 = h_1h_2h_3\,\mathrm{d}u_1\,\mathrm{d}u_2\,\mathrm{d}u_3$.

A.5.5 Transformation of vector components

Suppose that the vector $\boldsymbol{A}$ is denoted, in Cartesian coordinates, by

$$\boldsymbol{A} = A_1^C\hat{\boldsymbol{x}}_1 + A_2^C\hat{\boldsymbol{x}}_2 + A_3^C\hat{\boldsymbol{x}}_3 = \sum_i A_i^C\hat{\boldsymbol{x}}_i$$

(This is merely a different notation in which: A_1^C, A_2^C, A_3^C are equal to A_x, A_y, A_z, respectively, and $\hat{\boldsymbol{x}}_1$, $\hat{\boldsymbol{x}}_2$, $\hat{\boldsymbol{x}}_3$ are the same as $\hat{\boldsymbol{x}}$, $\hat{\boldsymbol{y}}$, $\hat{\boldsymbol{z}}$, respectively.) In curvilinear coordinates, we have

$$\boldsymbol{A} = A_1\hat{\boldsymbol{a}}_1 + A_2\hat{\boldsymbol{a}}_2 + A_3\hat{\boldsymbol{a}}_3 = \sum_k A_k\hat{\boldsymbol{a}}_k$$

where A_k is the vector component along the u_k-axis. The curvilinear unit vectors can be expressed in terms of the Cartesian unit vectors by

$$\hat{\boldsymbol{a}}_k = \sum_i \alpha_{ki}\hat{\boldsymbol{x}}_i \qquad k = 1, 2, 3; \quad i = 1, 2, 3$$

$$\text{i.e.} \qquad \begin{bmatrix} \hat{\boldsymbol{a}}_1 \\ \hat{\boldsymbol{a}}_2 \\ \hat{\boldsymbol{a}}_3 \end{bmatrix} = \begin{bmatrix} \alpha_{11} & \alpha_{12} & \alpha_{13} \\ \alpha_{21} & \alpha_{22} & \alpha_{23} \\ \alpha_{31} & \alpha_{32} & \alpha_{33} \end{bmatrix} \begin{bmatrix} \hat{\boldsymbol{x}}_1 \\ \hat{\boldsymbol{x}}_2 \\ \hat{\boldsymbol{x}}_3 \end{bmatrix}$$

and the transformation matrix $\boldsymbol{\alpha}$ is given by

$$\alpha_{ki} = \hat{\boldsymbol{a}}_k \cdot \hat{\boldsymbol{x}}_i = \frac{1}{h_k}\frac{\partial \boldsymbol{r}}{\partial u_k} \cdot \hat{\boldsymbol{x}}_i = \frac{1}{h_k}\frac{\partial x_i}{\partial u_k} \qquad (x_i \equiv x, y, z)$$

Also

$$A_k = \sum_i \alpha_{ki}A_i^C \quad \text{and} \quad A_i^C = \sum_k \alpha_{ki}^{-1}A_k$$

where α_{ki}^{-1} are the components of the inverse of $\boldsymbol{\alpha}$.

In generalized curvilinear coordinates:

(i) The gradient of a scalar field is given by

$$\mathbf{grad}\; V = \sum_k \hat{\boldsymbol{a}}_k \frac{1}{h_k}\frac{\partial V}{\partial u_k}$$

(ii) The divergence of a vector field is given by

$$\textbf{div}\ \boldsymbol{F} = \frac{1}{h_1h_2h_3}\left[\frac{\partial}{\partial u_1}(F_1h_2h_3) + \frac{\partial}{\partial u_2}(F_2h_3h_1) + \frac{\partial}{\partial u_3}(F_3h_1h_2)\right]$$

(iii) The curl of a vector field is given by

$$\begin{aligned}\textbf{curl}\ \boldsymbol{F} = {} & \frac{\hat{\boldsymbol{a}}_1}{h_2h_3}\left(\frac{\partial(h_3F_3)}{\partial u_2} - \frac{\partial(h_2F_2)}{\partial u_3}\right) + \frac{\hat{\boldsymbol{a}}_2}{h_1h_3}\left(\frac{\partial(h_1F_1)}{\partial u_3} - \frac{\partial(h_3F_3)}{\partial u_1}\right) \\ & + \frac{\hat{\boldsymbol{a}}_3}{h_1h_2}\left(\frac{\partial(h_2F_2)}{\partial u_1} - \frac{\partial(h_1F_1)}{\partial u_2}\right)\end{aligned}$$

(iv) The Laplacian of a scalar is given by

$$\begin{aligned}\textbf{div}\ (\textbf{grad}\ V) = {} & \frac{1}{h_1h_2h_3}\left[\frac{\partial}{\partial u_1}\left(\frac{h_2h_3}{h_1}\frac{\partial V}{\partial u_1}\right) + \frac{\partial}{\partial u_2}\left(\frac{h_3h_1}{h_2}\frac{\partial V}{\partial u_2}\right)\right. \\ & \left. + \frac{\partial}{\partial u_3}\left(\frac{h_1h_2}{h_3}\frac{\partial V}{\partial u_3}\right)\right]\end{aligned}$$

A.5.6 Cylindrical coordinates

In circular cylindrical coordinates (see Figure A.5a and b) we have:

$$u_1 = \rho, \quad u_2 = \phi, \quad u_3 = z$$

$$\hat{\boldsymbol{a}}_1 = \hat{\boldsymbol{a}}_\rho \quad (\text{or } \hat{\boldsymbol{\rho}}), \quad \hat{\boldsymbol{a}}_2 = \hat{\boldsymbol{a}}_\phi \quad (\text{or } \hat{\phi}), \quad \hat{\boldsymbol{a}}_3 = \hat{\boldsymbol{a}}_z \quad (\text{or } \hat{\boldsymbol{z}}\text{or } \hat{\boldsymbol{z}})$$

$$x = \rho\cos\phi, \quad y = \rho\sin\phi, \quad z = z$$

$$\begin{aligned}\Longrightarrow h_1 = h_\rho &= \sqrt{\left(\frac{\partial x}{\partial \rho}\right)^2 + \left(\frac{\partial y}{\partial \rho}\right)^2 + \left(\frac{\partial z}{\partial \rho}\right)^2} = \sqrt{\cos^2\phi + \sin^2\phi} = 1 \\ h_2 = h_\phi &= \sqrt{\left(\frac{\partial x}{\partial \phi}\right)^2 + \left(\frac{\partial y}{\partial \phi}\right)^2 + \left(\frac{\partial z}{\partial \phi}\right)^2} = \sqrt{(-\rho\sin\phi)^2 + (\rho\cos\phi)^2} = \rho \\ h_3 = h_z &= \sqrt{\left(\frac{\partial x}{\partial z}\right)^2 + \left(\frac{\partial y}{\partial z}\right)^2 + \left(\frac{\partial z}{\partial z}\right)^2} = \sqrt{1^2} = 1\end{aligned}$$

Differential displacement

$$\mathrm{d}\boldsymbol{r} = \sum_k \hat{\boldsymbol{a}}_k h_k\ \mathrm{d}\boldsymbol{u}_k = \hat{\boldsymbol{\rho}}\ \mathrm{d}\rho + \hat{\phi}\rho\ \mathrm{d}\phi + \hat{\boldsymbol{z}}\ \mathrm{d}z$$

$$\Longrightarrow |\mathrm{d}\boldsymbol{r}| = \sqrt{(\mathrm{d}\rho)^2 + (\rho\mathrm{d}\phi)^2 + (\mathrm{d}z)^2}$$

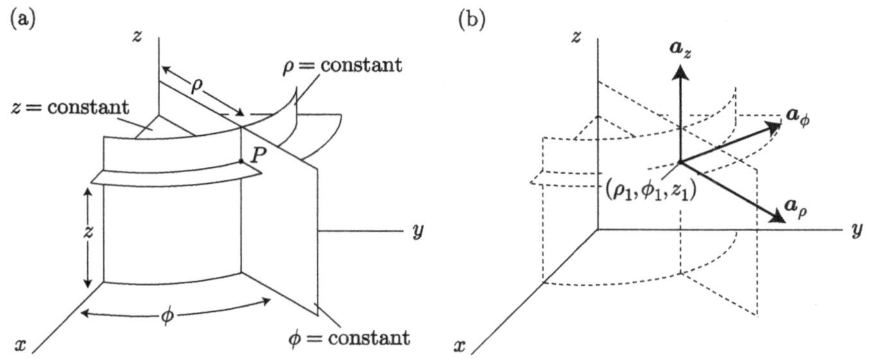

Figure A.5 Cylindrical coordinates. (a) The three mutually perpendicular surfaces of the circular cylindrical coordinate system. (b) The three direction vectors of the circular cylindrical coordinate system.

Differential surface and volume elements

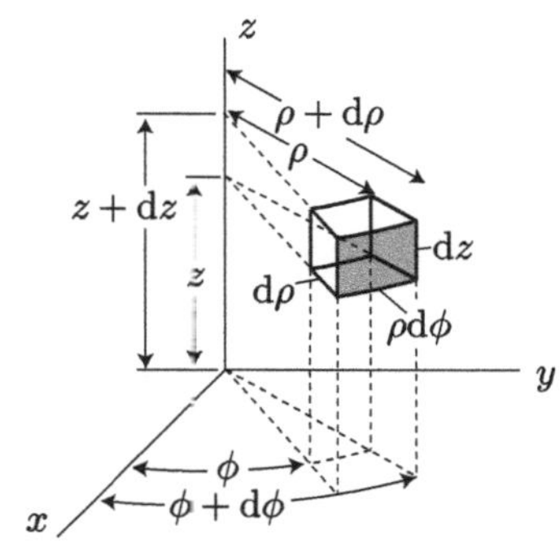

$$
\begin{aligned}
ds_1 &= \hat{\boldsymbol{\rho}} h_2 h_3 \, du_2 \, du_3 = \rho \, d\phi \, dz \\
ds_2 &= \hat{\boldsymbol{\phi}} h_3 h_1 \, du_1 \, du_3 = d\rho \, dz \\
ds_3 &= \hat{\boldsymbol{z}} h_1 h_2 \, du_1 \, du_2 = \rho \, d\rho \, d\phi
\end{aligned}
$$

$$dV = h_1 h_2 h_3 \, du_1 \, du_2 \, du_3 = \rho \, d\rho \, d\phi \, dz$$

Transformation matrix

$$
\boldsymbol{\alpha} = \left[\frac{1}{h_k}\frac{\partial x_i}{\partial u_k}\right] = \begin{bmatrix} \frac{1}{1}\frac{\partial x}{\partial \rho} & \frac{1}{1}\frac{\partial y}{\partial \rho} & \frac{1}{1}\frac{\partial z}{\partial \rho} \\ \frac{1}{\rho}\frac{\partial x}{\partial \phi} & \frac{1}{\rho}\frac{\partial y}{\partial \phi} & \frac{1}{\rho}\frac{\partial z}{\partial \phi} \\ \frac{1}{1}\frac{\partial x}{\partial z} & \frac{1}{1}\frac{\partial y}{\partial z} & \frac{1}{1}\frac{\partial z}{\partial z} \end{bmatrix} = \begin{bmatrix} \cos\phi & \sin\phi & 0 \\ -\sin\phi & \cos\phi & 0 \\ 0 & 0 & 1 \end{bmatrix}
$$

Unit vectors

Since

$$\hat{\boldsymbol{a}}_k = \sum_i \alpha_{ki} \hat{\boldsymbol{x}}_i$$

$$
\begin{aligned}
\Longrightarrow \quad \hat{\boldsymbol{\rho}} &= \hat{\boldsymbol{x}} \cos\phi + \hat{\boldsymbol{y}} \sin\phi \\
\hat{\boldsymbol{\phi}} &= \hat{\boldsymbol{x}}(-\sin\phi) + \hat{\boldsymbol{y}} \cos\phi \\
\hat{\boldsymbol{z}} &= \hat{\boldsymbol{z}}
\end{aligned}
$$

F_ρ, F_ϕ, F_z in terms of F_x, F_y, F_z

$$\begin{bmatrix} F_\rho \\ F_\phi \\ F_z \end{bmatrix} = \begin{bmatrix} \cos\phi & \sin\phi & 0 \\ -\sin\phi & \cos\phi & 0 \\ 0 & 0 & 1 \end{bmatrix} \begin{bmatrix} F_x \\ F_y \\ F_z \end{bmatrix}$$

$$\Longrightarrow \quad \begin{aligned} F_\rho &= F_x \cos\phi + F_y \sin\phi \\ F_\phi &= -F_x \sin\phi + F_y \cos\phi \\ F_z &= F_z \end{aligned}$$

$$\text{where } \cos\phi = \frac{x}{\sqrt{x^2+y^2}} \quad \text{and } \sin\phi = \frac{y}{\sqrt{x^2+y^2}}$$

F_x, F_y, F_z in terms of F_ρ, F_ϕ, F_z

The matrix for the inverse transformation is

$$\begin{bmatrix} \cos\phi & -\sin\phi & 0 \\ \sin\phi & \cos\phi & 0 \\ 0 & 0 & 1 \end{bmatrix}$$

so that

$$\Longrightarrow \quad \begin{aligned} F_x &= F_\rho \cos\phi - F_\phi \sin\phi \\ F_y &= F_\rho \sin\phi + F_\phi \cos\phi \\ F_z &= F_z \end{aligned}$$

grad, div, and curl

$$\nabla V = \textbf{grad } V = \hat{\boldsymbol{\rho}}\frac{\partial V}{\partial \rho} + \hat{\phi}\frac{1}{\rho}\frac{\partial V}{\partial \phi} + \hat{\boldsymbol{z}}\frac{\partial V}{\partial z}$$

$$\nabla \cdot \boldsymbol{F} = \textbf{div } \boldsymbol{F} = \frac{1}{\rho}\frac{\partial}{\partial \rho}(\rho F_\rho) + \frac{1}{\rho}\frac{\partial F_\phi}{\partial \phi} + \frac{\partial F_z}{\partial z}$$

$$\nabla \times \boldsymbol{F} = \textbf{curl } \boldsymbol{F} = \hat{\boldsymbol{\rho}}\left(\frac{1}{\rho}\frac{\partial F_z}{\partial \phi} - \frac{\partial F_\rho}{\partial z}\right) + \hat{\phi}\left(\frac{\partial F_\rho}{\partial z} - \frac{\partial F_\rho}{\partial z}\right) + \frac{\hat{\boldsymbol{z}}}{\rho}\left(\frac{\partial(\rho F_\phi)}{\partial \rho} - \frac{\partial F_\rho}{\partial \phi}\right)$$

$$\nabla^2 V = \textbf{div (grad } V) = \frac{1}{\rho}\frac{\partial}{\partial \rho}\left(\rho\frac{\partial V}{\partial \rho}\right) + \frac{1}{\rho^2}\frac{\partial^2 V}{\partial \phi^2} + \frac{\partial^2 V}{\partial z^2}$$

A.5.6.1 Spherical coordinates

In spherical coordinates, (see Figure A.6a and b) we have:

$$u_1 = r, \quad u_2 = \theta, \quad u_3 = \phi$$

$$\hat{\boldsymbol{a}}_1 = \hat{\boldsymbol{a}}_r \quad (\text{or } \hat{\boldsymbol{r}}), \quad \hat{\boldsymbol{a}}_2 = \hat{\boldsymbol{a}}_\theta \quad (\text{or } \hat{\boldsymbol{\theta}}), \quad \hat{\boldsymbol{a}}_3 = \hat{\boldsymbol{a}}_\phi \quad (\text{or } \hat{\phi})$$

$$x = r\sin\,\theta\cos\,\phi, \quad y = r\sin\theta\sin\phi, \quad z = r\cos\theta$$

$$\begin{aligned}
\Longrightarrow h_1 = h_r &= \sqrt{\left(\frac{\partial x}{\partial r}\right)^2 + \left(\frac{\partial y}{\partial r}\right)^2 + \left(\frac{\partial z}{\partial r}\right)^2} \\
&= \sqrt{\sin^2\theta\cos^2\phi + \sin^2\theta\sin^2\phi + \cos^2\theta} = 1 \\
h_2 = h_\theta &= \sqrt{\left(\frac{\partial x}{\partial \theta}\right)^2 + \left(\frac{\partial y}{\partial \theta}\right)^2 + \left(\frac{\partial z}{\partial \theta}\right)^2} \\
&= \sqrt{(r\cos\theta\cos\,\phi)^2 + (r\cos\theta\sin\phi)^2 + (r\sin\theta)^2} = r \\
h_3 = h_\phi &= \sqrt{\left(\frac{\partial x}{\partial \phi}\right)^2 + \left(\frac{\partial y}{\partial \phi}\right)^2 + \left(\frac{\partial z}{\partial \phi}\right)^2} \\
&= \sqrt{(-r\sin\theta\sin\phi)^2 + (r\sin\theta\cos\phi)^2 + 0} = r\sin\theta
\end{aligned}$$

Differential displacement

$$\mathrm{d}\boldsymbol{r} = \sum_k \hat{\boldsymbol{a}}_k h_k\,\mathrm{d}\boldsymbol{u}_k = \hat{\boldsymbol{r}}\,\mathrm{d}r + \hat{\boldsymbol{\theta}} r\,\mathrm{d}\theta + \hat{\phi} r\sin\theta\,\mathrm{d}\phi$$

$$\Longrightarrow |\mathrm{d}\boldsymbol{r}| = \sqrt{(\mathrm{d}r)^2 + (r\mathrm{d}\theta)^2 + (r\sin\theta\mathrm{d}\phi)^2}$$

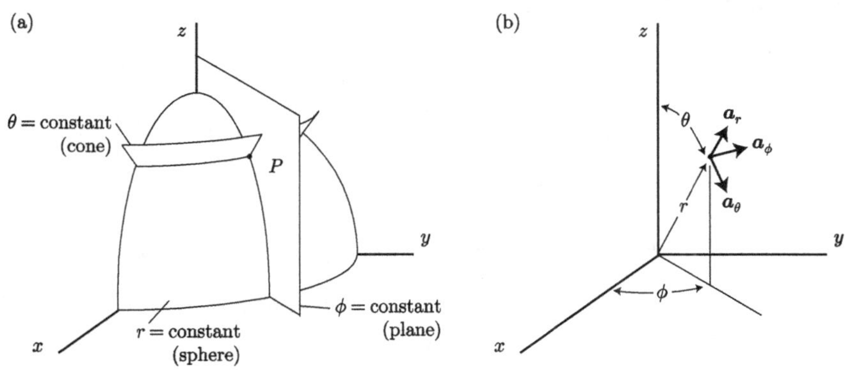

Figure A.6 Spherical coordinates. (a) The three mutually perpendicular surfaces of the spherical coordinate system. (b) The three direction vectors in spherical coordinates.

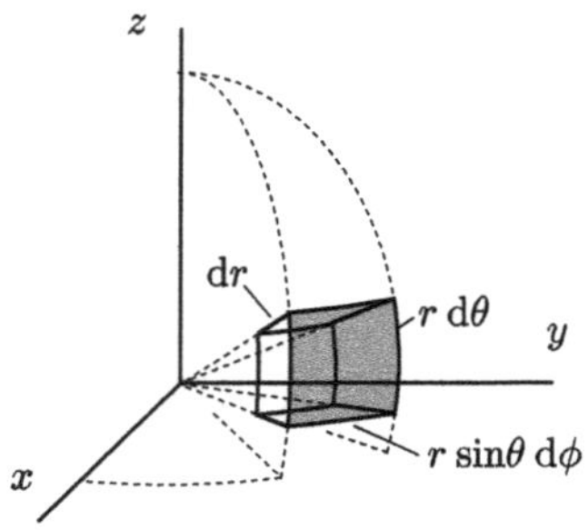

Differential surface and volume elements

$$\begin{aligned}
\mathrm{d}s_1 &= h_2h_3\,\mathrm{d}u_2\,\mathrm{d}u_3 = r^2\sin\theta\,\mathrm{d}\theta\,\mathrm{d}\phi\\
\mathrm{d}s_2 &= h_3h_1\,\mathrm{d}u_1\,\mathrm{d}u_3 = r\sin\theta\,\mathrm{d}r\,\mathrm{d}\phi\\
\mathrm{d}s_3 &= h_1h_2\,\mathrm{d}u_1\,\mathrm{d}u_2 = r\,\mathrm{d}r\,\mathrm{d}\theta\\
\mathrm{d}V &= h_1h_2h_3\,\mathrm{d}u_1\,\mathrm{d}u_2\,\mathrm{d}u_3 = r^2\sin\theta\,\mathrm{d}r\,\mathrm{d}\theta\,\mathrm{d}\phi
\end{aligned}$$

Transformation matrix

$$\boldsymbol{\alpha} = \left[\frac{1}{h_k}\frac{\partial x_i}{\partial u_k}\right] = \begin{bmatrix} \dfrac{1}{1}\dfrac{\partial x}{\partial r} & \dfrac{1}{1}\dfrac{\partial y}{\partial r} & \dfrac{1}{1}\dfrac{\partial z}{\partial r} \\[2ex] \dfrac{1}{r}\dfrac{\partial x}{\partial \theta} & \dfrac{1}{r}\dfrac{\partial y}{\partial \theta} & \dfrac{1}{r}\dfrac{\partial z}{\partial \theta} \\[2ex] \dfrac{1}{r\sin\theta}\dfrac{\partial x}{\partial \phi} & \dfrac{1}{r\sin\theta}\dfrac{\partial y}{\partial \phi} & \dfrac{1}{r\sin\theta}\dfrac{\partial z}{\partial r\sin\theta} \end{bmatrix}$$

$$\text{i.e.}\quad \boldsymbol{\alpha} = \begin{bmatrix} \sin\theta\cos\phi & \sin\theta\sin\phi & \cos\theta \\ \cos\theta\cos\phi & \cos\theta\sin\phi & -\sin\theta \\ -\sin\phi & \cos\phi & 0 \end{bmatrix}$$

Unit vectors

Since

$$\hat{\boldsymbol{a}}_k = \sum_i \alpha_{ki}\hat{\boldsymbol{x}}_i$$

$$\begin{aligned}
\implies\quad \hat{\boldsymbol{r}} &= \hat{\boldsymbol{x}}\sin\theta\cos\phi + \hat{\boldsymbol{y}}\sin\theta\sin\phi + \hat{\boldsymbol{z}}\cos\theta\\
\hat{\boldsymbol{\theta}} &= \hat{\boldsymbol{x}}\cos\theta\cos\phi + \hat{\boldsymbol{y}}\cos\theta\sin\phi - \hat{\boldsymbol{z}}\sin\theta\\
\hat{\boldsymbol{\phi}} &= -\hat{\boldsymbol{x}}\sin\phi + \hat{\boldsymbol{y}}\cos\phi
\end{aligned}$$

F_r, F_θ, F_ϕ ***in terms of*** *F_x, F_y, F_z*

$$\begin{bmatrix} F_r \\ F_\theta \\ F_\phi \end{bmatrix} = \begin{bmatrix} \sin\theta\cos\phi & \sin\theta\sin\phi & \cos\theta \\ \cos\theta\cos\phi & \cos\theta\sin\phi & -\sin\theta \\ -\sin\phi & \cos\phi & 0 \end{bmatrix} \begin{bmatrix} F_x \\ F_y \\ F_z \end{bmatrix}$$

$$\begin{aligned}
\implies\quad F_r &= F_x\sin\theta\cos\phi + F_y\sin\theta\sin\phi + F_z\cos\theta\\
F_\theta &= F_x\cos\theta\cos\phi + F_y\cos\theta\sin\phi - F_z\sin\theta\\
F_\phi &= -F_x\sin\phi + F_y\cos\phi
\end{aligned}$$

F_x, F_y, F_z in terms of F_ρ, F_ϕ, F_z

The matrix for the inverse transformation is

$$\begin{bmatrix} \sin\theta\cos\phi & \cos\theta\cos\phi & -\sin\phi \\ \sin\theta\sin\phi & \cos\theta\sin\phi & \cos\phi \\ \cos\theta & -\sin\theta & 0 \end{bmatrix}$$

Hence

$$\Longrightarrow \quad \begin{aligned} F_x &= F_r\sin\theta\cos\phi + F_\theta\cos\theta\cos\phi - F_\phi\sin\phi \\ F_y &= F_r\sin\theta\sin\phi + F_\theta\cos\theta\sin\phi + F_\phi\cos\phi \\ F_z &= F_r\cos\theta - F_\theta\sin\theta \end{aligned}$$

grad, div, and curl

$$\boldsymbol{\nabla} V = \textbf{grad } V = \hat{\boldsymbol{r}}\frac{\partial V}{\partial r} + \hat{\boldsymbol{\theta}}\frac{1}{r}\frac{\partial V}{\partial \theta} + \hat{\phi}\frac{1}{r\sin\theta}\frac{\partial V}{\partial \phi}$$

$$\boldsymbol{\nabla}\cdot\boldsymbol{F} = \textbf{div } \boldsymbol{F} = \frac{1}{r^2}\frac{\partial}{\partial r}(r^2F_r) + \frac{1}{r\sin\theta}\frac{\partial}{\partial\theta}(F_\theta\sin\theta) + \frac{1}{r\sin\theta}\frac{\partial F_\phi}{\partial\phi}$$

$$\begin{aligned} \boldsymbol{\nabla}\times\boldsymbol{F} = \textbf{curl } \boldsymbol{F} &= \frac{\hat{\boldsymbol{r}}}{r\sin\theta}\left(\frac{\partial(\sin\theta F_\phi)}{\partial\theta} - \frac{\partial F_\theta}{\partial\phi}\right) \\ &+ \hat{\boldsymbol{\theta}}\left(\frac{1}{r\sin\theta}\frac{\partial F_r}{\partial\phi} - \frac{1}{r}\frac{\partial(rF_\phi)}{\partial r}\right) + \frac{\hat{\phi}}{r}\left(\frac{\partial(rF_\theta)}{\partial r} - \frac{\partial F_r}{\partial\theta}\right) \end{aligned}$$

$$\begin{aligned} \nabla^2 V = \textbf{div (grad } V) &= \frac{1}{r^2}\frac{\partial}{\partial r}\left(r^2\frac{\partial V}{\partial r}\right) + \frac{1}{r^2\sin\theta}\frac{\partial}{\partial\theta}\left(\sin\theta\frac{\partial V}{\partial\theta}\right) \\ &+ \frac{1}{r^2\sin^2\theta}\frac{\partial^2 V}{\partial\phi^2} \end{aligned}$$

B

Power series

We list some results pertaining to power series. A good reference for this appendix is Knopp [10].

A *power series* in the variable x is an infinite summation of the form $\sum_{n=0}^{\infty} a_n(x-x_0)^n$, where x_0 is a constant (possibly zero) and a_n is the coefficient of the nth term in the series. An infinite sum has meaning only if it's *convergent*. By the ratio test (the simplest test for convergence), a series converges for x such that

$$\lim_{n\to\infty}\left|\frac{a_{n+1}(x-x_0)^{n+1}}{a_n(x-x_0)^n}\right| = |x-x_0| \lim_{n\to\infty}\left|\frac{a_{n+1}}{a_n}\right| < 1.$$

The series thus converges (absolutely) for $|x-x_0|<r$, where $r^{-1} \equiv \lim_{n\to\infty}|a_{n+1}/a_n|$ (which may be zero). The range of x for which $|x-x_0|<r$ is called the *interval of convergence*. The ratio test presumes that $\lim_{n\to\infty}|a_{n+1}/a_n|$ exists. There can be cases where there are two or more limit points of the sequence $|a_{n+1}/a_n|$, in which case a better definition of the interval of convergences is $r^{-1} = \overline{\lim}|a_n|^{(1/n)}$, where $\overline{\lim}$ means the maximum limit point [10, p. 154].

It can be shown that a power series defines a smooth function f for x within the interval of convergence:

$$f(x) \equiv \sum_{n=0}^{\infty} a_n(x-x_0)^n. \qquad (|x-x_0|<r) \tag{B.1}$$

(A *smooth function* has continuous derivatives of all orders.) A function is said to be *analytic* at x_0 if it has a power-series representation for $|x-x_0|<r$. An analytic function thus has derivatives of all orders; such derivatives are also analytic at x_0 and are represented by power series obtained by differentiating Eq. (B.1) term by term. Thus,

Mathematical Methods in Physics, Engineering, and Chemistry, First Edition.
Brett Borden and James Luscombe.

$$f'(x) = a_1 + 2a_2(x - x_0) + \cdots + na_n(x - x_0)^{n-1} + \cdots$$
$$f''(x) = 2a_2 + 6a_3(x - x_0) + \cdots + n(n-1)a_n(x - x_0)^{n-2} + \cdots$$
$$\vdots \tag{B.2}$$
$$f^{(n)}(x) = n!a_n + (n+1)!a_{n+1}(x - x_0) + \cdots + \frac{(n+k)!}{k!}a_{n+k}(x - x_0)^k + \cdots.$$

From Eqs. (B.2) and (B.1),

$$f(x_0) = a_0 \qquad f'(x_0) = a_1 \qquad \cdots \qquad f^{(n)}(x_0) = n!a_n. \tag{B.3}$$

A power series can therefore be expressed in terms of f and its derivatives at $x = x_0$:

$$f(x) = f(x_0) + f'(x_0)(x - x_0) + \cdots + \frac{1}{n!}f^{(n)}(x_0)(x - x_0)^n + \cdots . \tag{B.4}$$

Equation (B.4) is the familiar Taylor series for f.

Often we want to know whether a given function $f(x)$, defined in a neighborhood of x_0, is analytic at x_0. A practical definition is that *a function is analytic if its Taylor series about x_0 converges to the function in some neighborhood of x_0*. Analytic functions are infinitely differentiable, or smooth, but not all smooth functions are analytic. There are examples of non analytic smooth functions. Consider

$$f(x) \equiv \begin{cases} \mathrm{e}^{-(1/x)} & x > 0 \\ 0 & x \leq 0. \end{cases}$$

This function is infinitely differentiable for all $x > 0$. Precisely at $x = 0$, all derivatives are equal to zero. The Taylor series of f about $x = 0$ converges to the zero function,

$$\sum_{n=0}^{\infty} \frac{f^{(n)}(0)}{n!}x^n = \sum_{n=0}^{\infty} \frac{0}{n!}x^n = 0.$$

The function f is therefore not analytic at $x = 0$: The Taylor series does not converge to $f(x)$ for $x > 0$. While this example is pathological (meaning you won't run into it in physics), it illustrates the point. If f has derivatives of all orders at x_0, then, for finite n:

$$f(x) = f(x_0) + f'(x_0)(x - x_0) + \cdots + \frac{1}{n!}f^{(n)}(x_0)(x - x_0)^n + R_n(x),$$

where the "remainder" is given by (using the mean-value theorem of calculus)

$$R_n(x) = \frac{f^{(n+1)}(\xi)}{(n+1)!}(x - x_0)^{n+1},$$

where ξ is between x and x_0. Another way of characterizing whether a function is analytic is that the remainder must vanish as $n \to \infty$: $\lim_{n\to\infty} R_n(x) = 0$.

C

The gamma function, $\Gamma(x)$

The gamma function, $\Gamma(x)$, the graph of which is shown in Figure C.1, originated in the quest to find a generalization of factorials to real numbers. Can one find a function $\Gamma(x)$ such that $\Gamma(x+1) = x\Gamma(x)$? The answer is in the affirmative; it's defined by the *Euler integral*

$$\Gamma(x) \equiv \int_0^{\infty} t^{x-1} \mathrm{e}^{-t} \, \mathrm{d}t. \qquad (x > 0) \tag{C.1}$$

From direct integration of Eq. (C.1), we have the special values $\Gamma(1) = 1$ and $\Gamma(\frac{1}{2}) = \sqrt{\pi}$. For $x = n$ a positive integer, Eq. (C.1) can be evaluated by repeated integration by parts, with the result $\Gamma(n) = (n-1)!$ $(n = 1, 2, \ldots)$.

Recursion relation

One of the most useful properties of $\Gamma(x)$ is the recursion relation it satisfies. Integrate Eq. (C.1) by parts,

$$\Gamma(x) = -t^{x-1}\mathrm{e}^{-t}|_0^{\infty} + (x-1)\int_0^{\infty} t^{x-2}\mathrm{e}^{-t} \, \mathrm{d}t = (x-1)\Gamma(x-1).$$

This result is usually written

$$\Gamma(x+1) = x\Gamma(x). \tag{C.2}$$

Thus, $\Gamma(x)$ is the generalization of the factorial to real numbers.

Equation (C.2) is invaluable for numerical purposes; it also enables $\Gamma(x)$ to be analytically continued to negative values of x, as we'll show.

Mathematical Methods in Physics, Engineering, and Chemistry, First Edition.
Brett Borden and James Luscombe.

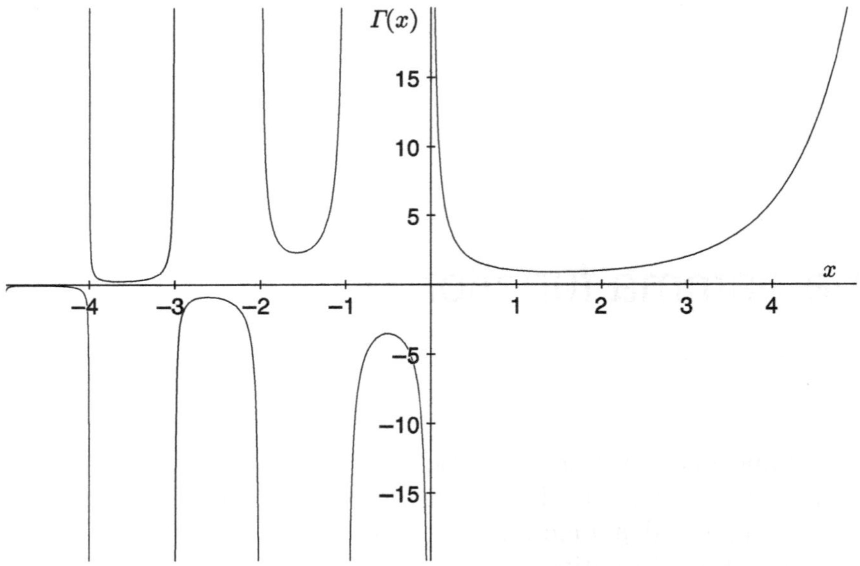

Figure C.1 $\Gamma(x)$ versus x.

We see from Eq. (C.2) that $\lim_{x\to 0} x\Gamma(x) = \Gamma(1) = 1$, and thus, as $x \to 0$, $\Gamma(x) \approx x^{-1}$, i.e. it has a simple pole[1] at $x = 0$. Moreover, it has simple poles at all negative integers. Let $x = -n$ in Eq. (C.2) and iterate:

$$\Gamma(-n) = \frac{1}{(-n)}\Gamma(-n+1) = \frac{1}{(-n)}\frac{1}{(-n+1)}\Gamma(-n+2)$$
$$= \cdots = \frac{(-1)^n}{n!}\Gamma(0). \tag{C.3}$$

Clearly, $\Gamma^{-1}(-n) = 0$.

Limit formula

The gamma function has other useful properties that can be derived starting from the Euler definition of the exponential function

$$\lim_{n\to\infty}\left(1 - \frac{t}{n}\right)^n = \mathrm{e}^{-t}. \tag{C.4}$$

Combining Eq. (C.4) with Eq. (C.1) suggests that we consider the limiting behavior of the integral

[1] A *pole* is a type of singularity that a function can have; see Chapter 8. Near a *simple pole* at $x = x_0$ (the types of poles possessed by $\Gamma(x)$), a function $f(x) \approx a/(x - x_0)$, where the constant a is known as the *residue* of the function. The residue is defined as $a = \lim_{x\to x_0}(x - x_0)f(x)$.

$$\Gamma_n(x) \equiv \int_0^n \left(1 - \frac{t}{n}\right)^n t^{x-1}\, \mathrm{d}t \tag{C.5}$$

as $n \to \infty$ for fixed x. Through repeated integration by parts, $\Gamma_n(x)$ has the value

$$\Gamma_n(x) = \frac{n! n^x}{x(x+1)\cdots(x+n)}. \tag{C.6}$$

It can be shown[2] that $\lim_{n\to\infty}\Gamma_n(x) = \Gamma(x)$. Thus, we have the *Euler limit formula*,

$$\Gamma(x) = \lim_{n\to\infty} \frac{n! n^x}{x(x+1)\cdots(x+n)}. \tag{C.7}$$

Reflection formula

An application of Eq. (C.7) is to derive the *Euler reflection formula*,

$$\Gamma(x)\Gamma(1-x) = \frac{\pi}{\sin \pi x} \qquad (x \neq 0, \pm 1, \pm 2, \ldots). \tag{C.8}$$

Using Eq. (C.7),

$$\begin{aligned}\frac{1}{\Gamma(x)\Gamma(1-x)} &= \lim_{n\to\infty}\left\{\frac{x(x+1)\cdots(x+n)}{n! n^x}\frac{(1-x)(2-x)\cdots(n+1-x)}{n! n^{1-x}}\right\} \\ &= x \lim_{n\to\infty}\left(\frac{n+1-x}{n}\right)\prod_{k=1}^{n}\left(1-\frac{x^2}{k^2}\right) = x\prod_{k=1}^{\infty}\left(1-\frac{x^2}{k^2}\right) \\ &= \frac{\sin \pi x}{\pi},\end{aligned} \tag{C.9}$$

where we've used Eq. (8.53). Inverting Eq. (C.9), we have Eq. (C.8). For $x = \frac{1}{2}$ in Eq. (C.8), we recover $\Gamma(\frac{1}{2}) = \sqrt{\pi}$, the same as we found from Eq. (C.1).

Digamma function

In Chapter 5, the derivative of $\Gamma^{-1}(x)$ is required, evaluated at integers. To derive this quantity takes a few steps, and it helps to first get the logarithmic derivative of $\Gamma(x)$. Using Eq. (C.2), $\Gamma(t) = (t-1)(t-2)\cdots(t-m)\Gamma(t-m)$. Take the logarithm, $\ln\Gamma(t) = \ln(t-1) + \ln(t-2) + \cdots + \ln(t-m) + \ln\Gamma(t-m)$. Now differentiate,

$$\frac{\mathrm{d}}{\mathrm{d}t}\ln\Gamma(t) = \frac{1}{t-1} + \frac{1}{t-2} + \cdots + \frac{1}{t-m} + \frac{1}{\Gamma(t-m)}\Gamma'(t-m). \tag{C.10}$$

[2]See for example Olver [47, p. 33].

Turn Eq. (C.10) around to isolate $\Gamma'(1)$ because that's a tabulated quantity:

$$\Gamma'(1) = -\sum_{k=1}^{t} \frac{1}{k} + \frac{\Gamma'(t+1)}{\Gamma(t+1)}. \tag{C.11}$$

We know that $\Gamma(n+1) = n!$. Using Stirling's approximation (Eq. (8.39)) for large n, $n! \overset{n\to\infty}{\sim} \sqrt{2\pi} n^{n+\frac{1}{2}} e^{-n}$. Differentiating Stirling's approximation[3] with respect to n, it's straightforward to show that

$$\frac{\Gamma'(n+1)}{\Gamma(n+1)} \overset{n\to\infty}{\sim} \ln n. \tag{C.12}$$

Substituting Eq. (C.12) into Eq. (C.11) and letting $n \to \infty$,

$$\Gamma'(1) = -\lim_{n\to\infty} \left[\sum_{k=1}^{n} \frac{1}{k} - \ln n \right] \equiv -\gamma \approx -0.5772, \tag{C.13}$$

where $\gamma = 0.57721\cdots$ is the *Euler–Mascheroni constant.* Combine Eq. (C.13) with Eq. (C.10),

$$\frac{\mathrm{d}}{\mathrm{d}t} \ln \Gamma(t) = \frac{\Gamma'(t)}{\Gamma(t)} \equiv \psi(t) = \begin{cases} -\gamma & t = 1 \\ -\gamma + \sum_{k=1}^{t-1} k^{-1}, & t \geq 2 \end{cases} \tag{C.14}$$

where $\psi(t)$ is the *digamma function* evaluated at positive integers.[4] Assembling the pieces (for $n \geq 1$):

$$\frac{\mathrm{d}}{\mathrm{d}t}\Gamma^{-1}(t)\bigg|_{t=n} = -\Gamma^{-2}(t)\Gamma'(t)|_{t=n} = -\frac{1}{\Gamma(t)}\psi(t)\bigg|_{t=n} = -\frac{1}{(n-1)!}\psi(n). \tag{C.15}$$

We'll require (Chapter 5) the derivative of $\Gamma^{-1}(x)$ evaluated at zero and negative integers. Let $t \to t+k+1$ in Eq. (C.10) and let $m = k+1$:

$$\frac{\Gamma'(t+k+1)}{\Gamma(t+k+1)} = \frac{1}{t+k} + \frac{1}{t+k-1} + \cdots + \frac{1}{t} + \frac{\Gamma'(t)}{\Gamma(t)}.$$

Using this formula, we have

$$\begin{aligned} \frac{\mathrm{d}}{\mathrm{d}t}\left(\frac{1}{\Gamma(t)}\right) &= -\frac{1}{\Gamma(t)}\frac{\mathrm{d}}{\mathrm{d}t}\ln\Gamma(t) \\ &= \frac{1}{\Gamma(t)}\left[\frac{1}{t+k} + \frac{1}{t+k-1} + \cdots + \frac{1}{t} - \frac{\Gamma'(t+k+1)}{\Gamma(t+k+1)}\right]. \end{aligned}$$

[3] In Stirling's approximation, we're treating n as a continuous quantity, valid for large n.
[4] The digamma function is defined as $\psi(z) \equiv \Gamma'(z)/\Gamma(z)$.

Now formally let $t \to -k$, where $k = 0, -1, -2, \cdots$; $\Gamma(t)$ in the denominator diverges at these points, eliminating all terms except the first. Using Eq. (C),

$$\begin{aligned} \frac{\mathrm{d}}{\mathrm{d}t}\left(\frac{1}{\Gamma(t)}\right)\bigg|_{t=-k} &= \lim_{t\to -k}\left(\frac{1}{\Gamma(t)(t+k)}\right) \\ &= (-1)^k k! \lim_{x\to 0}\left(\frac{1}{x\Gamma(x)}\right) = (-1)^k k!. \end{aligned} \tag{C.16}$$

D

Boundary conditions for Partial Differential Equations

As discussed in Chapter 9, the boundary conditions on Green functions must be chosen in conjunction with the boundary conditions on the solutions of the underlying partial differential equation (PDE). In this appendix, we examine the boundary conditions for each of the three possible types of linear PDEs (see Section 3.4).

All second-order linear PDEs in two variables can be reduced to one of three canonical forms, Eqs. (3.66), (3.69), or (3.72). To discuss the permissible boundary conditions on each type of PDE, consider Figure D.1 that indicates a domain of the xy-plane over which a PDE involving the function ψ holds, together with a curve Γ on which boundary conditions are applied. It might be supposed (as with ordinary differential equations [ODEs]) that at each point of Γ, one could specify ψ (Dirichlet boundary condition), its derivative (Neumann boundary condition), or a combination (mixed or Cauchy boundary conditions). As we'll see, the situation is more involved for PDEs.

One complication is that the domain might not be *bounded* (contrary to what's shown in Figure D.1). Another is that to speak of derivatives of functions of more than one independent variable, a direction must be specified. The change in $\psi(x, y)$ under an infinitesimal displacement $\mathrm{d}\boldsymbol{r}$ is obtained from the gradient, $\mathrm{d}\psi = \boldsymbol{\nabla}\,\psi \cdot \mathrm{d}\boldsymbol{r}$. The derivative of ψ in the direction of a general unit vector $\hat{\boldsymbol{u}}$ (the directional derivative) is $D_{\hat{\boldsymbol{u}}}\psi \equiv \hat{\boldsymbol{u}} \cdot \boldsymbol{\nabla}\,\psi$ (see Appendix A). For PDEs, Neumann boundary conditions consist of specifying the *normal derivative*, $\hat{\boldsymbol{n}} \cdot \boldsymbol{\nabla}\,\psi$, where $\hat{\boldsymbol{n}}$ is a unit vector locally perpendicular to the boundary.[1]

[1] A component of $\boldsymbol{\nabla}\,\psi$ *along* the boundary is essentially equivalent to a Dirichlet boundary condition: A tangential component of $\boldsymbol{\nabla}\,\psi$ can be integrated along the boundary yielding (up to a constant) the value of ψ on the boundary.

Mathematical Methods in Physics, Engineering, and Chemistry, First Edition.
Brett Borden and James Luscombe.

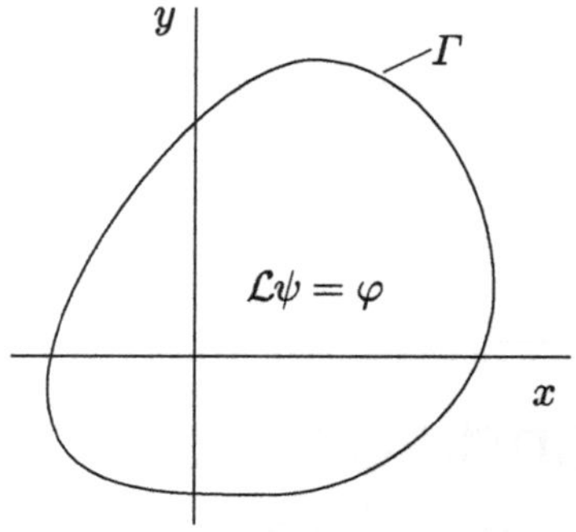

Figure D.1 Boundary value problem.

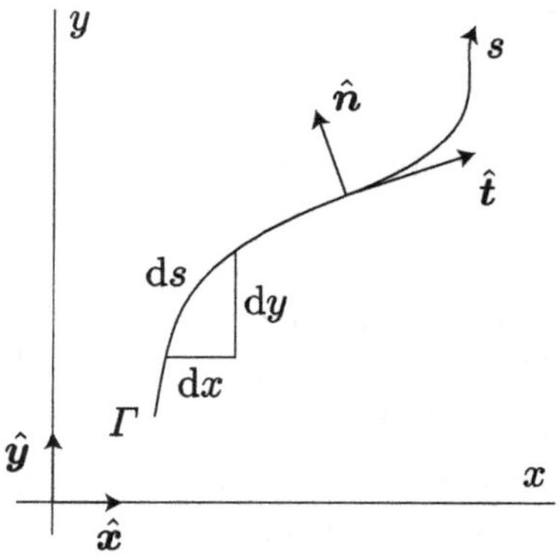

Figure D.2 Boundary curve Γ.

For second-order ODEs, the specification of ψ and ψ' at point x_0 is sufficient to determine the derivatives $\psi^{(n)}(x_0)$ for $n \geq 2$, thus ensuring the existence[2] of $\psi(x)$ in the form of a Taylor series for x near x_0. Is the same true for PDEs? Does specifying ψ and $\nabla\psi \cdot \hat{n}$ at a boundary suffice to determine ψ everywhere? As we'll see: It depends.

Let Γ be specified parametrically by functions $x = X(s)$ and $y = Y(s)$, where s is the arc length with $(\mathrm{d}s)^2 = (\mathrm{d}x)^2 + (\mathrm{d}y)^2$ (see Figure D.2). At any point of Γ, the *tangent vector* $\hat{t}$ has components (in the $\hat{x}$, $\hat{y}$ basis) $\hat{t} = (\mathrm{d}X/\mathrm{d}s,\ \mathrm{d}Y/\ \mathrm{d}s) \equiv (X', Y')$. The normal vector $\hat{n}$ thus has components[3] $\hat{n} = (-Y', X')$. The gradient $\nabla\psi$ has components (ψ_x, ψ_y) in the same basis. Thus, we have the normal derivative

$$\hat{n} \cdot \nabla\psi = -Y'\psi_x + X'\psi_y, \tag{D.1}$$

while the derivative of ψ along Γ is:

$$\frac{\mathrm{d}\psi}{\mathrm{d}s} \equiv \hat{t} \cdot \nabla\psi = X'\psi_x + Y'\psi_y. \tag{D.2}$$

[2]Second-order ODEs have power-series solutions determined by the value of a function and it's derivative at a point (Section 5.1.3). All higher derivatives can be found from the power series.

[3]You should be able to show that $\hat{t} \cdot \hat{n} = 0$. Moreover, you should see that $\hat{t} \cdot \hat{t} = \hat{n} \cdot \hat{n} = 1$.

Equations (D.1) and (D.2) can be solved[4] for ψ_x and ψ_y:

$$\psi_x = X'\frac{d\psi}{ds} - Y'\nabla\psi\cdot\hat{n} \qquad \psi_y = X'\nabla\psi\cdot\hat{n} + Y'\frac{d\psi}{ds}.$$

The derivatives ψ_x and ψ_y are therefore determined on the boundary by the boundary conditions.[5] What about second derivatives?

With ψ_x and ψ_y known, they can be differentiated along the boundary:

$$\frac{d}{ds}\psi_x = \psi_{xx}X' + \psi_{xy}Y' \qquad \frac{d}{ds}\psi_y = \psi_{xy}X' + \psi_{yy}Y', \tag{D.3}$$

where $\psi_{xy} \equiv \partial^2\psi/\partial x\partial y$, etc.[6] Between Eq. (D.3) and the original PDE (see Eq. (3.56)), we have three equations in three unknowns:

$$\begin{pmatrix} R & S & T \\ X' & Y' & 0 \\ 0 & X' & Y' \end{pmatrix}\begin{pmatrix}\psi_{xx}\\ \psi_{xy}\\ \psi_{yy}\end{pmatrix} = \begin{pmatrix} F \\ d\psi_x/\,ds \\ d\psi_y/\,ds\end{pmatrix}, \tag{D.4}$$

where all quantities on the right are known from boundary data. Solutions to Eq. (D.4) exist when the determinant of the matrix is nonzero. Evaluate the determinant:

$$\begin{vmatrix} R & S & T \\ X' & Y' & 0 \\ 0 & X' & Y' \end{vmatrix} = (X')^2\left(R\left(\frac{Y'}{X'}\right)^2 - S\frac{Y'}{X'} + T\right)$$
$$= (X')^2\left[R\left(\frac{dy}{dx}\right)^2 - S\frac{dy}{dx} + T\right], \tag{D.5}$$

where $Y'/X' = dy/\,dx$ on the boundary.

If the determinant is nonzero, the second derivatives of ψ on Γ are determined by the boundary data. In that case, all higher derivatives of ψ on Γ can be calculated. The value of ψ at points away from Γ can then be obtained from a Taylor series.[7] If the determinant is zero, however, the second derivatives of ψ on Γ are *not* determined by boundary data. Under what conditions does the determinant vanish? *The terms in square brackets of Eq.* (D.5) *vanish at points where Γ is tangent to a characteristic curve* (characteristics are discussed in Section 3.4). To see this, note from Eq. (3.65) that when $dy/\,dx = -\lambda_\pm$, the terms in square brackets vanish by virtue of Eq. (3.60). *Characteristics are curves* significance of characteristics

[4]The determinant of the coefficients is equal to -1 and so a solution always exists.

[5]Said differently, ψ_x and ψ_y cannot be specified independently: They're determined by $\psi(s)$ and $\nabla\psi\cdot\hat{n}$.

[6]Equation (D.3) presumes that the second-order derivatives ψ_{xx}, ψ_{xy}, ψ_{yy} exist – the very issue we're trying to establish.

[7]ψ is determined within a distance of the boundary equal to the radius of convergence of the series.

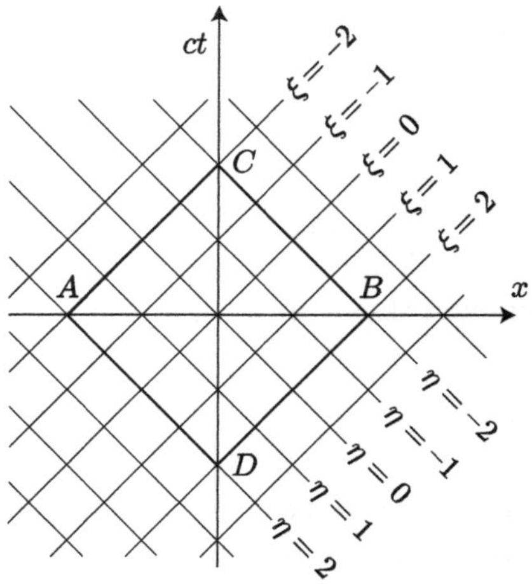

Figure D.3 Characteristic curves for the one-dimensional wave equation.

along which the second derivatives of ψ are incompletely specified by the PDE, Eq. (3.56). Characteristics thus complicate the issue of boundary conditions.

We can now state the boundary conditions (and boundaries!) suitable for PDEs.

Hyperbolic

The simplest hyperbolic PDE is the wave equation. Its characteristics are the straight lines $\xi = x - ct$ and $\eta = x + ct$. A point (x, ct) has coordinates (ξ, η) with $x = \frac{1}{2}(\eta + \xi)$ and $ct = \frac{1}{2}(\eta - \xi)$. Let the boundary conditions $\psi(x, 0)$ and $\nabla\psi \cdot \hat{\boldsymbol{n}} = (\partial\psi/\partial t)(x, t)|_{t=0}$ be given[8] on the line AB in Figure D.3. The wave equation has solutions of the form (see Section 3.4)

$$\psi(x, t) = f(\xi) + g(\eta), \tag{D.6}$$

where f and g are determined by the boundary conditions:

$$\begin{aligned} \psi(x, 0) &= f(x) + g(x) \\ \nabla\psi \cdot \hat{\boldsymbol{n}} &= -cf'(x) + cg'(x). \end{aligned} \tag{D.7}$$

Equation (D.7) can be solved[9] for f and g,

$$f(x) = \frac{1}{2}\psi(x, 0) - \frac{1}{2c}\int^{x} (\nabla\psi \cdot \hat{\boldsymbol{n}})(x')\mathrm{d}x' \quad g(x) = \frac{1}{2}\psi(x, 0) + \frac{1}{2c}\int^{x} (\nabla\psi \cdot \hat{\boldsymbol{n}})(x')\,\mathrm{d}x'. \tag{D.8}$$

[8]Note that the boundary conditions here, in space and time, are what are normally referred to as *initial conditions*. Because the value of ψ and its slope $\partial\psi/\partial t$ are both specified at $t = 0$, we can refer to these boundary conditions as *mixed initial conditions*.

[9]The integration constant is immaterial because it cancels in subsequent formulae.

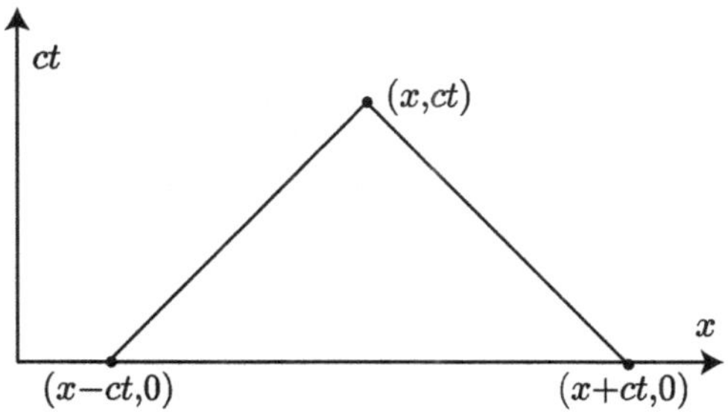

Figure D.4 Domain of dependence for $\psi(x, ct)$.

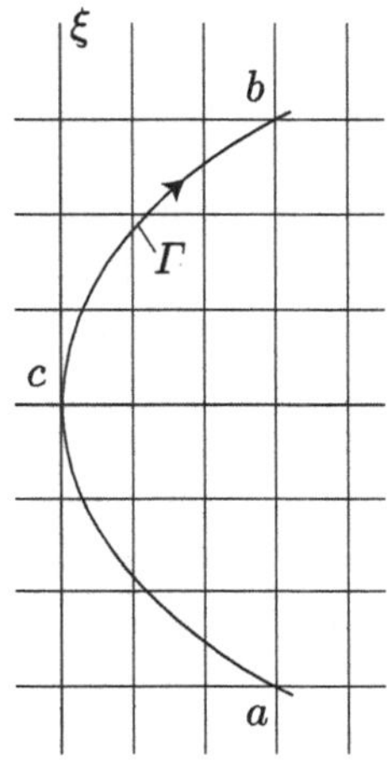

Figure D.5 Boundary curve tangent to a characteristic at point c.

From Eqs. (D.6) and (D.8),

$$\psi(x,t) = \frac{1}{2}[\psi(x-ct,0) + \psi(x+ct,0)] + \frac{1}{2c}\int_{x-ct}^{x+ct} (\nabla\,\psi \cdot \hat{\boldsymbol{n}})(x')\,\mathrm{d}x'. \tag{D.9}$$

At a given point (x, ct), $\psi(x, t)$ depends on the value of ψ at $(x - ct, 0)$ and $(x + ct, 0)$, and on the values of $\nabla\,\psi \cdot \hat{\boldsymbol{n}}$ along the interval $[x - ct, x + ct]$. The characteristics can thus be thought of as curves along which partial boundary information "propagates" to the point (x, ct). The interval $[x - ct, x + ct]$ is known as the *domain of dependence* of ψ at (x, ct), because $\psi(x, t)$ depends *only* on the boundary data in that interval.[10] Unless both characteristics that pass through a given point (x, ct) intersect the boundary curve, *the solution is not determined at that point.* Referring to Figure D.3, the solution is determined in the region $ACBD$.

Solutions are unambiguously determined by boundary data when the characteristics intersect the boundary curve only once. In Figure D.5, the boundary curve Γ becomes tangent to a characteristic at point c and "loops back" on itself, intersecting the family of ξ-characteristics twice.

---

[10] Readers familiar with the theory of relativity will recognize Figure D.4 as a spacetime diagram, with ξ and η the "lightcone coordinates."

Boundary conditions specified along portion ac of Γ suffice to determine the values of the ξ-characteristics that intersect the boundary between a and c *and* between c and b. Boundary conditions specified along portion cb of Γ then "overdetermine" the solution to the right of Γ, implying no solution at all. *Closed* boundaries, such as depicted in Figure D.1, cannot lead to well-defined solutions because every interior characteristic crosses the boundary twice. *Hyperbolic PDEs have solutions determined by boundary conditions on open boundaries that are nowhere tangent to characteristics.*

boundaries for hyperbolic PDEs

Elliptic

It turns out that *boundary conditions on open boundaries cannot be specified for elliptic PDEs.*[11] Boundary conditions on open boundaries, which work for hyperbolic PDEs,[12] fail for elliptic equations. Moreover, *mixed* boundary conditions cannot be specified on closed boundaries, as we show. *Unique solutions to elliptic PDEs are obtained for closed boundaries with Dirichlet or Neumman boundary conditions, but not both.*

boundaries for elliptic PDEs

We can see why mixed boundary conditions cannot be specified for elliptic PDEs by looking at the *uniqueness* of solutions. Consider the Poisson equation, $\nabla^2\psi(\boldsymbol{r}) = -\rho(\boldsymbol{r})$. Here, we seek a solution in a three-dimensional domain V enclosed by a surface S. Suppose there are *two* solutions ψ_1 and ψ_2 satisfying the Poisson equation, each satisfying the same type of boundary condition on S: Either both satisfy Dirichlet conditions or both satisfy Neumann conditions. The difference in solutions $\phi(\boldsymbol{r}) \equiv \psi_1(\boldsymbol{r}) - \psi_2(\boldsymbol{r})$ therefore satisfies the Laplace equation $\nabla^2\phi = 0$ with *homogeneous boundary conditions.* The function $\phi(\boldsymbol{r})$ satisfies homogenous boundary conditions because ψ_1, ψ_2 satisfy the same type of boundary conditions. Consider the construct $F \equiv \int_V \boldsymbol{\nabla}\phi \cdot \boldsymbol{\nabla}\phi \, \mathrm{d}^3r$, which is non-negative, $F \geq 0$. Using the identity $\boldsymbol{\nabla}\cdot(\phi\boldsymbol{\nabla}\phi) = \phi\nabla^2\phi + \boldsymbol{\nabla}\phi\cdot\boldsymbol{\nabla}\phi$, F is equivalent to (because $\nabla^2\phi = 0$) $F = \int_V \boldsymbol{\nabla}\cdot(\phi\boldsymbol{\nabla}\phi)\,\mathrm{d}^3r = \int_S \phi\boldsymbol{\nabla}\phi\cdot\hat{\boldsymbol{n}}\,\mathrm{d}s$. If $\phi = 0$ on S (Dirichlet condition), then $F = 0$. If $\boldsymbol{\nabla}\phi\cdot\hat{\boldsymbol{n}} = 0$ on S (Neumann condition), then $F = 0$. If $F = 0$, then[13] $\boldsymbol{\nabla}\phi = 0$, and $\phi =$ constant. For Dirichlet conditions, $\phi = 0$, and the solution is unique. For Neumann conditions, however, the solution is unique *up to a constant.* Can we have solutions satisfying mixed boundary conditions? Suppose $\phi + \beta\boldsymbol{\nabla}\phi\cdot\hat{\boldsymbol{n}} = 0$ on S, or that $\phi = -\beta\boldsymbol{\nabla}\phi\cdot\hat{\boldsymbol{n}}$ on S. If $\beta > 0$, then $F < 0$, which cannot be. For $\beta < 0$, $F > 0$, implying $\boldsymbol{\nabla}\phi \neq 0$, and thus the solution is not unique.

boundary conditions for the Poisson equation

[11] No attempt at completeness is made regarding this point as there is a substantial amount of mathematics underlying it. Reference can be made to Sneddon [48, pp. 110–114], who uses the method of characteristics; to Morse and Feshbach [18, section 6.2], who use numerical methods; and to Hadamard [49, chapters 1 and 2].

[12] When the boundary is not a characteristic curve, nor tangent anywhere to a characteristic.

[13] For $F = \int_V (\nabla\phi)^2\,\mathrm{d}^3r = 0$, the integrand must be zero everywhere.

Unique solutions of the Poisson equation require Dirichlet or Neumann conditions on closed boundaries, but not both.

Uniqueness of solutions for parabolic and hyperbolic PDEs

The new wrinkle in progressing from elliptic to parabolic or hyperbolic equations is *time*. The usual parabolic equation in physics is the diffusion equation, Eq. (3.70). For no variations in time, the diffusion equation reduces to the Laplace equation (Poisson if inhomogeneous); *ditto for the wave equation.* The diffusion and wave equations thus share with elliptic equations that unique solutions are obtained for closed *spatial* boundaries with either Dirichlet or Neumann conditions, but not both. One sees stated, however, that boundary conditions for parabolic and hyperbolic equations must be specified on *open* surfaces [18, p. 706]. An ambiguity arises when time is considered a dimension – what's meant by an *open* space-time volume? We don't ordinarily think of time as closing on itself, and thus, when space and time are taken together, a closed spatial boundary concatenated with the time dimension comprises an open boundary in spacetime.[14]

Diffusion equation

The inhomogeneous diffusion equation, Eq. (3.8) (for a system in spatial volume V that is bounded by surface S), can be written in terms of a *diffusion operator*

$$\mathcal{L}\psi \equiv \left(D\nabla^2 - \frac{\partial}{\partial t}\right)\psi(\boldsymbol{r},t) = -\rho(\boldsymbol{r},t).$$

Let there be functions ψ_1, ψ_2 satisfying $\mathcal{L}\psi_1 = -\rho(\boldsymbol{r},t)$ and $\mathcal{L}\psi_2 = -\rho(\boldsymbol{r},t)$, where each satisfies the same boundary conditions on S *and* are such that at time $t = t_0$, $\psi_1(\boldsymbol{r},t_0) = \psi_2(\boldsymbol{r},t_0)$, i.e. both satisfy the same *initial condition.* An initial condition is a "boundary condition" in time.[15] Form the difference $\phi(\boldsymbol{r},t) \equiv \psi_1(\boldsymbol{r},t) - \psi_2(\boldsymbol{r},t)$. The quantity $\phi(\boldsymbol{r},t)$ satisfies the homogeneous diffusion equation, $\mathcal{L}\phi = 0$, subject to homogeneous spatial boundary conditions *and* a homogeneous initial condition, $\phi(\boldsymbol{r},t_0) = 0$. initial condition

Consider $F \equiv \int_{t_0}^{T}\int_V \phi\mathcal{L}\phi \, \mathrm{d}^3r \, \mathrm{d}t = 0$, where $T > t_0$. (Note that F involves an integral over four-dimensional spacetime.) F is zero because

[14]What constitutes a surface? Ordinarily surfaces are two-dimensional structures that we see from three dimensions, such as the surface of an orange. Higher dimensional surface-like structures are referred to as *hypersurfaces*. A hypersurface is an embedding in n-dimensional space of an $(n-1)$-dimensional surface.

[15]Boundary conditions are the specification of quantities not involving the highest order derivatives in a PDE. Because the diffusion equation contains a first-order time derivative, the value of the function must be specified at a given instant of time, an initial condition. The wave equation, in contrast, has second-order time derivatives, and we must specify initial conditions on the function and its first time derivative.

$\mathcal{L}\phi = 0$. Make use of the identities $\phi\nabla^2\phi = \boldsymbol{\nabla}\cdot(\phi\,\boldsymbol{\nabla}\,\phi) - (\nabla\phi)^2$ and $\phi\partial\phi/\partial t = \frac{1}{2}(\partial/\partial t)(\phi)^2$. Using the divergence theorem,

$$\begin{aligned} F =& D\int_{t_0}^{T}\int_S \phi\,\boldsymbol{\nabla}\,\phi\cdot\hat{\boldsymbol{n}}\;\mathrm{d}s\;\mathrm{d}t \\ &-\left[D\int_{t_0}^{T}\int_V (\nabla\phi(\boldsymbol{r},t))^2\;\mathrm{d}^3r\;\mathrm{d}t + \frac{1}{2}\int_V (\phi(\boldsymbol{r},T))^2\;\mathrm{d}^3r\right] = 0. \quad \text{(D.10)}\end{aligned}$$

For either Dirichlet or Neumann conditions specified on S (but not both), the first integral on the right of Eq. (D.10) vanishes. Because $F = 0$, the terms in square brackets in Eq. (D.10) must vanish,

$$D\int_{t_0}^{T}\int_V (\nabla\phi(\boldsymbol{r},t))^2\;\mathrm{d}^3r\;\mathrm{d}t + \frac{1}{2}\int_V (\phi(\boldsymbol{r},T))^2\;\mathrm{d}^3r = 0.$$

Both terms are non negative ($D > 0$); they cannot cancel, each must separately vanish. The integrands, being non negative, must therefore vanish for all values of their arguments. The last integral in particular requires that $\phi(\boldsymbol{r},T) = 0$, for all $T > t_0$. *Unique solutions to the diffusion equation are obtained for $t > t_0$ from either Dirichlet or Neumann conditions, but not both, on closed spatial boundaries, and for an initial condition at $t = t_0$.* No further conditions may be imposed at a future time, as that would overdetermine the solution. Thus, one has boundary conditions specified on closed spatial surfaces, with time being an *open* boundary for $t > t_0$.

boundary conditions for the diffusion equation

Wave equation

Assume there are two solutions $\psi_1(\boldsymbol{r},t)$ and $\psi_2(\boldsymbol{r},t)$ of the inhomogeneous wave equation satisfying the same boundary conditions and the same initial conditions at $t = t_0$, $\psi(\boldsymbol{r},t_0)$, and $\partial\psi(\boldsymbol{r},t)/\partial t|_{t_0}$. Form the difference $\phi(\boldsymbol{r},t) \equiv \psi_1(\boldsymbol{r},t) - \psi_2(\boldsymbol{r},t)$, where $\phi(\boldsymbol{r},t)$ satisfies the homogeneous wave equation with homogeneous boundary conditions and homogeneous initial conditions. We show that $\phi(\boldsymbol{r},t) = 0$.

Consider the non negative quantity

$$H(\boldsymbol{r},t) \equiv \frac{1}{2}\left[\frac{1}{c^2}\left(\frac{\partial\phi}{\partial t}\right)^2 + \boldsymbol{\nabla}\,\phi\cdot\boldsymbol{\nabla}\,\phi\right].$$

As can be shown,

$$\frac{\partial H}{\partial t} = \frac{\partial\phi}{\partial t}\nabla^2\phi + \boldsymbol{\nabla}\,\phi\cdot\boldsymbol{\nabla}\,\frac{\partial\phi}{\partial t} = \boldsymbol{\nabla}\cdot\left(\frac{\partial\phi}{\partial t}\,\boldsymbol{\nabla}\,\phi\right),$$

where we've used that ϕ satisfies the homogeneous wave equation. Then, consider

$$\frac{\mathrm{d}}{\mathrm{d}t}\int_V H(\boldsymbol{r},t)\,\mathrm{d}^3r = \int_V \frac{\partial H}{\partial t}\,\mathrm{d}^3r = \int_V \boldsymbol{\nabla}\cdot\left(\frac{\partial \phi}{\partial t}\boldsymbol{\nabla}\,\phi\right)\,\mathrm{d}^3r$$
$$= \int_S \frac{\partial \phi}{\partial t}\boldsymbol{\nabla}\,\phi\cdot\hat{\boldsymbol{n}}\,\mathrm{d}s = 0.$$

The last equality follows because either $\phi = 0$ or $\boldsymbol{\nabla}\,\phi\cdot\hat{\boldsymbol{n}} = 0$ on S (homogeneous Dirichlet or Neumann conditions).

Thus, $\int_V H(\boldsymbol{r},t)\,\mathrm{d}^3r$ = constant. Evaluate the constant using the initial condition, $\phi(\boldsymbol{r},t_0) = 0$, implying that $\int_V H(\boldsymbol{r},t)\,\mathrm{d}^3r = 0$ for all t. We therefore have that

$$\frac{1}{c^2}\int_V \left(\frac{\partial \phi}{\partial t}\right)^2 \mathrm{d}^3r + \int_V (\nabla\phi)^2\,\mathrm{d}^3r = 0.$$

Each term is non negative; each must separately vanish, and the integrand for each is non negative, implying that ϕ is a constant independent of $\boldsymbol{r}$ and t. Because ϕ vanishes at $t = t_0$, we have $\phi(\boldsymbol{r},t) = 0$. *Solutions of the wave equation are unique for Dirichlet or Neumann conditions on S, but not both, and for mixed initial conditions throughout V.*

boundary conditions for the wave equation

Summary

- Elliptic PDEs have unique solutions for Dirichlet or Neumann conditions, but not both, on closed spatial boundaries.
- Parabolic PDEs have unique solutions for $t > t_0$, either for Dirichlet or Neumann conditions on closed spatial boundaries, but not both, and for an initial condition at one instant of time, $t = t_0$.
- Hyperbolic PDEs have unique solutions for boundary conditions specified on open boundaries (open in time, in both directions, past and future) that are nowhere tangent to characteristic curves. Solutions of the wave equation are unique for Dirichlet or Neumann conditions on S, but not both, and for mixed initial conditions throughout V.

References

[1] B. Borden and J. Luscombe. *Essential Mathematics for the Physical Sciences, Volume One.* IOP Concise Physics, 2017.
[2] G.E. Shilov. *Linear Algebra*. Prentice-Hall, 1971.
[3] P.R. Halmos. *Finite-Dimensional Vector Spaces*. D. Van Nostrand, 1958.
[4] G.H. Hardy, J.E. Littlewood, and G. Pólya. *Inequalities*. Cambridge University Press, 1952.
[5] F. Riesz and B. Sz.-Nagy. *Functional Analysis*. F. Ungar Publishing, 1955.
[6] P.A.M. Dirac. A new notation for quantum mechanics. *Mathematical Proceedings of the Cambridge Philosophical Society*, 35:416–418, 1939.
[7] R.P. Feynman. Simulating physics with computers. *International Journal of Theoretical Physics*, 21:467–488, 1982.
[8] G. Birkhoff and S. MacLane. *A Survey of Modern Algebra*. Macmillan, 1977.
[9] M.H. Stone. *Linear Transformations in Hilbert Space*. American Mathematical Society, 1932.
[10] K. Knopp. *Theory and Applications of Infinite Series*. Dover, 1990.
[11] N.M. Naimark. *Normed Rings*. P. Noordoff, 1959.
[12] A. Friedman. *Foundations of Modern Analysis*. Holt, Rinehart, and Winston, 1970.
[13] E.A. Coddington and N. Levinson. *Theory of Ordinary Differential Equations*. McGraw-Hill, 1955.
[14] R. Courant and D. Hilbert. *Methods of Mathematical Physics, Volume I*. Interscience Publishers, 1953.
[15] M.J. Lighthill. *Introduction to Fourier Analysis and Generalized Functions*. Cambridge University Press, 1958.
[16] O. Svelto. *Principles of Lasers*. Plenum Press, 1998.
[17] R. Courant and D. Hilbert. *Methods of Mathematical Physics, Volume II*. Interscience Publishers, 1962.
[18] P.M. Morse and H. Feshbach. *Methods of Theoretical Physics*. McGraw-Hill, 1953.
[19] G. Birkhoff and G.-C. Rota. *Ordinary Differential Equations*. Wiley, 1978.
[20] P.A.M. Dirac. *The Principles of Quantum Mechanics*. Oxford University Press, 1958.
[21] H. Goldstein. *Classical Mechanics*. Addison-Wesley, 1950.
[22] M. Abramowitz and I.A. Stegun. *Handbook of Mathematical Functions*. US Government Printing Office, 1964.
[23] J.W. Brown and R.V. Churchill. *Complex Variables and Applications*. McGraw-Hill, 2009.
[24] E.T. Whittaker and G.N. Watson. *A Course of Modern Analysis*. Cambridge University Press, 1950.
[25] A.I. Markushevich. *The of Functions of a Complex Variable, Volume I*. American Mathematical Society Chelsea Publishing, 1977.
[26] A.I. Markushevich. *Theory of Functions of a Complex Variable, Volume II*. American Mathematical Society Chelsea Publishing, 1977.
[27] R. Paley and N. Weiner. *Fourier Transforms in the Complex Domain*. American Mathematical Society, 1934.
[28] E.C. Titchmarsh. *Introduction to the Theory of Fourier Integrals*. Oxford University Press, 1947.
[29] G.N. Watson. *A Treatise on the Theory of Bessel Functions*. Cambridge University Press, 1952.
[30] L.I. Schiff. *Quantum Mechanics*. McGraw-Hill, 1968.
[31] R.G. Newton. *Scattering Theory of Waves and Particles*. Springer-Verlag, 1982.
[32] G. Strang. *Linear Algebra and Its Applications*. Harcourt Brace Jovanovich, 1980.
[33] A.N. Tikhonov and V.Y. Arsenin. *Solutions of Ill-Posed Problems*. Winston, 1977.
[34] D.J. Griffiths. *Introduction to Electrodynamics*. Pearson, 2013.
[35] L.D. Landau and E.M. Lifshitz. *Theory of Elasticity*. Pergamon Press, 1970.
[36] L.D. Landau and E.M. Lifshitz. *Fluid Mechanics*. Pergamon Press, 1987.
[37] J.F. Nye. *Physical Properties of Crystals: Their Represenation by Tensors and Matrices*. Oxford University Press, 1957.
[38] O. Veblen. *Invariants of Quadratic Differential Forms*. Cambridge University Press, 1927.

Mathematical Methods in Physics, Engineering, and Chemistry, First Edition.
Brett Borden and James Luscombe.

[39] J.A. Schouten. *Tensor Analysis for Physicists*. Dover, 1989.
[40] H.A. Lorentz, A. Einstein, H. Minkowski, and H. Weyl. *The Principle of Relativity*. Dover, 1952.
[41] I.M. Gelfand. *Lectures on Linear Algebra*. Interscience Publishers, 1961.
[42] R.L. Bishop and S.I. Goldberg. *Tensor Analysis on Manifolds*. Macmillan, 1968.
[43] G. Weinreich. *Geometrical Vectors*. University of Chicago Press, 1998.
[44] S. Weinberg. *Gravitation and Cosmology*. Wiley, 1972.
[45] A.C. Aitken. *Determinants and Matrices*. Oliver and Boyd, 1964.
[46] R.M. Wald. *General Relativity*. University of Chicago Press, 1984.
[47] F.W.J. Olver. *Asymptotics and Special Functions*. Academic Press, 1974.
[48] I.N. Sneddon. *Elements of Partial Differential Equations*. McGraw-Hill, 1957.
[49] J. Hadamard. *Lectures on Cauchy's Problem in Linear Differential Equations*. Dover, 2014.

Index

absolute tensor 351
addition theorem
  Bessel functions 177
  Legendre polynomials 165, 191
  spherical harmonics 160, 265, 269, 271
adjoint
  adjoint operator 55–56
  equation 68
  operator **27**
  self-adjoint equations 56
    integrating factor 56
  self-adjoint operator 28, 56, 251, 257, 298 *see also* Hermitian
    eigenvalues, 61
    eigenvectors 307, 308
    eigenvectors form orthonormal basis 51, 66
Airy function 233
alternating tensor 356
analytic functions 122, 195, *see also* complex analysis
  analytic continuation 177
  Cauchy integral formula 210
  Cauchy-Riemann conditions 200
  complex plane
    finite complex plane 202
    point at infinity 201
  contour integral 206
    Cauchy theorem 206
  differentiable in a region **198**
  entire function 202
  interval of convergence 122
  Laurent series 215
  Liouville's theorem 212
  Morera's theorem 208
  power series 402
  power series representations 214
  radius of convergence 122–123
  singularities 201
  zeros 215
analytic signal 238
anisotropic conductivity 324
antisymmetric
  matrix 30
  tensor 321
aperiodic 91
arrow of time 273
associated Legendre equation *see* Legendre equation
associated Legendre functions 157
  orthogonality 158
asymptotic notation 173
axial vector 361

balance equation *see* continuity equation
Bessel functions 81, 88, *138*, 173
  addition theorem 177
  Bessel equation 81, 121, 137
    power series solution 137
    self-adjoint form 180, 189
  Bessel series 182
  first kind 138
  generating function 175
  half-integer order *see* spherical Bessel functions
  Hankel functions 174
  integral representation 178
  large-argument form 174
  modified 194
  Neumann function 81, *140*
  order 83
  orthogonality 180
  properties 175
  radial equation 81
  recursion relations 177
  second kind 141
  small-argument form 173
  spherical *see* spherical Bessel functions
  Wronskian 139, 178, 259
Bessel series 182, 185
Bessel's inequality 16
big O notation 49
bilinear operator 342
binomial
  coefficient 149
  expansion 149
bivector 361
boundary conditions 56, 58, 60, 78–80, 86, 100, 149, 158, 166, 255, 267, 288, 289, 312, 409–417
  Dirichlet **58**, 60, 102, 181, 184, 261, 263, 272, 274, 279, 409

Mathematical Methods in Physics, Engineering, and Chemistry, First Edition.
Brett Borden and James Luscombe.

boundary conditions (*cont'd*)
  homogeneous **58**, 252, 253, 257
  mixed **58**, 409, 414
  Neumann **58**, 260–262, 272, 274, 279, 409
  nonhomogeneous **58**
  periodic 58, 60, 91
bounded operator **23**, 24, 28, 33, 37, 51, 303, 312
branch point 220, 221
Bromwich integral 244

Cartesian tensor 324
Cauchy boundary conditions *see* boundary conditions: mixed
Cauchy integral theorem 207
Cauchy-Riemann conditions **200**
characteristic curve 86
characteristic polynomial 37
Chebyshev differential equation 69
Christoffel symbols 368
  uniqueness 368
closed interval 4
closed set 7, 44
codomain of a mapping 8
coefficient functions 51, 121 *see also* power series solutions to ODEs
cofactor 31, 33
cogrediently 27
compact operator 303, 312
completeness **15**, 104
  completeness relation 67, 70, 260
    Bessel functions 193
    complex exponentials 97
    Legendre polynomials 151
    spherical Bessel functions 189
    spherical harmonics 160
    vector space 17
  complex exponentials 183
  Hilbert space 45
  spherical harmonics 264
  Sturm-Liouville system 69, 91, 257
complete orthonormal set **15**, 46, 67, 91, 96, 150, 257, 302 *see also* completeness
  completeness relation 98
  Hermitian kernels 299
complex analysis 195–244
  analytic signal 238
  antiderivative 212
  approximation of integrals 230
    method of stationary phase 235
    method of steepest descent 233
  branch point 201, 203, 207, 220, 240, 241
  Cauchy integral formula 210
  complex plane 195
  contour integration 208
    Bromwich integral 244
    infinite integrals 223
    integrands containing $\cos\theta$ and $\sin\theta$ 222
    Jordan's lemma 226
    poles on the contour 226
  deforming the contour 209
  Green's theorem in the plane 207
  Hilbert transform 238
  integration 202
  Laurent series 216
  meromorphic function 229
  Mittag-Leffler theorem 229
  Morera's theorem 208
  multivalued functions 220
  Paley-Wiener theorem 239
  power series representations of analytic functions 213
  residue *see* residue
  residue theorem 218
  singularities 217
    essential singularity 217
    residue of the pole 218
  Titchmarsh theorem 240
components of a vector 6
componentwise addition, multiplication 4
condition number 311
conductivity tensor 324
conic sections 85
connected set 196
conservative field **387**
conserved quantity 71
constant tensor 339, 358
continuity equation 72
contour integral **202**
  analytic functions 206
  Cauchy theorem 206
  Jordan curves 203
contravariant
  components 333, 339, 340
  indices 333
    convention 345
  tensor 333, 338
  transformation 339
  vector 327, 332, 337, 339, 349, 353, 358, 361
    geometric interpretation 347
convolution integral
  Fourier **111, 293**, 306
    convolution theorem 113, 238
    linear, shift-invariant operators 112
  Laplace **242**, 293
    convolution theorem 242
coordinate *see also* curvilinear coordinates
  basis 334
  curve 334
  surface 334
  transformation 328
countable set 46
countably infinite 46

covariant
divergence 369
components 337, 340
curl 369
derivative 364, 367
equation 339
indices 339
convention 345
tensor 339
transformation 27
vector 327, 336, 337, 341, 356, 361, 366
geometric interpretation 347
curl 363, 368, **380**
"circulation per unit area" 387
curvilinear coordinates 390–399
differential displacement 392
representation of vector operators (summary) 399
surface and volume elements 393
unit vectors 391
cyclic invariance of the trace 36
cylindrical coordinates 394

d'Alembertian operator 279
Darboux inequality 206
degenerate eigenvectors *see* eigenproblem
derivative of a determinant 370
determinant 29, 37, 38, 40, 42, 53
Laplace expansion in cofactors 33
Levi-Civita symbol 29
properties 32
diagonal matrix 20, **40**
differential operator 18, 51, 377–390
diffusion equation 72, **75**, 89
diffusion operator 272
inhomogeneous 74
diffusion operator 415
diGamma function **406**
dimension of a vector space (dim) **5**
dimension theorem 22
Dirac delta function **65**, 66, 97, 105, 254, 263, 277, 279
generalized function 66
integral representation 105
Legendre polynomial representation 151
multi dimensional 261, 266
properties 106
relation to step function 106
Dirac notation 5, 70, 323
bras 5, **11**, **28**
inner product 11
kets **5**, 28, 97
directional derivative **377**, 409
Dirichlet boundary condition *see* boundary conditions
discriminant 85
divergence 369, 370, **379**
"flux per unit volume" 386
divergence theorem 72, 261, 262, 266, 386, 416

domain 196
of a mapping 8
of dependence 413
dot product 11, 322
dual basis 324, 336
dual space **9**, 344
dyad **322**

eigenfunction *see* eigenproblem
eigenproblem
characteristic polynomial 37
degenerate eigenvectors 36
determinant 42
diagonalizable matrix 41
eigenequation **37**, 41
Hilbert-Schmidt theory 298
self-adjoint operator 299
eigenfunction **60**, 97, 147, 158, 162, 181, 284
eigenfunction expansion 66, 103
eigenvalue **38**, 60, 181, 185
multiplicty of an eigenvalue 36
spectrum 36
eigenvector **38**
Hermitian 52
diagonalizable 42
eigenvalues 39
eigenvectors 39
least-squares 307
Sturm-Liouville system 60, 91, 257
discrete spectrum 63
eigenfunction 181
eigenfunctions are orthogonal 62
eigenvalues are real-valued 61
weight function 58
trace 42
Unitary 52
zero eigenvalue 259
eigenvector *see* eigenproblem
Einstein summation convention 328
elastic modulus tensor 326
elementary vectors 2
elliptic PDE **87**, 261, 414
entire function 202
equation of a plane 347
equipotential surface 327
equivalence principle 367
Euclidean space, $\mathbb{E}^N$ **4**
Euler integral 403
Euler reflection formula 405
expansion theorem 68

finite support 239
flux **384**
Fourier analysis 91–115
Fourier series 92
completeness 97

Fourier analysis (*cont'd*)
complex exponential form 96
Fourier cosine and sine series 95
Gibbs phenomenon *94*
Parseval's theorem 103
Fourier transform 107
convolution theorem 113, 238
difference kernels 293
discrete 113
Fourier transform pair 111
integral equations 293
inverse transform 110
sinc function 109
Fourier coefficients 67
Fourier-Bessel transform 185
Fourier's law *see* heat conduction equation
Fredholm integral equation 291
free index 332
Fresnel integrals 235
Frobenius (method of) *see* power series solutions to ODEs
function space **4**
fundamental tensors 345
fundamental theorem of algebra 37, 212
fundamental theorem of Riemannian geometry 368

Gamma function 138, 234, **403**, 403–407
Euler limit formula 405
reflection formula 405
Gauss' theorem *see* divergence theorem
generalized function 66
generalized inverse 306
generalized Kronecker delta 359
generating function
Bessel functions 175
Legendre polynomials 152
geometric object 25, 333
gradient 72, 327, 336, 337, 347, 348, 364, **378**
Graf addition theorem 269
Gram-Schmidt process 12
Green function 251–282
boundary conditions 253
diffusion equation 272
free space 278
direct construction 255
eigenfunction expansion 257
free space 263
generalized 259
jump discontinuity 255
Kirchoff integral theorem 283
matching conditions 255
Poisson equation 261
reciprocity relation 253
regular part 263
three-dimensional Helmholtz equation 270
expansion in spherical harmonics 271
two-dimensional Helmholtz equation 267
wave equation 279
free-space 281
Green's identity 261
Green's theorem in the plane 207, 384

Hankel functions 174
heat conduction equation 73
Fourier's law 73
Heaviside function 106
Helmholtz equation **77**, 91, 121, 235, 256
cylindrical coordinates 80
radial equation 81
single-valued solution 80
fundamental importance 88, 89
green function expansion in spherical harmonics 271
Helmholtz operator 266
plane-wave expansion in spherical harmonics 190
rectangular coordinates 78
spherical Bessel functions 186
spherical coordinates 82
three-dimensional Green function 270
two-dimensional Green function 267
Helmholtz operator 266
Helmholtz theorem 389
Hermitian *see also* self-adjoint operator
conjugate 28
kernel 299
matrix **27**
diagonalizable 41
eigenproblem 39
operator 27, 59
eigenvectors form orthonormal basis 52
must be bounded 29
Hermitian conjugate 27
Hilbert-Schmidt theory *see* integral equations
Hilbert space 45, **68**
separable 46
Hilbert transform **238**
homogeneous differential equation **51**, 74
hyperbolic PDE 86, 279, 412, 415
hypersurface **415**

identity operator 17
ill-posed problems
condition number 311
regularization and Tikhonov regularization 313
image, of an operator 22
indicial equation 133
indicial roots 133
inertia tensor 324
infinite dimensional vector space 6
infinite product representation 247
inhomogeneous differential equation **51**, 74, 76 *see also* Green function

initial condition 415
inner product **10**, 55, 67, 298
  Dirac notation 11
inner product space 10
integral equations 287–315
  classification 290
  difference kernels 293
  equivalence to differential equations 288
  first and second kind 290
  Fourier kernels 294
  Fredholm 291
  generalized inverse 306
  Hilbert-Schmidt theory 298
  kernel **291**
    resolvent kernel 292
    self-adjoint kernel 297
    separable kernel 295
  matrix form 302
  measurement space 303
  Neumann series 291
  nullspace 304
  numerical approaches 302
  Volterra 290
integral operator 17
integrating factor 56
interior point 196
interval of convergence 401
invariant volume element 352
inverse of an operator 24
inversion of coordinate axes 361
irregular singular points *see* power series solutions to ODEs
isometry 34
isomorphism 8, 11, 337
isotherm 327

Jacobian
  determinant 163, 351
  matrix 328, 338, 379
Jordan curves 203
Jordan's lemma 226

kernel of a transformation 22
Kirchhoff integral representation 283
Kramers-Kronig relations 242
Kronecker delta **13**, 64, 331, 339

Lagrange's identity 56
Laplace equation 75, **76**, 121, 231, 263, 415
  cylindrical coordinates 183
  rectangular coordinates 79, 101
  spherical coordinates 166
Laplace expansion 31
Laplace transform 293
  inverse Laplace transform (Bromwich integral) 242
Laplacian 76, 78, **380**
Laurent series 216, 218
Legendre equation 83, 121, 142
  associated Legendre equation 83, 147, 157
Legendre polynomials *142*, 148
  addition theorem 160, 191
  completeness 151
  generating function 152
  integral representation 248
  Legendre series 151
  orthogonality 151
  parity 148
  properties 148
  recursion formula 156
  Rodrigues formula 150
Legendre series 151, 152
level set 327
Levi-Civita
  symbol 29, 357
  tensor 359
limit vector 44
line integrals 382
linear algebra 1–47
linear combination 1, 5
linear coordinate transformation 330
linear functional **8**
linear independence 5
linear operator **16**, 51
  adjoint **27**
  bound 24
  bounded 23, 28, 33, 37, 52, 303, 312
  differential 18
  integral 17
  inverse 24
  kernel 22
  matrix representation 18
  range 22
  unitary 33
linearity 1
Liouville theorem 212
lowering indices 344

mapping 8
matrix
  algebra 20–22
  change of basis 25
  determinant **29**
  diagonal
    diagonalizing 40
  eigenproblem 36
  Hermitian 27
  Hermitian conjugate 27
  matrix inverse **32**
  orthogonal 34
  real-valued symmetric always diagonalizable 42
  similarity transformation **27**
  symmetric 42

matrix (*cont'd*)
trace **35**, 42
transpose 21
unitary 34
matrix elements of an operator **19**, 55, 302
Maxwell stress tensor 320
measurement space 303
meromorphic function **229**
metric compatibility 368
metric tensor 342
Minkowski's inequality 4
minor of a matrix 32
Mittag-Leffler theorem 229
mixed boundary condition *see* boundary conditions
mixed initial conditions **412**
modified Bessel function 194
Morera's theorem 208, 212
multilinear function 354
multiplicative operator 17
multipole expansion 164

Neumann boundary condition *see* boundary conditions
Neumann function *see* Bessel functions
Neumann series 292
noise filtering 313
norm
of a vector 10
of an operator 23
normal derivative 409
normalized eigenfunction 62
normalized vector 14
null vector 3
nullity, dimension of the kernel 22
nullspace 22, 304

open set 196
operator *see* linear operator
operator norm 23
order of a differential equation 18
order of, notation 49
ordinary points *see* power series solutions to ODEs
orthogonal matrix **34**
orthogonal vectors 12
orthonormal basis 12, 66
orthonormal set of vectors 13
outer product 10, 28, 341

Padé approximant 228
Paley-Weiner theorem 239
parabolic PDE **87**, 272, 415
paraxial approximation 83
Parseval's theorem
Fourier series 103
Legendre polynomials 151
partial fraction expansion 117, 229, 243
partial sum 43, 67
partial wave expansion 297
passive coordinate transformation 329
Pauli matrices 47
periodic boundary condition **58**
periodic function **91**
permutation 30
definition 356
phase space 52
phasor 236
piecewise continuous 68, 92, 257
Poisson equation 74, **76**
polar vector 361
polynomial vector space 4
positive definite 10
positive definite metric 344
potential function
scalar potential 76, 153, 155, 164, 166, 380, 388
vector potential 389
power series 401
analytic 402
definition 401
interval of convergence 401
properties 122
uniqueness 124
power series solutions to ODEs 121–137
Bessel equation 137
Legendre equation 142
method of Frobenius 130
indicial equation 133
indicial root cases 133
indicial roots 133
method of undetermined coefficients 122
ordinary points 125
recursion relation 125
second solution 133, 135
Wronskian method 137
singular points 125, 128
regular and irregular 130, 131
variable coefficients 121
principle of covariance 340
pseudoscalar 362
pseudotensor 351, 362

quadratic form 85
quotient theorem 341

$r_<$, $r_>$ notation 269
radioactive decay 249
radius of convergence 122, 214
range of an operator 22
rank, dimension of the range 22
rank-nullity theorem 22
rank of a tensor 319, **333**
rational function 228
reciprocal basis 336
reciprocal space 276

recursion formula
  Bessel functions 177
  Gamma function 403
  Legendre polynomials 156
regularization of ill-posed problems 313
regular singular points *see* power series solutions to ODEs
relative tensors 351
representation, of a vector 5
residue
  finding residues
    differentiation method 219
    Laurent series method 218
  residue of the pole 218
  residue theorem 218
Rodrigues formula **150**

scalar **2**
scalar field 329
scattering amplitude 289
Schouten index notation 330
Schrödinger equation
  bound particle 74
  free particle 73
Schwarz inequality 11
self-adjoint operator *see* adjoint
self-consistency condition 285
separated boundary conditions 58
separation of variables 76, 77
  integral equations 295
  separable coordinate systems 88
  separation constant 77
sequence space 4
similarity transformation **27**
simple zero 215
simply connected 207
sinc function *105*, 109
  alternate definition 105
singularities 201
singular operator 24
singular points *see* power series solutions to ODEs
singular value decomposition (SVD) 310
smooth function 18, 122, 401
source function 74
span of a vector space 6
spectrum of an operator *see* eigenproblem
spherical Bessel functions 83, 88, 186
  completeness 190
  in terms of elementary functions 186
  integral representation 193
  large-argument forms 188
  orthogonality 189
  plane-wave expansion 190
  small-argument forms 188
  spherical Bessel series 189
  spherical Bessel-Fourier transform 189
spherical coordinates 397
spherical Hankel functions 186
  in terms of elementary functions 188
spherical harmonics 83, 88, 147–167, 264
  addition theorem 160, 265, 269, 271
  completeness on the sphere 159
  eigenfunctions of L^2 162
  multipole expansion 164
  plane-wave expansion 190
spherical symmetry 147
square integrable function 4, 64
standard basis 7
stationary phase method 235
steepest descent method 233
step function *see* Dirac delta function
Stirling approximation 235
Stokes' theorem 386
strain tensor 326
stress tensor 320, 350
Sturm-Liouville theory 51
  adjoint equation 69
  eigenproblem 39, 60
    continuous spectrum 63
    discrete spectrum 63
    eigenfunction expansion 66
    eigenfunctions are orthogonal 62
    eigenfunctions form a complete set 66
    eigenvalues are real-valued 61
  self-adjoint equations 56
  Sturm-Liouville system 58
    regular 59
Sturm separation theorem 55
subspace **7**
summation convention 328
superposition 1
surface integrals 383
Sylvester's criterion 344
symmetric
  matrix 42
  tensor 321
symmetry 161

target space of a mapping 8
Taylor/Maclaurin series 214
  radius of convergence 214
tensor 319–373, **327**
  absolute 351
  conductivity 324
  contravariant 327, 337, 339, 340, 350, 353, 355, 358, 362, 367
    geometric interpretation 347
    indices 345
  covariant 327, 336, 339, 340, 350, 353–356, 358, 362, 367
    geometric interpretation 347
    indices 345
  density 351
  inertia 324

tensor (*cont'd*)
  pseudotensor 351, 362
  rank 319, **334**
  relative 351
  stress 320
  symmetry operations 356
three-dimensional delta function 261
Tikhonov regularization 314
time-translation invariance 273, 280
Titchmarsh theorem 240
top hat function *109*
torsion tensor 367
totally symmetric, antisymmetric 356
trace of a matrix **35**, 42
transpose of a matrix 21
triangle inequality 12

uncertainty principle 110
uniqueness
  inhomogeneous ODE 253
  inhomogeneous PDEs 409
unit dyadic 325, 346
unit matrix 20
unitary matrix **34**
unitary operator **33**
  eigenvectors form orthonormal basis 52

vector calculus 377–390
vector components 6
vector, elementary 2
vector norm 10
vector space **3**
  l^2 and L^2 4
  basis
    non-orthogonal 6
    orthonormal 13
  dimension 5
  Dirac notation 5
  field 3
  function space 4
  Hilbert space 68
  infinite dimensional 6
  inner product 10
  inner product space 10
  linear independence 5
  linear operator 16
    matrix representation 19 *see also* matrix
  polynomial space 4
  sequence space 4
  square-integrable functions (L^2) 68
Volterra integral equations 291

wave equation **75**, 184, 236, 282, 292, 412, 415, 416
  homogeneous 75
  inhomogeneous 76
wave operator 279
wedge product 363
well-posed problems 312
Wronskian determinant 54, 61, 70, 137, 139, 178

zero matrix 21
zero of a function 55
zeros of analytic functions 215